MSP430 Microcontroller Basics

MSP430 Microcontroller Basics

John H. Davies

ELSEVIER

AMSTERDAM • BOSTON • HEIDELBERG • LONDON
NEW YORK • OXFORD • PARIS • SAN DIEGO
SAN FRANCISCO • SINGAPORE • SYDNEY • TOKYO

Newnes is an imprint of Elsevier

Newnes

Newnes is an imprint of Elsevier
30 Corporate Drive, Suite 400, Burlington, MA 01803, USA
Linacre House, Jordan Hill, Oxford OX2 8DP, UK

 Recognizing the importance of preserving what has been written, Elsevier prints its
books on acid-free paper whenever possible.

Library of Congress Cataloging-in-Publication Data
Application submitted

British Library Cataloguing-in-Publication Data
A catalogue record for this book is available from the British Library.

ISBN: 978-0-7506-8276-3

For information on all Newnes publications,
visit our Web site at: *http://www.books.elsevier.com*

Printed and bound in the United Kingdom

Transferred to Digital Print 2010

"To Elizabeth."

Contents

Preface

About a decade ago, I took over the teaching of a first-year, second-semester course on digital electronics. It covered flip-flops, counters, and state machines, all built from small-scale integrated circuits. One of the projects at the end was to build a digital die. In many ways it was an excellent exercise because there were so many feasible ways in which it could be approached—simple counters, Johnson counters, or state machines. My concern was that it was very close to the project that I had experienced in my first course on digital electronics, which was back in the mid-1970s. The technology was close to the state of the art then, but was it still appropriate after so many years? Another feature of our course is that it is taken not only by electronic engineers but also by students from the science faculty, mostly computer scientists. I wanted these students to leave with a feeling for what can readily be done with modern programmable electronics in smaller-scale systems. I therefore replaced the material in the second half of the course with microcontrollers. (Do not worry, state machines were not abandoned—they are taught with hardware description languages in the context of programmable logic devices.)

More recently, I thought that the time had come to review the choice of microcontroller. We traditionally used 8-bit processors because modern devices have versatile peripherals and sophisticated embedded emulation and are quite powerful enough for most applications. Then the Texas Instruments MSP430 caught my eye. A problem with 8-bit microcontrollers is that 8 bits are too few for addresses, which are typically 16 bits long, and this means that data and addresses cannot be treated on an equal footing. In contrast, the MSP430 has a uniform, 16-bit architecture throughout: The address bus, data bus, and registers in the CPU are all 16 bits wide. The CPU has a modern design with plenty of registers, most of which can be used equally for data or addresses. It has a small instruction set with orthogonal addressing and an ingenious constant generator, which is used to emulate many operations that would otherwise need their own, distinct instructions. In many ways these features make the 16-bit MSP430 simpler than a typical 8-bit processor.

Of course an elegant architecture does not generate many sales in the real world. More important are the range of peripherals and development tools. The MSP430 offers the usual selection of peripherals plus some less common modules, including sigma–delta analog-to-digital converters and operational amplifiers. Some devices include hardware multipliers and digital-to-analog converters, which provide a complete signal chain (although, of course, Texas Instruments also offers an enormous range of digital signal processors). There is a choice of two free development environments (always an important consideration in education). One is IAR Embedded Workbench, which is available for a wide range of microcontrollers. Another, Code Composer Essentials, is produced by Texas Instruments itself. A third option is the GCC toolchain for MSP430 at mspgcc.sourceforge.net.

I have not yet mentioned the major selling point of the MSP430, which is its low power consumption. Many microcontrollers are based on long-established designs with low-power modes grafted onto them. This means that returning to full power from a low-power mode is often awkward and in some cases is virtually a reset operation. The MSP430 is refreshingly different because it was designed from the outset for low-power operation. Entry to low-power modes and exit from them is straightforward, supported by a versatile clock system. For example, the clock module includes a digitally controlled oscillator that restarts at full speed from a low-power mode in less than 1 μs in newer devices. In many applications the MSP430 is put into a low-power mode, from which it is awakened by interrupts. These automatically restore full power for the interrupt service routine and return the processor to low power when it has finished. No extra code is needed for this: It is an intrinsic part of the interrupt mechanism. Most peripherals are designed for low power, although this can sometimes make them a little more complicated than would otherwise be necessary. The main point is that low-power modes are easy to use.

The quality of the data sheets and user's guides is another issue in education and those for the MSP430 are fine. Unfortunately one item was missing in the area of documentation: a suitable textbook in English. I wrote this book to fill the gap.

Outline

Most textbooks on microcontrollers follow one of two approaches. The first is to present a sequence of projects to explore successive aspects of the device. I think that this works well for simpler architectures, notably the 8-bit PICs, because it enables the reader to write functioning programs rapidly. This always feels good. Unfortunately I am not sure that it works as well for more advanced peripherals, which need considerable explanation before the reader can learn to use them fully.

The alternative approach is to describe each module in the microcontroller fully and in turn, starting with the CPU and instruction set and working out to the peripherals. This makes for a well-organized reference book but can be tedious as a textbook.

I tried to steer a course between these two. My inspiration is Kernighan and Ritchie's *The C Programming Language*, which starts with a "Tutorial Introduction" before exploring the language systematically in subsequent chapters. I think that it takes rather more introduction to a microcontroller so the "simple tour," which is my equivalent to the tutorial, does not start until Chapter 4. Before that, the first chapter contains a general introduction to embedded systems and microcontrollers. This sets the scene for Chapter 2, which focuses on the MSP430 and gives a broad view of its features. I include a chapter on hardware and software for developing applications, which I hope will be particularly useful for readers who are new to microcontrollers. It also contains some reminders of features of the C language that are more prominent in programs for microcontrollers than desktop computers—bit fields for instance. This leads into the tour, which runs through some simple programs to illustrate input and output, the inevitable flashing LEDs, and an introduction to one of the timers (the MSP430 has several).

The remainder of the book provides a more systematic description of the MSP430. I start with the CPU and instruction set, and show how the constant generator is used to provide further "emulated" instructions. The clock system is also described in this chapter. It is followed by Chapter 6 on subroutines, interrupts, and low-power modes. I already mentioned that a major feature of the MSP430 is the way in which low-power modes are handled automatically when interrupts are serviced.

Subsequent chapters are concerned with the most widely used peripherals. Chapter 7 on digital input and output starts with the usual parallel ports and goes on to describe liquid crystal displays, which many MSP430s can drive directly. There is a wide selection of timers in the MSP430, which are covered in the next chapter. This is followed by a lengthy chapter on analog input and output. The MSP430 offers many peripherals for analog-to-digital conversion, ranging from a simple comparator to a 16-bit sigma–delta module. I do not think that you can use any of these without some understanding of their characteristics, which explains the length of this chapter. Some MSP430s include operational amplifiers and digital-to-analog converters, which I described briefly. The final long chapter is on communication. I cover only three types of communication—serial peripheral interface, inter-integrated circuit bus, and asynchronous—but there are several peripherals for these in different variants of the MSP430, so there is a lot to explain.

The very last chapter provides an introduction to the MSP430X, an extended architecture with a 20-bit address bus that can handle 1 MB of memory. There is also an appendix to take the reader through the steps of editing, building, and debugging the first project, which can sometimes be a frustrating experience.

I find it annoying when books contain large chunks copied directly from data sheets and have tried to avoid this. You cannot hope to program a microcontroller without the data sheet at your side. Having said that, I start by going through each bit of the registers that control the peripherals used for the early programs. The idea is to explain how a typical peripheral is configured. After that I become more selective and concentrate on the overall function of the peripheral instead. Usually I pick out a few details that I think need extra explanation but skip the more mundane aspects. They are in the example programs in any case.

I include links to many of Texas Instruments' application notes because I can see no point in repeating material that has been thoroughly explained already. I find that many students are strangely reluctant to use this valuable resource. There are a few reminders about code examples for the same reason.

C or Assembly Language?

Most small microcontrollers are now programmed using the C language so the question might seem redundant. In fact often columns in newletters on embedded systems often carry articles with titles such as "Is Assembly Language Dead?" However, the answer seems to be clearly that assembly language is *not* dead for small microcontrollers, such as the MSP430. Most code is written in C but you may occasionally need to write a subroutine in assembly language to perform an operation that cannot be written out directly in C. Two examples are operations that require bitwise rotations rather than shifts and calculations that can be done more efficiently by exploiting special instructions of the CPU, such as binary-coded decimal arithmetic. Intrinsic functions often avoid the need for assembly language but not always.

More important, assembly language is often needed for debugging and this is the most compelling reason for describing it in a textbook. Small microcontrollers typically spend much of their time interacting with hardware by manipulating the registers that control the peripherals. Debugging may require stepping through lines of assembly language to check each step. You have to look at the manual to check the details of each instruction, but it helps to have a general idea of how the assembly language works.

From a pedagogical point of view, assembly language is useful to illustrate the architecture of the processor. In fact the MSP430 is simple enough that you can explore the thinking behind the design of the instruction set. Besides, assembly language can be fun (in small doses).

My approach is to develop the first, simple programs in Chapter 4 using both C and assembly language to show the relation between them. However, C dominates by the end of the chapter. Assembly language makes a strong showing in the next two chapters, which cover architecture, subroutines, and interrupts, including a section on mixing C and assembly language. Almost all remaining programs are in C, with assembly language reappearing only briefly for a function to convert numbers to binary-coded decimal. The listings in the text are read directly from the programs that I tested.

Companion Web Site

Please visit the companion Web site for this book at www.elsevierdirect.com/companions/9780750682763 and download the programs used as examples in the book. These programs were read into the text of the book from the workspaces that I used for testing, which means that the downloaded files should match the book perfectly. Links are also provided for data sheets, user's guides, and development tools. Solutions to the odd-numbered examples are freely available on the companion Web site but the remaining solutions are offered only to instructors.

Acknowledgments

It is a pleasure to thank numerous people who have helped me in various ways to write this book. Many are from Texas Instruments: Bonnie Baker, Jacob Borgeson, Andreas Dannenberg, Colin Garlick, Thomas Mitnacht, and Robert Owen. I am particularly grateful to Adrian Valenzuela for his comments on the final draft. Several engineers from other companies were kind enough to provide advice and assistance: Edward Gibbins and Steve Duckworth from IAR, Tom Baugh of SoftBaugh, Paul Curtis of Rowley Associates, David Dyer of Ericsson and Fernando Rodriguez while he was at Texas Instruments. Finally, I am grateful to colleagues and students at Glasgow University, from whom I have learnt an enormous amount over the years. I'd like to thank Fernando Rodriguez (not the same person who was at Texas Instruments) and David Muir in particular, with both of whom I have run a wide range of projects on embedded systems and microcontrollers—from tutor boxes with flip-flops to the electronic systems of a Formula Student racing car.

John Davies, Milngavie

CHAPTER 1

Embedded Electronic Systems and Microcontrollers

This chapter provides a short introduction to embedded electronic systems, where they are used, and ways in which they can be implemented. Microcontrollers were originally developed from microprocessors for use in embedded electronic control systems, as their name implies. They include a processor and most or all of the memory, clock, and other systems needed to support it. Everything is inside a single package, which is why a microcontroller is often described as a "computer on a chip." I review the main features of a typical small microcontroller before setting the scene for the rest of the book with the MSP430.

1.1 What (and Where) Are Embedded Systems?

Suppose that you asked people in the developed world to show you the products in their house that contained "computer chips." (Admittedly, this term is deliberately vague.) Probably they would point to a personal computer and stop there. If you tried harder, you might be offered a game console or personal digital assistant. It is unlikely that they would mention cellular phones, which contain a startling degree of processing power just for communication, to say nothing of taking photographs and playing games. There is hardly an electrical consumer product nowadays that does not rely on digital control. This seems reasonable for washing machines and video recorders, but one might wonder why a toaster or a kettle needs any digital electronics. These products contain *embedded* electronic systems: The processor supports the operation of the product but is not the main reason for purchasing it. There are said to be about 100 embedded processors for each computer, so high-profile, leading-edge microprocessors make up a small part of the market in terms of volume. Fancy modern cars have approaching 100 processors and even a personal computer has embedded processors in its keyboard, mouse, screen, disk drives, and so on. The snag for an engineer is that *embedded* seems synonymous with *invisible* and few people appreciate the extent to which they rely on electronics.

Embedded systems encompass a broad range of computational power. A crude classification is given by the number of bits that can be manipulated at a time. Many processors perform very simple tasks, for which 8 or even 4 bits are sufficient. For example, I have an electric toothbrush that pauses briefly every 30 s to remind me to move on to the next quarter of my mouth. The electronics in a remote control, kettle, or toaster need not be very sophisticated either. A bigger device that can handle 8 bits is needed for something like a washing machine. These may be comparable in power with the processors used in the first personal computers and, in fact, the descendents of the microprocessors of those days are still widely used. Digital control has been employed in car engines for many years, since the first legislation was introduced to reduce pollution and raise efficiency. These relied on 16-bit processors for a long time, but 32 bits are now needed to provide the necessary performance. A cellular phone also has a 32-bit processor as well as specialized hardware for digital signal processing. The subject of this book is the Texas Instruments MSP430, which is a straightforward, modern 16-bit processor designed specially for low-power applications.

1.2   Approaches to Embedded Systems

Many different approaches can be taken for the design of embedded systems. The general trend is toward digital systems and increasing integration: Systems that used analog electronics or small-scale integrated circuits (ICs) in the past are now more likely to use larger digital ICs. It is easier to follow this evolution for a simple system so we look at a couple of these first.

1.2.1   Timer for an Electric Toothbrush

First consider a very simple application, the timer for the toothbrush mentioned earlier. Here are a few possible ways of building a timer to give a 30 s delay.

Small-Scale Integration: The 555

Mention "timer" to many electronic engineers and they will immediately think of the 555. Introduced by Signetics in 1972, this has proven so versatile that complete books have been devoted to it. In a straightforward circuit it provides a square wave whose period is given very roughly by RC, the product of the values of a resistor and capacitor. (Really there are two resistors, and RC should be multiplied by a constant, but that does not affect the argument.) We want $RC \approx 30$ s and it is usually a good idea not to exceed $R = 1 \, M\Omega$, or leakage currents can become a problem. This means that $C \approx 33 \, \mu F$, which is a standard value. The snag is that capacitors of this value are usually electrolytic, which are physically large and leaky, but the circuit would probably work. This solution needs an eight-pin 555, two resistors, and one electrolytic capacitor.

Medium-Scale Integration: 4000 Series CMOS

A large capacitor is needed with the 555 because the period of oscillation is so long. A remedy is to reduce the period (increase the frequency) and feed the output into a counter, which could be selected from the 4000 series of small- and medium-scale ICs. This was the first family of CMOS logic and dates from 1968. They are slow but operate from a wide range of voltages and are particularly suitable for battery-powered circuits. Unlike the 7400 series, which provides mostly straightforward logic components, several unusual functions are available.

A suitable device for this application is the 4060. It contains a 14-bit ripple counter and an internal oscillator circuit to provide a clock, which needs either a crystal or two resistors and a capacitor. The outputs from ten stages of the counter are brought to pins so it need not just divide the clock by the maximum factor of $2^{14} = 16,384$. Without designing the system in detail, the oscillator can now run at a few hundred Hz. This permits a much smaller, nonelectrolytic capacitor. The snag is that the 4060 comes in a 16-pin package, so the overall circuit might take up more space than with the 555.

Large-Scale Integration: Small Microcontroller

I suspect that the real product contains a small microcontroller. This is a complete "computer on a chip" and is described in the next section. Several manufacturers produce these in eight-pin and even six-pin packages, only a few millimeters across. They have complete internal oscillators, so no external components are needed (except perhaps a decoupling capacitor). Even these tiny components can do far more than timing 30 s intervals. Perhaps they also supervise the charging of the battery, which is indicated by a light-emitting diode (LED). Maybe they could drive a small speaker to play a tune as well.

1.2.2   Electronic Dice

The project in my undergraduate course on digital electronics (a long time ago) was to build an electronic dice from gates and flip-flops. The top circuit in Figure 1.1 is more

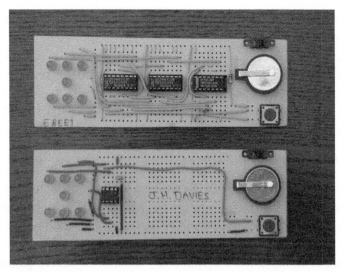

Figure 1.1: Electronic dice built using (top) *JK* flip-flops and gates and (bottom) an eight-pin microcontroller.

modern but uses 7400 series logic ICs and the same idea (it was formerly used in one of my department's courses). Two packages contain *JK* flip-flops and the third contains gates, which drive the flip-flops through the correct sequence and provide the clock. This is a simple example of a Moore state machine.

The lower photograph shows a circuit built on the same board using an eight-pin microcontroller. The economy in components and size is obvious. In fact, the on–off switch is superfluous because the microcontroller automatically switches off the LEDs after a few seconds and enters a low-power mode until the button is pressed again.

1.2.3   Larger Systems

Large embedded systems might contain fairly standard personal computers inside them. Many automatic teller machines (ATMs) are built like this. The advantage is that the hardware is standard and a huge range of software is available, including operating systems. On the other hand, the systems are large and consume a lot of power. Their reliability may also be questionable. Three general approaches can be taken between this extreme and small-scale integration.

Application-specific integrated circuits (ASICs): Specially designed for a particular application as their name implies. They provide the best performance but are extremely expensive to design and test. This restricts them to applications with very large volume or where performance must be bought at any price.

Field-programmable gate arrays (FPGAs) and programmable logic devices (PLDs): Essentially an array of gates and flip-flops, which can be connected by programming the device to produce the desired function. This is specified using a hardware description language such as VHDL or Verilog. Older programmable logic devices, such as the 22V10, contain a set of flip-flops (ten in this case) whose inputs come from an array of AND and OR gates. They are often used to provide the "glue" logic needed to support a large processor. Field-programmable gate arrays have a more versatile structure and may be enormous, with over a billion (10^9) transistors.

Microcontrollers: These have nearly fixed hardware built around a central processing unit (CPU). The CPU controls a range of peripherals, which may provide both digital and analog functions such as timers and analog-to-digital converters. Small devices usually include both volatile and nonvolatile memory on the chip but larger processors may need separate memory. Their operation is usually programmed using a language such as C or C++.

In practice the distinction between these is blurred, particularly for larger ICs. For example, part of a FPGA may be designed to act as a processor (FPGAs are often used to test the design of microcontrollers). Some devices include a fixed processor with something like a FPGA that can be configured to provide the desired digital peripherals.

Example 1.1

Estimate the number of embedded electronic systems in your living room.

1.3 Small Microcontrollers

A *microprocessor* contains a complete digital processor, which includes at least the arithmetic logic unit and associated registers. The earliest devices, such as the Intel 4004 and Texas Instruments' TMS1000, were introduced at the beginning of the 1970s. Their breathtaking evolution since then has been toward increasing computational power and complexity. They are also more powerful in an electrical sense. Large, modern microprocessors need huge heat sinks and fans and can draw over 100 A of current. The reduction of power dissipation is a major thrust of current development, now that so many microprocessors are used in portable equipment, whose battery should last for as long as possible. A microprocessor needs many other components to support it. These include a (large) external memory and the other components that can be found on the motherboard of a personal computer.

It was realized from the start that microprocessors would also be useful to control electronic equipment, such as photocopiers. Here the emphasis was less on computational power; the drive was more to reduce the complexity of the hardware and increase reliability. The trend was therefore to integrate as many functions as possible on to the same chip as the processor. This gave rise to the *microcontroller* (MCU or μC), which typically contains all of the functions needed to make a complete computer system, including memory. A microcontroller also contains a selection of peripheral modules to provide commonly needed digital functions, such as timers, and often analog functions as well. Inevitably the distinction between microprocessors and microcontrollers is blurred.

Although microprocessors have evolved almost entirely toward increasing computational power, this is not true of microcontrollers. One trend has been toward increasing integration so that everything is in one package, including the clock oscillator. Another trend has been the increasing integration of analog functions, so that it is often cheaper to buy a microcontroller with an analog-to-digital converter (ADC) than it is to buy a standalone ADC. Processing power has increased in some families. For instance the PowerPC processor, which powered Macintosh computers for many years, is now widely used in microcontrollers for engine management. However, there has also been vigorous development of smaller, cheaper devices. There is now a wide selection of microcontrollers in eight-pin and even six-pin packages, costing well under $1. These are aimed at applications that might previously have used discrete components and small-scale integration or, more likely, did not include electronics at all.

From now on, I shall consider only "small" microcontrollers (MCUs) although as usual this term cannot be defined precisely. Typically these devices can process 8 or 16 bits of data and have a 16-bit address bus, which means that they can address 64 KB of memory directly with no paging or banking. (The upper-case letter K stands for a "binary" kilo, meaning $2^{10} = 1024$. A lower-case k means the decimal value $10^3 = 1000$.) Their main function is likely to be sequential control rather than computation. They are designed for low power and low cost. There is a wide range of products despite these limitations in packages with 6 to over 100 pins. Many billions have been sold—this is a big business.

Another distinction between microprocessors and microcontrollers is the operating system. It is very unlikely that a modern microprocessor would be used without an operating system such as linux, MacOS, or Windows. In contrast, small microcontrollers are unlikely to use an operating system at all: Software is written to run directly on the hardware without any additional support. When an operating system is used, it is very different from that on a desktop computer. This is because microcontrollers are often used in *real-time* systems, where they are required to respond to external events within a prescribed time. A specialized *real-time operating system* (RTOS) is used in such applications. The relation between a user's program and an RTOS is quite different from that between a program and the operating system on a desktop computer. When you turn on a desktop computer, you have to wait for the operating system to load before you can do anything. In contrast, a microcontroller starts up with the user's program, whose first job is to start the RTOS, configure it, and launch the desired tasks. A more obvious difference is that an RTOS may occupy only a few hundred bytes of memory, not megabytes.

Given that the role of a microcontroller is to control the system in which it is embedded, its inputs and outputs are clearly important. Look briefly at a washing machine as a typical application to see what is likely to be required:

- The switches on the front panel are usually on–off buttons, which provide digital inputs to the MCU. On the other hand, there might be some knobs that offer continuous adjustment of the temperature or spin speed, in which case the MCU must also handle an analog input.

- Some sensors, such as one to check that the door is closed, also give digital inputs.

- Other sensors, such as those that measure the temperature and level of the water, require analog inputs.

- The front panel has some sort of display. This may be something very simple, such as a few light-emitting diodes (LEDs), which need only a digital output (on or off). Numerical seven-segment displays are also simple to drive, although they may be multiplexed to reduce the number of connections needed; see the section "Digital Input and Output: Parallel Ports" on page 208. On the other hand, specialized hardware is needed for a liquid crystal display (LCD), which may be either in the microcontroller or built into the display itself.

- Some components inside the washing machine, such as the valves for water, need only to be switched on or off. They can therefore be driven from a digital output, although additional components may be needed to provide the voltage and current needed by the valves.

- In other cases, it appears that the output can be varied continuously. For example, the motor rotates slowly for washing and through a range of increasing speeds for spinning. This appears to need an analog output, but in practice, this is simulated using *pulse-width modulation* (PWM), described in the section "Output in the Up Mode: Edge-Aligned Pulse-Width Modulation" on page 330.

How much processing power is needed for this washing machine? The most rapidly changing outputs are the LCD and motor using PWM, which may need to be driven at frequencies around a kilohertz, but this will not load the processor itself, because dedicated peripherals are used for these tasks. The inputs need to be scanned but this need be done only a few times per second. Each phase in the washing takes several seconds at least, and a timer is included in the hardware for this. It seems that the processor itself has almost nothing to do. Sophisticated machines may use fuzzy logic to adjust the washing cycle, but there clearly is plenty of time for this.

Suppose that we add an "out of balance" sensor to the washing machine, which monitors the force on the drum as it rotates. This is to prevent the machine vibrating dangerously if the load is unevenly distributed. Machines typically spin the clothes at around 1200 rpm or 20 revolutions per second. The force might be measured 10 times per revolution or 200 times per second. The output of the sensor probably is a continuously variable voltage, so the MCU needs to perform an analog-to-digital conversion. This is a relatively slow process and may take 50 clock cycles. The processor therefore needs about 10,000 clock cycles per second for this task. Typical clock frequencies are several megahertz, so this is unlikely to tax the microcontroller. The key point is that the demanding tasks are handled by special hardware, so the processor itself need not be particularly powerful.

Example 1.2

A bread maker is another common domestic appliance that probably contains a microcontroller. How much processing power is needed for this?

1.4   Anatomy of a Typical Small Microcontroller

We are now in a position to set out the functions required inside a practical micro-controller. The following features, shown in Figure 1.2, are essential.

   Central processing unit: I define this to include

- Arithmetic logic unit (ALU), which performs computation.
- Registers needed for the basic operation of the CPU, such as the program counter (PC), stack pointer (SP), and status register (SR).

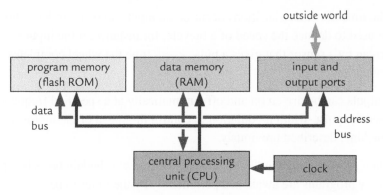

Figure 1.2: Essential components of a microcontroller.

- Further registers to hold temporary results.

- Instruction decoder and other logic to control the CPU, handle resets, and interrupts, and so on.

Memory for the program: Nonvolatile (read-only memory, ROM), meaning that it retains its contents when power is removed.

Memory for data: Known as random-access memory (RAM) and usually volatile.

Input and output ports: To provide digital communication with the outside world.

Address and data buses: To link these subsystems to transfer data and instructions.

Clock: To keep the whole system synchronized. It may be generated internally or obtained from a crystal or external source; modern MCUs offer considerable choice of clocks.

It is unlikely that any processor would lack any of these features, although their implementation may differ substantially. The big differences between devices comes from the range of *peripherals* included. These functions needed completely separate pieces of equipment long ago, but as technology improved, they could be included on the same printed circuit board as the processor. Now most peripherals are on the same integrated circuit as the processor and the nomenclature is no longer appropriate, but it has stuck. Here are some of the more common peripherals.

Timers: Most microcontrollers have at least one timer because of the wide range of functions that they provide. Here are just a few.

- The time at which transitions occur on an input can be recorded. This may be used to deduce the speed of a bicycle, for instance, if the input is driven by a sensor that gives a pulse every time the wheel completes a revolution.

- Outputs can be driven on and off automatically at a specified frequency. This is used for pulse-width modulation to control the speed of the motor in a washing machine, described previously.

- They provide a regular "tick" that can be used to schedule tasks in a program. Many programs are awakened periodically by the timer to perform some action—measure the temperature and transmit it to a base station, for example—then go to sleep (enter a low-power mode) until awakened again. This conserves power, which is vital in battery-powered applications.

Watchdog timer: This is a safety feature, which resets the processor if the program becomes stuck in an infinite loop.

Communication interfaces: A wide choice of interfaces is available to exchange information with another IC or system. They include serial peripheral interface (SPI), inter-integrated circuit (I^2C or IIC), asynchronous (such as RS-232), universal serial bus (USB), controller area network (CAN), ethernet, and many others.

Nonvolatile memory for data: This is used to store data whose value must be retained when power is removed. Serial numbers for identification and network addresses are two obvious candidates.

Analog-to-digital converter: This is very common because so many quantities in the real world vary continuously.

Digital-to-analog converter: This is much less common, because most analog outputs can be simulated using PWM. An important exception used to be sound, but even here, the use of PWM is growing in what are called *class D* amplifiers.

Real-time clock: These are needed in applications that must track the time of day. Clocks are obvious examples but data loggers are also an important case.

Monitor, background debugger, and embedded emulator: These are used to download the program into the MCU and communicate with a desktop computer during development. See the section "Access to the Microcontroller for Programming and Debugging" on page 57.

The processor communicates with these peripherals by reading from, and writing to, particular addresses in memory. These memory locations are called *special function registers* or *peripheral registers* to distinguish them from ordinary memories, which simply store data, but exactly the same commands are used—no special commands are needed. In practice, microcontrollers spend much of their time handling the peripheral registers. This shows the central role of memory in a microcontroller so I review its organization next.

1.5 Memory

Memory lies at the heart of any computer. It can be pictured as a pile (or piles) of pigeonholes. Each location can typically store 1 byte (8 bits or 1 B) of data and is often called a *register*, although this term is sometimes reserved for memories within the CPU. Memory is linked to the CPU by buses for data, address, and control as shown in Figure 1.2. Buses are shared sets of wires that join several components, rather like a multilane highway. Access to the bus must be controlled, as it is for the highway, to define when data are valid and ensure that two components do not try to write at the same time. This is the job of the control bus, which I have not drawn and shall ignore from now on. The number of wires in a bus defines its width, and processors are commonly characterized by width of the data bus, which is generally the same as the size of data that can be processed by the CPU. For example, an 8-bit processor has a data bus of this width and most operations in its CPU use 8 bits (although there may be further instructions to handle 16 bits as well). The address bus need not have the same width as the data bus and is often wider; an 8-bit address bus would only carry $2^8 = 256$ distinct addresses, which is too small to be useful. Many 8-bit processors have 16-bit address buses, for instance.

Addresses are always quoted in *hexadecimal* (hex) notation, where each digit has a range of 0–15. The values 10–15 are written as a–f or A–F. Hexadecimal values are distinguished by a prefix of "0x" in the C programming language and I follow this usage; other notations are often required for assembly language. The reason for using hexadecimal notation is that an 8-bit byte can be written as a two-digit hexadecimal number in the range 0x00–0xFF, corresponding to 0–255 in decimal. Similarly, a 16-bit address lies in the range 0x0000–0xFFFF. The four bits that correspond to each hexadecimal digit are (regrettably) known as nibbles.

The buses in a microprocessor are brought out to pins for access to external memory. Larger microcontrollers may also do this, but the buses usually remain hidden inside small microcontrollers. External memory can be added using a separate interface such as SPI.

1.5.1   Volatile and Nonvolatile Memory

Memory can be classified into two main varieties:

Volatile: Loses its contents when power is removed. It is usually called *random-access memory* or RAM, but the name is misleading because access to most other types of memory is equally random. The vital feature is that data can be read or written with equal ease. Volatile memory is used for data, and small microcontrollers often have very little RAM, sometimes only a few tens of bytes. The memory is usually *static* RAM, which means that it retains its data even if the clock is stopped (provided that power is maintained, of course). A single cell of static RAM needs six transistors. RAM therefore takes up a large area of silicon, which makes it expensive. Most memory in a desktop computer is *dynamic* RAM. This needs only one transistor per cell but must be refreshed regularly to maintain its contents, so it is not used in small microcontrollers.

Nonvolatile: Retains its contents when power is removed and is therefore used for the program and constant data. It is usually called *read-only memory* or ROM, but again this traditional name has become misleading. Most modern microcontrollers can write to their nonvolatile memory but it is much slower and more complicated than writing to RAM.

There are many types of nonvolatile memory in use:

Masked ROM: The data are encoded into one of the masks used for photolithography and written into the IC during manufacture. This memory really is read-only. It is used for the high-volume production of stable products, because any change to the data requires a new mask to be produced at great expense. Some MSP430 devices can be ordered with ROM, shown by a *C* in their part number. An example is the MSP430CG4619.

EPROM (electrically programmable ROM): As its name implies, it can be programmed electrically but not erased. Devices must be exposed to ultraviolet (UV) light for about ten minutes to erase them. The usual black epoxy encapsulation is opaque, so erasable devices need special packages with quartz windows, which are expensive. These were widely used for development before flash memory was widely available.

OTP (one-time programmable memory): This is just EPROM in a normal package without a window, which means that it cannot be erased. Devices with OTP ROM are still widely used and the first family of the MSP430 used this technology.

Flash memory: This can be both programmed and erased electrically and is now by far the most common type of memory. It has largely superseded electrically erasable, programmable ROM (EEPROM). The practical difference is that individual bytes of EEPROM can be erased but flash can be erased only in blocks. Most MSP430 devices use flash memory, shown by an *F* in the part number.

A higher voltage is needed to write to flash memory than is necessary for normal operation. This had to be supplied externally to early devices but modern components include a charge pump to generate the programming voltage internally. Flash memory is of course widely used in portable storage devices—memory cards, USB drives, and so on—but the technology is different. Microcontrollers use NOR flash, which is slower to write but permits random access. NAND flash is used in bulk storage devices and can be accessed only serially in rows.

Although flash currently dominates nonvolatile memory, new technologies may be on the way. There has been vigorous research into ferroelectric memory, silicon nanocrystals, and other approaches, which their promoters confidently expect to be adopted in the next few years.

1.5.2   Harvard and von Neumann Architectures

The two types of memory that we just reviewed, volatile and nonvolatile, can be treated in the two general ways as illustrated in Figure 1.3. General-purpose processors use almost exclusively the von Neumann architecture but both are used in microcontrollers.

Harvard Architecture

The volatile (data) and nonvolatile (program) memories are treated as separate systems, each with its own address and data bus. Many microcontrollers use this architecture, including Microchip PICs, the Intel 8051 and descendents, and the ARM9. The principal advantage is efficiency.

- It allows simultaneous access to the program and data memories. For instance, the CPU can read an operand from the data memory at the same time as it reads the next instruction from the program memory.

- The two systems can be separately optimized. For example, the PIC16 has a data memory with an 8-bit data bus and a 9-bit address bus. On the other hand, the program memory is 14 bits wide, so that each word holds a complete instruction, with a 13-bit address bus.

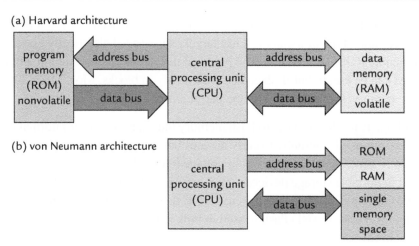

Figure 1.3: Harvard and von Neumann architectures for memory.

A problem with the Harvard architecture is that constant data (often lookup tables) must be stored in the program memory because it is nonvolatile. This means that constants cannot be read in the same way as volatile values from the data memory. Special "table read" instructions must therefore be provided or part of the program memory is mapped into data memory.

von Neumann Architecture

There is only a single memory system in the von Neumann or Princeton architecture. This means that only one set of addresses covers both the volatile and nonvolatile memories. The memory map, which shows the addresses at which each type of memory is located, becomes particularly important. The architecture is intrinsically less efficient because several memory cycles may be needed to extract a full instruction from memory. However, the system is simpler and there is no difference between access to constant and variable data. Microcontrollers with a von Neumann architecture include the MSP430, the Freescale HCS08, and the ARM7.

Example 1.3

What is the maximum number of locations that can be addressed in the data and program memories of the PIC16? Express your answer in decimal and hexadecimal notations.

Example 1.4

Suppose that the memory of a von Neumann processor is organized in bytes (as usual) and can store the same total number of *bits* as the PIC16. How many bytes are needed and what is the minimum width of the address bus?

1.6   Software

I have already mentioned the biggest difference between writing a program for a desktop computer and for a small microcontroller: the absence of an operating system (in most cases). The impact of this will become clear in Chapter 4. The tasks carried out by the program may be very different too. Desktop computers often spend considerable time on calculations, whether it is analyzing data or computing the next view in a game. The CPU in a microcontroller spends much of its time interacting with peripherals, although it may have to perform some calculations on the values received from a sensor, for instance.

Several languages may be used for programming a small microcontroller:

Machine code: The binary data that the processor itself understands. Each instruction has a binary value called an *opcode*. It is unrecognizable to humans, unless you spent a very long time on low-level debugging. Some very early computers had to be programmed in machine code, but that was long ago, thank goodness. You will see it, however, because the contents of memory are shown in the debugger and machine code is included in the "disassembly" (see later).

Assembly language: Little more than machine code translated into English. The instructions are written as words called *mnemonics* rather than binary values and a program called an *assembler* translates the mnemonics into machine code. It does a little more than direct translation, but not a lot, nothing like a compiler for a high-level language.

A major disadvantage of assembly language is that it is intimately tied to a processor and is therefore different for each architecture. Even worse, the detailed usage varies between development environments for the same processor. Most programming of small microcontrollers was done in assembly language until recently, despite these problems, mainly because compilers for C produced less-efficient code. Now the compilers are better and modern processors are designed with compilers in mind, so assembly language has been pushed to the fringes. A few operations, notably bitwise rotations, cannot be written directly in C, and for these assembly language may be much more efficient. However, the main argument for learning assembly language is for debugging. There is no escape if you need to check the operation of the processor, one instruction at a time. *Disassembly* is the opposite process to assembly, the translation of machine code to assembly language.

C: The most common choice for small microcontrollers nowadays. A compiler translates C into machine code that the CPU can process. This brings all the power of a high-level language—data structures, functions, type checking and so on—but C can usually be compiled into efficient code. Compilation used to go through assembly language but this is now less common and the compiler produces machine code directly. A disassembler must then be used if you wish to review the assembly language.

C++: An object-oriented language that is widely used for larger devices. A restricted set can be used for small microcontrollers but some features of C++ are notorious for producing highly inefficient code. Embedded C++ is a subset of the language intended for embedded systems. Java is another object-oriented language, but it is interpreted rather than compiled and needs a much more powerful processor.

BASIC: Available for a few processors, of which the Parallax Stamp is a well-known example. The usual BASIC language is extended with special instructions to drive the peripherals. This enables programs to be developed very rapidly, without detailed understanding of the peripherals. Disadvantages are that the code often runs very slowly and the hardware is expensive if it includes an interpreter.

1.7   Where Does the MSP430 Fit?

An enormous number of microcontrollers is available. They fill many pages of the distributors' catalog with a tiny typeface. Where does the subject of this book, the MSP430, fit into this spectrum?

The MSP430 was introduced in the late 1990s, although its ancestry goes back to the 4-bit TSS400. In summary, it is a particularly straightforward 16-bit processor with a von Neumann architecture, designed for low-power applications. The CPU is often described as a reduced instruction set computer (RISC) but this is debatable (if unimportant) and is considered in the section "Reflections on the CPU and Instruction Set" on page 153. Both the address and data buses are 16 bits wide. The registers in the CPU are also all 16 bits wide and can be used interchangeably for either data or addresses. This makes the MSP430 simpler than an 8-bit processor with 16-bit addresses. Such a processor must use its general-purpose registers in pairs for addresses or provide separate, wider registers.

In many ways, the MSP430 fits between traditional 8- and 16-bit processors. The 16-bit data bus and registers clearly define it as a 16-bit processor. On the other hand, it can address only $2^{16} = 64$ KB of memory. This is the same as the Freescale HCS08, an 8-bit family that also uses von Neumann memory and whose architecture goes back to the 6800 in the very early days of microprocessors. The corresponding 16-bit family is the HCS12, which uses paging to address up to 8 MB of memory. The absence of pages or banks in the memory makes the MSP430 very simple to use. (The picture is a little different with the MSP430X, which has extended registers and a wider address bus that can handle up to 1 MB of memory. I leave this until Chapter 11.)

Another feature of the MSP430 that stems from its recent introduction is that it is designed with compilers in mind. Most small microcontrollers are now programmed in C, and it is important that a compiler can produce compact, efficient code. The MSP430 has 16 registers in its CPU, which enhances efficiency because they can be used for local variables, parameters passed to subroutines, and either addresses or data. This is a typical feature of a RISC, but unlike a "pure" RISC, it can perform arithmetic directly on values in main memory. Microcontrollers typically spend much of their time on such operations.

The MSP430 is the simplest microcontroller in TI's current portfolio. Its more powerful siblings include the TMS470, which is based on the 32/16-bit ARM7, and the C2000, which incorporates a digital signal processor. Several features make the MSP430 suitable for low-power and portable applications:

- The CPU is small and efficient, with a large number of registers.

- It is extremely easy to put the device into a low-power mode. No special instruction is needed: The mode is controlled by bits in the status register. The MSP430 is awakened by an interrupt and returns automatically to its low-power mode after handling the interrupt.

- There are several low-power modes, depending on how much of the device should remain active and how quickly it should return to full-speed operation.

- There is a wide choice of clocks. Typically, a low-frequency watch crystal runs continuously at 32 KHz and is used to wake the device periodically. The CPU is clocked by an internal, digitally controlled oscillator (DCO), which restarts in less than 1 μs in the latest devices. Therefore the MSP430 can wake from a standby mode rapidly, perform its tasks, and return to a low-power mode.

- A wide range of peripherals is available, many of which can run autonomously without the CPU for most of the time.

- Many portable devices include liquid crystal displays, which the MSP430 can drive directly.

- Some MSP430 devices are classed as application-specific standard products (ASSPs) and contain specialized analog hardware for various types of measurement.

It is impossible to pick a single number to demonstrate the low-power consumption and great caution is needed when comparing different manufacturers' claims. For example, the F2013 draws around 4.5 mA when operating at its top speed of 16 MHz. This also needs its maximum supply voltage of 3.5 V. However, the supply can be reduced to 1.8 V and the current falls to 0.2 mA if a speed of 1 MHz is acceptable. In many applications, the microcontroller spends most of its time in standby mode, when a typical current is below 1 μA. Many batteries have a larger self-discharge current than this. The MSP430 can restart quickly because of its DCO, which may be an important factor in the overall power budget. The application note *MSP430 Competitive Benchmarking* (slaa205b) contains a comparison of the MSP430 with a range of other microcontrollers.

Currently four families of MSP430 are available. The letter after MSP430 shows the type of memory. Most part numbers include F for flash memory but some have C for ROM. There is a second letter for ASSPs to show the type of measurement for which they are intended: E for electricity, W for water, and G for signals that require a gain stage, provided by operational amplifiers. The next digit shows the family and the final two or three digits identify the specific device.

MSP430x1xx: Provides a wide range of general-purpose devices from simple versions to complete systems for processing signals. There is a broad selection of peripherals and some include a hardware multiplier, which can be used as a rudimentary digital signal processor. Packages have 20–64 pins.

MSP430F2xx: A newer, general-purpose family introduced in 2005. Its CPU can run at 16 MHz, double the speed of earlier devices, while consuming only half the current at the same speed. Some come in 14-pin packages, including a traditional plastic dual-in-line (PDIP) option, which is attractive for anybody who has to build circuits by hand. They do not require a crystal for their low-frequency clock. Pull-up or pull-down resistors are provided on the inputs to reduce the number of external components needed. There are many options for analog inputs. Even the smallest, 14-pin devices offer a 16-bit sigma–delta ADC.

This family is intended to supersede the MSP430x1xx over the next few years. Devices with the same pin-out and related model numbers are intended as drop-in replacements. For example, the F24x can replace the F14x.

MSP430x3xx: The original family, which includes drivers for LCDs. It is now obsolescent.

MSP430x4xx: Can drive LCDs with up to 160 segments. Many of them are ASSPs, but there are general-purpose devices as well. Their packages have 48–113 pins, many of which are needed for the LCD.

MSP430X: The original MSP430 architecture, extended to give the MSP430X in 2006, mainly so that it can address extra memory but with other improvements as well. Curiously, this is not marketed as a separate family: The devices are included in the MSP430F2xx and MSP430F4xx families with nothing in their part number to distinguish them. The CPU is a MSP430x if there is more than 64 KB of memory.

The letters MSP stand for *mixed signal processor*, which is a reminder that many practical applications require analog inputs. There is a selection of analog-to-digital converters with a resolution of up to 16 bits. An example of a system where this choice is important is the weighing machine shown in Figure 1.4, which is another project formerly used in my department. It includes the following functional blocks:

- The sensor has four resistive elements arranged as a Wheatstone bridge. Ideally, this is balanced when there is no load, giving $V_+ = V_-$. Two of the resistances increase and two decrease when a weight is placed on the scale pan, driving the bridge out of balance.

- A differential amplifier magnifies the difference in voltage between its input terminals, giving $V_{out} = A(V_+ - V_-)$, where A is the gain.

- The analog output of the amplifier is converted to a binary value in an analog-to-digital converter.

- The microcontroller multiplies the input by an appropriate factor so that the display gives the weight in grams or ounces and subtracts an offset so that the

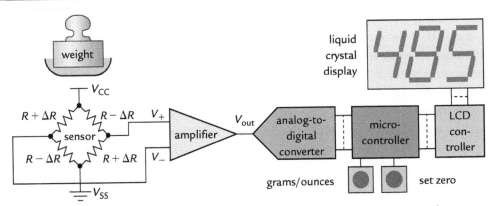

Figure 1.4: Weighing machine with a liquid crystal display, broken down into individual functions.

       display reads zero when no weight is present. It also reads the buttons and supervises the complete system.

- There is a serial interface between the microcontroller and the liquid crystal display, which has a built-in controller.

This system clearly needs a lot of components, including several integrated circuits. The project was designed this way for pedagogical reasons, to pull together functions that had been covered in other courses that students had taken on analog electronics, embedded systems, digital electronics, circuits, and so on. It was not intended as a practical product, although weighing machines would have been designed this way in the past.

In contrast, the whole system can be constructed from a sensor, an MSP430, a simple LCD without a controller, and a couple of decoupling capacitors. The MSP430x4xx family drives segmented LCDs directly, which eliminates the need for a controller. Several devices contain ADCs with high-resolution, differential inputs, which would work directly from the sensor without the need for an amplifer. The microcontroller can also manage the power drawn by the circuit so that the processor would be switched off when it was not needed and the whole system shut down after a period of inactivity. This is an ideal application for the MSP430—so well suited, in fact, that it is described in the application note *MSP430F42x Single-Chip Weight Scale* (slaa220). I focus on the MSP430 in the next chapter.

CHAPTER 2

The Texas Instruments MSP430

This chapter provides a review of the most important aspects of the hardware of the MSP430. It is a product of Texas Instruments, which I abbreviate to TI. The aim is to cover enough background for Chapter 4, where I tour through some simple programs that demonstrate the main functions. Individual peripherals are covered in detail later, as are some of the more intricate features of the CPU. The general plan follows roughly the layout of the data sheet. I concentrate on the original range of MSP430 devices, which can address 64 KB of memory. The extended MSP430X can address 1 MB but is otherwise similar. I point out the major differences but defer the details until Chapter 11.

I take as an example the MSP430F2013, or F2013 for short, which was introduced in 2005. It is one of the smallest MSP430s, with only 14 pins, but nevertheless contains a broad range of functions. It is also the target in TI's low-cost eZ430–F2013 development tool and its MSP430FG4618/F2013 Experimenter's Board. The F2003 is identical except for a smaller flash memory and a correspondingly lower price. The notation F20x3 refers to both the F2003 and F2013. It is common to find sets of devices that differ only in their memory; there are four variants of FG4616–FG4619, for instance. The F2003 and F2013 constitute one pair in a related set of devices (F2001, F2002, F2003, F2011, F2012, and F2013), known collectively as F20xx. The pairs differ mainly in their analog inputs.

2.1 The Outside View—Pin-Out

The *pin-out* shows which interior functions are connected to each pin of the package. There are several diagrams for each device, corresponding to the different packages in which it is produced. The F2013 is available in a traditional 14-pin plastic dual-in-line package (PDIP) with pins 0.1″ apart, which is a boon for hobbyists and students—a device that is large enough to solder easily by hand. There is also a plastic small-outline thin package (TSSOP). This has a similar shape but is a surface-mount device with pins 0.65 mm (about 0.025″) apart. A general warning: *Packages with the same shape do not always have the same pin-out.* This does not apply to the F2013, but I have been caught out by other companies' products. The third option is a tiny, quad flatpack no-lead (QFN) package, which is about 4 mm square with pads 0.65 mm apart. This has 16 leads instead of 14, which allows the analog and digital power connections to be brought out separately for better performance.

The pin-out for the PDIP and TSSOP packages is shown in Figure 2.1. Perhaps the most obvious feature is that almost all pins have several functions. This is typical of a modern, small microcontroller. Silicon is cheap but pins are expensive. The aim is therefore to integrate as many functions into as small a package as possible. Most applications do not use all the peripherals so, with luck, there is no conflict where a design needs more than one function on a pin simultaneously. A more detailed list of the functions on each pin is tabulated under *Terminal Functions* in the data sheet. Here is a sketchy description—the details are covered later in the section on each peripheral:

- V_{CC} and V_{SS} are the supply voltage and ground for the whole device (the analog and digital supplies are separate in the 16-pin package).

- P1.0–P1.7, P2.6, and P2.7 are for digital input and output, grouped into ports P1 and P2.

- TACLK, TA0, and TA1 are associated with Timer_A; TACLK can be used as the clock input to the timer, while TA0 and TA1 can be either inputs or outputs. These can be used on several pins because of the importance of the timer.

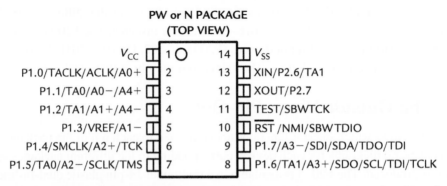

Figure 2.1: Pin-out of the MSP430F2003 and F2013, taken from the data sheet.

- A0−, A0+, and so on, up to A4±, are inputs to the analog-to-digital converter. It has four differential channels, each of which has negative and positive inputs. VREF is the reference voltage for the converter.

- ACLK and SMCLK are outputs for the microcontroller's clock signals. These can be used to supply a clock to external components or for diagnostic purposes.

- SCLK, SDO, and SCL are used for the universal serial interface, which communicates with external devices using the serial peripheral interface (SPI) or inter-integrated circuit (I^2C) bus.

- XIN and XOUT are the connections for a crystal, which can be used to provide an accurate, stable clock frequency.

- $\overline{\text{RST}}$ is an active low reset signal. *Active low* means that it remains high near V_{CC} for normal operation and is brought low near V_{SS} to reset the chip. Alternative notations to show the active low nature are _RST and /RST.

- NMI is the nonmaskable interrupt input, which allows an external signal to interrupt the normal operation of the program.

- TCK, TMS, TCLK, TDI, TDO, and TEST form the full JTAG interface, used to program and debug the device.

- SBWTDIO and SBWTCK provide the Spy-Bi-Wire interface, an alternative to the usual JTAG connection that saves pins.

A less common feature of the MSP430 is that some functions are available at several pins rather than a single one. This applies to Timer_A in the F2013. Channel 0 (TA0) is brought out to pins shared with P1.1 and P1.5 but their functionality is not identical: P1.1 can be used for both a "capture" input and a "compare" output but P1.5 is available only for output. Similarly TA1 is shared with P1.2, P1.6, and P2.6. All can provide an output, but only P1.2 can be configured for input to the timer. The details are set out in the table of *Terminal Functions* in the data sheet and explained further in the section on Timer_A2.

One of the first tasks in a program for a microcontroller is to configure the functions of each pin. All the pins can be used by a program except V_{CC} and V_{SS} for power and TEST/SBWTCK, which is reserved for debugging.

This is the view from *outside*: We next look *inside* to see how these pins are connected.

Example 2.1

Roughly, how many pins would the F2013 need if each had only one function?

2.2   The Inside View—Functional Block Diagram

Figure 2.2 shows a block diagram of the F2013. These are its main features:

- On the left is the CPU and its supporting hardware, including the clock generator. The emulation, JTAG interface and Spy-Bi-Wire are used to communicate with a desktop computer when downloading a program and for debugging.

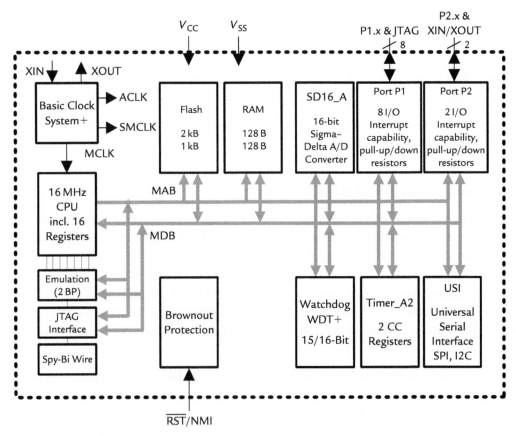

Figure 2.2: Block diagram of the MSP430F2003 and F2013, taken from the data sheet.

- The main blocks are linked by the *memory address bus* (MAB) and *memory data bus* (MDB).

- These devices have flash memory, 1 KB in the F2003 or 2 KB in the F2013, and 128 bytes of RAM.

- Six blocks are shown for peripheral functions (there are many more in larger devices). All MSP430s include input/output ports, Timer_A, and a watchdog timer, although the details differ. The universal serial interface (USI) and sigma–delta analog-to-digital converter (SD16_A) are particular features of this device.

- The brownout protection comes into action if the supply voltage drops to a dangerous level. Most devices include this but not some of the MSP430x1xx family.

- There are ground and power supply connections. Ground is labeled V_{SS} and is taken to define 0 V. The supply connection is V_{CC}. For many years, the standard for logic was $V_{CC} = +5$ V but most devices now work from lower voltages and a range of 1.8–3.6 V is specified for the F2013. The performance of the device depends on V_{CC}. For example, it is unable to program the flash memory if $V_{CC} < 2.2$ V and the maximum clock frequency of 16 MHz is available only if $V_{CC} \geq 3.3$ V.

TI uses a quaint notation for the power connections. The S stands for the source of a field-effect transistor, while the C stands for the collector of a bipolar junction transistor, a quite different device. The MSP430, like most modern integrated circuits, is built using complementary metal–oxide–silicon (CMOS) technology and field-effect transistors. I doubt if it contains any bipolar junction transistors except possibly in some of the analog peripherals.

There is only one pair of address and data buses, as expected with a von Neumann architecture. Some addresses must therefore point to RAM and some to flash, so it is a good idea to explore the memory map next.

2.3 Memory

The memory system can be viewed as a pile of pigeonholes, as we saw in the section "Memory" on page 11. Each register or pigeonhole holds 8 bits or 1 byte and this is the smallest entity that can be transferred to and from memory, as in most processors. The memory address bus is 16 bits wide so there are $2^{16} = 65,536 = 64\,\text{K} = 0\text{x}10000$ addresses. The first address is 0, like arrays in C, so the range is 0x0000 to 0xFFFF although not all these may be meaningful in a particular device. In other words, there may be no physical register at some addresses. On the other hand, 0 is a perfectly valid address (it is the Interrupt Enable 1 register, IE1), even though the NULL pointer in C usually has the value 0.

There are no connections for bringing the buses outside the chip to attach external memory. Nor is there any provision for increasing the amount of memory in the MSP430 by selecting banks or pages, as in some other processors. The MSP430X extends the range of memory by a factor of 16 to 2^{20} bytes by adding a further 4 bits to the address bus and the registers in the CPU.

The memory data bus is 16 bits wide and can transfer either a word of 16 bits or a byte of 8 bits. Bytes may be accessed at any address but words need more care. The address of a word is defined to be the address of the byte with the lower address, which must be even. Thus the two bytes at 0x0200 and 0x0201 can be considered as a valid word with address 0x0200, which may be fetched in a single cycle of the bus. This is shown in Figure 2.3. On the other hand, it is not possible to treat the two bytes at 0x0201 and 0x0202 as a single word because their address would be 0x0201, which is odd and therefore invalid. These two bytes straddle the boundary of two words. The memory system of some 16-bit processors can handle misaligned words like these, usually at the cost of an extra bus cycle, but the MSP430 does not. An important case is that instructions are composed of words and must therefore lie on even addresses.

Address	Bits				
0x0206	...				
0x0205	15, msb	14	... bits...	9	8
0x0204	7	6	... bits...	1	0, lsb
0x0203	byte				
0x0202	7, msb	6	... bits...	1	0, lsb
0x0201	more significant byte, MSB				
0x0200	less significant byte, LSB				
0x01FF	...				

Figure 2.3: Ordering of bits, bytes, and words in memory, adapted from the *MSP430x2xx Family User's Guide*. Addresses increase up the page.

Suppose that a word contains the hexadecimal value 0x1234. Its more significant (high-order) byte is 0x12 and the less significant (low-order) byte is 0x34. There are two ways in which these two bytes can be stored in the two bytes of a word in memory and both are in use.

Little-endian ordering: The low-order byte is stored at the lower address and the high-order byte at the higher address. This is used by the MSP430 and is the more common format.

Big-endian ordering: The high-order byte is stored at the lower address. This is used by the Freescale HCS08, for instance.

Little-endian ordering may appear more logical but has one awkward outcome. A debugger usually displays the contents of memory by showing the value of each byte by default. Addresses increase from left to right across each line. This means that the low-order byte is displayed first, followed by the high-order byte. Thus our value of 0x1234 is displayed as 34 12. It is easy to be puzzled by this. A simple solution is to display the contents of memory in words instead.

Bits within a byte or word are numbered with 0 being the least significant bit (lsb), conventionally written on the right, and 7 or 15 the most significant bit (msb). Unfortunately the same initials stand for most or least significant *byte* as well. I use LSB and MSB to distinguish these but TI use LSB and MSB for bits in their documents.

2.3.1 Memory Map

Figure 2.4 shows the memory map of the F2013. Maps are sometimes drawn with addresses increasing up the page, as one would normally draw the vertical axis on a graph. Otherwise they are drawn in the opposite sense, as one would write a list of numbers. I have followed TI's usage with addresses increasing upward.

Most MSP430 devices have a similar memory map, differing only in the size of the regions for RAM and code. The exceptions are devices with particularly large memory, such as the F1611 with 10 KB of RAM. On a large scale, the memory map shows the type of memory at each address or that an address is not used. On a finer scale, it shows the address and function of each register. These are listed in the data sheet for each device in the section "Peripheral File Map" on page 29. The CPU itself cares only about the numerical address of each register, but standard names are defined in the data sheet to make it easier for the programmer. These names are also understood by the development software.

Most devices have ranges of addresses that are not used, meaning that there are no registers at such addresses. These are shown in gray in Figure 2.4. In fact the largest regions are unused in devices with small memories, including this one.

Here is a brief description of each region:

Special function registers: Mostly concerned with enabling functions of some modules and enabling and signalling interrupts from peripherals.

Peripheral registers with byte access and peripheral registers with word access: Provide the main communication between the CPU and peripherals. Some must be accessed as words and others as bytes. They are grouped in this way to avoid wasting

Address	Type of memory
0xFFFF	interrupt and reset
0xFFC0	vector table
0xFFBF	flash code memory
0xF800	(lower boundary varies)
0xF7FF	
0x1100	
0x10FF	flash
0x1000	information memory
0x0FFF	*bootstrap loader*
0x0C00	*(not in F20xx)*
0x0BFF	
0x0280	
0x027F	RAM
0x0200	(upper boundary varies)
0x01FF	peripheral registers
0x0100	with word access
0x00FF	peripheral registers
0x0100	with byte access
0x000F	special function registers
0x0000	(byte access)

Figure 2.4: Memory map of the MSP430F2013, based on the data sheet and the *MSP430x2xx Family User's Guide.* **Addresses increase up the page and are not drawn to scale. Gray regions are unused and their size varies considerably between devices. The F2013 does not have a bootstrap loader but I have shown its location because it is present in most variants of the MSP430.**

addresses. If the bytes and words were mixed, numerous unused bytes would be needed to ensure that the words were correctly aligned on even addresses.

Random access memory: Used for variables. This always starts at address 0x0200 and the upper limit depends on the size of the RAM. The F2013 has 128 B.

Bootstrap loader: Contains a program to communicate using a standard serial protocol, often with the COM port of a PC. This can be used to program the chip but improvements in other methods of communication have made it less important than in the past, particularly for development. Details are given in the application note *Features of the MSP430 Bootstrap Loader* (slaa089). All MSP430s had a bootstrap loader until the F20xx, from which it was omitted to improve security.

Information memory: A 256 B block of flash memory that is intended for storage of nonvolatile data. This might include serial numbers to identify equipment—an address for a network, for instance—or variables that should be retained even when power is removed. For example, a printer might remember the settings from when it was last used and keep a count of the total number of pages printed. The information memory is laid out with smaller segments of flash than the code memory, which makes it more convenient to erase and rewrite. Segment A contains factory calibration data for the DCO in the MSP430F2xx family and is protected by default.

Code memory: Holds the program, including the executable code itself and any constant data. The F2013 has 2 KB but the F2003 only 1 KB.

Interrupt and reset vectors: Used to handle "exceptions," when normal operation of the processor is interrupted or when the device is reset. This table was smaller and started at 0xFFE0 in earlier devices.

The range of addresses has been extended from 64 KB to 1 MB in the MSP430X. This means that addresses require 20 bits rather than 16 and the MAB is therefore 4 bits wider. The bottom 64 KB of memory from 0x00000 to 0x0FFFF is laid out in exactly the same way as in the original MSP430. The additional memory, from 0x10000 to 0xFFFFF, is available for additional ROM. This permits larger programs and tables to be stored.

Example 2.2

What are the maximum sizes of RAM and flash memory if they expand to fill the adjacent unused regions? Include the vector table and information memory for the flash memory.

2.4   Central Processing Unit

The central processing unit (CPU) executes the instructions stored in memory. It steps through the instructions in the sequence in which they are stored in memory until it encounters a branch or when an *exception* occurs (interrupt or reset). It includes the arithmetic logic unit (ALU), which performs computation, a set of 16 registers designated R0–R15 and the logic needed to decode the instructions and implement them. The CPU can run at a maximum clock frequency f_{MCLK} of 16 MHz in the MSP430F2xx family and some newer MSP430x4xx devices, and 8 MHz in the others. It is built using *static* logic, which means that there is no minimum frequency of operation: The CPU can be stopped and will retain its state until it is restarted. This is essential for low-power operation to be straightforward.

The registers are shown in Figure 2.5. Each can hold a word of 16 bits. They do not have addresses in the main memory map, unlike some other designs of processors. The first four registers have dedicated functions with alternative names, while the remaining 12 R4–R15 are working registers for general purposes. Either words or bytes can be written to the CPU registers but the behavior for bytes is different from main memory: The destination is always the low byte and the high byte is cleared (reset to 0).

The generous set of 16 registers is characteristic of a reduced instruction set computer (RISC). This contrasts with the PIC16 family, for instance, where most operations on data must go through the working register W and indirect addressing is restricted to the single register FSR. The Freescale HCS12 is an example of a traditional, accumulator-based, 16-bit architecture in a complex instruction set computer (CISC). Its CPU contains the usual stack pointer, program counter, and condition code (status) register. In addition, there is a single accumulator for data and two index registers for addresses. These three dedicated registers can be compared with the 12 general-purpose registers in the MSP430, which can be used for either addresses or data. Moreover, we shall see in the section

15	... bits ...	0
R0/PC	program counter	0
R1/SP	stack pointer	0
R2/SR/CG1	status register	
R3/CG2	constant generator	
R4	general purpose	
	⋮	
R15	general purpose	

Figure 2.5: Registers in the CPU of the MSP430.

"Addressing Modes" on page 125 that the same addressing modes can be used with all registers, including the four dedicated ones, which greatly simplifies the instruction set.

I look briefly at the functions of the registers in the CPU. Full details are given in the section "Central Processing Unit" on page 119 after tour of the MSP430.

Program counter, PC: This contains the address of the next instruction to be executed, "points to" the instruction in the usual jargon. Instructions are composed of 1–3 words, which must be aligned to even addresses, so the lsb of the PC is hard-wired to 0.

Stack pointer, SP: When a subroutine is called, the CPU jumps to the subroutine, executes the code there, then returns to the instruction after the call. It must therefore keep track of the contents of the PC before jumping to the subroutine, so that it can return afterward. This is the primary purpose of the *stack*. Some processors use separate hardware for the stack but the MSP430 uses the top (high addresses) of the main RAM. The stack pointer holds the address of the most recently added word and is automatically adjusted as the stack grows downward in memory or shrinks upward.

Status register, SR: This contains a set of *flags* (single bits), whose functions fall into three categories. The most commonly used flags are C, Z, N, and V, which give information about the result of the last arithmetic or logical operation. The Z flag is set if the result was zero and cleared if it was nonzero, for instance. Decisions that affect the flow of control in the program can be made by testing these bits.

Setting the GIE bit enables maskable interrupts, which is explained in the section "Interrupts" on page 186. We do not use interrupts for the time being.

The final group of bits is CPUOFF, OSCOFF, SCG0, and SCG1, which control the mode of operation of the MCU. All systems are active when all bits are clear. Setting various combinations of these bits puts the MCU into one of its low-power modes, which is described in the section "Low-Power Modes of Operation" on page 198. We keep the CPU active for now.

Constant generator: This provides the six most frequently used values so that they need not be fetched from memory whenever they are needed. It uses both R2 and R3 to provide a range of useful values by exploiting the CPU's addressing modes. This will be explained in the section "Constant Generator and Emulated Instructions" on page 131 but the compiler or assembler handles the details automatically.

General purpose registers: The remaining 12 registers, R4–R15, are general working registers. They may be used for either data or addresses because both are 16-bit values, which simplifies the operation significantly.

The registers in the MSP430X have been widened to 20 bits (except SR) so that they can hold its longer addresses but otherwise they work in the same way.

2.5 Memory-Mapped Input and Output

Simple digital input and output takes place through sets of pins on the package of the integrated circuit called *ports*. Each port has up to 8 pins although not all may be available in a particular package. Many manufacturers label the ports with letters but TI uses numbers and the ports are called P1, P2, and so on (there is no P0 in current devices). For example, the F2013 has all pins of port P1 available, labeled P1.0–P1.7, but only pins P2.6 and P2.7 of port P2.

Typical pins can be configured for either input or output and some inputs may generate interrupts when the voltage on the pin changes. This is useful to awaken a system that has entered a low-power mode while it waits for input from a (slow) human. Pins usually have further functions as well as described in the section "The Outside View—Pin-Out" on page 21.

This is the view of the ports as seen from outside the package: How do they appear to the CPU inside? The MSP430, in common with most processors, uses *memory-mapped* input and output. This means that the ports simply appear to the CPU as particular memory registers called *peripheral registers*. Each port is associated with a byte and each bit corresponds to a pin on the package (if implemented). These registers can be read, written, and modified in almost the same way as simple registers in RAM. You can even do arithmetic with them, with some restrictions. For example, you simply read the register P1IN in the usual way to find the logical values on the inputs to port P1. Input voltages near ground, V_{SS}, give a logical 0 while voltages near the supply, V_{CC}, give a logical 1. The details of what is meant by *near* are discussed in the section "Analog Aspects of 'Digital' Inputs" on page 221. Similarly, writing to P1OUT drives the pins to the specified values if they are configured as outputs. As you might expect, writing a 0 drives the pin toward V_{SS} while a 1 drives it toward V_{CC}. Easy!

A different approach was taken in early Intel chips and their descendents, with a separate set of addresses for input and output. Different instructions were therefore needed for the ports, such as in and out.

Each port has several registers that are used to read, write, and configure it. Port P1 on the MSP430F2013 has eight, for instance (the maximum), whose details are given in the *MSP430x2xx Family User's Guide*. Each pin or bit can be controlled individually; it is not necessary to make a whole port behave in the same way. These are the three most important registers for port P1; the remainder are described in the section "Digital Input and Output: Parallel Ports" on page 208.

Port P1 input, P1IN: Reading returns the logical values on the inputs if they are configured for digital input and output. This register is read-only. It is also *volatile*, which means that it may change at a time that a program cannot predict. This is of course why the input port is there, so that the MCU can react to its surroundings. You want to know when a user presses a button, for instance. This is a reminder that P1IN is not just a simple memory and that some care is needed when programming in C (see the section "Definition of Peripheral Registers in C" on page 115).

Port P1 output, P1OUT: Writing sends the value to be driven onto the pin if it is configured as a digital output. If the pin is not currently an output, the value is stored in a buffer and appears on the pin if it is later switched to be an output.

Port P1 direction, P1DIR: A bit of 0 configures the pin as an input, which is the default. Writing a 1 switches the pin to become an output.

Example 2.3

Why is it useful to provide separate registers for input and output? Many microcontrollers have only one.

2.5.1   Other Peripherals

A list of the peripherals that can be found in typical microcontrollers, such as timers, was given in the section "Anatomy of a Typical Small Microcontroller" on page 8. They are controlled in the same way as the ports, through memory-mapped peripheral registers. Some of these registers hold words rather than bytes, which must be checked when using assembly language. A few larger devices have a direct memory access (DMA) controller, which enables peripherals to use the main memory without intervention from the CPU. This allows operations to proceed while the CPU is shut down to save power.

2.6   Clock Generator

We can leave most of the peripherals until later, but a clock is essential for every synchronous digital system. Full details follow in the section "Clock System" on page 163.

Basically the clock signal is a square wave whose edges trigger hardware throughout the device so that the changes in different components are synchronized.

Clocks for microcontrollers used to be simple. Usually a crystal with a frequency of a few MHz would be connected to two pins. It would drive the CPU directly and was typically divided down by a factor of 2 or 4 for the main bus. Unfortunately, the conflicting demands for high performance and low power mean that most modern microcontrollers have much more complicated clocks, often with two or more sources. In many applications the MCU spends most of its time in a low-power mode until some event occurs, when it must wake up and handle the event rapidly. It is often necessary to keep track of real time, either so that the MCU can wake periodically (every second or minute, for instance) or to time-stamp external events. Therefore, two clocks with quite different specifications are often needed:

- A fast clock to drive the CPU, which can be started and stopped rapidly to conserve energy but usually need not be particularly accurate.

- A slow clock that runs continuously to monitor real time, which must therefore use little power and may need to be accurate.

Several types of oscillator are used to generate the clock signal. These are the two extreme types, which are also the most common:

Crystal: Accurate (the frequency is close to what it says on the can, typically within 1 part in 10^5) and stable (does not change greatly with time or temperature). Crystals for microcontrollers typically run at either a high frequency of a few MHz to drive the main bus or a low frequency of 32,768 Hz for a real-time clock. The disadvantages are that crystals are expensive and delicate, the oscillator draws a relatively large current, particularly at high frequency, and the crystal is an extra component and may need two capacitors as well. Crystal oscillators also take a long time to start up and stabilize, often around 10^5 cycles, which is an unavoidable side effect of their high stability.

Resistor and capacitor (*RC*): Cheap and quick to start but used to have poor accuracy and stability. The components can be external but are now more likely to be integrated within the MCU. The quality of integrated *RC* oscillators has improved dramatically in recent years and the F20xx provides four frequencies calibrated at the factory to within ±1%.

There are also ceramic resonators that lie between these extremes, being cheaper but less accurate and stable than crystals. This list may change in the future with the rise of micro-electro-mechanical systems (MEMS). It may become possible to integrate a mechanical resonator on the same silicon chip as the MCU, which will combine high accuracy with low power.

Example 2.4

Why is a crystal with the strange-looking frequency of 32,768 Hz convenient for a real-time clock? (This is called a watch crystal.)

The MSP430 addresses the conflicting demands for high performance, low power, and a precise frequency by using three internal clocks, which can be derived from up to four sources. These are the internal clocks, which are the same in all devices:

- **Master clock, MCLK**, is used by the CPU and a few peripherals.

- **Subsystem master clock, SMCLK**, is distributed to peripherals.

- **Auxiliary clock, ACLK**, is also distributed to peripherals.

Typically SMCLK runs at the same frequency as MCLK, both in the megahertz range. ACLK is often derived from a watch crystal and therefore runs at a much lower frequency. Most peripherals can select their clock from either SMCLK or ACLK. The available sources vary considerably between families and individual devices. Rather than run through all of these now, here are the default configurations for the families. These are suitable for many applications with only minor changes. First, the MSP430x1xx and MSP430F2xx families:

- ACLK comes from a low-frequency crystal oscillator, typically at 32 KHz.

- Both MCLK and SMCLK are supplied by an internal *digitally controlled oscillator* (DCO), which runs at about 0.8 MHz in the MSP430x1xx and 1.1 MHz in the MSP430F2xx.

Note that ACLK requires an external crystal, which is supplied on most demonstration boards but not TI's basic development kits nor the eZ430–F2013. No crystal means no ACLK with most MSP430s. However, devices in the MSP430F2xx family (except the F21x1, which were first to be introduced) include an internal, very low-power, low-frequency oscillator (VLO) that can be selected for ACLK if there is no crystal.

One of the most important characteristics of the DCO is that it starts very rapidly at full speed, taking less than 1 μs in the MSP430F2xx. This is a critical feature for a low-power system.

The default MCLK and SMCLK in the MSP430x4xx family are different but ACLK is the same.

- ACLK comes from a low-frequency crystal oscillator, typically at 32 KHz.

- MCLK and SMCLK are supplied from the DCO, which is controlled by a frequency-locked loop (FLL). This locks the frequency at 32 times the ACLK frequency, which is close to 1 MHz for the usual watch crystal.

Again, no crystal means no ACLK. It appears that the MCU is paralyzed if no crystal is attached. However, the clock generator will detect a fault on ACLK and allow the DCO to run at its lowest frequency. Thus MCLK and SMCLK remain available.

Many applications do not require precise timing, and the DCO is satisfactory for the main clock. At present, there is no alternative to a crystal for an accurate frequency so many systems use a watch crystal for ACLK. Therefore, the default configuration is widely applicable, although the DCO often needs to be configured to a higher frequency.

2.7 Exceptions: Interrupts and Resets

Execution of a program usually proceeds predictably, but there are two classes of *exception* to this rule: interrupts and resets:

Interrupts: Usually generated by hardware (although they can be initiated by software) and often indicate that an event has occurred that needs an urgent response. A packet of data might have been received, for instance, and needs to be processed before the next packet arrives. The processor stops what it was doing, stores enough information (the contents of the program counter and status register) for it to resume later on and executes an *interrupt service routine* (ISR). It returns to its previous activity when the ISR has been completed. Thus an ISR is something like a subroutine called by hardware (at an unpredictable time) rather than software. A second use of interrupts, which is particularly important in the MSP430, is to wake the processor from a low-power state.

Resets: Again usually generated by hardware, either when power is applied or when something catastrophic has happened and normal operation cannot continue. This can happen accidentally if the watchdog timer has not been disabled, which is easy to forget. A reset causes the device to (re)start from a well-defined state.

The CPU must be told where to fetch the next instruction following an interrupt or reset. The address of this instruction is called a *vector* and can be specified in different ways. The simplest is to have a single vector for resets and another for all interrupts. This is adopted by the Microchip PIC16, for instance, which executes the instruction at address 0x0000 following a reset and that at 0x0004 following an interrupt. A similar approach is used in the ARM7TDMI core, although this has two interrupts (normal and fast). The problem with this method is that the interrupt service routine has no idea what caused the interrupt and must first check all the flags to determine the source. This can be a lengthy process, which conflicts with the purpose of using interrupts for urgent tasks.

The MSP430 uses *vectored interrupts* instead. Each ISR has its own vector, which is stored at a predefined address in a *vector table* at the end of the program memory (addresses 0xFFC0–0xFFFF). The address for each interrupt vector is listed in the data sheet. The vector table is at a fixed location, but the ISRs themselves can be located anywhere in memory. There is no need to hunt for the source of the interrupt in most cases, although a few interrupts have multiple sources. For example, the interrupt vector for the watchdog timer is stored at address 0xFFF4. When the watchdog timer requests an interrupt, the processor fetches the vector from the word stored at 0xFFF4, loads it into the program counter, and fetches its next instruction from this address to start the ISR. Most processors use a similar scheme and a vectored interrupt controller is added to the ARM7TDMI when it is packaged into a complete microcontroller.

The same method is used for a reset, whose vector is stored in the very last word of memory at 0xFFFE. When a reset occurs, most registers in the device are loaded with the initial conditions specified in the family user's guide. The program counter is then loaded with the reset vector. This is the address of the first instruction, which is next fetched from memory to start normal operation. Thus the first instruction can be stored anywhere in memory. This is not all the initialization that is needed; the user's program must do the rest. In particular, the watchdog timer must be configured or disabled before it times out and resets the device. The stack should be initialized if subroutines or interrupts are used, which applies to all but the most trivial programs. This is followed by code to configure the other peripherals used, notably the ports and clock.

2.8   Where to Find Further Information

Even a small, cheap microcontroller is a complicated system and a great deal of information is needed to exploit it fully. One of the pleasures of working in electronics is the high quality of the information provided by most manufacturers and TI is no exception. You will certainly need the data sheet for the particular device you are using and the corresponding family user's guide. Make sure that you have the up-to-date versions of these from the TI Web site, which should be your first point of reference.

2.8.1   Data Sheet

You will soon become familiar with the layout of this. I have referred to several sections already, but here is a brief summary of how the data sheet is organized, taking the MSP430F2013 as an example again. This data sheet, like most, covers a range of related devices: MSP430x20x1, MSP430x20x2, and MSP430x20x3. Here are the main sections:

Front page: Gives a brief overall description. This is useful for picking out the crucial features of a device when you look for particular set of peripherals, such as an analog-to-digital converter and an I^2C interface. There is inevitably some "cherry-picking" to show the device at its best. For example, the current in active mode is quoted as $220\,\mu A$ at 1 MHz and 2.2 V; later it is mentioned that the CPU can run up to 16 MHz. Obviously you do not get the low current at the same time as the fast clock. This might sound a bit "economical with the truth" but is not really: It is only fair to highlight the critical features on which a choice of device might be based. In any case, the full data are in the document, and it is often the case that you are looking for one aspect in particular.

Device pin-out: Shows the style of packages and their connections (Figure 2.1).

Functional block diagram: Shows the main systems within the integrated circuit; see the description of Figure 2.2.

Terminal functions: Show what peripherals can be connected internally to each pin and expands the information shown on the pin-out.

Short-form description: Gives a brief summary of the CPU, instruction set, operating (low-power) modes, and interrupt vector addresses.

Special function registers: Mainly control interrupts from the central functions rather than the peripherals.

Memory organization: Gives the main features of the memory map.

Flash memory: Gives further information on the segments of flash memory, which is important if you wish to erase it.

Peripherals: A brief summary for most modules, because they are described fully in the family user guide, although it is not always obvious how data are split between the two.

This section is most important for peripherals that have general-purpose inputs and outputs, which may be connected internally or externally in different ways. An example is to Timer_A. It has a choice of four clock inputs, for instance. Two of these are the general clocks, ACLK and SMCLK, but the others, INCLK and TACLK, may be derived from different sources. In some devices they are brought out independently but here INCLK = $\overline{\text{TACLK}}$. Either one of these may be connected to P1.0 if desired. This allows the timer to be clocked by either the positive or the negative edge of the input clock.

Peripheral file map: Lists the peripheral registers with their standard names, which should be recognized by the compiler or assembler.

Electrical characteristics: Cover a vast range of information. Keep safely within the "absolute maximum ratings" unless you wish to damage your device. You probably have to concentrate on one or more sections closely for a particular application. For example, suppose that the lifetime of a battery is most important. Several tables and plots show how the current drawn by the device depends on its operating mode, frequency, supply voltage, temperature, and so on.

Application information: Particularly lengthy in this data sheet due to separate sections for the three pairs of devices included. The most helpful parts are the tables that show how to configure each pin for its various functions (see the section "Memory-Mapped Input and Output" on page 32). The detailed circuitry associated with the pins is useful for hardware designers. For example, it shows that the inputs have Schmitt triggers, which are useful to reduce noise (see the section "Analog Aspects of 'Digital' Inputs" on page 221).

The detail in a good data sheet can seem overwhelming and it requires practice to select the critical information. Ron Mancini wrote an excellent article, "How to Read a Semiconductor Data Sheet" [51]. It focuses on operational amplifiers but the principles apply equally to microcontrollers. I want to emphasize one issue. Many quantities have *minimum*, *typical*, and *maximum* values tabulated, although often not all three. For example, the supply current in active mode I_{AM} under certain conditions has the following values listed:

- Minimum, unspecified.

- Typical, 220 μA.

- Maximum, 270 μA.

The *typical* value was found by measuring a large number of devices and taking an average. It is a fair bet that you would find a similar value if you were to buy a device and measure its current under the same conditions, which are carefully set out in the data sheet.

Now, suppose that this parameter were crucial to your design, which had to meet a challenging specification for the lifetime of a battery. Should you rely on the typical value? The answer would definitely be "No" if you were planning to make a large number of products. You should always design on the basis of the worst case, which in this example is the *maximum* supply current. You may need to read the data sheet closely to determine whether this parameter is tested. In other words, does the manufacturer guarantee that every device will have $I_{AM} \leq 270\,\mu\text{A}$ or is this merely a prediction based on testing one batch?

On the other hand, you might be building only a single system. In this case, you could probably afford to buy a few extra MCUs, test them, and select the chip with the lowest current. It would probably be safe to base your design on the typical value. Finally, no *minimum* value is specified. This is mainly because it would be useless for designing, which should normally be based on the worst case. Besides, the ideal value would be zero.

The figures quoted by reputable companies, such as TI, are usually conservative and apply over the full range of specified operating conditions. For example, the output available from the ports is specified over the full range of operating temperatures, -40 to $+85°C$. In many applications, your device is unlikely to experience the extremes of this range and you may feel able to take a more optimistic figure than the worst case. Of course you will be in trouble if a user takes your product to Antarctica.

Always check for *errata* to the data sheets. Some of the "errors" are long standing and effectively built into the device. For example, the CPU4 bug affects the (push) instruction but a workaround has been built into development software for years and the errata states "No fix planned." Other issues may concern specific devices. I flagged a few of the errata that affect programs and devices in this book.

2.8.2   Family User's Guide

The data sheet for a particular device should be read in conjunction with the corresponding family user's guide. The MSP430F2013 needs the *MSP430x2xx Family User's Guide* (slau144), for instance. This provides a detailed description of all the modules within the family, including the CPU. The separation keeps the individual data sheets shorter and is convenient because modules are implemented in the same way in most devices. As an example, all MSP430s have an identical Timer_A except for the number of channels. Highly specialized modules, such as the operational amplifiers, vary more between devices.

Near the front of the family user's guide is a helpful glossary and table of the notation used for registers. The general format is that each module has an introduction followed by a description of its function. A list of the associated registers is at the end. You will spend a lot of time reading this while you work out how to configure each peripheral. It also gives the values of each bit after reset to show how the device starts up.

2.8.3   FET User's Guide

It might seem odd to list this specially, but the *MSP-FET430 Flash Emulation Tool (FET) User's Guide* (slau138) is a mine of information even if you do not use the specific hardware or software for which it is intended. The main reason is the list of frequently asked questions (FAQs) on hardware, program development, and debugging. These are extremely helpful and I suggest that you print them for easy reference.

This document also describes the hardware built into the MSP430 for debugging and emulation. For a long time, it was not mentioned in the family user's guides but a section *Embedded Emulation Module (EEM)* has recently been added. It is helpful to know how many breakpoints are available when debugging and how much information about execution is stored on the chip. There is more in the application note *Using the MSP430 EEM Debug Feature with IAR 3.30* (slaa263).

2.8.4   Application Notes

Well over 100 application notes for the MSP430 are on the TI Web site at the time of writing. Some are fairly general, such as *MSP430 Software Coding Techniques* (slaa294), but most show how to solve specific problems. It is always worth looking through the list to see if somebody has solved your problem already.

I must pick out one of these application notes for special mention: Lutz Bierl's *Application Reports* (slaa024). Do not hit the print button automatically because this runs to 1088 pages. It is unfortunately somewhat out of date, being based on the obsolescent MSP430x3xx family, but the detail is remarkable. It covers software and hardware, including the CPU and peripherals. The author is one of the architects of the MSP430 and there is some overlap with his more recent book [2].

2.8.5   Code Examples

These are available for individual devices and the range of flash emulation tools, usually in both C and assembly language. For example, slac080 (C) and slac081 (assembly) are for the MSP430F2013 and related devices. They start with a basic program to toggle an LED using a software delay loop, `msp430x20x3_1.c` in this case, which I investigate in the section "Automatic Control Flashing Light by Software Delay" on page 91. The other programs illustrate the full range of peripherals. There are nearly 20 examples for Timer_A, for instance. (Remember not to try those that need ACLK if there is no crystal in your system unless your device includes the VLO.)

2.8.6   Training

TI offers various forms of training on the MSP430. The 430 Day is an introductory course while the MSP430 MCU Advanced Technical Conference (ATC) lasts for 2.5 days and offers in-depth training on many aspects of the MSP430. It is not cheap but the material is excellent. The community is small enough that you get to meet many of the people who designed the chip, wrote the application notes, and so on. There are also online courses, which are listed on the TI Web site.

2.8.7   Other Resources

Many other resources are listed on the TI Web site, including the third party network and discussion groups. You may wish to join the Yahoo MSP430 User's Group.

Example 2.5

Suppose that a F2013 needs to carry out a sequence of tasks that require a certain number of clock cycles. How should the system be designed for the minimum drain from a battery (current $\times$ time)? Is it best to use the maximum clock frequency, $f_{MCLK} = 16\,\text{MHz}$, to run for the shortest time? The current is lower if f_{MCLK} is reduced, but does this compensate for the longer time required to perform the tasks? Remember that the maximum frequency of MCLK depends on V_{CC}. Look at the figures in the data sheet for supply current in active mode. You might consider (a) 1 MHz as an example of a low frequency, (b) the maximum frequency permitted at the lowest value of V_{CC}, and (c) the maximum frequency permitted, 16 MHz. Alternatively, if you are more analytically minded, find an expression for the current as a function of f_{MCLK} and V_{CC}.

CHAPTER 3

Development

It has become dramatically easier and cheaper to develop programs for microcontrollers in the past decade or so. There are several reasons for this. Perhaps the most significant is the prevalence of flash memory, which makes it easy to download successive versions of a program into the MCU in its demonstration board or final system. Before that it was possible to download programs quickly into EPROM, although this often required a separate programmer, but it took a quarter of an hour to erase the chip under an ultraviolet light before a new program could be written. Development required much more complicated systems before EPROM was widely available.

The second advance is the provision of hardware for debugging and emulation on-chip. This is due to the steady reduction in the cost of transistors on silicon; it is cheaper to add a few thousand transistors than an extra pin (although silicon is not free). Even the cheapest MSP430, the 55¢ F2001, has hardware to set breakpoints on-chip, although it lacks the more comprehensive hardware of the larger devices. Advanced debugging formerly required expensive, complicated in-circuit emulators but on-chip emulation has largely made these redundant for small microcontrollers.

The third step forward is that almost all manufacturers now offer free development software. Inevitably the features are limited, but it is more than adequate for learning how to use the chip. The quality of this software has also improved, particularly debuggers.

Some aspects of developing software for embedded systems are much the same as for a desktop computer. You still have to write and compile the code, for instance. The processes diverge when it comes time to run and, inevitably, debug the system. Software for a desktop computer is usually developed on the same type of computer but this is not true for an embedded system. The code must be downloaded into the target (microcontroller), which resides in the system that it is intended to control. What do you do when it fails? How can you see what is going on inside the MCU? I suspect that we have all been reduced to staring at a chip in desperation, hoping that it will come to life and tell us what is wrong. There is nothing like the screen of a desktop computer; typically you might have a few light-emitting diodes (LEDs). If you really want to find out what is happening on the pins you need an oscilloscope or logic analyzer. This is where debugging an embedded system diverges seriously from a pure software project. Of course, the emulation and debugging hardware and software provide a good view of what is happening inside the chip, but the challenge is usually to localize the problem or devise a test that will stop the system when the fault occurs.

3.1 Development Environment

A development system includes some or all of these functions. The editor, compilers, and linker are usually combined into an *integrated development environment* (IDE), which keeps track of all the files required for a complete project—source files, headers, linker scripts, and so on—and looks much like you would expect for a desktop computer.

Editor: Used to write and edit programs (usually C or assembly code). A good editor helps you to lay out the code logically and use colors to highlight syntax. It is very helpful to have a quick way of locating the definition of symbols in header files, because these are used heavily in embedded systems.

Assembler or compiler: Produces executable code and checks for errors, preferably providing helpful messages. There may be add-ons to provide extra checking, such as MISRA C or "lint" for C programs. The degree of optimization can be changed; typically, you want limited optimization or none at all during debugging. Different dialects of C may be available with extensions for embedded systems.

Linker: Combines compiled files and routines from libraries and arranges them for the correct types of memory in the MCU.

Stand-alone simulator: Simulates the operation of the MCU on a desktop computer without the real hardware. Many simulators model only the CPU and memory but some also include peripherals. Simulation has the advantage that you have complete control and view of the MCU; the disadvantage is that it is tricky to simulate inputs and interrupts, which are an important part of most systems.

Embedded emulator/debugger: Allows software to run on the MCU in its target system under the control of a debugger running on a desktop computer, which usually has the same front end as the simulator. The computer and MCU communicate through a special interface, JTAG for the MSP430. This needs special emulation hardware on the MCU, which improves all the time.

In-circuit emulator: Specialized and expensive ($1000s) hardware that emulates the operation of the MCU under the control of debugging software running on a desktop computer. It does not use the usual version of the MCU but a special "bond-out" chip that provides access to the internal hardware of the emulated device. It is difficult even to connect an emulator in place of a modern MCU, whose pins (if any) may be only 0.5 mm apart. Highly sophisticated breakpoints and other traces can be made, but this equipment is now largely obsolete for small microcontrollers, which contain embedded emulation for free.

Flash programmer: Downloads ("burns") the program into flash memory on the MCU. This is done automatically by the debugger in development but a dedicated programmer is used in production.

A good choice of development systems for the MSP430 is listed on TI's Tools and Software page. Two are available free from TI themselves:

IAR Embedded Workbench Kickstart: The free version of IAR Embedded Workbench for MSP430 (EW430). It is limited to 4 KB of C code and has some other restrictions. There are several commercial versions with fewer restrictions and an evaluation version with a time limit of 30 days. Most of the documentation and examples provided by TI are currently written for EW430 and I have therefore used EW430 for this book. There is a detailed "walk through" in Appendix A. It is a fairly mainstream product, and you will probably not find any surprises if you have used another IDE. An advantage for learners is that it is available for a broad range of MCUs: Your hard-earned experience will not be wasted if you are subsequently assigned to a different device.

Code Composer Essentials Evaluation: The free version of Code Composer Essentials for the MSP430 (CCE), limited to 16 KB of C code. It was developed by TI itself but is less closely related to Code Composer for its DSP chips than you might expect from the name. It is based on the open-source Eclipse SDK. A few documents on the TI Web site support CCE, notably a special version of the *MSP-FET430 FLASH*

Emulation Tool (FET) User's Guide (slau157). Some of the usage, particularly for assembly language, is entirely different from EW430.

Several third-party IDEs are listed as well. Most offer an evaluation version so that you can try them out. Their look and feel can be distinctive, so try to find one in which you feel comfortable.

Typical IDEs offer at least two languages, C and assembly. I use both of these throughout the book and now review them briefly. There may also be support for C++ but I do not use this except the symbol for comments, mentioned later.

3.2   The C Programming Language

This is now the language of choice for small microcontrollers, for many good reasons. Its built-in and user-defined types, data structures, and flexible control flow make it far more efficient and reliable to write in C than assembly language. At the same time, it is a sufficiently low-level language that compilers can use to produce efficient code. This is aided by the architecture of modern processors, such as the MSP430, which are designed with compilers in mind rather than for hand-crafted assembly code.

I assume that you have a basic, working knowledge of C, which can be learned from any of the standard textbooks. A few features are more common in programs for embedded systems than those for desktop computers, which I review briefly.

3.2.1   Aspects of C for Embedded Systems

Programs for small embedded systems tend not to contain a lot of complicated manipulation of complex data objects. Instead, much code is usually devoted to the control of peripherals through their special registers. This means that the details of individual bits, bytes, and words are often important.

Declarations

The `const` and `volatile` qualifications are often critical, particularly to define special function registers. Their addresses must be treated as constant but the contents are often volatile, so this is a good exercise in the meaning of these key words.

const: Means that the value should not be modified: it is constant.

volatile: Means that a variable may appear to change "spontaneously," with no direct action by the user's program. The compiler must therefore not keep a copy of the variable in a register for efficiency (like a cache). Nor can the compiler assume that the variable remains constant when it optimizes the structure of the program—rearranging loops, for instance. If the compiler did either of these, the program might miss externally induced changes to the contents of the memory associated with the variable.

The peripheral registers associated with the input ports must obviously be treated as volatile in an embedded system. The values in these depend on the settings of the switches or whatever is connected to the port outside the MCU. Clearly the compiler must not assume that these never change. Here is a simple example, where the program waits for the value on its input port P1IN to change from the value OldP1IN. The loop would be pointless, and would be optimized out of existence, if P1IN were not declared volatile.

```
while (P1IN == OldP1IN) {                    // wait for change in P1IN
}
```

The use of these key words will be illustrated at length in the section "Definition of Peripheral Registers in C" on page 115. Special definitions are often set up with typedef to save repeating the details.

Shifts

It is sometimes necessary to shift bits in a variable, most commonly when handling communications. For example, serial data arrives 1 bit at a time and needs to be assembled into bytes or words for further processing. This would be done in hardware with a shift register and similarly by a shift operation in software. Assignment operators can also be used for shifts so you will see expressions like value <<= 1 to shift value left by one place.

Right-shifts work differently for signed and unsigned numbers. The operations are called *arithmetic* and *logical shifts*, and I explain the difference in the section "Shift and Rotate Instructions" on page 137.

Multiplication and division by power of 2 can be implemented efficiently as shifts but there is little point in writing this out yourself: Modern compilers do this sort of optimization automatically.

Low-Level Logic Operations

It is essential to distinguish between the Boolean logic operations and their bitwise counterparts:

- The *Boolean* form, such as the AND operation in `if (A && B)`, treats A and B as *single* Boolean values. Typically A $= 0$ means false and A $\neq 0$ means true. (A `bool` type in C++ and C99 can be used by including the header file `stdbool.h`.)

- In contrast, the *bitwise* operator `&` acts on each corresponding pair of bits in A and B *individually*. This means that eight operations are carried out in parallel if A and B are bytes. For example, if A = `10101010` and B = `11001100`, then A & B = `10001000`.

Similarly, the bitwise ordinary (inclusive) OR operation gives A | B = `11101110` and exclusive-or (XOR) gives A ^ B = `01100110`.

Masks to Test Individual Bits

The smallest entity that can be handled by the CPU is a byte but it is often necessary to test or modify individual bits in an embedded system. This can be done with bitwise logic operations. Suppose that we want to know the value of bit 3 on the input of port 1, for instance. This means bit 3 of the byte P1IN, abbreviated to P1IN.3, for which we can use the standard definition BIT3 = `00001000`. A pattern used for selecting bits is called a *mask* and has all zeroes except for a 1 in the position that we want to select. Now consider the bitwise AND operation TestP13 = P1IN & BIT3. All bits of TestP13 will be zero except for bit 3 because x & 0 = 0 for either value of the bit x. Bit 3 of TestP13 follows from x & 1 = x and is therefore made equal to bit 3 of P1IN. Overall, TestP13 is zero if P1IN.3 = 0 and nonzero if P1IN.3 $\neq 0$. This is often used in decisions, as shown next:

```
if ((P1IN & BIT3) == 0) {          // Test P1.3
    // Actions for P1.3 == 0
} else {
    // Actions for P1.3 != 0
}
```

The parentheses around (P1IN & BIT3) are vital because of the rules for precedence in C. See the section "Read Input from a Switch" on page 80 for this control flow in action. Always test masked expressions against 0. It is tempting to write `if ((P1IN &`

BIT3) == 1) for the opposite test to that above. Unfortunately it is never true because the nonzero result is BIT3, not 1. Use if ((P1IN & BIT3) ! = 0) instead.

Masks to Modify Individual Bits

Masks can also be used to modify the value of individual bits. Start with the inclusive OR operation. Its truth table can be expressed as x | 0 = x and x | 1 = 1. In words, taking the OR of a bit with 0 leaves its value unchanged, while taking its OR with 1 sets it to 1. Thus we could set (force to 1) bit 3 of port 1 with P1OUT = P1OUT | BIT3, which leaves the values of the other bits unchanged. This is usually abbreviated to P1OUT |= BIT3.

Clearing a bit (to 0) is done using AND, as in the previous section, but is slightly more tricky. The truth table can be expressed as x & 0 = 0 and x & 1 = x. Thus we AND a bit with 1 to leave it unchanged and clear it with 0. In this case, the mask should have all ones except for the bit that we wish to clear, so we use ~BIT3 rather than BIT3. Thus we clear P1OUT.3 with P1OUT &= ~BIT3.

Finally, bits can be toggled (changed from 0 to 1 or 1 to 0) using the exclusive-or operation. For example, P1OUT ^= BIT3 toggles P1OUT.3.

There is no need for the mask to contain only a single 1; masks with several nonzero bits can be used to operate on multiple bits. On the other hand, it seems a bit primitive to use masks at all, and bit fields offer another approach.

Bit Fields

Bit fields resemble structures except that the number of bits occupied by each member is specified. Here is an example for the Timer_A Control register TACTL, which I describe in the section "Automatic Control: Flashing a Light by Polling Timer_A" on page 105. The definition is taken from the standard EW430 header file io430x11x1.h, slightly reformatted to fit. Concentrate on struct {...} TACTL_bit for now; I explain the rest shortly.

```
__no_init volatile union {
    unsigned short TACTL;          // Timer_A Control
    struct {
        unsigned short TAIFG  : 1;  // Timer_A counter interrupt flag
        unsigned short TAIE   : 1;  // Timer_A counter interrupt enable
        unsigned short TACLR  : 1;  // Timer_A counter clear
        unsigned short        : 1;
        unsigned short TAMC   : 2;  // Timer_A mode control
        unsigned short TAID   : 2;  // Timer_A clock input divider
```

```
        unsigned short  TASSEL  : 2;  // Timer_A clock source select
        unsigned short          : 6;
    } TACTL_bit;
} @ 0x0160;
```

The number after the colon shows the number of bits in the field. Each field often holds a single bit but there are several exceptions here; the mode is controlled by the pair of bits TAMC, for instance. The unimplemented bits of the register must be included in the structure to keep the alignment correct but do not need names. Now we can use a field instead of a mask to test the TAIFG bit with if (TACTL_bit.TAIFG == 0). Similarly, this bit can be

- **Set** with TACTL_bit.TAIFG = 1.

- **Cleared** with TACTL_bit.TAIFG = 0.

- **Toggled** with TACTL_bit.TAIFG ^= 1.

Nothing is magical about bit fields: The compiler produces the same code whether you use masks or fields. Usually it is more straightforward to use fields, so I suggest that you do this and let the compiler convert them into masks. It is much more reliable than us.

There is a pitfall if you define your own bit fields, which is their ordering. It is not specified in C whether bits should be allocated from left to right or right to left. In other words, is the first bit the msb or the lsb? This is up to the compiler. The allocation is right to left, starting with the lsb, in EW430 but another compiler may use the reverse ordering.

Unions

Fields are convenient for manipulating a single bit or group of bits but not for writing to the whole register. It is easier to do this by composing a complete value using masks, like this:

```
    TACTL = MC_2 | ID_3 | TASSEL_2 | TACLR;
```

The section of the header file for TACTL includes a set of masks that correspond to the bit fields. We can therefore choose whether to treat the register as a single object (byte or word) or as a set of bit fields. This requires a *union*. A union may be used to store different types of data but here it provides two (or more) different ways of looking at the same data, either as a single 16-bit word, unsigned short TACTL, or as a set of bit fields, struct TACTL_bit.

The final line of the definition of TACTL contains @ 0x0160, which is an extension to standard C that fixes the address of the variable. I say more about this in the section "Definition of Peripheral Registers in C" on page 115.

3.2.2   Sizes and Types of Variables

It is often important to know the precise size of a variable—how many bytes it occupies and what range of values it can hold. The MSP430 is a 16-bit processor so it is likely that a char will be 1 byte and an int will be 2 bytes. There is still room for ambiguity, though: Is a plain char signed or unsigned? In other words, if you use it to hold a small integer, will it be interpreted as lying in the range 0..255 or −128..127? Always qualify a definition with signed or unsigned if it matters. Better still, use the definitions in stdint.h to remove any uncertainty. This provides types such as int8_t and uint8_t for signed and unsigned 8-bit integers, respectively.

Most variants of the MSP430 have no hardware for multiplication or division, so these operations are slow. Floating-point arithmetic is very expensive on a small microcontroller in terms of both storage and execution. It is better avoided where possible.

3.2.3   Coding Guidelines for C

TI recommends the following coding style guidelines for C, taken from the readme file in the code examples for the MSP430x11x1 (slac011):

1. No line should exceed 80 characters.

2. Use macros provided in the MSP430 header file.

3. Comments start in column 45, first word is capitalized.

4. For multiline comments, additional lines are *not* capitalized.

5. Use two-space indentation.

I have tried to follow these guidelines in the examples but I am pressed for space with room for only 72 characters across the width of the page. I therefore am relaxed about the rules on spacing.

TI provides no guidance about the format of long names for functions and variables, which is perhaps wise in view of the heated reactions this issue excites. Do you prefer a long_name, longName, or LongName? I personally prefer the middle format, but this may be no more than laziness and I have generally used `LongName` in this book because it seems to be most common in TI's examples.

You should use a standard layout for programs, which keeps them easy to read. There is good advice on this from Ganssle [15], Hills [29], and Labrosse [21] among many others. I mention just one point here: Always use braces `{}` after statements such as `if ()` or `while ()`, even if there is only one line of code and they are not strictly necessary. It is very easy to make mistakes where the program looks correct and is even indented to show the programmer's intention but the flow of control is wrong.

Every program should have a complete introduction and liberal comments to explain its operation to the person who has to update it in a few years. Good documentation is essential. I have kept the explanations brief in the examples because it would only repeat the text, and paper costs money (yours).

3.2.4   Initialization and Structure of Programs

The C language requires static and global variables to be initialized before a program runs. Most are cleared to 0 if explicit values are not specified. Occasionally this is an option in an IDE to save space if things are really tight. Disable the startup code at your peril.

The IDE also performs other initialization quietly. For example, it initializes the stack pointer before it calls the function to initialize your variables. This code is usually in a file called something like `startup.c`. The debugger runs this automatically and halts at the beginning of your `main()` function.

What this startup code does *not* do is initialize any peripherals, notably the watchdog timer and clock generator (look back at the section "Exceptions: Interrupts and Resets" on page 36). It is usually the first job of a program to set these up. The outline of a typical program is described in the application note MSP430 software coding techniques (slaa294).

3.2.5   Extensions and Restrictions to Standard C

Most code can be written in "standard" C. This is often called ANSI C but the formal title is ISO-C90. There is a more recent standard, ISO-C99, and most modern compilers support some aspects of it. Hills [29] describes the different standards and gives abundant advice on the safe and efficient use of C in embedded systems.

There is also an industry-standard Embedded C++ language (EC++), which is a subset of C++ intended for embedded systems programming. It is supported by EW430 but I have not used it except for comments. These can be introduced with // and continue to the end of the line with no need for a closing delimiter. This is part of C99 and is often more convenient than the traditional pair of delimiters /* and */ in C.

It is not possible to write a program for an embedded system in C without a few extensions. The most common extension is the @ operator, which defines the address of a variable. This is necessary because the addresses of peripheral registers are fixed in the MCU so the IDE cannot allocate them freely as it would for ordinary variables. See the section "Definition of Peripheral Registers in C" on page 115. There is often an extension to express numbers in binary notation, such as 0b10100101 by analogy with 0xA5 for hexadecimal notation, but this is not available in EW430.

Some operations are impossible within C and *intrinsic* functions must be supplied. These are typically operations on registers in the CPU. For example, the GIE bit in the status register (SR) is used to turn interrupts on and off, and there are intrinsic functions __enable_interrupt() and __disable_interrupt() in EW430 for this. The processor can also add binary-coded-decimal (BCD) variables, an operation which is not defined in C, so intrinsic functions are again provided. Of course, these are utterly nonportable but it is safer to use intrinsic functions than assembly language.

Finally, there are recommended restrictions to C as well. One of the reasons for the popularity of C is its flexibility, which is also a source of trouble. Many embedded systems are critical for safety, and their programs must therefore be reliable. There are recommended subsets of "safe C" that aim to eliminate unreliable constructions and usage. EW430 includes the widely used MISRA-C rules, developed by the Motor Industry Software Reliability Association. The software itself is included only in the more expensive versions but even Kickstart has the manual. The original version of MISRA-C has 127 rules in all, some required and some advisory. Many are straightforward rules for good programming practice, such as "All automatic variables shall be assigned a value before being used." Others are more specific to embedded systems, such as "Dynamic heap memory allocation shall not be used," which is a wise precaution in systems with limited memory. I strongly recommend that you read these rules and try to follow their spirit, even if you do not have to comply with them formally.

3.2.6 Traps, Pitfalls, and How to Avoid Them

This is a good place to mention the list of books in "Further Reading" that provide good advice on programming embedded systems. Jack Ganssle's books are excellent and cover the whole process of designing, building and debugging embedded systems. His columns *Embedded Pulse* and *Break Points* in *Embedded Systems Design* at www.embedded.com and his own newsletter at www.ganssle.com are a good read too. Dan Saks also writes an illuminating column in *Embedded Systems Design, Programming Pointers* [61]. Simon's *An Embedded Software Primer* [25] is a clear and authoritative introduction to the higher-level aspects of software in embedded systems.

Returning to C, Koenig's book *C Traps and Pitfalls* [31] has been rendered partly obsolete by newer standards of C but many of his examples remain valid. I have often found Steele, McMeniman, and Harbison's *C: A Reference Manual* [35] useful to clear up tricky points of the language. I learned C back in 1980 from the first edition of Kernighan and Ritchie's *The C Programming Language*, which I think is a wonderful book [30]—but not for beginners. There is a vast number of introductory books on C; find one whose style suits you.

3.2.7   Beyond Hand-Crafted C

My wife looked over my shoulder while I was writing a program for the MSP430 one day and remarked on the primitive method of development—leafing through the reference manual and working out the appropriate value to put in each register to configure all the peripherals that I was using. Now she uses FORTRAN, so it is serious to be accused of primitive work practices. Is there a better way? The object-oriented nature of C++ certainly assists some types of program but suppose that we stick with plain C. Several applications generate parts of embedded software automatically.

The more straightforward products are essentially aids to configuring the peripherals. Suppose that you want an interrupt from the timer with a particular period. Normally you must choose the clock for the timer, decide whether to prescale it, and finally work out the period of the timer in clock cycles. Most of this could be done automatically, and Keil's Startup Wizard is an example of this approach. VisSim from Visual Solutions (www.vissim.com) is an example of a more specialized product, which simulates a control system and converts the optimized algorithm into a C program.

A more comprehensive approach is provided by code generators, such as Processor Expert from Unis, which produce the whole framework of the project. Not only would the timer be configured, but the outlines of interrupt service routines would be created as well. Complete modules are available in software to build higher-level functions on top of the peripherals built into the hardware of the MCU. It sounds as though this should be the future of programming, but my impression is that the reception has been fairly cool. Maybe this is just conservatism or perhaps more development is needed. I had little success with this approach but put it down to lack of experience.

3.3   Assembly Language

Not many years ago, small microcontrollers were usually programmed in assembly language. There are a few cases, notably interrupt service routines, where it may still be worthwhile to use assembly for its efficiency. A few operations cannot be represented directly in C and may therefore be much faster in assembly, including bitwise rotations, binary-coded-decimal arithmetic and detailed manipulation of registers. On the other hand, most C compilers offer intrinsic functions for these operations.

There are many good reasons for avoiding assembly language. It is specific to a processor or at least to a family and therefore lacks portability. A complete program needs directives to the assembler, as in C, and these often vary between vendors' assemblers for the same device. Variables are just bytes or words and have no type, so it is up to the programmer to keep track of the significance of the value: Is an integer signed or unsigned, for instance? You must find a routine in a library or write your own to handle anything that goes beyond the instruction set of the processor, such as arithmetic with long integers. The assembler itself converts the mnemonics of assembly language directly into machine code and calculates a few addresses but that is about all—nothing like a compiler.

After that diatribe you might wonder whether it is worth studying assembly language at all. I think that it is still useful for debugging, particularly the detailed interaction between the program and hardware—peripherals and input/output ports. It shows exactly what the processor does and is therefore a good education in computer architecture. It is also fun when it works (and frustrating when it does not). On the other hand, the days of writing substantial programs in assembly are surely behind us.

We see in the section "Light LEDs in Assembly Language" on page 72 that we must do many tasks ourselves with assembly language that the development environment does automatically for C. For example, we must specify where to store the program in memory, where memory for variables should be allocated, and so on. It is even necessary to store the reset vector explicitly so that execution starts at the correct instruction (look back to the section "Exceptions: Interrupts and Resets" on page 36).

3.3.1 Layout of Assembly Language

The layout of assembly language is much more rigid than for C. A typical line has four parts:

```
RESET:  mov.w       #WDTPW|WDTHOLD,&WDTCTL   ; Stop watchdog timer
label:  operation   operands                ; Comment
```

The four parts are

1. **label**—starts in the first column and may be followed by a colon (:) for clarity.

2. **operation**—either an *instruction*, which is translated into binary machine code for the processor itself, or a *directive*, which controls the assembler.

3. **operands**—data needed for this operation (not always required).

4. **comment**—the rest of the line, following a semicolon (;).

Instructions and directives are not usually case sensitive but it is a good idea to follow a specific usage, such as TI's guidelines given later. Comments *must* be used freely in assembly language because it is practically incomprehensible without them.

It is easy to be caught out by the default base (radix) of numbers in assembly language. This was often hexadecimal in the past but modern assemblers are usually decimal. The C-style notation 0xA5 for hexadecimal numbers is now widely accepted by assemblers. Other common notations include $A5, h'A5' and 0A5h. TI often use the last form, where the leading zero is necessary because A5h could be the name of a variable. Binary numbers can similarly be written as 10100101b, which I use, or b'00011000'.

TI recommend the following coding style guidelines for assembly language. Again they are taken from the readme file in the code examples for the MSP430x11x1 (slac010).

1. No line should exceed 80 characters.

2. Use macros provided in the MSP430 header file.

3. Labels start in column 1 and are 10 characters or fewer.

4. Instructions / DIRECTIVES start in column 13.

5. Instructions are lower case and DIRECTIVES are UPPER CASE.

6. Operands start in column 21.

7. Comments start in column 45, the first word is capitalized.

8. For multiline comments, additional lines are *not* capitalized.

```
Column:     13      21                          45
            |       |                           |
Label       in/DIR  short_operand               ; Comment
                                                ; additional  comment  line
```

I have tried to follow these but am again hampered by the limit of 72 characters, so I had to relax the advice on spacing.

3.4   Access to the Microcontroller for Programming and Debugging

We clearly need a way to download a program into the MCU at the very least. Increasingly this is a two-way communication so that information can be retrieved from the chip during debugging. Three interfaces are used in the MSP430 range, two of which are available on most variants:

- Bootstrap loader (serial interface).

- Conventional JTAG (four wires).

- Spy-Bi-Wire JTAG (two wires).

The bootstrap loader (BSL) was built into all MSP430s until the F20xx in 2005 and is described in the application note *Features of the MSP430 Bootstrap Loader* (slaa089). It uses a standard serial protocol and needs only a level shifter for connection to the RS-232 (COM) ports of a computer. The BSL is permanently stored in masked ROM in the MSP430 and is executed instead of the user's program if particular signals are applied to the debugging pins of the MCU when it is reset. The flash memory can then be erased and programmed. A password protects against illegal access. Thus the device can be programmed for development or production or subsequently updated in the field. It does not provide for debugging. Serial monitors were widely used for debugging in the past but had the great disadvantage of being intrusive: The user's program had to halt while the debugger took control.

The use of JTAG for programming is described in the application note *Programming a Flash-Based MSP430 Using the JTAG Interface* (slaa149) but the interface can do far more than this. The abbreviation stands for Joint Test Action Group and JTAG is an IEEE standard, 1149.1. It was originally intended for testing circuit boards rather than integrated circuits and is based on the idea of a boundary scan. Very roughly, a shift register with a flip-flop is connected to each pin. The values on the pins can be loaded all at once into the shift register and read out serially through JTAG. The shift register is controlled by a test access port state machine (TAP controller). Thus JTAG needs at least four connections, labeled TDI, TDO, TMS, and TCK for the input, output, control, and clock of the shift register and state machine. Like all good standards, JTAG has been extended greatly and now does far more than boundary scans. It is widely used for debugging microcontrollers, although some manufacturers retain proprietary interfaces. Berger gives a good description in his book [9].

TI use a 14-pin connector for JTAG. This carries the four standard connections, power, ground, and $\overline{RST}$/ NMI, which may be needed to reset the chip during debugging. Some devices need further signals. The details are in the *MSP-FET430 FLASH Emulation Tool (FET) User's Guide* (slau138). An annoyance with the standard JTAG interface is that four pins is a serious loss on a small device. Another pin called TEST is therefore provided as well to switch the usual four pins between their input/output function and JTAG. For comparison, Freescale's background debug mode (BDM) interface needs only one pin, which carries data in both directions. The latest MSP430 devices, such as the F20xx, use a less extravagant Spy-Bi-Wire interface that needs only two pins instead of the standard four or five. All current MSP430 devices have a dedicated $\overline{RST}$/ NMI pin, which cannot be shared with digital input and output as in some other MCUs. It can be reconfigured to carry a nonmaskable interrupt (NMI) instead of reset ($\overline{RST}$, with a bar because it is active low—a logical value of 0 resets the device).

The JTAG signals are connected to a desktop computer through an interface. This is often called a *pod* or *wiggler* but TI calls it a *flash emulation tool* (FET). Pods often worked with the parallel port of a PC in the past but now mostly use USB or ethernet. They cost about $100 and are available from TI or any of the tool suppliers. Many of these can also blow a fuse inside the MSP430 to disable the JTAG interface and prevent malicious access.

The JTAG interface gives full access to the MCU when it has control. A debugger can display the value in almost any register—CPU, RAM, or ROM—and modify the value in registers where this is permitted. This sounds perfect but there is always a catch. There are side effects of reading some registers associated with interrupts and the debugger triggers these; it does not have privileged access in the MSP430. This can make it tricky to debug interrupt service routines as we see later.

JTAG also controls the on-chip debug logic, whose capability varies considerably between variants of the MSP430. In all cases, the program can be stopped, started, and single-stepped. The most basic devices, such as the F1121A, have two breakpoints that can be set in hardware. These are most simply used separately to stop execution at a selected statement. This means that the break is triggered when the address of the instruction is placed on the memory address bus (MAB). Alternatively a breakpoint can be set when a particular value appears on the memory data bus (MDB). The pair of breakpoints can be combined to stop when a variable in RAM (whose address is on the MAB) has a specified value (on the MDB).

Larger devices have an enhanced emulation module (EEM) where more sophisticated breakpoints can be defined. They also have a trace buffer, which stores a short record of events around the trigger. The details are given in the *MSP430 IAR Embedded Workbench IDE User Guide* and the application note *Using the MSP430 EEM Debug Feature with IAR 3.30* (slaa263).

3.5 Demonstration Boards

Most people start programming embedded systems with a commercial demonstration board. A good selection is listed on the TI Web site for the MSP430 and I'll briefly mention a few. A demonstration board has a selection of the following components:

- The MSP430 itself with the essential supporting hardware: crystal (if necessary), decoupling capacitor, pull-up resistors, socket for JTAG interface.

- Headers or sockets to provide access to the MCU's pins.

- Power supply, either from a battery or an external supply, in which case a voltage regulator is needed; perhaps a charge pump to permit operation from a single cell.

- Output display, always some plain LEDs, perhaps seven-segment LEDs or a segmented LCD for MCUs that can drive them directly. (I like to have a complete set of eight LEDs on one port but few boards offer this.)

- Input devices, almost always some simple push buttons and perhaps a potentiometer for analog input.

- RS-232 transceiver so that the MSP430 can communicate with a computer using its serial interface.

- Flash memory chip, particularly for data logging.

- Other input and output devices, such as microphone and speaker, radio or infrared transmitter and receiver.

- Sensors, depending on the intended application, which might measure temperature, pressure, light and the like.

- Connections for external peripherals, usually with standard interfaces for digital components, such as serial, SPI, I²C, Dallas 1-wire, or iButton.

- Interface for communication such as USB, ethernet or Zigbee.

Boards cost $20–$200 or more depending on the components (the MSP430 itself may well be cheaper than the PCB). In most cases, you need a separate JTAG pod but it is integrated onto a few demonstration boards. You also need a computer, of course. Most software runs only under Windows but there is increasing support for linux. I briefly describe a few demonstration boards to give you a feel for their features.

3.5.1   Texas Instruments eZ430–F2013

This is a tiny but complete development tool for the MSP430, shown in Figure 3.1. It plugs into a USB port, which is used for both communication and power, and that is it. In fact, it looks much like a USB memory drive. The USB–JTAG interface with an MSP430F169 and TUSB3410 takes up most of the tool. The target device is a F2013, which sits on a removable board at the tip. It has a single LED plus the essential decoupling capacitor and pull-up resistor on the reset. No crystal is provided because this

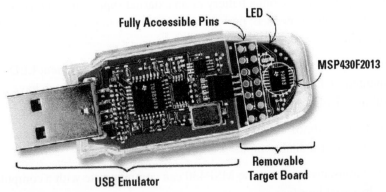

Figure 3.1: The TI eZ430–F2013 USB stick development tool.

MCU has an internal oscillator (VLO) that can be used for ACLK. There is a four-pin connector that carries power, ground, and two JTAG signals. A footprint for a 14-pin header is provided, which gives access to all pins of the MCU. Details are in the *eZ430–F2013 Development Tool User's Guide* (slau176).

The eZ430–F2013 is undoubtedly cute but I am not convinced of its usefulness, particularly for beginners. A good feature is that it is easy to see what is happening on each pin by poking an oscilloscope probe into the holes for the 14-pin header. On the other hand, it is difficult to get a good ground connection. There is no switch for an input, and the single LED is not on a pin that can be driven directly by the timer, which is a shame. I am tempted to remove the target board and use the rest as an interface to a more comprehensive demonstration board, which could be constructed by hand using a PDIP F20xx. The only snag is that the connector is delicate and not easy to buy. Another possibility would be to attach a 14-pin header to the target board. This would give access to all the pins but wreck the appearance and the board would no longer fit into the case. A neat idea by Peter Darnell of Visual Solutions Inc. is to mill a slot in the case so that it can be put back together with the header exposed.

3.5.2 Texas Instruments MSP430FG4618/F2013 Experimenter's Board

This is a large and remarkably complete demonstration board, shown in Figure 3.2. It includes two variants of the MSP430, one of the largest (FG4618)* and one of the smallest (F2013). Between them they illustrate almost all of the peripherals of the MSP430. The FG4618 drives a SoftBaugh LCD, which I describe in Chapter 7. There are different types of analog-to-digital converters in the two devices, both of which are covered in Chapter 9. The two MSP430s can communicate with each other using SPI and I²C, and the FG4618 is connected to an RS-232 interface, all of which are ideal for Chapter 10 on communication. The FG4618 also includes operational amplifiers and can provide a complete audio signal chain but I have no space for most of this nor for the optional ChipCon RF module. There is a JTAG interface for each of the MSP430s so you will need a separate JTAG–USB interface (or maybe two). The F2013 is wired only for the Spy-Bi-Wire interface, which is not supported by older JTAG pods.

You need to solder a pin or two to the board for a ground connection if you plan to use an oscilloscope to probe the board, which is a minor nuisance. A problem with all large

---

* This was a FG4619 in early boards, including mine.

Figure 3.2: The TI MSP430FG4618/F2013 Experimenter's Board.

demonstration boards is that they have a lot of jumpers that must be placed correctly. Check the diagrams carefully in the *User's Guide* (slau213) if some feature appears not to be working. Another helpful accessory is a set of wires with sockets that fit over the pins of the headers. These are used to join pins that are not next to each other, in which case jumpers will not work. They are hard to find and are jealously guarded by their owners. You can make them, although it is a little awkward, or buy them (SoftBaugh sells them as prototyping wires, for instance).

3.5.3   Texas Instruments Development Kits

These are more conventional kits, comprising a separate USB–JTAG interface and target board to suit a particular package of MSP430. The board has a zero insertion force (ZIF) socket so that devices can be inserted and removed safely. Typically there is an LED and a crystal, although the latter is not always installed. These kits are really intended for small-scale production rather than learning about the MSP430.

3.5.4   SoftBaugh

SoftBaugh (www.softbaugh.com) makes a large selection of boards. These include comprehensive evaluation systems, such as the ES437 shown in Figure 3.3, and more

Figure 3.3: The SoftBaugh ES437 evaluation system, taken from www.softbaugh.com.

modest demonstration boards, which are more focused on a particular aspect of the MSP430.

3.5.5   Olimex

Olimex (www.olimex.com) also sells a selection of boards. I based the introductory tour of the MSP430 in Chapter 4 on its starter kit for the F1121A (MSP430-1121STK), shown in Figure 4.1. This is a cheap and well-designed board that exercises most features of its small MSP430. Its MSP430-easyWeb3 board is an economical introduction to ethernet with the MSP430.

3.5.6   Others

Many other companies produce demonstration boards and are listed on the TI Web site. There are boards to demonstrate ethernet, ZigBee, USB, flash memory cards, and other technologies that are important in modern embedded systems.

Of course, it is fun to build your own board. Most variants of the MSP430 have their pins so densely crowded together that they are difficult to solder by hand, if they have pins at all. Only recently has the F20xx become available in a traditional, plastic dual-in-line package (PDIP) with pins 0.1″ apart, making manual assembly straightforward. You can even build a simple circuit on stripboard, and I made a couple to test programs for this book.

3.6   Hardware

I just mentioned that it is very difficult to assemble circuits with modern components because of their small packages with closely spaced pins, which are intended for surface mounting. The increasing difficulty in obtaining traditional, through-hole packages with pins on a 0.1″ pitch is a major headache for educators and hobbyists. Fortunately, several companies have addressed this need for the MSP430. One solution is to buy a *header board* or *breadboard module*. These are small PCBs that typically carry the microcontroller, its essential supporting components, and a JTAG socket; sometimes there is a voltage regulator as well to protect the MCU. The tiny connections of the MCU are brought out to sturdy pins on a 0.1″ pitch so the board looks like a large dual-in-line package, which can be plugged into a breadboard or a socket on a PCB. Alternatively, larger *prototype boards* have the same components mounted on a PCB with plenty of space to lay out the user's components.

There are are a few important points to watch if you build your own PCB.

- Do not forget decoupling or bypass capacitors on the power supply. The usual recommendation is to put a 0.1 μF ceramic capacitor as close to the power pins as possible to suppress noise, with a further 10 μF tantalum capacitor to provide greater charge storage. Put another 0.1 μF capacitor across the analog power pins if these are separate. The reference voltages on some analog modules require external capacitors as well.

- The layout of the crystal oscillator is particularly critical with modern, low-power MCUs. The value of the load capacitors, if needed, should be adjusted to account for the capacitance of tracks on the PCB. The whole circuit for the oscillator should be kept as compact as possible and should not overlap tracks that carry signals. TI recommends that the can of the crystal be soldered to ground. There is further guidance in the books by Catsoulis [10] and Wilmshurst [26].

- Make debugging easy for yourself. Install test points where you need to monitor the signal and provide plenty of pins for the ground clip of your oscilloscope probes.

- Finally, check that the $\overline{\text{RST}}$/NMI and other pins are connected as specified in the *MSP-FET430 FLASH Emulation Tool (FET) User's Guide* (slau138). Unused digital input/output pins may be left unconnected but must then be configured internally as advised in the section "Configuration of Unused Pins" on page 215.

3.7 Equipment

The basic equipment needed to get started with the MSP430 is just a PC, interface pod, and demonstration board. Most development software does not make serious demands on a modern PC. If you have some spare money, I suggest that you spend it on a big screen or a second one. Debuggers spawn a large number of windows and a large screen is a great boon. The debugger and editor are poorly integrated in some older software, in which case two screens are needed for efficient debugging, but this does not apply to any of the products I used for the MSP430. On the other hand, you frequently have to refer to data sheets and user's guides while writing and debugging, which would warrant a second screen if you do not have the documents on paper. A serial interface or USB–serial adapter may be useful if your demonstration board has an RS-232 socket.

The development of embedded systems becomes dramatically different from pure software projects when you start to look at real electrical signals. A digital multimeter (DMM) is the most basic tool. This can be used to check that the correct voltages are present on V_{CC} and V_{SS}, for instance. It is a good idea to do this before plugging in your precious microcontroller. A DMM can also measure the average current drawn from the supply, although this can be tricky if the system spends most of its time in a low-power mode. If you are more ambitious, Ganssle has an entertaining article in which he shows how a cheap analog multimeter can be used as a performance analyzer [59]. There are a lot of good ideas in his book too [15]. For example, try to reserve an input/output pin to use as a debugging output. I show an example of this in "A Software UART Using Timer_A" on page 590.

A DMM is of little use for signals that change rapidly: For these you need an oscilloscope, the tool that marks you out as an electronic engineer. A serious problem is that many signals show only infrequent events, nothing like the continuous sine wave that you might hope to find in an analog circuit. A digital storage oscilloscope (DSO) is the best tool for intermittent signals but unfortunately these are seriously expensive. Even the base models from companies like Agilent and Tektronix start at around $1000. Consider yourself lucky if your employer or school has a DSO available. Cheaper, analog scopes are of limited use but better than nothing. A DSO has versatile triggering, stores the signals, can perform calculations on the traces (usually including a Fourier transform to give the power spectrum), and downloads the data to a memory card or computer. They are wonderful.

Any oscilloscope is useful for the basic check that a signal is getting out of the MCU. This is often helpful with communications. You configure the appropriate peripheral, check its operation in loopback mode, where the signal is routed internally back from the transmitter to the receiver, and all is fine. Finally you connect the external equipment—and nothing happens. The problem is often that the ports of the MCU have not been configured properly so that the signals are not connected from the peripheral to the pins. Embarrassing, but we have all done it. Similar issues can arise with pulse width modulation, where again the port must be set up for output from the timer.

Fortunately there is a cheaper alternative to a real DSO, which is a data acquisition unit or "USB oscilloscope." These are something like the front end of an oscilloscope, with amplifiers and analog-to-digital converters. The digitized signals are transferred, usually over USB, to a desktop computer where they are displayed. The instrument is also controlled using a visual interface. A good place to look for these is in *Circuit Cellar* magazine. The selection and performance of these devices is improving continually. Some are for analog signals, some are like logic analyzers, and a few are mixed-signal systems that combine both functions. They cost a tiny fraction of bespoke instruments. I have been impressed by the quality of data but they are much less convenient to use than a genuine oscilloscope. It is hard to beat a simple knob for changing the gain or time base of a display, particularly when you already have one hand holding a probe onto an awkward component.

That reminds me—please be careful with oscilloscope probes; good ones cost at least $100 and are wrecked by wheeling a chair over them. Do not forget to clip the ground lead onto the system under test or the signals will not make any sense. Make life easy for yourself by providing several ground pins when you design a PCB.

Now that the hardware is constructed, it is time to write some programs. The next chapter takes you on an elementary tour of the MSP430.

CHAPTER 4

A Simple Tour of the MSP430

This chapter provides an introductory tour of the MSP430 and its basic features. The programs are selected to run on the simplest demonstration board currently available, the Olimex 1121 starter kit (MSP430-1121STK). A photograph is shown in Figure 4.1 with a simplified circuit in Figure 4.2. This uses a 20-pin MSP430F1121A and provides the following inputs and outputs:

- LEDs active low, LED1 on P2.3 (TA1 out) and LED2 on P2.4 (TA2 out).

- Push buttons with pull-up resistors, active low, B1 on P2.1 (INCLK in) and B2 on P1.2 (CCI1A in).

- Piezo sounder between P2.0 and P2.5.

Figure 4.1:   Olimex MSP430-1121STK starter kit.

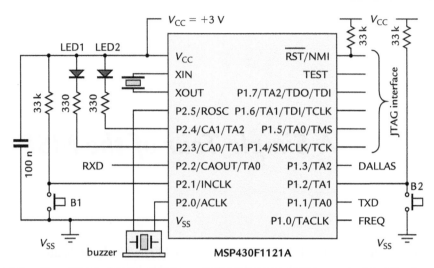

Figure 4.2: Simplified circuit of Olimex MSP430-1121STK starter kit. The layout of the pins is chosen for convenience and does not match the physical pinout of the device.

- Frequency input on P1.0 (TACLK in).

- Dallas iButton interface on P1.3 (CCI2A in).

- Asynchronous serial interface (RS-232) on P1.1 (transmit) and P2.2 (receive).

Many of these pins can be used for digital input and output or by Timer_A, whose connections are shown in parentheses in this list, but I do not cover the timer until Chapter 8. Instead I concentrate on functions offered by the core of the chip rather than the peripherals. I also avoid interrupts to keep things simple. The general approach is to write programs in C followed by assembly language, although the two programs are not always equivalent. The aims are to write outputs (light LEDs), read inputs (pushbutton), and control something automatically (flash LEDs). The main obstacles are behind us after these have been accomplished.

4.1   First Program on a Conventional Desktop Computer

The introductory program in a course on C for a desktop computer, shown in Listing 4.1, has hardly changed since I learned the language in 1980. It simply displays the text `hello, world` on the "console" and exits.

Listing 4.1: The classic "hello, world" program in C.

```
#include <stdio.h>
void main (void)
{
    printf("hello, world\n");
}
```

The first program on a simple embedded system is equivalent but apparently less ambitious: to illuminate one or more light-emitting diodes (LEDs). The main hurdle, as for the desktop computer, is likely to be set by the development software. There is the further problem, however, of finding whether the fault lies in the software or hardware if the program is built successfully but nothing happens. Before embarking on this route, two features of hello, world are worth emphasizing:

- The program needs substantial support from an operating system. This is most obvious for the printf() function, which sent the characters to a printer back in 1980 but is now more likely to display the text in a "console" window on the screen. However, the operating system must also load the program into memory, start execution at the correct point, and resume control when the program has finished. There is no operating system, printer, or screen in a small embedded system, which is why simple LEDs are used for the output. A C compiler ensures that the program is stored in the correct part of memory and that execution starts at the correct point, but these details must be handled by a user if the program is written in assembly language. Finally, a user's program must never exit, because there is no operating system to take over: It must retain control of the processor until the power is switched off.

- Another important influence of the operating system is that it insulates the user from the details of the system on which the program runs: The code is the same whether the message appears on a printer or in a window and whether the operating system is linux, MacOS, or Windows. In contrast, a program for an embedded system is utterly dependent on the processor and its surrounding hardware. Even the simple program to illuminate a LED cannot be written until we know the pin to which the LED is connected and whether it is active high or low (Figure 4.3). The first step must therefore be to understand the hardware. Read the manual for your development board. It probably has an example program as well.

4.2   Light LEDs in C

The simplest program is to light the LEDs on the demonstration board in a fixed pattern. LEDs can be connected in two standard ways, shown in Figure 4.3. The resistor limits the current: A few mA are sufficient for a modern LED. In the active high circuit, the LED illuminates if the pin is driven high (logic 1) and extinguishes if the pin is driven low (logic 0). The Olimex 1121STK uses the opposite configuration, active low. This may seem perverse, but the active low configuration was common in the past because outputs were better at sinking current (allowing it to flow to ground, V_{SS}) than sourcing current (supplying it from V_{CC}). This asymmetry goes back to the fundamental physics of silicon, and most modern circuits are designed to compensate for it.

Example 4.1

A trivial calculation for the electrical engineers: What value of R is needed to limit the current to 4 mA assuming that $V_{CC} = 3$ V and the LED drops 1.8 V?

We saw in the section "Memory-Mapped Input and Output" on page 32 that input and output is *memory mapped*, which means that nothing unusual is needed to control output pins: We simply write in the normal way to the register that controls them, P2OUT here. The complete program needs to carry out the following tasks:

1. Configure the microcontroller.

2. Set the relevant pins to be outputs by setting the appropriate bits of P2DIR.

3. Illuminate the LEDs by writing to P2OUT.

4. Keep the microcontroller busy in an infinite, empty loop.

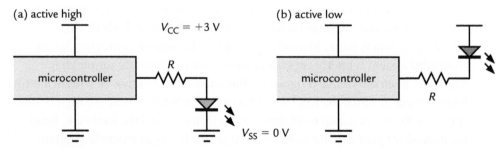

Figure 4.3: Connection of an LED to a pin of a microcontroller: (a) active high and (b) active low.

The program is shown in Listing 4.2.

Listing 4.2: Program `ledson.c` in C to light LEDs with a constant pattern.

```c
// ledson.c - simple program to light LEDs
// Sets pins to output, lights pattern of LEDs, then loops forever
// Olimex 1121STK board with LEDs active low on P2.3,4
// J H Davies, 2006-05-17; IAR Kickstart version 3.41A
//-----------------------------------------------------------------
#include <msp430x11x1.h>            // Specific device

void main (void)
{
    WDTCTL = WDTPW | WDTHOLD;        // Stop watchdog timer
    P2DIR = 0x18;           // Set pins with LEDs to output, 0b00011000
    P2OUT = 0x08;           // LED2 (P2.4) on, LED1 (P2.3) off (active low!)
    for (;;) {              // Loop forever...
    }                       // ...doing nothing
}
```

Let us look through this in detail.

1. The file starts with comments that explain the purpose of the program, the hardware on which it runs, and the software used for development. I kept these brief because they would only repeat the text of the book. I use the year–month–day format for the date to avoid confusion between American and European usage.

2. As usual, there is an `#include` directive for a header file. However it is not the familiar `stdio.h`, because there is no standard input and output. Instead it is a file that defines the addresses of the special function and peripheral registers and other features specific to the device being used.

3. The first line of C stops the watchdog timer (see the section "Anatomy of a Typical Small Microcontroller" on page 8), which would otherwise reset the chip after about 32 ms. This is explained fully in the section "Watchdog Timer" on page 276. A complete program would configure many other aspects of the chip but the remaining defaults are satisfactory here.

4. The two pins of port P2 that drive the LEDs are set to be outputs by writing to P2DIR. For safety the ports are always inputs by default when the chip starts up (power-up reset).

5. The LEDs are illuminated in the desired pattern by writing to P2OUT. Remember that a 0 lights an LED and 1 turns if off because they are active low.

6. The final construction is an empty, infinite `for` loop to keep the processor busy. It could instead be written as `while (1) {}` but some compilers complain that the condition in the `while()` statement is always true.

The infinite loop is needed because the processor does not stop of its own accord: It goes on to execute the next instruction and so on, until it tries to read an instruction from an illegal location and cause a reset. This is different from introductory programming courses, where you learn to write programs that perform a definite task, such as displaying `hello, world`, and stop. In these cases, an infinite loop is disastrous. In contrast, there is an infinite loop in every interactive program, such as a word processor, or one that runs continuously, such as an operating system. For example, the computer on which I am writing this book spends most of its time in an infinite loop waiting for me to hit the next key. The empty loop is a waste of the MCU but keeps it under control. A better approach is to put the processor to sleep—into a low-power mode—which we do later.

Example 4.2

Try this on your demonstration board. You may need to make some small modifications if the LEDs are connected to different pins and if the configuration is active high rather than active low. The program may be trivial but getting it to work in the development environment probably is more challenging. Tips for getting started with EW430 and various development kits are in Appendix A.

You may find a minor flaw in this program if you experiment with lighting the LEDs in different patterns. How can it be corrected?

4.3   Light LEDs in Assembly Language

Listing 4.3 shows the equivalent program in assembly language. It clearly takes a great deal more effort than in C: There is no = sign for assignment, for example. We also handle "housekeeping" issues explicitly, which the development system does automatically for C, such as where the program should be stored in memory and where execution should start.

Listing 4.3: Program `ledsasm.s43` in assembly language to light LEDs with a constant pattern.

```
; ledsasm.s43 - simple program to light LEDs, absolute assembly
; Lights pattern of LEDs, sets pins to output, then loops forever
; Olimex 1121STK board with LEDs active low on P2.3,4
```

```
; J H Davies, 2006-05-17; IAR Kickstart version 3.41A
;-----------------------------------------------------------------
#include <msp430x11x1.h>          ; Header file for this device

    ORG     0xF000               ; Start of 4KB flash memory
Reset:                           ; Execution starts here
    mov.w   #WDTPW|WDTHOLD,&WDTCTL ; Stop watchdog timer
    mov.b   #00001000b,&P2OUT
                    ; LED2 (P2.4) on, LED1 (P2.3) off (active low!)
    mov.b   #00011000b,&P2DIR     ; Set pins with LEDs to output
InfLoop:                         ; Loop forever...
    jmp     InfLoop              ; ...doing nothing
;-----------------------------------------------------------------
    ORG     0xFFFE               ; Address of MSP430 RESET Vector
    DW      Reset                ; Address to start execution
    END
```

The file starts with comments as usual, which are introduced with semicolons in assembly language. A convenient feature of EW430 is that the same header file is used as in C, but everything is different after that. The suffix 'b' denotes a binary number, as in 00001000b. Let us look at the program in detail.

4.3.1 Assignment: The Move Instruction

We are so familiar with statements like a = b in C that it is easy to forget how much is implied in an assignment:

- It does not depend on the type of data, whether a and b are char, int, or other types.

- We do not have to worry how a and b are stored and whether b is a constant, variable, or expression.

- The compiler converts the data if a and b are of different types (within the rules, of course).

None of this is true in assembly language:

- The CPU can transfer only a byte or a word and must be told which.

- We must specify the location and nature of the source and destination explicitly.

- There is no conversion of the data (unless you do something sneaky).

The basic instruction in assembly language is mov.w *source,destination* to move a word or mov.b for a byte. If you omit the suffix and just use mov the assembler assumes that you mean `mov.w`. I recommend that you always specify `.w` or `.b` explicitly for safety.

This instruction *copies* a word or byte from one location to another. It leaves the source as it was, despite the name *move*. This usage is unfortunate but universal. Watch the order of operands: Many assembly languages are written the other way around. The locations are specified using various *addressing modes*. There are seven of these in all but we look at only the simplest three for now and do not worry about how they are implemented. A single character denotes the mode in the operand.

- **immediate, #:** The value itself (word or byte) is given and stored in the word following the instruction. This is also known as a *literal* value.

- **absolute, &:** The address of a register in memory space is given and stored in the word following the instruction.

- **register, R:** This specifies one of the 16 registers in the CPU.

The immediate mode is clearly useful only for the source but the other two can be used for either the source or destination, provided that it is possible to write to the destination—it is not in ROM, for instance. It is not necessary for all moves to go via a register in the ALU as in some other processors, notably RISCs. For example, two operations are required to load a literal value into a register in the PIC16: The value is first loaded into the working register with `movlw`, then stored in a file register with `movwf`. The MSP430 can do this with a single instruction.

Listing 4.3 contains only the form `mov #immediate,&absolute`. The processor itself needs purely numerical addresses or data so we could control the LEDs connected to port P2 by writing

```
mov.b    #00001000b,&0x0029   ; LED2 (P2.4) on, LED1 (P2.3) off
```

A byte with *immediate value* (#) 00001000b is copied to the register at *address* (&) 0x0029. If you check the memory map you can confirm that this is the port P2 output register P2OUT. Clearly this style is unhelpful. Fortunately the assembler allows us to use symbolic constants as in C, which are much clearer to understand. It substitutes their values from the header file, like the Find/Replace operation in an editor. Therefore we can write

```
mov.b    #00001000b,&P2OUT    ; LED2 (P2.4) on, LED1 (P2.3) off
```

which is much clearer. Please use symbols instead of numerical values wherever possible. The header file includes a set of constants such as BIT3, which could be used instead of 00001000b. I also used constants in the instruction to stop the watchdog timer.

Warning: In many assembly languages it is not necessary to write an & before absolute addresses. Unfortunately the MSP430 uses a different addressing mode called *symbolic* if you omit the &. It should make no practical difference on the MSP430, but you need to be more careful with the MSP430X (see the section "Instruction Set of the MSP430X" on page 607).

Later in this chapter we encounter other ways of assigning values. For example, the clear instruction clr.w or clr.b puts the value of the destination to 0. In many processors this is a distinct instruction but not in the MSP430: The assembler translates clr.b P2OUT to mov.b #0,P2OUT. You can see this in the Disassembly window of the debugger. This is an example of an emulated instruction. It is implemented in a clever way that is explained in the section "Constant Generator and Emulated Instructions" on page 131, which avoids having to store the immediate value of 0 in the instruction.

The program in assembly language writes to P2OUT before P2DIR, the opposite order from the program in C. This ensures that the correct values appear on the pins as soon as they are made into outputs. If the pins are switched to output first, the outputs initially are driven to the values that happen to be sitting in P2OUT. These are random if power has just been turned on or are left over from the previous program. You should have seen this if you experimented with different patterns of LEDs. It is perfectly legal to write to P2OUT while the pin is configured as an input: The value waits in a buffer until the pins are enabled for output.

4.3.2   Infinite Loop

All the instructions in Listing 4.3 are moves except for the infinite, empty loop at the end. Assembly language offers nothing like the flexible constructions for loops in C. The flow of control is based on *jump* instructions, which can be either conditional or unconditional (always taken). Unconditional jumps are equivalent to goto in C. You are strongly advised not to use such instructions in high-level programming but you have no choice in assembly language.

Normally the CPU executes instructions from memory one after the other (see the section "Central Processing Unit" on page 30). The program counter (PC) is automatically incremented so that it is ready to fetch the next instruction as soon as it is needed. This sequence is broken by a jump, which causes a new address to be loaded into in the PC. Most jumps require only a small change in the PC and it requires less space to store the *change* rather than the complete, final value itself. This is how `jmp` works: Both the instruction itself and the change in PC fit into a single word of memory provided that the jump is less than 1KB. Fortunately we do not have to calculate the change in PC ourselves: The assembler does it for us. Put a label at the destination of the jump, `InfLoop:` in Listing 4.3, and use this as the target of the `jmp` instruction.

4.3.3   Housekeeping

We have now dealt with all the assembly language itself, but must still address several issues that the C compiler and linker would handle for us. This requires a number of *directives* to the assembler. These tell the assembler how it should process the code and are much like lines that start with # in C; they are not translated into instructions for the processor itself.

Where Should the Program Be Stored in Memory?

We must tell the assembler where it should store the program in memory. Generally the code should go into the flash ROM and variables should be allocated in RAM. We put the code in the most obvious location, which is at the start of the flash memory (low addresses). The *Memory Organization* section of the data sheet shows that the main flash memory has addresses 0xF000–FFFF. We therefore instruct the assembler to start storing the code at 0xF000 with the organization directive `ORG 0xF000`.

Where Should Execution of the Program Start?

Which instruction should the processor execute first after it has been reset (or turned on)? Some processors, such as the PIC16F, start with the instruction at address 0. You might expect the MSP430 to do something similar and start with the instruction at the lowest address in its flash memory. However, this would not always work because we might have stored the program elsewhere in memory. Even in the PIC16F, the first instruction is usually something like `goto start`, so a more versatile scheme would clearly be useful. In the MSP430, the *address* of the first instruction to be executed is stored at a specific location in flash memory, rather than the instruction itself. This address, called the *reset vector*, occupies the highest 2 bytes of the vector table at 0xFFFE:FFFF. It is set up as follows:

1. Put a label in front of the first instruction to be executed. I use `Reset` in Listing 4.3.

2. Use an `ORG 0xFFFE` directive to tell the assembler where the reset vector should be stored. It is at the same address in all current MSP430 devices.

3. This is followed by a `DW Reset` directive for "define word," which tells the assembler to store the following word (2 bytes) in memory. This is the address of the first instruction to be executed and the assembler replaces the label with its numerical value.

Many processors use reset vectors in a similar way.

Whew—that is it. The final directive is `END`, whose purpose is obvious. The assembler complains if you omit it.

Example 4.3

Try this on your demonstration board. Some tips and suggestions are in the section "Developing a Project in Assembly Language" on page 633.

4.3.4 Machine Code

Perhaps assembly language is now so clear that you want something more obscure. Both the C compiler and assembler ultimately produce binary machine code for the CPU to process. This is shown in the Disassembly window of the debugger or you can inspect the contents of the flash memory. The first three active lines of Listing 4.3 produce the machine code shown as comments:

```
mov.w    #WDTPW|WDTHOLD,&WDTCTL    ;  40B2  5A80  0120
mov.b    #00001000b,&P2OUT         ;  42F2  0029
mov.b    #00011000b,&P2DIR         ;  40F2  0018  002A
```

I have broken the machine code into 16-bit words (four hexadecimal digits). The first word on each line is the instruction itself. The most significant hexadecimal digit is the operation code, or *opcode*, which is a 4 in all three lines because all these are moves. The remainder of the instruction word describes the nature of the operands: whether they are bytes or words and how the addresses are specified. In the first case, the operand is a word, the source is an immediate value, and the absolute address of the destination is provided. The instruction is followed by a second word with the immediate data and a third word with the address of the destination.

The third line is similar but the instruction is slightly different, because a byte is moved rather than a word. The immediate data are just the hexadecimal representation of the binary value given in the assembly code, padded out to fill a word. The final word is the address of the P2DIR register.

You might reasonably expect the second line to be very similar to the third, but in fact it is quite different: There are only two words instead of three and the instruction is not the same. The reason is that the immediate operand of 0x0008 is one of the frequently used values provided by the constant generator built in to the CPU, mentioned in the section "Central Processing Unit" on page 30. There is no need to include the value in the instruction, which saves both memory and time.

Make sure that memory is displayed in words rather than bytes if you wish to look at instructions. You may get an unpleasant surprise if you use bytes instead, because `40B2 5A80 0120` will be shown as `B2 40 80 5A 20 01`. This is an unfortunate consequence of little-endian ordering and I warned you in the section "Memory" on page 25.

This potential confusion emphasizes the point that instructions and data are indistinguishable at this level. Everything is just made of binary numbers, ones and zeros. Suppose that something goes wrong and a word of data is loaded into the CPU by mistake instead of an instruction. There is no way of telling that this is not intended to be an instruction and the CPU will (probably) execute it because most possible words are valid instructions.

4.3.5 Relocatable Assembly Language

The style of program shown in Listing 4.3 is known as *absolute assembly* because the memory addresses are given explicitly in the source using `ORG` directives. Many of TI's examples are written in this way. It is straightforward but inelegant because it is poor practice to "hardwire" constants into a program. What would happen if you wanted to run the program on a different device, for instance? There is a good chance that you might remember to change the `include` file but you could easily forget the address in one of the `ORG` directives.

One solution would be to use symbolic constants such as RomStart but there are no definitions like this in the header file msp430x11x1.h. This is because the header file is really intended for C, where the allocation of memory is done automatically by the linker. The addresses are therefore stored in the linker control file instead.

It is possible to use the linker with assembly language so that addresses need not be written into the program. This is called *relocatable assembly*. It is often avoided because linkers have such a bad reputation: Most linker control scripts make assembly language look friendly. Fortunately it is straightforward for simple programs like ours. Listing 4.4 shows the relocatable version of the program to light LEDs. It is clearer and no more complicated than the absolute assembly in Listing 4.3.

Listing 4.4: Relocatable assembly language `ledsrel.s43` to light LEDs.

```
;  ledsrel.s43 - relocatable program to light LEDs
;  Lights pattern of LEDs, sets pins to output, then loops forever
;  Olimex 1121STK board with LEDs active low on P2.3,4
;  J H Davies, 2006-06-01; IAR Kickstart version 3.41A
; -----------------------------------------------------------------
#include <msp430x11x1.h>               ; Header file for this device

        RSEG    CODE                   ; Program goes in code memory
Reset:                                 ; Execution starts here
        mov.w   #WDTPW|WDTHOLD,&WDTCTL  ; Stop watchdog timer
        mov.b   #00001000b,&P2OUT
                          ; LED2 (P2.4) on, LED1 (P2.3) off (active low!)
        mov.b   #00011000b,&P2DIR      ; Set pins with LEDs to output
InfLoop:                               ; Loop forever...
        jmp     InfLoop                ; ...doing nothing
; -----------------------------------------------------------------
        RSEG    RESET                  ; Segment for reset vector
        DW      Reset                  ; Address to start execution
        END
```

The basic action of the linker is to group parts of the program that use the same type of memory into *segments* and allocate these to appropriate addresses of the MCU. Here we use only two types of memory, the executable code and the reset vector, whose segments have obvious names. The directive RSEG CODE tells the assembler that the following instructions should be put in the CODE segment, which the linker then puts at the correct address in flash memory. RSEG stands for "relocatable segment," meaning that the address is assigned by the linker (the alternative is ASEG for "absolute segment," in which case we must provide the address). The relation between segments and addresses is defined in the linker control script, lnk430F1121A.xcl for this device, but you probably do not need to look at it.

Unfortunately there is a general problem with assembly language and relocatable assembly in particular. We made it easier to change device but are instead tied to particular software. The instruction set of the processor is defined by TI so you can be reasonably sure that the language is the same in all assemblers for the same chip (although the layout and notation for constants may be different). Directives, however, vary wildly between development environments. Some, like ORG, are widely used, but even these are not universal. In particular, Code Composer Essentials uses an entirely different approach from EW430. The format of the header file is different and so is the directive to include it, it does not permit absolute assembly at all (no ORG directives) and executable code is introduced with. text instead of RSEG CODE. There are numerous other differences, which are described in Appendix D of the *User's Guide* (slau157).

4.4   Read Input from a Switch

Now that we can write output to LEDs, the next step is to read input from a switch. The standard way of connecting a simple push button is shown in Figure 4.4. The *pull-up resistor* R_{pull} holds the input at logic 1 (voltage V_{CC}) while the button is up; closing the switch shorts the input to ground, logic 0 or V_{SS}. The input is therefore *active low*, meaning that it goes low when the button is pressed. You might like to think "button down $\rightarrow$ input down."

A wasted current flows through the pull-up resistors to ground when the button is pressed. This is reduced by making R_{pull} large, but the system becomes sensitive to noise if this is carried too far and the Olimex 1121STK has $R_{pull} = 33\,k\Omega$, which is typical. This circuit is so common that most MCUs offer internal pull-ups to reduce the number of

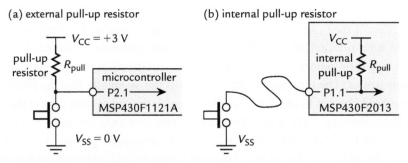

Figure 4.4: Standard, active low connection of a push button to an input with an (a) external and (b) internal pull-up resistor R_{pull}.

external components needed. TI has been curiously late in adding these to the MSP430 but they are available in the MSP430F2xx family and recent MSP430x4xx devices. Pull-up or pull-down resistors can be activated by setting bits in the PxREN registers, provided that the pin is configured as an input. The MCU behaves randomly if you forget this step because the inputs floats; see the section "Configuration of Unused Pins" on page 215.

Example 4.4

Compare the current wasted by the pull-up resistor with the operating current drawn by the MCU. (Of course, the button is not down all the time. Programming tricks can be used to reduce this current if it is important.)

A program can respond to inputs in two ways. For a simple analogy, suppose I am waiting for my daughter to come home while writing this book. She has forgotten her key so I need to unlock the door for her.

- I could go regularly to the door and look out to see whether she has returned, after each paragraph for instance. This is *polling*.

- I could carry on writing continuously until I receive an *interrupt* from the doorbell. I then finish the sentence to leave the work in a well-defined state and go to the door.

This example is trivial but contains the main points that concern a program. Polling is simple but carries the overhead of regular checking, which is high if checks must be made frequently. An interrupt needs special hardware (the doorbell) and requires tidying up so that normal operation can be resumed after the interrupt has been serviced. This makes interrupts somewhat tricky to program and I therefore defer them to Chapter 6.

The next program lights an LED when a button is pressed—the MCU acts as an expensive piece of wire. I assume that we use LED1 (P2.3) and button B1 (P2.1) on the Olimex 1121STK. There are many ways of doing even this straightforward task and two possible flowcharts are shown in Figure 4.5.

4.4.1   Single Loop with a Decision

The first approach has an infinite loop that tests the state of the button on each iteration. It turns the LED on if the button is down and turns it off if the button is up. The only

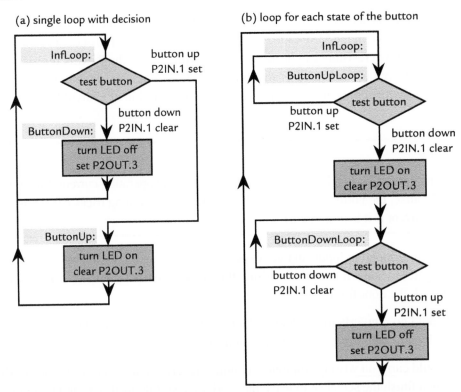

Figure 4.5: Two flow diagrams for lighting LED1 when button B1 is pressed: (a) single loop containing a decision, (b) loop for each state of the button.

awkward feature is that both the button and LED are active low. Listing 4.5 shows the details. It is similar to `msp430x11x1_P2_poll` in the code examples for the F1121A.

Listing 4.5: Program `butled1.c` in C to light LED1 when button B1 is pressed. This version has a single loop containing a decision statement.

```
// butled1.c - press button to light LED
// Single loop with "if"
// Olimex 1121STK board, LED1 active low on P2.3,
//    button B1 active low on P2.1
// J H Davies, 2006-06-01; IAR Kickstart version 3.41A
// -------------------------------------------------------------------
#include <msp430x11x1.h>          // Specific device

// Pins for LED and button on port 2
#define LED1      BIT3
#define B1        BIT1
```

```
void main (void)
{
    WDTCTL = WDTPW | WDTHOLD;      // Stop watchdog timer
    P2OUT |= LED1;                 // Preload LED1 off (active low!)
    P2DIR = LED1;                  // Set pin with LED1 to output
    for (;;) {                     // Loop forever
        if ((P2IN & B1) == 0) {    // Is button down? (active low)
            P2OUT &= ~LED1;        // Yes: Turn LED1 on (active low!)
        } else {
            P2OUT |= LED1;         // No: Turn LED1 off (active low!)
        }
    }
}
```

There are a few new points about this program:

1. I defined symbolic constants for B1 and LED1, which is better practice than using their bit patterns explicitly. The constants BIT1 and BIT3 are defined in the header file.

2. Bitwise logical operations are used to set or reset the output LED1 rather than writing to the whole byte P2OUT. It makes no difference here because only one bit is used for output but this method will be essential later. See the section "Aspects of C for Embedded Systems" on page 46 for a reminder of how this works.

3. A mask is also used inside the if statement to pick out the value of the bit associated with the push button. *You must include the parentheses* around P2IN & B1 or the program does not work as intended. The reason is that the == operator has a higher precedence than &. This is painfully easy to forget. (Of course you could be a real C hacker and write just if (P2IN & B1) with other changes to match. Good luck.)

4. It is not necessary to put braces around the statements that follow if and else because they are only single lines. However, I strongly recommend that you *always* do this. If not, you will forget the braces around multiple lines sooner or later and it can be very difficult to pin down the problem when the program does not work as intended.

Example 4.5

Try this. In the simulator you can mimic the input from the button by changing the value of P2IN.P1 in the Register > Port 1/2 window. With an emulator you should see the value of

P2.1 change in the Register window when you press the button. If not, check that pull-ups are installed or enabled.

Modify your program so that the LED lights when the button is up rather than down.

Example 4.6

Extend the program so that button B1 controls LED1 and button B2 controls LED2 independently. Button B2 is connected to P1.2 on port P1 rather than port P2.

Example 4.7

Light LED1 only while *both* buttons B1 and B2 are pressed.

4.4.2   Single Loop with Decision in Assembly Language

Listing 4.6: Program butasm1.s43 with single loop in assembly language to light LED1 when button B1 is pressed.

```
; butasm1.s43 - press button to light LED
; Single loop with decision
; Olimex 1121STK, LED1 active low on P2.3, B1 active low on P2.1
; J H Davies, 2006-06-01; IAR Kickstart version 3.41A
; -------------------------------------------------------------------
#include <msp430x11x1.h>              ; Header file for this device

; Pins for LED and button on port 2
LED1       EQU       BIT3
B1         EQU       BIT1

      RSEG        CODE                ; Program goes in code memory
Reset:                               ; Execution starts here
      mov.w      #WDTPW|WDTHOLD,&WDTCTL  ; Stop watchdog timer
      bis.b      #LED1,&P2OUT         ; Preload LED1 off (active low!)
      bis.b      #LED1,&P2DIR         ; Set pin with LED1 to output
InfLoop:                             ; Loop forever
      bit.b      #B1,&P2IN            ; Test bit B1 of P2IN
      jnz        ButtonUp             ; Jump if not zero, button up
ButtonDown:                          ; Button is down
      bic.b      #LED1,&P2OUT         ; Turn LED1 on (active low!)
      jmp        InfLoop              ; Back around infinite loop
ButtonUp:                            ; Button is up
      bis.b      #LED1,&P2OUT         ; Turn LED1 off (active low!)
      jmp        InfLoop              ; Back around infinite loop
; -------------------------------------------------------------------
      RSEG        RESET               ; Segment for reset vector
      DW          Reset               ; Address to start execution
      END
```

Listing 4.6 shows the corresponding program in assembly language. It contains several new directives, instructions, and structures:

1. The two directives with `EQU` are equivalent to `#define` in C and provide the same symbols.

2. The instruction `bis` stands for "bit set" and is equivalent to `|=` in C. The first operand is the mask that selects the bits and the second is the register whose bits should be set, `P2OUT` in the first case. The characters `#` and `&` are necessary to show that the values are immediate data and an address, as in the `mov` instruction. Similarly, `.b` is again needed because `P2OUT` is a byte rather than a word.

3. The complementary instruction `bic` stands for "bit clear." It clears bits in the destination if the corresponding bit in the mask is set. For example, `bic.b #0b00011000,&P2OUT` clears bits 3 and 4 of `P2OUT`. There is no need to take the complement of the mask as in the C program, where `&=~ mask` is used to clear bits.

4. Inevitably nothing is as sophisticated as an `if-else` construction in assembly language. The only conditional instructions are jumps that depend on bits in the status register. Therefore the general approach is to carry out an operation that affects the status register and jump according to the result. Here we want to test the state of the push button so we use `bit.b #B1,&P2IN`. This instruction stands for "bit test" and is the same as the `and` instruction except that it affects only the status register; it does not store the result. In this case it calculates `B1 & P2IN` and either sets the `Z` bit in the status register if the result is zero or clears `Z` if the result is nonzero. It does not affect `P2IN`. See the section "Arithmetic and Logic Instructions with Two Operands" on page 133 for more details.

5. The next step is to test the `Z` bit and jump if the bit is clear. This means that the result of the `bit` operation is nonzero, which in turn means that the `B1` bit must be set and the button is up. You can imagine how easy it is to muddle this train of thought. It is best to work through the instructions by hand, "playing computer" with pencil and paper, to check that intricate code like this works as expected. The two complementary branch instructions that test the `Z` bit are

    - **jz—jump if zero**, meaning that the `Z` bit is set.

    - **jnz—jump if nonzero**, meaning that the `Z` bit is clear (reset).

In both cases the jump is taken if the condition is true, otherwise the next instruction is executed in the usual way. Here control jumps to `ButtonUp` if the button is up, B1 is set, the result is nonzero, and the Z bit is clear.

6. If the button is down, the jump is not taken and the next instruction is executed. The label `ButtonDown` is redundant but useful to show why this part of the code is being executed. Anything helps to make assembly language clearer.

Some processors have instructions to test bits directly, although these may be restricted to a limited range of addresses. They may even have a single instruction to test a bit and jump according to the result. Indeed the only conditional instructions in a PIC16F are the pair "skip if set" and "skip if clear," which can be used for any bit in any register. In contrast, the MSP430's tests work only on the status register. Therefore an explicit test is needed to set up the status register before a conditional branch, as previously, unless an arithmetic or logical operation does this automatically.

4.4.3 Two Loops, One for Each State of the Button

Figure 4.5b and Listing 4.7 show another way of controlling the LED. In this case two `while` loops are inside an infinite loop. The program is trapped inside the first loop while the button is up and in the second while it is down. The actions to be taken when the button is pressed or released—turning the LED on and off—are put in the transitions between the loops, not within the loops themselves.

Listing 4.7: Program `butled2.c` in C to light LED1 when button B1 is pressed. This version has a loop for each state of the button.

```
// butled2.c - press button to light LED
// Two loops, one for each state of the button
// Olimex 1121STK board, LED1 active low on P2.3,
//   button B1 active low on P2.1
// J H Davies, 2006-06-01; IAR Kickstart version 3.41A
//------------------------------------------------------------------
#include <msp430x11x1.h>              // Specific device

// Pins for LED and button on port 2
#define LED1      BIT3
#define B1        BIT1

void main (void)
{
    WDTCTL = WDTPW | WDTHOLD;         // Stop watchdog timer
    P2OUT = LED1;                     // Preload LED1 off (active low!)
    P2DIR = LED1;                     // LED1 pin output, others input
```

```
    for (;;) {                          // Loop forever
        while ((P2IN & B1) != 0) {      // Loop while button up
        }                               // (active low) doing nothing
// Actions to be taken when button is pressed
        P2OUT &= ~LED1;                 // Turn LED1 on (active low!)
        while ((P2IN & B1) == 0) {      // Loop while button down
        }                               // (active low) doing nothing
// Actions to be taken when button is released
        P2OUT |= LED1;                  // Turn LED1 off (active low!)
    }
}
```

What is the difference between these two approaches?

1. The LED is continually being switched on or off in the first version. Of course this has no visible effect but the important point is that the action is repeated continually.

2. The LED is turned on or off only when necessary in the second version, just at the points when the button is pressed or released.

There is no practical difference between these for the simple task of lighting the LED while the button is down, but it makes a big difference for the following examples.

Example 4.8

Write a program to toggle the LED each time the button is pressed: Turn it on the first time, off the second, on the third, and so on.

Example 4.9

Write a program to count the number of times the button is pressed and show the current value on the LEDs. Of course this works better on a demonstration board with more LEDs. On the Olimex 1121STK, you should set the pins for both LEDs to outputs and turn the LEDs off initially. The statement P2OUT -= LED1 causes the displayed value to be incremented by 1. Why does this work? (It would be clearer if the LEDs were active high and connected to bits 0 upward.) This is a reminder of memory-mapped input and output: We can do arithmetic on the output register P2OUT as if it were a simple location in RAM.

You may find that the display sometimes jumps by more than 1 when the button is pressed. The reason is explained in the section "Switch Debounce" on page 225.

Example 4.10

Suppose that you have a complete set of LEDs on port P1. Write a program to measure the time for which a button on P2.7 is pressed and display it as a binary value on the LEDs (arbitrary units). This can be done by incrementing P1OUT in a loop while the button is down. (You may wish to reduce the frequency of MCLK.) The main point is to select the correct format of program.

4.4.4 Addressing Bits Individually in C

It seems rather primitive that we have to pick out the value of individual bits by masking them explicitly in Listings 4.5 and 4.7. The processor may not be able to address bits directly but why can the compiler not do the dirty work for us? Why can we not write something like 'P2OUT.3 = 1' to set the bit, for instance? The good news is that we can, but it needs a header file in which the registers are defined as *bit fields*, mentioned in the section "Aspects of C for Embedded Systems" on page 46. EW430 provides such definitions in a second set of header files, whose names begin with io430 rather than msp430. We can then write P2OUT_bit.P2OUT_3 = 1 to set bit 3 of P2OUT, and so on. The usage is a bit clumsy so I defined more user-friendly symbols in Listing 4.8. The io430 headers can be used for C programs instead of the msp430 files but do not work for assembly language.

Listing 4.8: Program `butled3.c` in C to light LED1 when button B1 is pressed. This version uses bit fields rather than masks.

```
// butled3.c - press button to light LED
// Two loops, one for each state of the button
// Header file with bit fields
// Olimex 1121STK board, LED1 active low on P2.3,
//    button B1 active low on P2.1
// J H Davies, 2006-06-01; IAR Kickstart version 3.41A
//-------------------------------------------------------------
#include <io430x11x1.h>           // Specific device, new format header

// Pins for LED and button
#define LED1      P2OUT_bit.P2OUT_3
#define B1        P2IN_bit.P2IN_1

void main (void)
{
    WDTCTL = WDTPW | WDTHOLD;         // Stop watchdog timer
```

```
      LED1 = 1;                             // Preload LED1 off (active low!)
      P2DIR_bit.P2DIR_3 = 1;                // Set pin with LED1 to output
      for (;;) {                            // Loop forever
          while (B1 != 0) {                 // Loop while button up
          }                                 // (active low) doing nothing
// actions to be taken when button is pressed
          LED1 = 0;                         // Turn LED1 on (active low!)
          while (B1 == 0) {                 // Loop while button down
          }                                 // (active low) doing nothing
// actions to be taken when button is released
          LED1 = 1;                         // Turn LED1 off (active low!)
      }
}
```

4.4.5 Two Loops in Assembly Language

Listing 4.9 shows the corresponding program in assembly language. It is very similar to Listing 4.6 and there are no new features.

Listing 4.9: Program `butasm2.s43` in assembly language to light LED1 when button B1 is pressed. There is a loop for each state of the button.

```
; butasm1.s43 - press button to light LED
; Two loops, one for each state of the button
; Olimex 1121STK, LED1 active low on P2.3, B1 active low on P2.1
; J H Davies, 2006-06-01; IAR Kickstart version 3.41A
; --------------------------------------------------------------------
#include <msp430x11x1.h>             ; Header file for this device

; Pins for LED and button on port 2
LED1      EQU       BIT3
B1        EQU       BIT1

      RSEG       CODE                 ; Program goes in code memory
Reset:                                ; Execution starts here
      mov.w      #WDTPW|WDTHOLD,&WDTCTL  ; Stop watchdog timer
      bis.b      #LED1,&P2OUT         ; Preload LED1 off (active low!)
      bis.b      #LED1,&P2DIR         ; Set pins with LED1 to output
InfLoop:                              ; Loop forever
ButtonUpLoop:                         ; Loop while button up
      bit.b      #B1,&P2IN            ; Test bit B1 of P2IN
      jnz ButtonUpLoop                ; Jump if not zero, button up
; Actions to be taken when button is pressed
      bic.b      #LED1,&P2OUT         ; Turn LED1 on (active low!)
ButtonDownLoop:                       ; Loop while button down
      bit.b      #B1,&P2IN            ; Test bit B1 of P2IN
      jz  ButtonDownLoop              ; Jump if zero, button down
; Actions to be taken when button is released
      bis.b      #LED1,&P2OUT         ; Turn LED1 off (active low!)
```

```
    jmp      InfLoop                ; Back around infinite loop
; - - - - - - - - - - - - - - - - - - - - - - - - - - - - - - - - - - - - - - -
    RSEG     RESET                  ; Segment for reset vector
    DW       Reset                  ; Address to start execution
    END
```

4.4.6 Digital Die

A good exercise is to design a digital die (singular of dice to be pedantic). Ideally this should display a random number from 1 to 6 on a display when the button is pressed (or released). Unfortunately this needs more than two LEDs so the Olimex 1121STK is not ideal. Suppose that we have a board with six LEDs available to make a useful display. (Seven LEDs in the traditional pattern would make a better product.)

The first issue is to find a way of generating "random" numbers and there are many possibilities. Here are three:

- We could take a "snapshot" of a fast clock, such as the timer module on the MCU. A more sophisticated approach is to use the difference between two clocks. This is described in the application note *Random Number Generation Using the MSP430* (slaa338).

- There are algorithms for generating pseudo-random numbers from a shift register with feedback.

- The simplest approach is to write a loop that cycles rapidly around the six states while the button is down and stops when the button is released.

Taking the third approach, suppose that the program shows a fixed display when the button is released and cycles rapidly through the range of values while the button is pressed to "roll" the die. How fast a loop do we need? Human reaction times are around 0.1 s and are perhaps reproducible to 0.01 s (a guess). This means that the system should make several cycles through 1–6 in 0.01 s so that a user cannot predict the value when the button is released. Let us say ten cycles, so each must take less than 1 ms. It is unlikely that the complete cycle will need more than 100 clock cycles of the processor, so the clock must run at 100 KHz or faster. This is no problem if we use the digitally controlled oscillator (DCO), which typically runs at around 1 MHz. On the other hand, a watch crystal at 32 KHz might be a bit slow, although the estimates are very conservative.

The program can be based on the "two loop" structure by extending it to six states. The main loop works like this:

1. Show the display for a value of 1. Test the state of the button and trap the program in a loop if the button is up; go on to the next number if the button is down.

2. Do the same for values 2–5.

3. Do the same for 6, except that this time the program returns to the beginning of the main loop to display a 1 again.

The tricky aspect is to ensure that the die is fair or unbiased, which means that all values should be equally likely. This in turn means that must be an equal delay between the tests of the button for each value. It is hard to ensure this if the program is written in C, so assembly language is the better choice. The problem is handling the final value of 6, because this must return to the beginning of the main loop rather than passing on directly to the next value. This may require an extra `jmp` instruction, which breaks the symmetry of the code for each value. Here are two possible solutions:

- Write the code for each value in the same way, which probably means that a `jmp` instruction is used to go to the next value.

- Compensate for the extra two cycles taken in the `jmp` instruction by padding the code for the other values with *no operation* or `nop` instructions. A `nop` consumes one clock cycle without performing any operation and is used when short delays are needed.

Example 4.11

Try this within the constraints of your demonstration board. Count the number of cycles spent in each state using the simulator to ensure that there is no bias. (It is hard to check this by repeatedly rolling the die and counting the number of occurrences of each value because of the intrinsic randomness. A χ^2 test might be appropriate if you know enough statistics.)

4.5 Automatic Control: Flashing Light by Software Delay

The next step is to get the MCU to control something by itself and the simplest task is to flash an LED on and off. Let us do this with a period of about 1 Hz, which needs a delay of 0.5 s while the LED remains on or off. Timing is a very common function of MCUs so almost all contain specialized hardware to act as a timer—many devices have several timers with different characteristics. We explore the timer in the section "Automatic Control: Flashing a Light by Polling Timer_A" on page 105, but use a simpler approach now. This is a software timing loop, which just means an empty loop that consumes cycles of the processor. (We could use nops for short delays but not for 0.5 s.) The idea is shown in Listing 4.10. One of the two LEDs on the Olimex 1121STK is initially lit and the display is toggled (complemented) by the xor operation. The delay between changes is set by the for loop, which simply goes around doing nothing for DELAYLOOPS iterations. This is essentially the same as the simplest program in TI's set of code examples, msp430x11x1_1.c.

Listing 4.10: Program flashled.c to flash LEDs with a frequency of roughly 1 Hz using a software delay.

```
// flashled.c - toggles LEDs with period of about 1s
// Software delay for() loop
// Olimex 1121STK, LED1,2 active low on P2.3,4
// J H Davies, 2006-06-03; IAR Kickstart version 3.41A
//-------------------------------------------------------------
#include <msp430x11x1.h>              // Specific device

// Pins for LEDs
#define LED1     BIT3
#define LED2     BIT4
// Iterations of delay loop; reduce for simulation
#define DELAYLOOPS  50000

void main (void)
{
    volatile unsigned int LoopCtr;   // Loop counter: volatile!

    WDTCTL = WDTPW | WDTHOLD;         // Stop watchdog timer
    P2OUT = ~LED1;                    // Preload LED1 on, LED2 off
    P2DIR = LED1|LED2;                // Set pins with LED1,2 to output
    for (;;) {                        // Loop forever
        for (LoopCtr = 0; LoopCtr < DELAYLOOPS; ++LoopCtr) {
        }                             // Empty delay loop
        P2OUT ^= LED1|LED2;           // Toggle LEDs
    }
}
```

Perhaps the most important aspect of this program is the declaration of the variable LoopCtr:

- The critical feature is the volatile key word. This essentially tells the compiler not to perform any optimization on this variable (look back at the section "The C

Programming Language" on page 46). If it were omitted and optimization were turned on, the compiler would notice that the loop is pointless and remove it, together with our delay.

- It is `unsigned` to give a range of 0–65,535, which is needed for 50,000 iterations.

- A less obvious pitfall is the assumption that the size of an integer is 16 bits so that it can hold the desired range of values. Yes, I did check the manual for the compiler but it would be more professional to specify the size explicitly by replacing `unsigned int` with the key word `uint16_t`. This is defined in `stdint.h`, which must be included.

Example 4.12

Try this program and measure the duration of the delay. This can be done in real time using a watch. It is also interesting to check the number of clock cycles per iteration. This can be found from the CYCLECOUNTER in the Register > CPU Registers window of the EW430 simulator. The default clock in the F1121A is the DCO at about 800 KHz but this may be different in other devices.

It is easy to monitor the value of a variable such as `LoopCtr` in the debugger. Hovering the mouse over the name of a variable in the source code reveals its type and value (you must first make the window active by clicking in it). It may be more convenient to add the variable to the Watch window. This can be done by right-clicking on the name of the variable in the Code window and selecting Add to Watch in the contextual menu.

You might like to set a *breakpoint* on the line that toggles the LEDs. This is a point at which the debugger halts execution of the program in either the simulator or emulator. Breakpoints can be set in either the editor or debugger. Click in this line and choose Edit > Toggle Breakpoint from the menu. A red blob appears in the margin and the debugger highlights the line in red as well, both in the source and disassembly. You can then select Debug > Go and the program runs until it reaches the breakpoint. A shortcut is to right-click on the line after the loop and select Run to Cursor from the contextual menu.

Try writing the loop differently to see whether the delay changes. You could use `while` or `do-while`, count down rather than up, and compare `--LoopCtr` and `LoopCtr--` (these can be substantially different).

Example 4.13

Take out the `volatile` key word to test the warning. The effect depends on the degree of optimization for speed, which is selected in Project > Options > C/C++ Compiler > Optimizations.

This approach is obviously crude and is a gross waste of the MCU but may be used for short, uncritical delays. For example, a delay is sometimes needed while starting up the chip so that a peripheral can stabilize. A problem with this approach, as you found in the examples, is that the delay is hard to predict because it depends on the way in which the C code is compiled. This can be avoided by using assembly language, which we do in the next section.

4.5.1   Delay Loop in Assembly Language

Listing 4.11 shows the corresponding program in assembly language. The major new feature is that we need a variable for the loop counter. This raises the question, What is meant by a variable in assembly language? There is nothing like the declaration of `LoopCtr` in C. A variable is really no more than a memory location that we choose to use for a particular purpose. It is as if we stick a label `LoopCtr` on a pigeonhole to remind us of what is stored in it. In other words, the name of a variable is just a symbol that will be translated to the address of its memory location.

Where should the memory be located? Clearly it cannot be ROM or we will be unable to store a new value. One option is RAM, and we look at this later, but the simplest option for the delay loop is to use one of the CPU's internal registers. This might not be a good approach in a larger program because the usage could conflict with other routines, but there is no problem here. I chose the register R4 and used its name explicitly in the code but it would be better practice to assign a symbol such as `LoopCtr` again.

Listing 4.11: Program `flshasm1.s43` in assembly language to flash LEDs with a frequency of roughly 1 Hz using a software delay. A CPU register is used for the loop counter.

```
; flshasm1.s43 - toggles LEDs with period of about 1s
; Loop counter in CPU register R4, up loop with compare
; Olimex 1121STK, LED1,2 active low on P2.3,4
; J H Davies, 2006-06-06; IAR Kickstart version 3.41A
; -----------------------------------------------------------------
```

```
#include <msp430x11x1.h>              ; Header file for this device

; Pins for LED on port 2
LED1        EQU BIT3
LED2        EQU BIT4
; Iterations of delay loop; reduce for simulation
DELAYLOOPS  EQU 50000

        RSEG    CODE                  ; Program goes in code memory
Reset:                                ; Execution starts here
        mov.w   #WDTPW|WDTHOLD,&WDTCTL ; Stop watchdog timer
        mov.b   #LED2,&P2OUT          ; Preload LED1 on, LED2 off
        bis.b   #LED1|LED2,&P2DIR     ; Set pins with LED1,2 to output
InfLoop:                              ; Loop forever
        clr.w   R4                    ; Initialize loop counter
DelayLoop:                            ; [clock cycles in brackets]
        inc.w   R4                    ; Increment loop counter     [1]
        cmp.w   #DELAYLOOPS,R4        ; Compare with maximum value [2]
        jne     DelayLoop             ; Repeat loop if not equal   [2]
        xor.b   #LED1|LED2,&P2OUT     ; Toggle LEDs
        jmp     InfLoop               ; Back around infinite loop
; ----------------------------------------------------------------
        RSEG    RESET                 ; Segment for reset vector
        DW      Reset                 ; Address to start execution
        END
```

The construction of the loop is slightly different from that in the C program. Most loops in C are for or while loops, in which the test takes place at the start of each iteration. The only type of test in assembly language is a conditional jump, so it is easier to place the test at the end of the loop, where there has to be a jump in any case. This is like a do-while loop in C:

```
LoopCtr = 0;
do {
    ++LoopCtr;
} while (LoopCtr != DELAYLOOPS);
```

The significant difference is that the body of a do-while loop is always executed at least once, even if the test is never true. In contrast, a while loop is not executed at all if the test fails on the first iteration. This means that the assembly language would malfunction if DELAYLOOPS were 0: It would give the maximum possible delay instead of no delay at all. A test could be included to avoid this. Alternatively, we could enter the loop by jumping to the test rather than starting at the top, as we see later.

The test at the end of the loop is done by the compare instruction cmp. This computes the difference R4 - DELAYLOOPS and sets the status register but does not store the result, unlike a real subtraction. It is only a test, like `bit`. It is followed by the conditional jump `jne`, which stands for "jump if not equal." A great advantage of assembly language is that we can work out the length of the delay exactly: Each iteration takes five clock cycles.

Nothing is wrong with this program, but loops are rarely written like this in assembly language unless it is essential that the variable count upward. It is shorter to write the loop so that the variable counts down to 0 because the result of the `dec` instruction can be tested directly in a conditional jump. This is shown in Listing 4.12. There is no need for an extra comparison because the z flag will be set when the result of `dec` is 0. This construction is so common that many processors have a special "decrement and branch if nonzero" instruction.

Listing 4.12: Part of program `flshasm2.s43` to show a typical loop counting down to 0 in assembly language.

```
       mov.w    #DELAYLOOPS,R4    ; Initialize loop counter
DelayLoop:                        ; [clock cycles in brackets]
       dec.w    R4                ; Decrement loop counter  [1]
       jnz      DelayLoop         ; Repeat loop if not zero [2]
```

The loop now takes only three clock cycles per iteration. This is 150,000 cycles for the complete delay, which gives about 0.19 s with an 800 KHz clock. Thus the LEDs flash at about 2.7 Hz, considerably faster than the target of 1 Hz. We could try to remedy this by increasing DELAYLOOPS but an integer cannot be much larger than the current value. Better solutions would be to slow down the loop by inserting some nops or putting a further loop around it.

4.5.2   Assembly Language with Variables in RAM

A register in the CPU is a good place to store a variable that is needed rapidly for a short time but RAM is used for variables with a longer life. We might as well use the first address available and there are several ways in which this can be allocated. The data sheet shows that RAM lies at addresses from 0x0200–02FF so we should use 0x0200 and 0x0201 for LoopCtr, which needs 2 bytes. The most obvious approach is to assign the address explicitly:

```
LoopCtr       EQU 0x0200       ; 2 bytes for loop counter
```

This shows immediately that the variable is just a name for a memory location, which is instructive, but it is easy to make mistakes. Suppose that we want another variable in the address that follows `LoopCtr`. We must remember to add 2 to the address, because `LoopCtr` is a word rather than a byte, so the next address free is 0x0202. The potential for error is obvious. The assembler therefore provides a directive `DS` to *define storage*. We must tell it where this storage is to be defined, which needs an `ORG` statement in absolute assembly. Thus we could allocate storage for `LoopCtr` and another variable as follows:

```
        ORG     0x0200          ; Start of RAM
LoopCtr     DS  2               ; 2 bytes for loop counter
Another     DS  1               ; 1 byte for another variable
```

In relocatable assembly we specify the segment rather than the absolute address. Several segments are used for variables in C, depending on how they are declared and initialized. I use DATA16_N, which is intended for uninitialized variables:

```
        RSEG    DATA16_N        ; Memory for variables
LoopCtr     DS  2               ; 2 bytes for loop counter
Another     DS  1               ; 1 byte for another variable
```

These are something like static variables in C in the sense that they remain in existence and retain their value until the MCU is reset, even if they are not being used. They are not initialized automatically, however. Another important distinction is that variables do not have a "type" in most assembly languages. The two bytes we have reserved could hold a signed integer, an unsigned integer, a unicode character, and so on. It is up to the programmer to keep track of the meaning of the data and the assembler provides no checks. A few assemblers have special directives to define the type of variable held in storage, but this is rare.

The complete program is shown in Listing 4.13. The `dec` operation takes four cycles for a variable in RAM rather than in a CPU register, which doubles the length of the delay.

Listing 4.13: Program `flshasm3.s43` in assembly language to flash LEDs with a frequency of roughly 1 Hz using a software delay. The loop counter is stored in RAM.

```
; flshasm3.s43 - toggles LEDs with period of "about" 1s
; Loop counter in RAM, counts loop to zero
; Olimex 1121STK, LED1,2 active low on P2.3,4
; J H Davies, 2006-06-07; IAR Kickstart version 3.41A
; -------------------------------------------------------------------
```

```
#include <msp430x11x1.h>              ; Header file for this device

; Pins for LED on port 2
LED1         EQU  BIT3
LED2         EQU  BIT4
; Iterations of delay loop; reduce for simulation
DELAYLOOPS   EQU  50000
;-----------------------------------------------------------------------
     RSEG    DATA16_N                  ; Memory for variables
LoopCtr      DS   2                    ; 2 bytes for loop counter
;-----------------------------------------------------------------------
     RSEG    CODE                      ; Program goes in code memory
Reset:                                 ; Execution starts here
     mov.w   #WDTPW|WDTHOLD,&WDTCTL    ; Stop watchdog timer
     mov.b   #LED2,&P2OUT             ; Preload LED1 on, LED2 off
     bis.b   #LED1|LED2,&P2DIR        ; Set pins with LED1,2 to output
InfLoop:                               ; Loop forever
     mov.w   #DELAYLOOPS,&LoopCtr     ; Initialize loop counter
DelayLoop:                             ; [clock cycles in brackets]
     dec.w   &LoopCtr                 ; Decrement loop counter  [4]
     jnz     DelayLoop                ; Repeat loop if not zero [2]
     xor.b   #LED1|LED2,&P2OUT        ; Toggle LEDs
     jmp     InfLoop                  ; Back around infinite loop
;-----------------------------------------------------------------------
     RSEG    RESET                     ; Segment for reset vector
     DW      Reset                     ; Address to start execution
     END
```

Example 4.14

Try this in the debugger. You can monitor the value of LoopCtr as in C, except that the debugger has no way of knowing that it is unsigned, so this must be set by hand in the Watch window. It is assumed that the variable is a signed, 16-bit integer if you hover the mouse over the name of the variable in the (active) source window, so the value may not always be what you expect.

It is also worth opening the Memory window and selecting RAM. Change the width to 2x Units to see words rather than bytes, which avoids trouble with little-endian ordering. The RAM is initially filled with random values, but you will see the contents of the word at 0x0200 change when LoopCtr is initialized and subsequently decremented.

If you look at the disassembly for the C program, you find that the compiler takes neither of the two approaches we used for assembly language. Instead the address of the loop counter appears as @SP or 0x0(SP). This is because automatic variables are normally created on the stack. See the section "Storage for Local Variables" on page 179. Sometimes the compiler creates a subroutine for the loop, which makes the addressing even more obscure.

4.6   Automatic Control: Use of Subroutines

It is good practice to package distinct functions into subroutines or functions because this makes code easier to write, debug, and reuse. This is also more reliable because individual subroutines can be tested before being built into a large program. Moreover, we can make the function more versatile. For example, we could write a delay loop with a duration of 0.1 s and use it to build a subroutine to give an overall delay of $0.1n$ s, where the multiplier n is passed as a parameter. This is trivial in C but needs new material in assembly language.

Example 4.15

Rewrite the program in C that flashes the LEDs at 1 Hz to use a function, where the delay in units of 0.1 s is passed as a parameter.

Listing 4.14 shows a program to flash the LEDs at 1 Hz. It uses a subroutine `DelayTenths`, which is called with the `call` instruction. This sounds obvious, but it is painfully easy to forget the # sign before the name. This unusual requirement arises because `call` in the MSP430 uses the same addressing modes as the arithmetic and logic instructions. Here we provide an immediate value, the address given by the label `DelayTenths`. Therefore the # sign must be inserted, just as in statements like `mov.b #0,P2OUT`. It is a nuisance in these circumstances but the advantage is that subroutines can be called in versatile ways. You can set up arrays of subroutines, for instance.

No special structure starts a subroutine in assembly language, nothing like the declaration in C with a list of parameters and the type of return value. Instead there is only a label, just like the target of a jump. The subroutine itself is based around the delay loop in Listing 4.12, whose range has been adjusted to give a delay of about 0.1 s. This is packaged within an outer loop, which is executed n times. The main program and subroutine must agree on how to pass the desired delay in units of 0.1 s. Here we use the CPU register R12. This must be documented clearly so that the parameter does not get lost in transit when the subroutine is called. Here it is described in the comment before the subroutine, which also warns the user that the value is destroyed; the subroutine uses it as the outer loop counter. Parameters are passed in the same way in C provided that they are small enough to fit in registers R12–R15. This is not enforced by the hardware and is purely a convention defined in the reference manual for the compiler.

There is another important feature about the subroutine. It is possible that it might be called with a parameter of 0 and must behave correctly in this limit. I mentioned earlier that loops in assembly language are usually written with the test at the end, like do-while loops in C. Here the test must be evaluated first so that no delay loops are performed if R4 = 0. The solution is to jump into the loop just before the test rather than entering at the top. This is why the first statement in the subroutine is jmp LoopTest. The compiler uses this construction for while loops in C, which you can see by stepping through the code in the disassembly window of the debugger.

Finally, the subroutine ends with a return instruction, which restores control to the calling function. You must always enter a subroutine through a call and leave with a ret. Disaster is guaranteed if you ignore this. (The MSP430X can use calla and reta instead so you see these in the disassembly if your target is the F64618 on an MSP430FG4618/F2013 Experimenter's Board, for instance. There is no problem with call and ret for small programs, such as ours.)

Listing 4.14: Program flsubasm.s43 in assembly language to flash LEDs with a frequency of roughly 1 Hz using a subroutine and a passed parameter.

```
; flsubasm.s43 - flash LEDs at 1Hz with 800KHz MCLK
; Software delay subroutine based on down loop for 0.1s
; Multiple of 0.1s passed to subroutine in R12
; Olimex 1121STK, LED1,2 active low on P2.3,4
; J H Davies, 2006-06-20; IAR Kickstart version 3.41A
; ------------------------------------------------------------
#include <msp430x11x1.h>              ; Header file for this device

; Pins for LED on port 2
LED1           EQU        BIT3
; Iterations of delay loop for about 0.1s (3 cycles/iteration)
DELAYLOOPS   EQU       27000
; ------------------------------------------------------------
     RSEG    CSTACK                   ; Create stack (in RAM)
; ------------------------------------------------------------
     RSEG    CODE                     ; Program goes in code memory
Reset:                               ; Execution starts here
     mov.w   #SFE(CSTACK),SP          ; Initialize stack pointer
main:                                ; Equivalent to start of main() in C
     mov.w   #WDTPW|WDTHOLD,&WDTCTL   ; Stop watchdog timer
     bis.b   #LED1,&P2OUT             ; Preload LED1 off
     bis.b   #LED1,&P2DIR             ; Set pin with LED1 to output
InfLoop:                             ; Loop forever
     mov.w   #5,R12                   ; Parameter for delay, units of 0.1s
     call    #DelayTenths             ; Call subroutine: don't forget #
     xor.b   #LED1,&P2OUT             ; Toggle LED
     jmp     InfLoop                  ; Back around infinite loop
; ------------------------------------------------------------
; Subroutine to give delay of R12*0.1s
; Parameter is passed in R12 and destroyed
; R4 is used for loop counter but is not saved and restored
; Works correctly if R12 = 0: the test is executed first as in while(){}
; ------------------------------------------------------------
```

```
DelayTenths:
    jmp      LoopTest                    ; Start with test in case R12 = 0
OuterLoop:
    mov.w    #DELAYLOOPS,R4              ; Initialize loop counter
DelayLoop:                               ; [clock cycles in brackets]
    dec.w    R4                          ; Decrement loop counter   [1]
    jnz      DelayLoop                   ; Repeat loop if not zero [2]
    dec.w    R12                         ; Decrement number of 0.1s delays
LoopTest:
    cmp.w    #0,R12                      ; Finished number of 0.1s delays?
    jnz      OuterLoop                   ; No: go around delay loop again
    ret                                  ; Yes: return to caller
; ---------------------------------------------------------------------
    RSEG     RESET                       ; Segment for reset vector
    DW       Reset                       ; Address to start execution
    END
```

The difference between calling a subroutine and a simple jump is that control returns to the calling routine when a subroutine has finished. Here this means that control returns to xor.b #LED1,&P2OUT after DelayTenths has completed its work, illustrated in Figure 4.6. How does the processor keep track of this and find its way back?

When the subroutine is called, the CPU stores a copy of the current value of the program counter (PC), which contains the address of the next instruction that would normally be executed—the xor.b here. This is called the *return address*. It then loads the address of the subroutine into the PC, DelayTenths here, and starts executing that. At the ret instruction it retrieves its copy of the return address, places this in the PC, and resumes execution. This is why you must never enter a subroutine without a call, simply by walking down the code into it: No address will be stored for ret to retrieve.

The return address is stored in the *stack*, which grows down from the highest address in RAM. The CPU keeps track of the stack through the stack pointer (SP) register and the

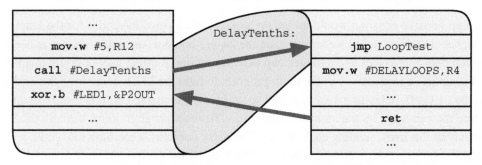

Figure 4.6: Flow of control when a subroutine is called and returns.

process described in detail in the section "Stack Pointer (SP)" on page 120. The vital point is that the stack pointer must be initialized to the first address beyond the RAM before any subroutine is called (it is decremented before it is first used). RAM extends from 0x0200–0x02FF in the F1121A so the initialization can be done with the following line in absolute assembly language:

```
mov.w    #0x0300,SP        ; Initialize stack pointer
```

It is slightly more complicated in relocatable assembly and needs two steps. First we must tell the linker to create a segment for the stack with RSEG CSTACK. The stack pointer can then be initialized to the top of this segment with mov.w #SFE(CSTACK),SP, where SFE calculates the ending address of a segment plus 1. This is just what we want. I initialized SP as the very first instruction and added a label main after it as a reminder that this is where the main() function starts in a C program. Including the label conveniently suppresses a complaint from the EW430 debugger as well.

4.6.1   Morse Code Flasher

More interesting patterns can be produced on the LEDs now that the delay can be controlled through the parameter passed to the subroutine. A simple example is to send a few characters of Morse code. This represents common characters—letters, numbers, and a few further characters—by sequences of long and short pulses. Long pulses are called *dashes* and are conventionally three times the length of the short pulses, called *dots*. There are short pauses between dots and dashes, longer ones between letters, and so on. Morse code uses short sequences for common letters, such as *dot* for E and *dash* for T, with longer sequences for numbers and infrequently used letters. Abbreviations are used to send common messages, of which the most famous is *dot-dot-dot dash-dash-dash dot-dot-dot* for the distress signal SOS, which achieved fame after the sinking of the *Titanic*.

A comprehensive program would include a lookup table to give the codes for the letters of the alphabet and so on, but we simply take the dashes and dots as input. A pause is needed between dots and dashes and a longer gap between letters. A crude way of doing this is to put a long list of instructions in the main program to light the LED, delay for a dot or dash, turn the LED off, delay to give a gap, and so on for all the characters. It is more elegant to store the pattern to be transmitted in an array. The main program loops through the elements of the array, sending each in turn. The array contains something like DOT, LETTER, DASH, DOT, ENDTX, where LETTER stands for the longer pause between letters and ENDTX marks the end of the message. The names DOT and DASH are just symbols for the appropriate delay in units of 0.1 s. A straightforward way of getting the gap between letters is to send a flash of zero duration, which appears as two successive pauses between symbols. This has the minor bug that it gives a short flash while the subroutine is called for zero delay but it will be invisible in practice. (Feel free to provide a better solution.)

Example 4.16

Extend the previous exercise to send the "word" SMS using C. The code for M is two dashes and you should recognize the pattern.

A program in assembly language is shown in Listing 4.15. Several new features are associated with the array that stores the message. This message is a constant and should therefore be stored in the flash memory, not RAM. We already saw how to do this for the reset vector, which was stored using a DW directive for *define word*. The delays are short here so it makes sense to store them in bytes and the corresponding directive is DB. This can be followed by one byte, a sequence of bytes, or a string. A label before the first DB acts as the name of the array, Message here. I spread the values over several lines for clarity. We need to tell the linker to store these data in the flash memory and a suitable standard segment is DATA16_C, which is used for constant data in C. This is allocated after the code, as you can see by opening the Memory window in the debugger and choosing FLASH.

Listing 4.15: Program `morsasm2.s43` in assembly language to send a short message in Morse code on an LED.

```
; morsasm2.s43 - sends dots and dashes
; Software delay subroutine with down loop for 0.1s
; Bug: gives very brief flash between letters
; Message stored as string in ROM with loop in main routine
; Registers in main routine: R5 used as loop counter,
;   R12 to pass delay to subroutine
; Olimex 1121STK, LED1,2 active low on P2.3,4
; J H Davies, 2006-06-20; IAR Kickstart version 3.41A
;-------------------------------------------------------------------
#include <msp430x11x1.h>    ; Header file for this device

; Pins for LED on port 2
LED1          EQU     BIT3
; Iterations of delay loop for about 0.1s (3 cycles/iteration)
DELAYLOOPS    EQU     27000
; Durations of symbols for morse code in units of 0.1s
; LETTER gives gap between letters; ENDTX terminates message
DOT           EQU     2
DASH          EQU     6
SPACE         EQU     2
```

```
LETTER        EQU     0
ENDTX         EQU     0xFF
; -----------------------------------------------------------------
      RSEG    CSTACK          ; Create stack (in RAM)
; -----------------------------------------------------------------
      RSEG    CODE            ; Program goes in code memory
Reset:                        ; Execution starts here
      mov.w   #SFE(CSTACK),SP ; Initialize stack pointer
main:                         ; Equivalent to start of main() in C
      mov.w   #WDTPW|WDTHOLD,&WDTCTL  ; Stop watchdog timer
      bis.b   #LED1,&P2OUT    ; Preload LED1 off
      bis.b   #LED1,&P2DIR    ; Set pin with LED1 to output
      clr.w   R5              ; Initialize counter to step through message
      jmp     MessageTest     ; Jump to test so it is evaluated first
MessageLoop:
      bic.b   #LED1,&P2OUT      ; LED1 on
      mov.b   Message(R5),R12   ; Load duration of delay as parameter
      call    #DelayTenths      ; Call subroutine: don't forget the #
      bis.b   #LED1,&P2OUT    ; LED1 off
      mov.w   #SPACE,R12      ; Load duration of delay (space)
      call    #DelayTenths    ; Call subroutine
      inc.w   R5              ; Next symbol to send
MessageTest:
      cmp.b   #ENDTX,Message(R5) ; Reached end of message?
      jne     MessageLoop        ; No: continue around loop
InfLoop:                      ; Yes: loop forever
      jmp     InfLoop         ; around infinite, empty loop
; -----------------------------------------------------------------
; Subroutine to give delay of R12*0.1s
; Parameter is passed in R12 and destroyed
; R4 is used for loop counter but is not saved and restored
; Works correctly if R12 = 0: the test is executed first as in while(){}
; -----------------------------------------------------------------
DelayTenths:
      jmp     LoopTest        ; Start with test in case R12 = 0
OuterLoop:
      mov.w   #DELAYLOOPS,R4  ; Initialize loop counter
DelayLoop:                    ; [clock cycles in brackets]
      dec.w   R4              ; Decrement loop counter   [1]
      jnz     DelayLoop       ; Repeat loop if not zero [2]
      dec.w   R12             ; Decrement number of 0.1s delays
LoopTest:
      cmp.w   #0,R12          ; Have we finished number of 0.1s delays?
      jnz     OuterLoop       ; No: go around delay loop again
      ret                     ; Yes: return to caller
; -----------------------------------------------------------------
      RSEG    DATA16_C        ; Segment for constant data in ROM
Message:                      ; Message to send (dots and dashes)
      DB      DOT,DOT,DOT,LETTER
      DB      DASH,DASH,DASH,LETTER
      DB      DOT,DOT,DOT,ENDTX
; -----------------------------------------------------------------
      RSEG    RESET           ; Segment for reset vector
      DW      Reset           ; Address to start execution
      END
```

Having stored this array, how do we retrieve the values of individual elements? In C we use an array, something like Message[i] with i as the index. I used the register R5 for the index and it turns out that the usage in assembly language is very similar: Message(R5). This is called *indexed* addressing. The CPU adds the constant Message to the value in R5 and uses the sum as the address of the desired data—simple. (It would be a little more complicated if the elements of the array were larger than bytes.)

Example 4.17

Write a program in assembly language for a die with a lookup table for the patterns. Assume that you have LEDs on all bits of port P1 and a button on P2.7. The table can be treated in the same way as Message in Listing 4.15.

4.7 Automatic Control: Flashing a Light by Polling Timer_A

Software delay loops are a waste of the processor because it is not available for more useful actions. There are other serious problems too, such as the unpredictability of delays written in C. All microcontrollers therefore have special hardware to act as a timer, often many of them. The heart of a timer is just a counter fed from a clock, so the idea is little different from the software loop. In particular, a timer has no idea of real time: It just counts clock cycles. It is up to the programmer to ensure that the value held in the counter is meaningful. Also, the recorded time is only as accurate as the frequency of the clock, which therefore needs a crystal if something like a reliable time of day is needed.

There are several types of timer in different variants of the MSP430. We use Timer_A, which is available in all devices. Chapter 8 is devoted to timers so only a brief description is given here. Timer_A is based on a 16-bit register TAR, shown in Figure 4.7, which counts clock pulses. There are also several capture/compare channels, which are usually used to trigger events and drive outputs directly, but we bypass them and take a more basic approach. The simplest way of generating a fixed delay with Timer_A is to wait for TAR to overflow. This is normally done by requesting an interrupt but we instead use polling as in the section "Read Input from a Switch" on page 80.

The operation of Timer_A is controlled by reading from and writing to memory-mapped registers, as for all peripherals. These are described thoroughly in the user's guide for each

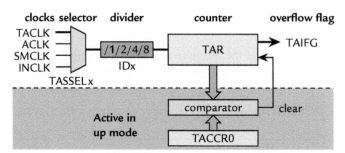

Figure 4.7: Simplified structure of part of Timer_A. Only the clock and counter TAR, shown above the broken line, are used in Continuous mode. The comparator and capture/compare register TACCR0 are also needed in Up mode.

family. We need only the Timer_A Control Register TACTL. The layout of the individual bits of this register is shown in Figure 4.8. The address of the register is given in the data sheet but symbols for this and the bits are defined in the header files. You would not use numbers when symbols are available, would you?

Three items are given for each bit:

- Its position in the word, which should not be needed (use symbolic names instead).

- Its name, which is defined in the header file and should be known to the debugger; some bits are not used, which I show by a gray fill.

- The accessibility and initial condition of the bit; here they can all be read and written with the exception of TACLR, where the missing r indicates that there is no meaningful value to read. The (0) shows that each bit is cleared after a power-on reset (POR).

15					10	9	8
						TASSELx	
rw–(0)	rw–(0)	rw–(0)	rw–(0)	rw–(0)	rw–(0)	rw–(0)	rw–(0)

7	6	5	4	3	2	1	0
IDx		MCx			TACLR	TAIE	TAIFG
rw–(0)	rw–(0)	rw–(0)	rw–(0)	rw–(0)	w–(0)	rw–(0)	rw–(0)

Figure 4.8: The Timer_A Control Register TACTL. Gray bits are unused.

The user's guide goes on to describe the function of each bit or group of bits:

Timer_A clock source select, TASSELx: There are four options for the clock: the internal SMCLK or ACLK or two external sources. We use SMCLK because it is always available, which needs TASSELx = 10.

Input divider, IDx: The frequency of the clock can be divided before it is applied to the timer, which extends the period of the counter. We use the maximum factor of eight, which needs IDx = 11.

Mode control, MCx: The timer has four modes. By default it is off to save power. We first use the simplest Continuous mode, in which TAR simply counts up through its full range of 0x0000–0xFFFF and repeats. This needs MCx = 10.

Timer_A clear, TACLR: Setting this bit clears the counter, the divider, and the direction of the count (it can go both up and down in up/down mode). The bit is automatically cleared by the timer after use. It is usually a good idea to clear the counter whenever the timer is reconfigured to ensure that the first period has the expected duration.

Timer_A interrupt enable, TAIE: Setting this bit enables interrupts when TAIFG becomes set. We do not use this.

Timer_A interrupt flag, TAIFG: This bit can be modified by the timer itself or by a program. It is raised (set) by the timer when the counter becomes 0. In continuous mode this happens when the value in TAR rolls over from 0xFFFF to 0x0000. An interrupt is also requested if TAIE has been set. The program must clear TAIFG so that the next overflow can be distinguished.

The sub-main clock SMCLK runs at the same speed as MCLK by default, which is 800 KHz in the F1121A. If this were used to clock the timer directly, the period would be $2^{16}/800\,\text{KHz} \approx 0.08\,\text{s}$. We want about 0.5 s and therefore divide the frequency of the clock by 8 using IDx. This gives a delay of about 0.66 s, close enough.

Listing 4.16: Program `timrled1.c` to Flash LEDs with a frequency of roughly 1 Hz by polling free-running Timer_A.

```
// timrled1.c - toggles LEDs with period of about 1.3s
// Poll free-running timer A with period of about 0.65s
// Timer clock is SMCLK divided by 8, continuous mode
// Olimex 1121STK, LED1,2 active low on P2.3,4
// J H Davies, 2006-06-12; IAR Kickstart version 3.41A
//-------------------------------------------------------------------
```

```
#include <io430x11x1.h>         // Specific device

// Pins for LEDs
#define LED1    BIT3
#define LED2    BIT4

void main (void)
{
    WDTCTL = WDTPW|WDTHOLD;              // Stop watchdog timer
    P2OUT = ~LED1;                      // Preload LED1 on, LED2 off
    P2DIR = LED1|LED2;                  // Set pins for LED1,2 to output
    TACTL = MC_2|ID_3|TASSEL_2|TACLR;   // Set up and start Timer A
// Continuous up mode, divide clock by 8, clock from SMCLK, clear timer
    for (;;) {                          // Loop forever
        while (TACTL_bit.TAIFG == 0) {  // Wait for overflow
        }                              //   doing nothing
        TACTL_bit.TAIFG = 0;           // Clear overflow flag
        P2OUT ^= LED1|LED2;            // Toggle LEDs
    }                                  // Back around infinite loop
}
```

Listing 4.16 shows the program. Timer_A is set up by ORing the desired options, using the symbols defined in the header file. Unfortunately these are not particularly user-friendly; it would be a lot clearer if we could write TASSEL_SMCLK instead of TASSEL_2, for instance. I set the TACLR bit to ensure that the counter is clear before it starts. This is good practice although it is not strictly necessary here.

The first construct within the infinite for loop is another loop that polls the timer. This simply looks at the TAIFG flag and waits for it to become set, indicating that TAR has rolled over to 0. The next step is to clear this flag so that the next rollover is recognized. This is followed by the main action of the program—just the toggling of the display here. More tasks could be added here, provided that they do not take longer than the period of the timer. The result is a *paced loop*, a straightforward structure for a program that carries out a sequence of tasks at regular intervals. Nowadays it would be unusual to pace the loop by polling the timer; instead the MCU would save energy by entering a low-power mode after it had completed the tasks and wait for the timer to wake it again.

4.7.1   Timer_A in Up Mode

Finer control over the delay is obtained by using the timer in Up mode rather than continuous mode. The maximum desired value of the count is programmed into another register, TACCR0. In this mode TAR starts from 0 and counts up to the value in TACCR0, after which it returns to 0 and sets TAIFG. Thus the period is TACCR0 + 1 counts. Here the clock has been divided down to 100 KHz so we need 50,000 counts for a delay of 0.5 s and should therefore store 49,999 in TACCR0. (Of course this level of precision is pointless given the limited accuracy of the DCO.) Only a small change to the program is needed, shown in Listing 4.17. It is safer to configure the timer while it is not running so TACTL is written last, because the timer starts as soon as its mode is changed from stop.

Listing 4.17: Configuration of Timer_A for up mode in program `timrled2.c`.

```
    TACCR0 = 49999;                        // Upper limit of count for TAR
    TACTL = MC_1 | ID_3 | TASSEL_2 | TACLR;    // Set up and start Timer A
//  "Up to CCR0" mode, divide clock by 8, clock from SMCLK, clear timer
```

Example 4.18

Try this on your demonstration board. How accurate is the delay?

Debugging is much more challenging when peripherals are used. Unfortunately it is not possible to simulate the complete program because EW430 does not include peripherals: It is a CPU core simulator only. If you try, you will see that TAR never leaves 0. It is possible to mimic the effect of the polling loop by setting TAIFG = 1 manually but this does not test the timing. (It is also possible to simulate interrupts but that must wait for a later chapter.)

Emulation also gets interesting because the timer may keep running even when the CPU has been stopped to examine the registers. The emulation hardware in newer devices provides individual control over the clocks (use Emulation > Advanced > Clock Control… in the EW430 debugger) but this does not apply to the F1121. I experimented and found that SMCLK (taken from the DCO) stops when the CPU is halted but that ACLK runs continuously (see the next example).

Example 4.19

Timer_A is often clocked from ACLK, which is typically provided by a 32 KHz watch crystal. Try this if your board is so equipped, such as the Olimex 1121STK. (TI development kits may come with a loose crystal, which needs to be soldered.) Set the source appropriately (TASSELx bits) and remove the division by 8 (IDx bits). In continuous mode the delay will be 2 s, too long, so use up mode instead. The length of the delay should now be as accurate as your watch.

4.7.2   Morse Code on the Buzzer

The Olimex 1121STK includes a piezoelectric buzzer, which would be useful for the Morse code transmitter. It does not respond to a steady (DC) voltage, like the LED, but needs to be driven with an alternating voltage at the desired frequency. It is therefore connected between two pins of the microcontroller, as shown in Figure 4.9, rather than between one pin and either ground or V_{CC}. There is no voltage across the buzzer if both pins are at the same voltage. If one pin is raised to $V_{CC} = 3\,V$ and the other is lowered to $V_{SS} = 0\,V$, then a potential difference of $+3\,V$ appears across the buzzer. Reversing the voltages on the pins puts $-3\,V$ across the buzzer, measured in the same direction. Thus we can produce an alternating voltage of $\pm 3\,V$ by periodically toggling the outputs between 10 and 01.

Example 4.20

Extend your C program for Morse code to try this. Two delays are now needed, one for the frequency of the buzzer and one for the duration of the dots and dashes. You could use a software delay to drive the buzzer, placed inside the loop that polls the timer. A better approach is to use the timer to provide the frequency for the buzzer. Toggling the pins at 1 KHz gives a note of 500 Hz, which is not too irritating. Count the number of times that the timer overflows to give delays for the dots and dashes.

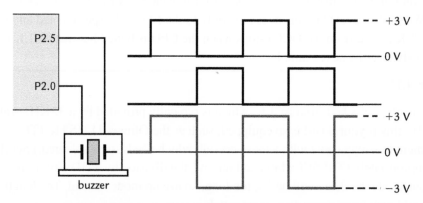

Figure 4.9: Connection of a piezoelectric buzzer between two output pins, which are driven in antiphase to produce an alternating voltage of $\pm 3\,V$ across the buzzer.

4.7.3 Random Light Display

A pretty application of the delay is a random light show on the LEDs. Of course this is rather limited with only two LEDs but the principle can be applied to bigger displays. This again uses a delay set by the timer but requires a calculation for the next pattern to display.

Pseudorandom sequences of bits are widely used, particularly to encrypt transmissions sent over insecure media such as the Internet. They are "pseudorandom" rather than truly random because they can be reproduced exactly if you know the algorithm and starting condition. (They would be of little use for communication if the receiver could not decode the signal.) A simple way of generating a pseudorandom sequence uses hardware based on a shift register is shown in Figure 4.10. I drew the flip-flops back to front compared with the usual symbol so that the least significant bit is on the right, which is how binary numbers and registers are usually laid out.

The circuit without the exclusive-OR gate and its connections is a plain *shift register*. A D flip-flop simply reads the value on its *D* input at a clock transition and transfers it to its *Q* output. Thus the value in flip-flop 0 is transferred to flip-flop 1 after a clock transition. At the same time the value in flip-flop 1 is transferred to flip-flop 2 and so on. The pattern of bits simply shifts one place to the left in each clock cycle. An input is applied to the first flip-flop, 0.

A *ring counter* is made by connecting the output of the last flip-flop to the input of the first flip-flop. Now the pattern in the register rotates indefinitely. It repeats after *N* cycles, where *N* is the number of flip-flops. Both shifts and rotations can be applied to registers in a MCU; see the section "Shift and Rotate Instructions" on page 137.

A pseudorandom sequence requires more complicated feedback. The simplest method, shown in the figure, is to take the feedback from an exclusive-OR gate connected to the

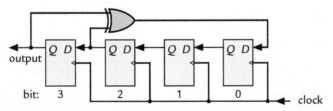

Figure 4.10: A shift register with feedback through an exclusive-OR gate from the last two stages, used to generate a pseudorandom stream of bits.

outputs of the last two stages. Ideally this gives a sequence of $2^N - 1$ states before it repeats. A register of N flip-flops has 2^N states in all so one is missing. This is the state with all flip-flops 0, which feeds back a 0 and therefore does not change after a clock transition. The counter must therefore be "seeded" with a nonzero value. The counter in Figure 4.10 with $N = 4$ gives the sequence 0001, 0010, 0100, 1001, 0011, 0110, 1101, 1010, 0101, 1011, 0111, 1111, 1110, 1100, 1000 and repeats. Unfortunately this simple circuit does not give the sequence of maximal length for all values of N and more complicated feedback may be needed. Annoyingly it fails for the convenient values $N = 8$ and 15 but the simple approach works for $N = 4, 7$, and 15. There is a good description in Section 9.33 of *The Art of Electronics* [42].

Example 4.21

Write a spreadsheet to simulate the shift register with feedback and confirm that the sequence is correct. What happens if you add another flip-flop so that $N = 5$?

Listing 4.18: Program to produce a pseudorandom bit sequence by simulating a shift register with feedback.

```
// random1.c - pseudorandom sequence on LEDs
// Poll timer A in Up mode with period of about 0.5s
// Timer clock is SMCLK divided by 8, up mode, period 50000
// Olimex 1121STK, LED1,2 active low on P2.3,4
// J H Davies, 2006-06-16; IAR Kickstart version 3.41A
//-----------------------------------------------------------------------
#include <io430x11x1.h>                    // Specific device
#include <stdint.h>                        // For uint16_t

// Pins for LEDs
#define LED1        BIT3
#define LED2        BIT4
// Parameters for shift register; length <= 15 (4 is good for testing)
#define REGLENGTH   15
#define LASTMASK    ((uint16_t) (BIT0 << REGLENGTH))
#define NEXTMASK    ((uint16_t) (BIT0 << (REGLENGTH-1)))

void main (void)
{
    uint16_t pattern;                       // next pattern to be displayed

    WDTCTL = WDTPW|WDTHOLD;                  // Stop watchdog timer
    P2OUT = LED1|LED2;                       // Preload LEDs off
    P2DIR = LED1|LED2;                       // Set pins with LEDs to output
    TACCR0 = 49999;                          // Upper limit of count for TAR
    TACTL = MC_1|ID_3|TASSEL_2|TACLR;        // Set up and start Timer A
// "Up to CCR0" mode, divide clock by 8, clock from SMCLK, clear timer
    pattern = 1;                             // Initialize pattern (nonzero)
```

```
    for (;;) {                          // Loop forever
        while (TACTL_bit.TAIFG == 0) {  // Wait for timer to overflow
        }                               //    doing nothing
        TACTL_bit.TAIFG = 0;            // Clear overflow flag
        P2OUT = pattern;                // Update pattern (lower byte)
        pattern <<= 1;                  // Shift for next pattern
// Mask two most significant bits, simulate XOR using switch, feed back
        switch (pattern & (LASTMASK|NEXTMASK)) {
        case LASTMASK:
        case NEXTMASK:
            pattern |= BIT0;            // XOR gives 1
            break;
        default:
            pattern &= ~BIT0;           // XOR gives 0
            break;
        }
    }                                   // Back around infinite loop
}
```

Listing 4.18 shows a program to simulate the shift register with feedback. The timer is polled in the Up mode to give a duration of about 0.5 s for each display. The unsigned 16-bit integer pattern represents the shift register. Feedback is taken from two adjacent bits, which correspond to flip-flops in the counter. The effective length of the counter is set by REGLENGTH so the active bits are in positions 0 to (REGLENGTH − 1). Two masks are constructed to pick out the final two bits after they have been shifted one position to the left by the <<= 1 operation. Remember that all the bits of BIT0 are 0 except for position 0. The result of BIT0 << REGLENGTH therefore has all bits 0 except for position REGLENGTH. The most significant active bit of the shift register is here after it has been shifted so LASTMASK picks it out. Similarly, NEXTMASK picks out the next to last active bit.

It would be possible to use an explicit exclusive-OR operation to find the feedback but I used a more straightforward if inelegant method. The two relevant bits are masked and tested in a switch statement. If either one is set then bit 0 is set, otherwise bit 0 is cleared. (The clearing is not really necessary because bit 0 should be clear after the left-shift operation.)

The usual output from this system is a pseudorandom stream of bits. This can be taken from any position or flip-flop, although it is conventional to use the most significant bit, as shown in Figure 4.10. The random light display uses the two LEDs on bits 3 and 4 as a small window into the register itself. The display looks prettier with a shorter register (smaller REGLENGTH) because there are no long runs of zeros or ones, although the pattern repeats more quickly.

Example 4.22

Find a more elegant way of performing the exclusive-OR operation. It need not be fast—0.5 s is available between updates of the display.

4.8   Header Files and Issues Brushed under the Carpet

Most of the programs described in this chapter are unusual for microcontrollers because the actions are performed by software. The examples on Timer_A are more typical because hardware is configured to provide the heartbeat of the program. The majority of the code in many programs interacts with the peripherals through the registers that control them: configuring timers to drive outputs with pulse-width modulation, sending bytes to external memory through a serial peripheral interface, and so on. The interface between the program and these registers is provided by the header files, which therefore play a major role in any project. Typically a programmer sits in front of a computer with the target hardware on one side and the reference manual on the other (on paper or on a second screen if you are so lucky). It is therefore important to understand what information is contained in the header files. We also saw that many tasks must be carried out explicitly when using assembly language but happen "by magic" in C, and look at these briefly too.

Two sets of header files are provided with EW430: `msp430xyyy.h` can be used with both C and assembly, while `io430xyyy.h` can be used only with C but offers better access to the fields within each register. Their contents can be divided into four main parts.

- A set of standard masks for individual bits, `BIT0–BITF`, which we already used.

- Definitions of bits in the status register and intrinsic functions for changing them in C. These are particularly important for entering low power modes, as we see in the section "Low-Power Modes of Operation" on page 198.

- Definitions of special function and peripheral registers, which fill most of the file.

- Definitions of interrupt vectors, described in the section "Interrupts" on page 186.

Much less information is needed for assembly language so we review this first.

4.8.1   Header Files for Assembly

Variables do not have a type in assembly language so the only vital information needed is the address of each register. You might therefore expect that the header file would contain a list of equates such as this:

```
P2OUT     EQU      0x0029
```

Many IDEs do indeed provide such files, but those in EW430 are rather more complicated because files like msp430x11x1.h are intended for both C and assembly. They are therefore written using macros DEFC(name, address) that expand in different ways for the two languages. In assembly they become sfrb name = address, where sfrb is a specific directive for defining special function byte registers. It associates the address with the name, like the equate directive, but also provides further information for the assembler and debugger.

The other information that we need is the significance of each bit within the registers. Symbols are therefore defined to provide masks with the names given in the reference manuals. We already used these to configure the timer. There are no definitions for individual bits of the port registers; use BIT0–BITF for these.

4.8.2   Header Files for C

These are more complicated because they must define both the type of value held in the registers and their addresses. Individual bits can also be addressed as bit fields in C, which avoids the need to use masks.

Definition of Peripheral Registers in C

It is instructive to see how this might be done in standard C before looking at how it is implemented in the files provided. This is also a good illustration of many of the aspects of C mentioned in "Aspects of C for Embedded Systems" on page 46.

The first problem is that you cannot define the address of a variable in C. The & operator takes the address of a variable, so it would be wonderful if we could write something like this:

```
uint8_t &P2OUT = 0x0029;      // define address of P2OUT
```

In fact this is close to a *reference* in C++ but is not allowed in standard C. However, addresses are used extensively in C: A pointer holds the address of a variable. So we could be slightly less ambitious and define pP2OUT as a pointer to the register rather than the register itself:

```
uint8_t * pP2OUT = (uint8_t *) 0x0029;      // pointer to P2OUT
```

I have used uint8_t to make it clear that this is an unsigned 8-bit integer but you may well see unsigned char. This definition works but is dangerous because pP2OUT is a variable and there is nothing to stop us changing its value by mistake. This can be prevented by using the const qualifier but it must be placed carefully:

```
uint8_t * const pP2OUT = (uint8_t *) 0x0029; // constant pointer to P2OUT
```

Putting const after the * forbids the program from changing the value of the pointer pP2OUT but does not restrict access to the register to which this points, P2OUT itself. This is what we want. In contrast, const uint8_t * pP2OUT would wrongly apply the qualifier to *pP2OUT. This would stop us modifying P2OUT but would not protect pP2OUT.

This definition is complete for P2OUT, which is rather like an ordinary register in RAM: It can be read and written and is purely under the control of software. Most registers that control peripherals can also be modified by the hardware. One example is TACTL, which we used to configure Timer_A. The Timer_A interrupt flag TAIFG is set by the hardware when the counter returns to 0 and we polled this to wait for a delay. The definition must therefore include the volatile qualifier to warn the compiler that the register may be changed by the hardware in a way that is not under direct control of the program:

```
volatile uint16_t * const pTACTL = (uint16_t *) 0x0160;
```

This prevents the polling loop from being optimized out of existence. Most peripheral registers have definitions like this but some need a little more. The port input registers, such as P2IN, are not only volatile (or they would be of little use) but are also read-only. We can prevent the program writing to them by adding another const qualifier:

```
const volatile uint8_t * const pP2IN = (uint8_t *) 0x0028;
```

This now has two const keywords because neither the pointer nor the register should be modified. It seems bizarre to see both const and volatile together but it makes sense:

const says that the program should not modify the register while volatile tells us that the hardware may do so.

It looks wasteful to define 'variables' in this way, although a modern compiler will almost certainly take advantage of the const qualifier to treat pP2IN as a constant stored in ROM rather than as a variable in RAM. However, this possibility can be avoided by defining a symbolic constant instead:

```
#define pP2IN ((const volatile uint8_t *) 0x028)
```

A definition may sometimes produce more efficient code but its major advantage is that we can go on to declare a name for the register itself, not a pointer:

```
#define myP2IN (*((const volatile uint8_t *) 0x028))
```

Now myP2IN can be used in exactly the same way as P2IN from the usual header file. Dan Saks [61] has written an instructive series of articles on memory-mapped device registers, which describes other ways in which such registers can be defined and the ways in which compilers interpret the definitions.

Having gone through all this, I have to admit that this is not how the registers are defined in the real header files. Instead, P2IN itself is defined in io430x11x1.h like this:

```
_no_init _READ volatile unsigned char P2IN @ 0x0028;
```

The significant difference is the @ notation, which is a widely used extension to standard C that specifies the address of the variable. The full description of a peripheral register in an io430 header is elaborate because it allows us to treat the register either as a set of bit fields or as a single entity with a collection of masks to manipulate individual bits or fields. I described this for TACTL in the section "Aspects of C for Embedded Systems" on page 46.

4.8.3 Other Issues

There is another header file, intrinsics.h, which you might wish to include in more advanced C programs. It contains declarations of functions that cannot be written in standard C. Many of these concern low-power modes and interrupts, which are covered in Chapter 6.

It is clear from a comparison of the programs in this chapter that the C compiler and linker do a great deal of work that we have to do by hand in assembly. Much of this is done when the program starts execution and is contained in `cstartup.c`. Here is a summary of the main tasks:

- The reset vector is defined to point to the first executable statement.

- The stack pointer is initialized.

- Static and global variables are initialized. Many variables should be 0 initially so the compiler allocates them to their own segment DATA16_Z in RAM and the startup code runs through a loop to clear them. Other variables that need to be initialized to nonzero values go into a separate segment in RAM, DATA16_I. Their initial values are stored in flash memory in a segment DATA16_ID, which is copied into DATA16_I by the startup code.

All this happens invisibly. The debugger automatically executes the startup code when it loads a program and stops when it encounters the `main` function. (This can be changed if you want to examine the startup process.)

It is vital to emphasize that the startup code does *not* initialize or configure any of the functions of the microcontroller, with the single exception of the stack pointer. *This is all left to you.* In particular you must handle or stop the watchdog timer or your chip will keep resetting in a puzzling manner.

The segments I just mentioned are defined in the linker script with the extension `.xcl`, which was described in the section "Relocatable Assembly Language" on page 78. The simple memory model of the MSP430, a single, linear 16-bit address space, means that the linker script is straightforward. Many microcontrollers have their memory arranged in banks to extend the range of addresses beyond the width of the address bus. Usually restrictions are placed on where the startup and interrupt vectors can point and so on, which makes the linker files spectacularly incomprehensible.

This completes our introductory tour of the MSP430. We next take a more detailed look at the processor itself before exploring the most widely used peripherals for the remainder of the book.

CHAPTER 5

Architecture of the MSP430 Processor

The chapter describes the central processing unit (CPU) of the MSP430 and its most closely associated modules, the clock generator and reset circuitry.

The earlier parts of the chapter, which cover the CPU and its instruction set, will be of most interest to assembly language programmers. On the other hand, it is helpful to be acquainted with this material for several reasons. Foremost is debugging. Also, many of the examples in TI's application notes are written in assembly language. In any case the MSP430 is not complicated. The CPU has only 27 instructions, most of which can use all appropriate addressing modes (up to 7). Its large number of general-purpose registers can hold many variables and addresses so there is less use of the stack, which can be confusing.

We first take a closer look at the CPU, filling in the details skipped in the section "Central Processing Unit" on page 30. As usual I concentrate on the original MSP430 and leave the MSP430X to Chapter 11.

5.1 Central Processing Unit

I repeat the sketch of the registers in Figure 5.1 for convenience. There are four special-purpose and 12 general-purpose registers, all of which can be addressed in the same ways. Let us look at these in detail.

5.1.1 Program Counter (PC)

This contains the address of the next instruction to be executed—"points to" the instruction in the usual jargon. Instructions are composed of 1–3 words, which must be aligned to even addresses, so the lsb of the PC is hardwired to 0.

15	... bits...	0
R0/PC	program counter	0
R1/SP	stack pointer	0
R2/SR/CG1	status register	
R3/CG2	constant generator	
R4	general purpose	
	⋮	
R15	general purpose	

Figure 5.1: Registers in the CPU of the MSP430.

The usual cycle of execution is that the contents of the PC are placed on the address bus and the next instruction is fetched from this address. The value in the PC is automatically increased by 2 after each fetch so that it is ready for the next word. One or two further words may be fetched if the instruction needs them. The PC is now ready with the address of the next instruction when the current one has been executed. Thus instructions are executed sequentially unless there is an explicit jump. In this case the address of the new instruction is produced as a result of the current operation and is written over the value in the PC. Subroutines and interrupts also modify the PC but in these cases the previous value is saved on the stack and restored later.

5.1.2   Stack Pointer (SP)

When a subroutine is called the CPU must jump to the subroutine, execute the code there, and finish by returning to the instruction after the call. It must therefore keep track of the contents of the PC before jumping to the subroutine so that it can return afterward. This is done with a *stack*, which is also known as a last in–first out (LIFO) data structure. It is something like a spring-loaded plate dispenser in a cafeteria. You can only take the plate off the top of the pile. If a new plate is added, the old ones disappear out of sight and you must take the most recently added plate. The analogy is not exact because a processor can read any memory location on the stack, not just its top. On the other hand, it is definitely a problem if the CPU tries to read from an empty stack, just as it is for a cafeteria to run out of plates.

In some processors, such as the PIC16, the stack is implemented in hardware. This has the advantage of speed, because the PC can be copied to the stack in parallel with fetching the next instruction. The stack is also protected from accidental damage by the user's program. A serious disadvantage is that the stack cannot be used for anything other than return addresses. In modern practice the stack is also heavily used for temporary variables, passing parameters to subroutines and returning the result, particularly for compiled languages such as C.

The MSP430 is conventional and allocates the stack in general-purpose RAM. The plates move up and down in the dispenser but data are not moved around like this in memory. In the MSP430, as in many other processors, the stack is allocated at the top of RAM and grows *down* toward low addresses. The stack pointer holds the address of the top of the stack. The MSP430 has a last used or full descending stack, meaning that the SP contains the address of the most recently added word. (The alternative is a next available or empty descending stack, in which case the SP holds the address of the next available word.) The lsb of the stack pointer is hardwired to 0 in the MSP430, which guarantees that it always points to valid words. A byte is therefore wasted to preserve the alignment to words if a single byte is placed on the stack. The wasted byte is unchanged; it is not cleared as in a CPU register following a byte operation.

The operation of the stack is illustrated in Figure 5.2. The specific addresses are for the F2013 or another device with 128 B of RAM. Generally, 0x027F should be replaced with the highest address in RAM. Stack operation follows these steps:

(a)   The stack is empty after the processor has been reset. The stack pointer should contain the address of the last word used, but there is none yet. It must therefore be initialized to the first address beyond the top of RAM, 0x0280 in the F2013. The hardware of the processor does *not* do this itself. For programs written in C, the compiler initializes the stack automatically as part of the startup code, which runs silently before the program starts, but *you must initialize SP yourself in assembly language*. This was shown in the section "Automatic Control: Use of Subroutines" on page 99.

(b)   The word 0x1234 has been added or *pushed* on to the stack. (The details of the instructions are explained later; recall that # means an immediate or literal value.) The value of SP is first decreased by 2 so that it points to the new location on the stack, then the value is copied to this address. This is called *predecrement addressing*.

(c)   The byte 0x56 has been written to the stack. It goes into the lower byte of the next word, whose upper byte is wasted. A further word of 0x789A is written afterward and SP now points to this value.

(d)   A word has been removed, pulled or *popped* from the stack into the register R15. This retrieves the most recently added value, 0x789A, and copies it into R15. The

(a) Stack after initialization.

(b) Stack after `push.w #0x1234`.

(c) Stack after `push.b #0x56` followed by `push.w #0x789A`.

(d) Stack after `pop.w R15`.

Figure 5.2: Operation of the stack in the MSP430F2013, whose RAM lies from 0x0200 to 0x027F.

stack pointer is increased by 2 afterward (postincrement addressing) and now points to the previously added value, the byte 0x56. Effectively the word 0x789A has been removed from the stack because it is now at a lower address than SP. The value remains in RAM until further values are pushed onto the stack, which grows downward and overwrites the old contents.

This shows how the stack operates as a last in–first out structure, which is essential where one subroutine calls another (the calls are nested) so that their return addresses are retrieved in the correct sequence. Note that the "top" of the stack is actually at the lowest active address in memory and therefore at the bottom of the sketches.

15	...	9	8	7	6	5	4	3	2	1	0
reserved			V	SCG1	SCG0	OSC OFF	CPU OFF	GIE	N	Z	C

Figure 5.3: Individual bits in the status register.

5.1.3   Status Register (SR)

This contains a set of *flags* (single bits) shown in Figure 5.3, whose functions fall into three categories. The reserved bits are not used in the MSP430 but some have been taken up in the MSP430X to extend the length of addresses. The status register is known as the condition code register in some processors.

Result of Arithmetic or Logic Operations

The C, Z, N, and V bits are affected by many of the operations performed by the ALU, but not all—see the section "Instruction Set" on page 132 for details:

- The main function of the *carry* bit C is to flag that the result of an arithmetic operation is too large to fit in the space allocated. In other words, an overflow occurred. Here is an example with bytes for simplicity. The hexadecimal sum $0x75 + 0xC7 = 0x13C$, where the result is too large to be held in a single byte. The processor would put 0x3C in the destination and set the carry bit to show that the result had overflowed and that a 1 should be carried into the next more significant byte. The carry flag also takes part in rotations and shifts. It is sometimes used as temporary storage to pass a bit from one register to another or to a subroutine.

- The *zero* flag Z is set when the result of an operation is 0. A common application is to check whether two values are equal: They are subtracted and the Z bit is tested to see whether the result is 0, which shows that the values are the same. This is used in Listing 4.6.

- The *negative* flag N is made equal to the msb of the result, which indicates a negative number if the values are signed.

- The *signed overflow* flag V is set when the result of a signed operation has overflowed, even though a carry may not be generated. An example is the sum $0x75 + 0x67 = 0xDC$. There are no problems if the variables are all unsigned.

However, if they are signed, the two operands have their msbs clear and are therefore positive, but the result has its msb set and therefore is interpreted as the negative number −0x24. (Remember that a byte can hold the values 0 to 0xFF if it is unsigned or −0x80 to 0x7F if it is signed.) The V bit would be set in this case to show that an overflow has occurred if the values are signed.

It is rarely necessary to test these flags explicitly because the instruction set contains a set of decisions, such as "jump if greater or equal," which make the appropriate tests. See Maxfield and Brown [37] in "Further Reading" for more background.

Enable Interrupts

Setting the *general interrupt enable* or GIE bit enables maskable interrupts, provided that the individual sources of interrupts have themselves been enabled. Clearing the bit disables all maskable interrupts. There are also nonmaskable interrupts, which cannot be disabled with GIE. This is explained in the section "Interrupts" on page 186.

Control of Low-Power Modes

The CPUOFF, OSCOFF, SCG0, and SCG1 bits control the mode of operation of the MCU. All systems are fully operational when all bits are clear. Setting combinations of these bits puts the device into one of its low-power modes, which is described in the section "Low-Power Modes of Operation" on page 198.

5.1.4   Constant Generators

Both R2 and R3 are used to provide the 6 most commonly used constants. This saves storing the values in the program and having to fetch them each time. The operation depends on the addressing modes and is therefore described in the section "Constant Generator and Emulated Instructions" on page 131 after these have been explained.

5.1.5   General-Purpose Registers

The remaining 12 registers R4–R15 have no dedicated purpose and may be used as general working registers. They may be used for either data or addresses because both are 16-bit values, which simplifies the operation significantly. Certain conventions should be followed if subroutines are written in assembly language, which specify the registers that should be used for passing parameters and returning a result. All this is handled by the compiler for programs written in C but is important if you mix C with assembly language—see the section "Conversion from Binary to Binary-Coded Decimal" on page 270.

5.2 Addressing Modes

A key feature of any CPU is its range of addressing modes, the ways in which operands can be specified. The MSP430 has four basic modes for the source but only two for the destination in instructions with two operands. These modes are made more useful by the way in which they interact with the CPU's registers. All 16 of these are treated on an almost equal basis, including the four special-purpose registers R0–R3 or PC, SP, SR/CG1, and CG2. The combination of the basic addressing modes and the registers gives the seven modes listed in the user's guides, although eight could reasonably be claimed.

An instruction itself fits into a single word of 16 bits, although it may be followed by further words to provide addresses or an immediate value. There are three formats of instruction. I shall use TI's standard abbreviations of src for source and dst for destination. It is important to distinguish between these because the destination has fewer addressing modes.

Double operand (Format I): Arithmetic and logical operations with two operands such as add.w src, dst, which is the equivalent of dst += src in C. Note the different ordering of src and dst in C and assembly language. Both operands must be specified in the instruction. This contrasts with accumulator-based architectures, where an accumulator or working register is used automatically as the destination and one operand.

Single operand (Format II): A mixture of instructions for control or to manipulate a single operand, which is effectively the *source* for the addressing modes. The nomenclature in TI's documents is inconsistent in this respect.

Jumps: The jump to the destination rather than its absolute address, in other words the offset that must be added to the program counter.

The "return from interrupt" instruction reti is unique in requiring no operands. This would usually be described as *inherent* addressing but TI curiously classifies it as Format II without data. (This follows from its binary opcode.)

We are now in a position to examine the range of addressing modes. The reason for the asymmetry between source and destination is straightforward: Not enough bits are available in an instruction (a 16-bit word). I return to this in the section "Reflections on the CPU and Instruction Set" on page 153.

5.2.1   Register Mode

This uses one or two of the registers in the CPU. It is the most straightforward addressing mode and is available for both source and destination. For example,

```
mov.w    R5,R6    ; move (copy) word from R5 to R6
```

The registers are specified in the instruction word; no further data are needed. It is also the fastest mode and this instruction takes only 1 cycle. Any of the 16 registers can be used for either source or destination but there are some special cases:

- The PC is incremented by 2 while the instruction is being fetched, before it is used as a source.

- The constant generator CG2 reads 0 as a source.

- Both PC and SP must be even because they address only words, so the lsb is discarded if they are used as the destination.

- SR can be used as a source and destination in almost the usual way although there are some details about the behavior of individual bits.

For byte instructions,

- Operands are taken from the lower byte; the upper byte is not affected.

- The result is written to the lower byte of the register and the upper byte is cleared.

The upper byte of a register in the CPU cannot be used as a source. If this is needed, the 2 bytes in a word must first be swapped with swpb.

5.2.2   Indexed Mode

This looks much like an element of an array in C. The address is formed by adding a constant base address to the contents of a CPU register; the value in the register is not changed. Indexed addressing can be used for both source and destination. For example, suppose that R5 contains the value 4 before this instruction:

```
mov.b    3(R5),R6     ; load byte from address 3+(R5)=7 into R6
```

The address of the source is computed as $3 + (R5) = 3 + 4 = 7$. Thus a byte is loaded from address 7 into R6. The value in R5 is unchanged. There is no restriction on the address for a byte but remember that words must lie on even addresses.

Indexed addressing can be used for the source, destination, or both. The base addresses, just the single value 3 here because only one address is indexed, are stored in the words following the instruction. They cannot be produced by the constant generator.

The instruction is not normally used like this with numerical constants. More typically the base is the address of the first element of an array or table and the register holds the index. Thus the indexed address Message(R5) is equivalent to the element Message[i] of an array of characters in C, assuming that R5 is used for the index and R5 = i. Look back at Listing 4.15.

However, there is an important difference between C and assembly. The CPU always calculates the indexed address in *bytes*, while C takes account of the size of the object when calculations are performed with pointers. Suppose that Words[] is an array of words. In C the two expressions Word[i] and *(Word+i) are equivalent. The corresponding indexed address would be Word(R5) with R5 = 2i because each word is 2 bytes long.

These examples use general-purpose registers but the full power of indexed addressing is released when the special registers are used as well.

Symbolic Mode (PC Relative)

In this case the program counter PC is used as the base address, so the constant is the offset to the data from the PC. TI calls this the symbolic mode although it is usually described as PC-relative addressing. It is used by writing the symbol for a memory location without any prefix. For example, suppose that a program uses the variable LoopCtr, which occupies a word. The following instruction stores the value of LoopCtr in R6 using symbolic mode:

```
mov.w    LoopCtr,R6      ; load word LoopCtr into R6, symbolic mode
```

The assembler replaces this by the indexed form

```
mov.w    X(PC),R6              ; load word LoopCtr into R6, symbolic mode
```

---

where X = $LoopCtr - PC$ is the offset that needs to be added to PC to get the address of LoopCtr. This calculation is performed by the assembler, which also accounts for the automatic incrementing of PC.

This seems a complicated way of specifying the address and is not appropriate for addresses that are fixed in the memory map, such as those of peripheral registers. In a few cases, PC-relative addressing is essential to produce *position-independent code*, which can be moved in memory without affecting its function. Symbolic addressing is useful for such applications but it is highly specialized, such as bootloaders where the code runs from RAM rather than ROM. It is hard to see the attraction of symbolic addressing in the MSP430 apart from this, because absolute addressing can also reach the whole memory map. It is more important with 20-bit addresses in the MSP430X.

Absolute Mode

The constant in this form of indexed addressing is the absolute address of the data. This is already the complete address required so it should be added to a register that contains 0. The MSP430 mimics this by using the status register SR. It makes no sense to use the real contents of SR in an address so it behaves as though it contains 0 when it is used as the base for indexed addressing. This is one of its roles as constant generator CG1.

Absolute addressing is shown by the prefix & and should be used for special function and peripheral registers, whose addresses are fixed in the memory map. This example copies the port 1 input register into register R6:

```
mov.b    &P1IN,R6              ; load byte P1IN into R6, absolute mode
```

The assembler replaces this by the indexed form

```
mov.b    P1IN(SR),R6           ; load byte P1IN into R6, absolute mode
```

---

where P1IN is the absolute address of the register.

SP-Relative Mode

TI does not claim this as a distinct mode of addressing, but many other companies do! The stack pointer SP can be used as the register in indexed mode like any other. Recall from the section "Stack Pointer (SP)" on page 120 that the stack grows down in memory and that SP points to (holds the address of) the most recently added word. Suppose that we wanted to copy the value that had been pushed onto the stack before the most recent one. The following instruction will do this:

```
mov.w    2(SP),R6          ; copy most recent word but one from stack
```

For example, suppose that the stack were as shown in Figure 5.2(d) with $SP = 0x027C$. Then the preceding instruction would load 0x1234 into R6. This type of manipulation is important in subroutines and interrupts and is illustrated in Chapter 6.

5.2.3   Indirect Register Mode

This is available only for the source and is shown by the symbol @ in front of a register, such as @R5. It means that the contents of R5 are used as the *address* of the operand. In other words, R5 holds a pointer rather than a value. (The contents of R5 would be the operand itself if the @ were omitted.) Suppose that R5 contains the value 4 before this instruction:

```
mov.w    @R5,R6   ; load word from address (R5)=4 into R6
```

The address of the source is 4, the value in R5. Thus a word is loaded from address 4 into R6. The value in R5 is unchanged. This has exactly the same effect as indexed addressing with a base address of 0 but saves a word of program memory, which also makes it faster. This is very loosely the equivalent of `r6 = *r5` in C, without worrying about the types of the variables held in the registers.

Indirect addressing cannot be used for the destination so indexed addressing must be used instead. Thus the reverse of the preceding move must be done like this:

```
mov.w    R6,0(R5)     ; store word from R6 into address 0+(R5)=4
```

The penalty is that a word of 0 must be stored in the program memory and fetched from it. The constant generator cannot be used.

5.2.4   Indirect Autoincrement Register Mode

Again this is available only for the source and is shown by the symbol @ in front of a register with a + sign after it, such as @R5+. It uses the value in R5 as a pointer and automatically increments it afterward by 1 if a byte has been fetched or by 2 for a word. Suppose yet again that R5 contains the value 4 before this instruction:

```
mov.w    @R5+,R6
```

A word is loaded from address 4 into R6 and the value in R5 is incremented to 6 because a word (2 bytes) was fetched. This is useful when stepping through an array or table, where expressions of the form *c++ are often used in C.

This mode cannot be used for the destination. Instead the main instruction must use indexed mode with an offset of 0, followed by an explicit increment of the register by 1 or 2. The reverse of this move therefore needs two instructions:

```
mov.w    R6,0(R5)      ; store word from R6 into address 0+(R5)=4
incd.w   R5            ; R5 += 2
```

This is undoubtedly a bit clumsy.

Autoincrement is usually called *postincrement* addressing because many processors have a complementary *predecrement* addressing mode, equivalent to *--c in C, but the MSP430 does not.

An important feature of the addressing modes is that all operations on the first address are fully completed before the second address is evaluated. This needs to be considered when moving blocks of memory. The move itself might be done by a line like this:

```
mov.w    @R5+,0x0100(R5)
```

Suppose as usual that R5 initially contains the value 4. The contents of address 4 is read and R5 is double-incremented to 6 because a word is involved. Only now is the address for the destination calculated as 0x0100 + 0x0006 = 0x0106. Thus a word is copied from address 0x0004 to 0x0106; the offset is not just the value of 0x0100 used as the base address for the destination. The compiler takes care of these tricky details if you write in C, but you are on your own with assembly language.

Immediate Mode

This is a special case of autoincrement addressing that uses the program counter PC. Look at this example:

```
mov.w    @PC+,R6 ; load immediate word into R6
```

The PC is automatically incremented after the instruction is fetched and therefore points to the following word. The instruction loads this word into R6 and increments PC to point to the next word, which in this case is the next instruction. The overall effect is that the word that followed the original instruction has been loaded into R6. This is how the MSP430 handles immediate or literal values, which are encoded into the stream of instructions. It is the equivalent of r6 = constant.

This is available only for the source of an instruction because it is a type of autoincrement addressing but it is hard to see when it would be useful for the destination.

5.3   Constant Generator and Emulated Instructions

A complex instruction set computer (CISC) has special instructions for performing many common operations. For example, there may be a "clear" operation to reset a register to 0 and an "increment" operation to increase its value by 1. In contrast, a reduced instruction set computer (RISC) uses general instructions for these operations. Instead of clearing a register it stores 0 into it, in the same way as it would store any other number. Similarly, it increments a value by using the standard addition instruction to add 1. This could require commonly used values like 0 and 1 to be stored frequently in the code, which would be wasteful of both memory and time because the values would have to be fetched from memory whenever they were needed.

To improve efficiency, most RISCs therefore have one or more registers that are "hardwired" to commonly used values. Typically there is only one value per register but the designers of the MSP430 cleverly exploited its four addressing modes for a source to get four values from its dedicated constant generator R3/CG2. The status register R2/SR/CG1 can be read and written almost normally in Register mode but it would make no sense to use its value as part of an address. It therefore acts as constant generator CG1 when it is used as a source in the other three addressing modes. Thus seven constants are available from the two registers. These provide the base of 0 for absolute addressing, already described in the section "Absolute Mode" on page 128, and the six immediate values 0, 1, 2, 4, 8, and 0xFFFF $= -1$ for signed values. The chosen values are frequently used in comparison, arithmetic, and logic operations and for masks to pick out bits 0–3. It is no accident that the most commonly tested bits in the status register are in these positions (Figure 5.3) and the same is true for registers in other modules.

For example, direct (register) addressing of R3/CG2 returns the value 0, so mov.w R3,R5 clears R5. Similarly, mov.w @R2,R5 uses indirect addressing on R2/SR /CG1, which returns the value 8 and this will be stored in R5.

These constants are combined with many of the 27 native instructions to provide a further 24 *emulated* instructions. They can be written in the same way as "real" instructions and the assembler converts them to native instructions with the appropriate constant from CG1 or CG2. A problem with this approach is that the constants can be used only as the source. The other operand is therefore a destination with restricted addressing modes, which can sometimes be a nuisance.

Do not attempt to use any of the constant generators as a destination. This applies even in cases where the destination is not changed. It is tempting to try this for instructions such as comparison, but it is not allowed. Of course R2 may be used as a destination in register mode, when it acts as SR rather than CG1. Absolute addressing can definitely be used for the destination, which effectively uses CG1 as base. Neither mode works for CG2 as destination.

5.4 Instruction Set

The instructions are thoroughly documented in the section "RISC 16-Bit CPU" of the family user's guides. The MSP430 has 27 native instructions, and a further 24 emulated instructions are defined to make life easier for the programmer. These include common operations such as "clear," which is implemented as an ordinary move with a value of 0 provided by the constant generator. I list all instructions for completeness but concentrate on the unusual features and traps for the unwary. The instruction set is *orthogonal* with few exceptions, meaning that all addressing modes can be used with all instructions and registers. I show the .w form for operations that can use either bytes or words.

Aside: It sounds as though the MSP430 has fewer instructions than the PIC16 with 35, but trivial comparisons of radically different processors are always misleading. It might be more accurate to say that the PIC has 28 instructions with up to three addressing modes. For example, the operand for arithmetic and logic instructions can be a literal value, taken from a register whose address is given explicitly, or in a register whose address is specified indirectly in FSR.

5.4.1   Movement Instructions

There is only the one mov instruction to move data. It can address all of memory as either source or destination, including both registers in the CPU and the whole memory map. This is an excellent feature. Some processors have distinct instructions for loading a CPU register from memory, storing it to memory, and memory-to-memory moves if these are available at all:

```
mov.w    src,dst ; move (copy)                dst = src
```

Note the order of the operands, which is the opposite of the equivalent statement in C (and some other assembly languages).

Peculiarity: The status bits are not affected by mov. Strings in C end with the null character \0 and other lists are often terminated by 0, so it would be helpful to detect this. The Z flag is affected by the move itself in many processors but an explicit test must be used in the MSP430.

Stack Operations

These push data onto the stack and pop them off, as described in the section "Stack Pointer (SP)" on page 120:

```
push.w   src      ; push data onto stack      *--SP = src
pop.w    dst      ; pop data off stack        dst = *SP++      emulated
```

The SP is fixed to be even, so a word of stack space is always consumed, even if only a byte is added. The pop operation is emulated using postincrement addressing but push requires a special instruction because predecrement addressing is not available.

5.4.2   Arithmetic and Logic Instructions with Two Operands

Binary Arithmetic Instructions with Two Operands

These are fairly standard. The carry bit should be interpreted as "not borrow" for subtraction:

```
add.w    src,dst ; add                        dst += src
addc.w   src,dst ; add with carry             dst += (src + C)
adc.w    dst     ; add carry bit              dst += C          emulated
sub.w    src,dst ; subtract                   dst -= src
subc.w   src,dst ; subtract with borrow       dst -= (src + ~C)
```

```
sbc.w   dst      ; subtract borrow bit        dst -= ~C        emulated
cmp.w   src,dst  ; compare, set flags only    (dst - src)
```

The compare operation cmp is the same as subtraction sub except that only the bits in SR are affected; the result is not written back to the destination. There are many examples of operations on more than one word with the carry/borrow bit in Section 5.1 of *Application Reports* (slaa024). Maxfield and Brown [37] give an entertaining account of binary arithmetic.

Arithmetic Instructions with One Operand

All these are emulated, which means that the operand is always a destination:

```
clr.w    dst     ; clear                      dst = 0          emulated
dec.w    dst     ; decrement                  dst--            emulated
decd.w   dst     ; double decrement           dst -= 2         emulated
inc.w    dst     ; increment                  dst++            emulated
incd.w   dst     ; double increment           dst += 2         emulated
tst.w    dst     ; test (compare with 0)      (dst - 0)        emulated
```

The test operation is the special case of comparison with 0. In many processors the clear operation differs from a move with the value 0 because a move sets the flags but a clear does not. This does not apply to the MSP430 because a move does not set the flags.

Decimal Arithmetic

These instructions are used when operands are binary-coded decimal (BCD) rather than ordinary binary values. This means that the value of each nibble is restricted to the range of unsigned, decimal integers 0–9 instead of the full hexadecimal range 0–F. BCD is often used for values to be displayed in decimal form because it saves having to convert the binary value to a set of decimal digits. This is useful in a clock, for instance, as we see in the section "Simple Applications of the LCD" on page 264. Maxfield has several articles explaining why BCD is important and how to use it—even signed BCD [60, 63]:

```
dadd.w   src,dst  ; decimal add with carry    dst += src + C
dadc.w   dst      ; decimal add carry bit      dst += C         emulated
```

There is only one native instruction for decimal arithmetic, dadd. This adds its source *plus the carry bit* decimally to its destination. The result is a BCD number provided that the operands were valid BCD numbers themselves. Some processors have a "decimal adjust" instruction instead, which converts the value in a register from binary to BCD.

The emulated instruction dadc adds only the carry bit decimally to the destination. The

CPU itself provides no other operations on nibbles, but there are routines for converting between BCD and binary numbers in Section 5.5 of *Application Reports* (slaa024) and I show an example in the section "Conversion from Binary to Binary-Coded Decimal" on page 270.

Peculiarity: The mnemonic dadd is misleading: It would better have been called daddc for "decimal add with carry." Make sure that you set or clear the carry bit before using dadd unless the carry has been determined by a previous operation.

Example 5.1

Is it possible to construct an emulated decinc instruction to increment a register decimally, assuming that it already contains a BCD value? Or is more than one instruction needed? What about a decrement?

Logic Instructions with Two Operands

These are not quite the same as in many other processors:

```
and.w   src,dst ; bitwise and                     dst &= src
xor.w   src,dst ; bitwise exclusive or            dst ^= src
bit.w   src,dst ; bitwise test, set flags only    (dst & src)
bis.w   src,dst ; bit set                         dst |= src
bic.w   src,dst ; bit clear                        dst &= ~src
```

The MSP430 has the usual and and exclusive-OR xor instructions but not an explicit inclusive-OR. The and and bitwise test operations are identical except that bit is only a test and does not change its destination.

Peculiarity: The Z bit is affected in the usual way by these operations and the carry bit is given by $C = \sim Z$. The idea of this is that the carry bit can subsequently be rotated into another register to ease serial–parallel conversion.

The bit set bis and bit clear bic instructions are used with masks to set and clear bits. The bis operation is very similar to inclusive-OR and bic mask is likewise related to and ~mask.

Peculiarity: The bis and bic operations do not affect the status bits. Therefore bis is not quite a substitute for the usual inclusive-OR, which would be expected to affect SR. This is not a serious loss because the effect of an inclusive-OR operation on the status bits is largely predictable.

Example 5.2

Why did the designers provide a `bic` instruction as well as `and`?

Bit operations are called *read–modify–write* operations because the CPU cannot operate on bits individually: It must read the register into the ALU, perform the operation, and write the result back. This can have unwanted side effects with some special registers. One example is registers associated with interrupts, where a read may automatically clear flags. Trouble can also arise with input/output ports in some processors but this is not a problem on the MSP430 with its separate input and output registers.

Logic Instructions with One Operand

There is only one of these, the invert `inv` instruction, also known as *ones complement*, which changes all bits of 0 to 1 and those of 1 to 0:

```
inv.w    dst      ; invert bits              dst = ~dst      emulated
```

It is emulated using `xor` and inherits its peculiarity C = ~Z. Its operand is a destination. It is not the same as changing the sign of a number, which is the twos complement.

Example 5.3

Why is there no emulated instruction to change the sign of a twos-complement number?

Byte Manipulation

These instructions do not need a suffix because the size of the operands is fixed:

```
swpb     src      ; swap upper and lower bytes (word only)
sxt      src      ; extend sign of lower byte (word only)
```

The swap bytes instruction `swpb` swaps the two bytes in a word; there is no corresponding swap nibbles for the nibbles in a byte. The sign extend instruction `sxt` is used to convert a signed byte into a signed word. It copies the value of bit 7, which gives the sign of the lower byte, into bits 8–15 and gives flags C = ~Z. The opposite operation, truncation from a word to a byte, can be done with `mov.b`.

Multiplication, if present in hardware at all, is performed by a peripheral—it is not implemented in the ALU and does not appear in the instruction set. There are routines for multiplication in software in Section 5.1 of *Application Reports* (slaa024). Further methods are given in the application note *Efficient Multiplication and Division Using MSP430* (slaa329).

Operations on Bits in Status Register

There is a set of emulated instructions to set or clear the four lowest bits in the status register, those that can be masked using the constant generator:

```
clrc        ; clear carry bit              C = 0        emulated
clrn        ; clear negative bit           N = 0        emulated
clrz        ; clear zero bit               Z = 0        emulated
setc        ; set carry bit                C = 1        emulated
setn        ; set negative bit             N = 1        emulated
setz        ; set zero bit                 Z = 1        emulated
dint        ; disable general interrupts   GIE = 0      emulated
eint        ; enable general interrupts    GIE = 1      emulated
```

The carry bit should be set or cleared before instructions that take it as input unless it is a result of a previous operation. This applies to adc, addc, sbc, subc, dadc, dadd (particularly easy to forget because of the mnemonic), and the rotations rlc and rrc. The GIE flag affects only maskable (general) interrupts; see the section "Interrupts" on page 186.

5.4.3   Shift and Rotate Instructions

Processors often offer three types of shifts and rotations as illustrated in Figure 5.4, although the treatment of the carry bit varies. They differ in the treatment of the bits that are shifted out of and into the register:

- **Logical shift** inserts zeroes for both right and left shifts.

- **Arithmetic shift** inserts zeroes for left shifts but the most significant bit, which carries the sign, is replicated for right shifts.

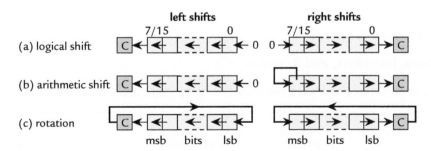

Figure 5.4: Left and right logical shifts, arithmetic shifts, and rotations on an 8- or 16-bit register.

- **Rotation** does not introduce or lose any bits; bits that are moved out of one end of the register are passed around to the other.

Usually the carry bit is included in rotations and it may gain the bit that is shifted out by arithmetic or logical shifts.

The MSP430 has arithmetic shifts and rotations, all of which use the carry bit. The right-shifts are native instructions but the left shifts are emulated, so the left- and right-shifts have different addressing modes available:

```
rla     dst     ; arithmetic shift left              emulated
rra     src     ; arithmetic shift right
rlc     dst     ; rotate left through carry          emulated
rrc     src     ; rotate right through carry
```

Peculiarities: The mnemonics for the arithmetic shifts imply that they are rotations, which is misleading. There are no logical shifts in the MSP430 but they have been added to the MSP430X. A logical shift left is the same as an arithmetic shift left so there is no problem there. A logical shift right can be emulated by first clearing the carry bit and making a rotation right.

Example 5.4

How else might a logical shift right be emulated?

The rotation operation is not available in C so assembly language may be needed if this instruction is critical. Shifts in C are always logical for unsigned values but the nature of shifts for signed values is undefined. They are arithmetic in EW430 to match the instruction set. Multiword shifts can be constructed using the carry bit in much the same way as multiword arithmetic.

5.4.4 Flow of Control

Subroutines, Interrupts, and Branches

These are mainly straightforward but there is a tricky point about addresses:

```
br      src     ; branch (go to)              PC = src    emulated
call    src     ; call subroutine
ret             ; return from subroutine                  emulated
reti            ; return from interrupt
nop             ; no operation (consumes single cycle)    emulated
```

Peculiarity: Both br and call can use the full range of addressing modes for a source. The most common elementary use of call is for a subroutine that begins at a particular label. This label is translated by the assembler to the address of the first instruction in the subroutine: direct addressing. This is the value that should be loaded into the PC to call the subroutine and is therefore like immediate data. It must consequently be given the prefix # like any other immediate value. For example, call #DelayTenths. This is *very* easy to forget. We used this in the section "Automatic Control: Use of Subroutines" on page 99 and I will remind you of this pitfall again.

The behavior is easier to understand with br, which is emulated. The instruction br label is translated into mov.w label, PC. This means that label is used as an absolute *address* so the contents of the word whose address is label are fetched and loaded into PC. It is more likely that we want to load the *value* label itself into the PC, which needs mov.w #label, PC. The call instruction must be handled in the same way. There is no problem with the jump instructions because they use offsets rather than full addresses and the compiler or assembler calculates these automatically.

The good side to this is that it is easy to select a branch or subroutine from a table by using indexed, indirect, or even autoincrement addressing. The interrupt handling for Timer_A is designed with this in mind.

Be sure that subroutines end with ret and interrupt service routines end with reti; the extra letter is crucial. The MSP430X uses calla and reta for subroutines.

The standard no-operation instruction nop is emulated to waste one cycle of the processor. There are further suggestions in the family user's guides for instructions to use more cycles but their side effects may need care.

Jumps, Unconditional and Conditional

The unconditional jump instruction is less tricky:

```
jmp      label    ; unconditional jump
```

The target is a straightforward label: It does not have the peculiarity (or versatility) of br. The difference between them is that:

- **jmp** fits in a single word, including the offset, but its range is limited to about ±1 KB from the current location.

- **br** can go anywhere in the address space and use any addressing mode but is slower and requires an extra word of program storage.

The nomenclature varies between manufacturers; jump and branch have the opposite meaning in Freescale processors, for instance. The symbol $ stands for the current value of the program counter in the assembler so jmp $ is a concise way of getting an empty, infinite loop.

The conditional jumps are the "decision-making" instructions and test certain bits or combinations in the status register. It is not possible to jump according to the value of any other bits in SR or those in any other register. Typically a bit test instruction bit is used to detect the bit(s) of interest and set up the flags in SR before a jump.

Many branches have two names to reflect different usage. For example, it is clearer to use jc if the carry bit is used explicitly—after a rotation, for instance—but jhs is more appropriate after a comparison:

```
jc    label   ; jump if carry set,        C = 1    same as jhs
jnc   label   ; jump if carry not set,    C = 0    same as jlo
jn    label   ; jump if negative,         N = 1
jz    label   ; jump if zero,             Z = 1    same as jeq
jnz   label   ; jump if nonzero,          Z = 0    same as jne
```

Assume that the "comparison" jumps follow cmp.w src,dst, which sets the flags according to the difference dst - src. Alternatively, tst.w dst sets the flags for dst - 0:

```
jeq   label   ; jump if equal,            dst = src    same as jz
jne   label   ; jump if not equal,        dst != src   same as jnz
jhs   label   ; jump if higher or same,   dst >= src   same as jc
jlo   label   ; jump if lower,            dst < src    same as jnc
```

A further pair of conditional jumps also tests the V bit for signed values:

```
jge    label   ; jump if greater or equal, dst >= src   signed values
jl(t)  label   ; jump if less than,         dst < src    signed values
```

Both mnemonics jl and jlt are used. It is up to the programmer to select the correct instruction. For example, suppose that two bytes contain 0x99 and 0x01. They are related by 0x99 > 0x01 if the values are unsigned but 0x99 < 0x01 if they are signed, twos complement numbers because 0x99 is the representation of −0x67.

Peculiarities: There are tests for the conditions $<$ and $\geq$ but not for $\leq$ nor $>$. It may be possible to choose the source and destination in a comparison to avoid this problem. Unfortunately the asymmetric addressing modes often prevent this, particularly if one value is immediate. Two tests may then be necessary.

5.4.5 Instruction Timing

The number of MCLK cycles required for most instructions is limited by access to memory. This is a typical feature of a RISC-like CPU with a von Neumann architecture and also applies to the ARM7, for instance. Values for *typical* instructions are listed in Table 5.1 but there are several exceptions, including instructions that change the flow of control and those where the destination is PC. The general principle for Format I instructions (two operands) is as follows. Most of these must read the instruction and two operands from memory and write the result back. The duration is set by the modes used to address memory for the operands.

- It takes one cycle to fetch the instruction word itself. This is all if both source and destination are in CPU registers. Values from the constant generators are effectively in registers.

- One more cycle is needed to fetch the source if it is given indirectly as @Rn or @Rn+, in which case the address is already in the CPU. This includes immediate data.

- Alternatively, two more cycles are needed if one of the indexed modes is used. The first is to fetch the base address, which is added to the value in a CPU register to get the address of the source. A second cycle is necessary to fetch the operand itself. This includes absolute and symbolic modes.

Table 5.1: Number of MCLK cycles required for typical instructions. It applies only to logical and arithmetic instructions and when the destination is not PC.

Format I destination	Rs	Source @Rs, @Rs+	S(Rs)
Rd	1	2	3
D(Rd)	4	5	6
Format II	1	3	4

- Two more cycles are needed to fetch the destination in the same way if it is indexed.

- A final cycle is needed to write the destination back to memory if required; no allowance is needed for a register in the CPU.

These are illustrated in Figure 5.5. It is amusing that no dedicated cycles are needed for the computation itself. A similar principle applies to the arithmetic operations with a single operand but most instructions with Format II change the flow of control and have individual timings. Jumps always use two cycles, whether conditional jumps are taken or not.

There is no difference in the timing between bytes and words so it is pointless to squeeze variables into bytes in the hope of gaining speed. Of course it is an equally bad idea to waste memory by storing large tables as words when the elements would fit into bytes.

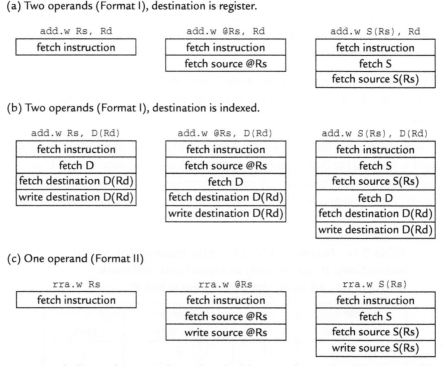

(a) Two operands (Format I), destination is register.

Figure 5.5: Cycle-by-cycle operation of typical instructions, showing the traffic with memory.

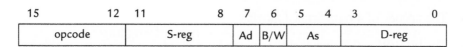

Figure 5.6: Breakdown of a Format I (double operand) instruction.

A few instructions have been speeded up in the MSP430X. These include `mov`, which does not need to fetch its destination, and `bit` and `tst`, which do not produce a result that needs to be written back. The number of cycles for these instructions with addressing modes as in `mov.w Rs,D(Rd)` has been reduced from four to three. The interfaces to subroutines and interrupt service routines have also been made faster.

5.4.6   Machine Code

Occasionally you may need to decode the binary machine code to deduce the instruction to be executed. This is an unfulfilling activity and fortunately is rarely needed with modern debuggers.

The layout of the bits within a Format I instruction (two operands) instruction is shown in Figure 5.6, taken from the family user's guide. These are the individual fields:

- **opcode** (4 bits) is the operation code. The highest 12 values are used for Format I instructions, the remainder for jumps and Format II.

- **S-Reg** and **D-Reg** (4 bits each) specify the CPU registers associated with the source and destination; the registers either contain the operands themselves or their contents are used to form the addresses.

- **As** (2 bits) gives the mode of addressing for the source, which has four basic modes.

- **Ad** (1 bit) similarly gives mode of addressing for the destination, which has only two basic modes.

- **B/W** (1 bit) chooses whether the operand is a byte (1) or a word (0).

Here is a trivial example of a move from register to register with the resulting machine code:

```
mov.w    R5,R6    ; 4506
```

The instruction can be broken into its fields of opcode = 4, S-reg = 5, Ad = 0, B/W = 0, As = 0, D-reg = 6. What do these mean?

- The opcode of 4 represents a move.

- The bit B/W = 0 shows that the operand is a word.

- The addressing mode for the source is As = 0, which is register. The register is S-reg = 5, which is R5 as expected.

- Similarly, the addressing mode for the destination is Ad = 0, which again means register. The register is D-reg = 6 = R6.

Here is an addition rather than a move:

```
add.w    R5,R6    ; 5506
```

The machine code is identical except for the opcode, which is now 5 rather than 4. The specification of the operands is unchanged. This is because of the orthogonality: All instructions use the same addressing modes.

Let us move an immediate value instead of a register:

```
mov.w    #5,R6    ; 4036 0005
```

Now there are two words. The fields of the instruction are opcode = 4, S-reg = 0, Ad = 0, B/W = 0, As = 3 = 11b, D-reg = 6. The difference is in the specification of the source, which means autoincrement. The register is Sreg = 0, which is the PC. Autoincrement addressing on the PC is the way in which immediate values are implemented. The value itself is contained in the second word.

Next look at a value of 4 instead, which can be supplied by the constant generator. I use a byte rather than a word for a change:

```
mov.b    #4,R6    ; 4266
```

This breaks into opcode = 4, S-reg = 2, Ad = 0, B/W = 1, As = 2 = 10b, D-reg = 6. The B/W bit flags a byte rather than a word. The source appears to have indirect register mode on R2/SR/CG1, but this is translated by the constant generator into a value of 0x0004, as required. Only a single word is needed for the instruction.

Next, let us return to the first three active lines of Listing 4.3, which were reviewed in the section "Machine Code" on page 77.

```
mov.w    #WDTPW|WDTHOLD,&WDTCTL    ; 40B2  5A80  0120
mov.b    #00001000b,&P2OUT         ; 42F2  0029
mov.b    #00011000b,&P2DIR         ; 40F2  0018  002A
```

These are a bit more complicated because the source is given as an absolute address rather than a register. The first instruction breaks into opcode = 4, S-reg = 0, Ad = 1, B/W = 0, As = 3 = 11b, D-reg = 2. The source has autoincrement addressing on PC, which means an immediate value. The destination has Ad = 1, which means indexed. The register is Dreg = 0010b = R2/SR/CG1, which means absolute addressing (the register acts like a base value of 0). The instruction is followed by words for the immediate value and absolute address. The third line is very similar but the B/W bit is set to indicate a byte rather than a word. The second line looks rather different because the constant generator is used to provide the value of 4, as described earlier.

5.4.7   Illegal Operations

TI's documentation says little about illegal operations. In newer devices, such as the F2013, a reset is generated if the CPU tries to fetch an instruction from the range of memory allocated to the special function and peripheral registers (addresses below 0x0200). That is about it. There are many other possible illegal operations so I experimented. Needless to say, you should never write code that relies on any form of illegal operation and the effects will be different in other variants.

Fetching Data

Here are three possible illegal operations.

- Data can be fetched from a nonexistent address: empty regions of the memory map. Execution proceeds as normal but the "fetched" value is random.

- Data for a word should be aligned to even addresses but you could attempt to fetch a word from an odd address. I found that the lsb of the address was ignored and the word was fetched from the resulting even address.

- A related case of this is the range of peripheral registers with word access. It appears that the lsb of their addresses is ignored because they are intended to be read only as words. It is possible to read the lower byte alone but not the upper byte: An attempt to read the upper byte returns the lower byte instead.

Illegal Instructions

Instructions are 16-bit words in the MSP430 but not all 2^{16} possible binary values are used; roughly one eighth have no operations associated with them. The instruction map is shown in a table near the end of the section "RISC 16-bit CPU" in the family user's guides (the first four rows of this table are labeled incorrectly at the time of writing). The ranges 0x0000–0x0FFF and 0x1400–0x1FFF are unused and there are some gaps in the row for Format II instructions (one operand), 0x1000–0x13FF.

I tried a few of the unused values found that the CPU executed the "instructions" happily, although it was not clear what it did. This causes a reset in many processors, which have an illegal opcode trap. In practice the most likely unexpected instruction is 0xFFFF, which is read if the program wanders into erased, unprogrammed Flash memory (see the next section). It disassembles to the following instruction:

```
and.b    @R15+,0xFFFF(R15)    ; "instruction" in unprogrammed Flash
```

In most cases this simply increments R15 by 1 (why?) but it causes a reset if it attempts to modify the watchdog timer control register because it is protected by a password (see the section "Watchdog Timer" on page 276).

The MSP430X uses most of the unused opcodes for its extra instructions and very few unimplemented values are left.

Fetching Instructions

Instructions are usually fetched from Flash memory but there is nothing wrong with reading them from RAM instead. This is used in applications that write to the main Flash memory such as bootloaders or updaters. Nor is it an error to read from unprogrammed Flash memory, which contains all bits set (0xFFFF). On the other hand, it is surely an error to read an instruction from the peripheral registers and this causes a reset in the F2013 and other newer devices. The same is true if an attempt is made to load an instruction from unused ranges of memory. This is good behavior for safety.

5.5   Examples

I picked a few examples to highlight aspects of the instruction set.

5.5.1   Copying Strings

The first example is based on the standard library function strcpy(destination, source), which copies a null-terminated string from source to destination. A basic version could be implemented as follow. This is a classic example where it is most convenient to embed all the action in the test of the while loop, a good or bad feature of C depending on your point of view. There are no checks for the length of the destination, unterminated strings, or overlap of the source and destination, so this is a dangerous routine. The library version returns the value dest but I have not bothered with this.

```
void MyStrcpy(char * dest, const char * src)
{
    while ((*dest++ = *src++) != '\0') {
    }                              // empty body of loop
}
```

Now write this in assembly language. Suppose that the address of the destination is passed in R12 and the source in R14. First, suppose that autoincrement addressing were available for both source and destination and that the move instruction affected SR. The translation from C to assembly would then be trivial:

```
; Copy source string starting in R14 to destination starting in R12
MyStrCpy:
CopyLoop:
        mov.b    @R14+,@R12+      ; can't be done - dream on
        jnz      CopyLoop         ; continue if byte != '\0'
        ret
```

Unfortunately life is not so simple because autoincrement addressing is not available for the destination. The structure of the loop is a little unusual because the copy operation must be carried out once to terminate the destination after the test for a nonnull character has failed. This must be done explicitly if a simple condition is used in the while statement. (The structure does not suit a do–while loop either because the test must be carried out before the pointer is incremented.)

```
void MyStrCpy(char * dest, const char * src)
{
    while (*src != '\0') {         // test for end of string
        *dest++ = *src++;          // copy nonnull character
    }
    *dest = '\0';                  // terminate destination string
}
```

This can be translated into assembly without too much modification. As usual control jumps initially to the test, which is put at the end of the loop for convenience:

```
; Copy source string starting in R14 to destination starting in R12
MyStrCpy:
        jmp     CopyTest
CopyLoop:
        mov.b   @R14+,0(R12)    ; [2 words, 5 cycles]
        inc.w   R12             ; [1 word,  1 cycle]
CopyTest:
        tst.b   0(R14)          ; [2 words, 4 cycles]
        jnz     CopyLoop        ; [1 word,  2 cycles]
        clr.b   0(R12)          ; terminate string
        ret
```

Autoincrement addressing is used for the source of mov.b. This is not available for the destination, which must instead use indexed mode with a base of 0 and a separate instruction to increment the pointer R12. Both tst.b and clr.b also use indexed mode with a base of 0 because they are again destinations of emulated instructions. This is expensive in terms of clock cycles.

The loop can be made a little neater by exploiting the feature that moves do not affect SR. We can test the source, perform the move afterward without affecting the flags, and finally make the conditional jump. The increment cannot be moved, however, because it would overwrite SR. The complete program including a test harness is shown in Listing 5.1.

The string for the source is stored in the DATA16_C segment of Flash. The double quotation marks around the text instruct EW430 to interpret it as a C string and append a null character '0'. Use single quotation marks if you do not want the null character.

When testing this function, we want to confirm that all this string and no more gets copied. I therefore filled all words of RAM with the pattern 0xA5A5 as soon as the watchdog is stopped—even before initializing the stack. This shows another form of loop that duplicates a fixed source, which I load into a register for speed. I use words rather than bytes for speed because RAM contains an even number of bytes. The directives SFB and SFE provide the addresses of the beginning and end of RAM. It is a little tricky because the beginning is in the segment DATA16_N and the end is in CSTACK.

Another use of filling the RAM with a known pattern is to monitor the stack, which overwrites the words of A5A5 as it grows downward. This can be seen in action when the subroutine MyStrCpy is called and the return address is pushed on to the stack. The obvious pattern also makes it easy to check that there is a safe buffer between the variables at the bottom and the stack at the top. It would be trivial to rewrite the program to clear RAM instead.

Listing 5.1: Program `teststrcpy1.s43` to exercise `MyStrCpy`. The RAM is initially filled with the pattern A5A5 so that its use can be monitored.

```
; teststrcpy1.s43 - test MyStrCpy, copying from ROM to RAM
; Fills RAM with 0xA5A5 before creating stack
;   (so this cannot be called as a subroutine)
; TI eZ430, F2013, LED active high on P1.0
; J H Davies, 2006-07-28; IAR Kickstart version 3.41A
; ------------------------------------------------------------------
#include <msp430x20x3.h>           ; Header file for this device
; ------------------------------------------------------------------
        RSEG    DATA16_N           ; Memory for variables
DestStr DS      16                 ; 16 bytes string for destination in RAM
; ------------------------------------------------------------------
        RSEG    CSTACK             ; Create location for stack
; ------------------------------------------------------------------
        RSEG    CODE
Reset:
        mov.w   #WDTPW|WDTHOLD,&WDTCTL   ; Stop watchdog timer
        mov.w   #SFB(DATA16_N),R4   ; Pointer to start of RAM
        mov.w   #0xA5A5,R5          ; Pattern for fill
FillRAM:
        mov.w   R5,0(R4)           ; Write to RAM, clumsy destination
        incd.w  R4                 ; Step pointer to next word
        cmp.w   #SFE(CSTACK),R4    ; Set flags for (R4 - end of RAM)
; Not SFE(DATA16_N), which would include only the declared variables
        jlo     FillRAM            ; Repeat loop while R4 < end of RAM
        mov.w   #SFE(CSTACK),SP    ; Initialize stack pointer
main:                              ; Roughly like start of main() in C
        mov.w   #SourceStr,R14     ; Load address of source
        mov.w   #DestStr,R12       ; Load address of destination
        call    #MyStrCpy          ; Copy string (don't forget #!)
        jmp     $                  ; Infinite, empty loop
; ------------------------------------------------------------------
; Copy source string starting in R14 to destination starting in R12
; Both registers overwritten; no local registers used
; No checks for overlap, space in destination, unterminated source...
MyStrCpy:
        jmp     CopyTest
CopyLoop:
        inc.w   R12                ; [1 word,  1 cycle] inc dst address
CopyTest:
        tst.b   0(R14)             ; [2 words, 4 cycles] test source
        mov.b   @R14+,0(R12)       ; [2 words, 5 cycles] copy src -> dst
        jnz     CopyLoop           ; [1 word,  2 cycles] continue if not \0
        ret
; ------------------------------------------------------------------
        RSEG    DATA16_C           ; Segment for constant data in ROM
SourceStr:                         ; Source string
        DB      "hello, world\n"   ; "" causes a '\0' to be appended
; ------------------------------------------------------------------
        RSEG    RESET              ; Segment for reset vector
        DW      Reset              ; Address to start execution
        END
```

Example 5.5

Try this in the debugger. Open a window with View > Memory and select RAM. You should first see the loop fill the RAM with bytes of A5. Use Run to Cursor to jump to the end of the loop or you will get very bored. Watch the stack pointer change and the return address appear at the highest address in RAM when the subroutine is called. Characters are then copied one by one into RAM until the string is complete, including the newline and null characters.

Unfortunately the loop inside MyStrCpy in Listing 5.1 is neither smaller nor faster than the first version in assembly language; it still occupies 6 words and takes 12 cycles per iteration. For comparison, here is the assembly code that the EW430 compiler produced from the function in C:

```
; Copy source string starting in R14 to destination starting in R12
MyStrCpy:
        mov.b    @R14+,R15          ; [1 word,  2 cycles]
        mov.b    R15,0x0(R12)       ; [2 words, 4 cycles]
        inc.w    R12                ; [1 word,  1 cycle]
        tst.b    R15                ; [1 word,  1 cycle]
        jne      MyStrcpy           ; [1 word,  2 cycles]
        ret
```

The moved byte is stored temporarily in R15. This eliminates the second of the costly, indexed addresses for the destination and cuts the loop down to ten cycles per iteration. A possible disadvantage is that an extra register is used.

5.5.2   Copying Blocks of Data

Copying becomes easier if the offset between the source and destination is a known constant. This occurs when data are read from ROM into RAM during the initialization of a C program, for instance. An example is shown in Listing 5.2. This copies the same string for simplicity but can be used for any data, not necessarily terminated by a null byte. The address of the end must therefore be specified instead. The rest of the program is copied from Listing 5.1.

Listing 5.2: Part of program blockcpy1.s43, which copies a block of memory between fixed addresses.

```
; Copy data between fixed addresses BeginSource and BeginDest
; Address EndSource is 1 beyond the final byte of the source
    mov.w    #BeginSource,R14       ; Load address of source
    jmp      CopyTest
```

```
CopyLoop:
; This ought to be sufficient but the assembler complains...
    mov.b   @R14+,BeginDest-BeginSource-1(R14)   ; -1 allows for @R14+
; ...so I had to handle the wrapping around myself
    mov.b   @R14+,BeginDest-BeginSource+0xFFFF(R14)
                                    ; 0xFFFF allows for increment in @R14+
CopyTest:
    cmp.w   #EndSource,R14          ; Set flags for (R14 - end of source)
    jlo     CopyLoop
    jmp     $                      ; Infinite, empty loop
; --------------------------------------------------------------------
    RSEG    DATA16_C               ; Segment for constant data in ROM
BeginSource:                       ; Source data
    DB      "hello, world\n"       ; "" causes a '\0' to be appended
EndSource:                         ; Address beyond data to be copied
```

This loop uses autoincrement addressing on the source. The address of the destination is calculated from the same register, which is added to the displacement between source and destination. The tricky feature is that the autoincrement is performed before the address of the destination is calculated, so we must subtract 1 (or 2 if words are moved) to compensate. There is a further complication in Listing 5.2, because the destination has a lower address than the source, so the displacement is negative. This upset the EW430 assembler so I had to add 0xFFFF rather than subtract 1.

If the number of bytes to copy were specified instead of the final address, a further counter could be initialized and decremented to 0. That could be done here too. The loop would be faster but would require an extra register.

Example 5.6

Rewrite the program in Listing 5.2 to use a counter rather than a comparison with EndSource.

5.5.3   Recording the State of a Push Button

I mentioned in the section "Read Input from a Switch" on page 80 that mechanical devices such as the contacts in push buttons tend to *bounce*, meaning that they do not give clean transitions on the input to a microcontroller. Instead the input may make several transitions between 0 and 1 before settling at its final value. All inputs are susceptible to noise, which may give short spikes of the wrong value. Some sort of filtering is therefore necessary to produce a clean signal for the main function. This requires a record of the input from the button with time. One way of doing this is to read the input at regular intervals and record the values in a shift register. The latest input becomes the msb and the other values are

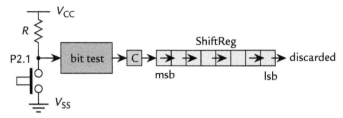

Figure 5.7: Use of a shift register to record the most recent eight states of a push button.

shifted right. The oldest value, which was in the lsb, is lost. The idea is shown in Figure 5.7. This can be done very simply because the bit operations give $C = {\sim}Z$ and the carry bit can be rotated into the shift register:

```
bit.b    #BIT1,&P2IN         ; Test button on bit 1 of P2IN; C = ~Z
rrc.b    ShiftReg            ; Insert C into msb of shift register
```

For example, suppose that P2IN.1 is high. The bit test gives nonzero so $Z = 0$, $C = 1$, and a 1 is inserted into the lsb of the byte ShiftReg. The inversion in $C = {\sim}Z$ conveniently gives the correct value of the inserted bit. We see how to use this in the section "Debouncing in Software" on page 233. Remember to initialize ShiftReg before use. It could be made a word instead of a byte or even longer if desired.

It is trickier to do this in C because the rotate instruction is not available. Instead the register can be shifted and its msb set or cleared according to the input:

```
ShiftReg >>= 1;                     // Shift register right by 1 position
if (P2IN_bit.P2IN_1 == 1) {  // Test button on bit 1 of P2IN
    ShiftReg |= BIT7;               // Set msb of byte
} else {
    ShiftReg &=~BIT7;               // Clear msb of byte
}
```

It is not necessary to clear the msb explicitly if the shift is logical rather than arithmetic. Define ShiftReg to be unsigned to ensure this.

Example 5.7

Combine this shift register with a delay, such as the polling loop on the timer in the section "Automatic Control: Flashing a Light by Polling Timer_A" on page 105, to make a delay line. The LED shows the state of the button after it has passed through the delay line rather than its state at the same time. Feed the button in to the shift register and use the output to drive the LEDs.

If you use C, you could try writing the code for the shift register in a more elegant way than the preceding snippet. This treats the register as a byte for the shift but as bits to insert the new value. Would a union of a byte and a bit field be better?

5.6   Reflections on the CPU and Instruction Set

This is not a book on the architecture of computers, and the huge variation between microcontrollers shows that there is not even remote agreement on the optimum solution to their design. However, the MSP430 has a simple enough CPU that it is worth spending a little time pondering its design. There is an introduction by one of the architects, Lutz Bierl, in chapter 1 of his *Application Reports* (slaa024) and more detail in his book [2].

5.6.1   Simplicity

The MSP430 was designed from the start for low power consumption. It should therefore complete the desired operations in as few clock cycles as possible, which requires a good match between programs and the instruction set. This might encourage a large set of instructions but a conflicting requirement is to use as little energy as possible in each clock cycle, which favors a small CPU with only a few instructions.

Most programs today are written in C so the processor is designed for efficient compilation rather than hand-crafted assembly code. This explains the small number of general-purpose instructions, the large number of registers, and the highly orthogonal nature of the instruction set, meaning that all addressing modes can be used with all instructions (with only a few exceptions). Conveniently, these features also require less logic to control the CPU. A further advantage of the orthogonality is that it is easy to address the stack, which is heavily used for parameters and local variables in subroutines.

An 8-bit processor might appear simpler at first glance but an advantage of a 16-bit processor is that registers can be used for either data or addresses. Moreover, most analog-to-digital converters now produce more than 8 bits of output, which also applies to peripherals like the timer, so it is more efficient to handle words rather than bytes.

5.6.2   Registers of the CPU

The programmer's model is simple and versatile: a straightforward set of 16 registers that includes all the special-purpose registers as well as the general-purpose registers for addresses and data. The combination of this feature with orthogonal addressing is powerful. For example, a subroutine can be selected from an array in the same way as data.

The constant generator is a particularly creative feature. It is common to have a fixed value of 0 available but the four addressing modes of CG2 and the three modes of addressing the status register that would otherwise be meaningless give seven values. This means that the "move" instruction can emulate "clear" without extra data or cycles. Similarly, "add" can emulate "increment," and so on. This greatly increases the speed and density of the code, which can be poor in processors with a reduced instruction set. That raises the vexed question of whether the MSP430 is a RISC.

5.6.3   Is the MSP430 a RISC—and Should It Be?

TI describes the MSP430 as having a RISC 16-bit CPU although it is sometimes more cautious and merely calls it RISC-like. A discussion of the classification alone would be sterile but it is interesting to see what features of a RISC were adopted by the designers of the MSP430, which they rejected, and why. Furber [36] has a fascinating and accessible discussion of RISC and CISC processors focused on the ARM family, which is usually considered to be one of the first RISC processors although it does not fit the usual definition fully. Four main features characterize a RISC:

Small set of general-purpose instructions: In contrast to the large number of instructions with different formats in a CISC (over 250 in the Freescale HCS08, for instance).

Large bank of general-purpose registers: Used for both data and addresses, in contrast to the small number of specialized registers in a CISC (the HCS08 has a single accumulator for data and a separate index register for addresses).

Load–store architecture: In which instructions that process data use only the registers in the CPU and cannot operate on memory directly; separate instructions *load* data from memory into the CPU and *store* the result back into memory.

Single-cycle execution: Unlike the variable number of cycles required by a CISC.

The MSP430 displays the first two features but not the last two, and it is worth considering why these were rejected. The basic reason is that small microcontrollers spend much of their time performing simple operations on registers in the main address space. These often reconfigure a peripheral. For example, suppose that we wish to stop Timer_A, which requires bits MC0 and MC1 of TACTL to be cleared. It can be done in a single instruction in the MSP430:

```
bic.w    #MC0|MC1,&TACTL ; stop timer [3 words, 5 cycles]
```

This requires three words to hold the instruction itself, the immediate data for the source, and the absolute address for the destination. It takes five cycles to fetch the instruction, fetch the immediate value, fetch the address of the destination, fetch the original value of the destination, and write back the modified value of the destination. How would this be done in a "pure" RISC processor? Here is a possibility, making minimal changes from the MSP430:

```
load.w   #TACTL,R4      ; load address of TACTL   [2w, 2c]
load.w   @R4,R5         ; load value of TACTL     [1w, 2c]
load.w   #MC0|MC1,R6    ; load immediate operand  [2w, 2c]
bic.w    R6,R5          ; perform operation       [1w, 1c]
store.w  R5,@R4         ; store result for TACTL  [1w, 2c]
```

I assume that all addresses must reside within the CPU, so that `load.w &TACTL,R5` is not permitted; instead we must first store the address as an immediate value into a register and use indirect addressing. As usual the number of cycles is set by the access to memory. The bad news is that this requires five instructions, seven words, and nine cycles, roughly twice the storage and time of the MSP430. This shows immediately why RISC processors have a reputation for poor code density. The RISC also is slower unless the processor is simple enough that the speed of the clock could be raised, but in practice this is probably limited by the speed of the memory rather than the CPU. It may be possible to embed a byte of immediate data in the instruction, which would assist operations on peripheral registers with byte access.

Another benefit of the MSP430's single instruction is that it cannot be broken by an interrupt—instructions are always completed before an interrupt is taken. The jargon is that it is *atomic*. In contrast, an interrupt could occur between the five instructions of the RISC processor. The programmer may need to guard against potential side effects; see the section "Issues Associated with Interrupts" on page 196.

Of course it is easy to find examples where the RISC performs better. The inner loop in MyStrCpy could be rewritten like this:

```
CopyLoop:
    load.b   @R14+,R15     ; [1 word,  2 cycles]
    store.b  R15,@R12+     ; [1 word,  2 cycles]
    jnz      CopyLoop      ; [1 word,  2 cycles]
```

This requires only half the storage and cycles of the version in Listing 5.1. The main saving comes from autoincrement addressing in the store operation, which is not permitted for a destination in the MSP430. It is likely that a RISC would offer further addressing modes, such as predecrement to complement postincrement, and that these could be used with a base address to provide further indexed modes. Small microcontrollers usually spend more time writing to peripheral registers than copying or manipulating blocks of data, so the first example is more relevant. Therefore the designers of the MSP430 chose not to adopt a strict load–store architecture but to allow operations on main memory instead.

5.6.4   Addressing Modes

A Format I instruction with two operands must fit into a 16-bit word. Between 8 and 16 such operations are likely to be required, which needs 4 bits to specify the operation. Four bits are needed to select CPU registers for source and destination and a further bit indicates bytes or words. This leaves only three bits for the addressing modes, of which two are used for the source and one for the destination. This sounds severely limited but clever use of the constant generator gives the seven or eight effective modes available for the source. Unfortunately the modes for the destination are restricted.

The lack of indirect or autoincrement modes for the destination can be frustrating when writing assembly language. In the example of MyStrCpy we had to replace the "dream" instruction mov.b @R14+,@R12+ with two, mov.b @R14+,0(R12) and inc.w R12. The most annoying feature is that a word of 0 must be stored in the instruction to provide the base for the indexed address 0(R12) to act like @R12. However, there are simply no spare bits in a word to accommodate these modes. Two bits could be released if there were only 8 registers instead of 16 but this would be a serious sacrifice given that 4 are special. We would probably have to give up CG2 and many of the constants.

Another missing mode for addressing is predecrement, which would be useful for pushing data on to the stack. In C this would be written as *--SP = PushedData. There would not need to be a native push instruction if this mode were available, but again there simply is no space for it in the instruction word.

The indexed and immediate addressing modes require extra data, which means that the overall instruction occupies more than one word and takes several cycles of the bus to fetch from memory. The decision not to adopt a load–store architecture therefore means that it is not possible to have single-cycle execution of all instructions. Instead, the number of cycles is set by accesses to memory. This is an intrinsic limitation of the von Neumann architecture and also applies to the ARM7, for instance (although that raises its computational performance with a pipeline). This could be avoided only by adopting a Harvard architecture, as in the ARM9 or PIC24, at the cost of considerable extra complexity.

5.6.5   Conclusions on the Instruction Set

I have generally found the MSP430 comfortable to program in assembly language. The orthogonality is wonderful: There is no need to consult the manual constantly to find which addressing modes are permitted with each instruction. There are few enough instructions that I can remember them all. Access to memory is simple and uniform, with no distinction between short and long addresses. The constant generator makes programs faster and more compact with no effort from the programmer, because the assembler replaces emulated instructions, numerical constants, and absolute addresses with code that use the constant generators. There are other helpful little features, such as $C = {\sim}Z$ after bitwise operations.

Nothing is perfect, of course, and the limited addressing modes for the destination hindered me most. It is occasionally a nuisance that mov does not set the status flags, requiring a separate test. Sometimes it is clumsy to work around the missing tests in the jump instructions. A few features of the assembly language have caught me out on numerous occasions, so I list them here:

- It is extremely easy to forget the # when calling subroutines in the most straightforward way and the results are disastrous.

- It is also easy to forget the & in front of absolute addresses, in which case the processor uses symbolic mode instead.

- This is less common, but you must remember to clear the carry bit before the decimal addition instruction dadd, which is really "add with carry."

- Be careful to distinguish between the tests jhs and jlo for *unsigned* numbers and jge and jl/jlt for *signed* numbers.

5.7   Resets

A *reset* is a sequence of operations that puts the device into a well-defined state, from which the user's program may start. This is obviously necessary when power is first applied. A reset is also generated if the device detects a serious fault in hardware or software from which the user's program cannot be expected to recover. A peculiarity of the MSP430 is that it has two levels of reset, depending on whether the reset was caused by hardware or software.

Power-on Reset (POR)

This is generated by the following severe conditions related to *hardware*:

- The device is powered up. More generally, a POR is raised if the supply voltage drops to so low a value that the device may not work correctly: a brownout. Powering up the device is just an extreme case of this. Almost all current MSP430s include a brownout detector.

- A low external signal on the $\overline{\text{RST}}$/NMI pin resets the device if the pin is configured for the reset function rather than the nonmaskable interrupt. The reset function $\overline{\text{RST}}$ is active by default.

- Larger variants have a more comprehensive supply voltage supervisor (SVS). This is configurable, unlike the brownout detector. It sets the SVSFG flag if the voltage falls below the programmed level and can optionally reset the device.

Power-up Clear (PUC)

This always follows a power-on reset. It is also generated when *software* appears to be out of control in the following ways:

- The watchdog timer overflows in watchdog mode. Remember that the watchdog is active by default and must either be disabled or regularly cleared before it rolls over.

- An attempt is made to write to the watchdog control register WDTCTL without the correct password 0x5A (available as the symbol WDTPW) in the upper byte. A reset is triggered even if the watchdog is disabled or in interval timer mode.

- The registers for the flash memory controller, FCTL*n*, are protected by a password in the same way as WDTCTL. The value is 0xA5, available as the constant FWKEY. This is to protect runaway software from corrupting the stored program.

- In newer devices, a PUC is triggered by an attempt to fetch an instruction from the range of addresses reserved for peripheral registers or to read unimplemented memory.

5.7.1 Conditions after Reset

The initial conditions for all registers and peripherals after a POR and PUC are specified in the family user's guides. These are the general effects:

- The $\overline{\text{RST}}$/NMI pin is configured for reset. It is also used for the two-wire JTAG interface (Spy-Bi-Wire) in some devices, such as the F2013.

- Most input/output pins are configured as digital inputs. There are a few exceptions, such as those pins that are shared with the crystal in the F2013.

- Other peripheral modules and registers are initialized as set out in the guide. The value is shown under each bit in the description of the registers. For example, rw–0 means that a bit can be both read and written and is initialized to 0 after a PUC. The slightly different notation rw–(0) shows that a bit is initialized to 0 only after a POR; it retains its value through a PUC.

- The status register is cleared. This means that the device operates at full power, even though it might have been in a low-power mode before the reset occurred.

- The watchdog timer starts in watchdog mode. This is essential because it is a safety feature but means that you must either service or disable it before it times out and resets the chip (continually).

- The program counter is loaded with the the reset vector, which is stored in the word at 0xFFFE. This provides the address of the first instruction to be executed.

Small programs usually (re)start in the same way, regardless of what caused the reset. This may not be true for large programs or networked systems, where it may be necessary to report the problem. It is also important to identify the source of a reset when debugging. No single register holds all the relevant flags but most are in the interrupt flag register 1, IFG1. This may contain the following flags, depending on the variant:

- **WDTIFG** shows that the watchdog timed out or its security key was violated.

- **OFIFG** indicates an oscillator fault (this causes a nonmaskable interrupt, not a reset).

- **RSTIFG** shows a reset caused by a signal on the $\overline{\text{RST}}$/NMI pin.

- **PORIFG** is set for a power-on reset.

- **NMIIFG** flags a nonmaskable interrupt (not reset) caused by a signal on the $\overline{\text{RST}}$/NMI pin.

For example, the WDTIFG bit is set when the watchdog times out. It is not cleared by a power-up clear and can therefore be tested to identify the source of the PUC.

Other flags are elsewhere. A flash security key violation is indicated by KEYV in the TCTL3 register. The supply voltage supervisor control register SVSCTL contains SVSFG. This is initialized to 0 only after a brownout reset (which includes powering up the device) so this reset can be identified too. There does not appear to be a flag for illegal instruction fetches.

5.7.2   Configuration of the Device by the User

The operations performed during the reset process are enough to get the processor operating in a well-defined state but the user must configure other features of the device and set up the peripherals as desired.

The registers that control critical features are protected against accidental writes from runaway code in many microcontrollers. Often the user may be permitted to write only once to each register, after which the value is locked. In this case it is important to write a value to each register to activate the protection, even if it only repeats the default value. Having said this, I do not believe that the MSP430 has any registers protected in this way. This allows any aspect of it to be reconfigured at any time.

It is important to realize that many registers are *not* initialized by a reset. For example, the output registers such as P1OUT are not cleared and contain either random values or the data from before the device was reset.

A typical program performs the following initialization before starting on its main activity. The first two steps are vital.

- Initialize the stack pointer to the top of RAM (or 1 byte beyond to be precise) before any subroutines are called. This is done automatically for you in C but not if you use assembly language.

- Configure the watchdog before it times out and resets the device.
  I simply stop it for most of the programs in this book.

- Set up the clock, which is described in the section "Clock System" on page 163.

- Configure all ports. Unused pins should never be left as unconnected (floating) inputs: See the section "Configuration of Unused Pins" on page 215 for different ways of handling this.

- Configure other peripherals, such as Timer_A.

- Finally, enable interrupts if you use them.

5.7.3   Hardware Issues

Two aspects of the hardware associated with resets may need care.

External Connection to the Reset Pin

The $\overline{\text{RST}}$/NMI pin must not be left floating and needs to be decoupled from noise. A typical connection recommended in the *MSP-FET430 Flash Emulation Tool (FET) User's Guide* (slau138) is a 47 kΩ pull-up resistor (remember that the input is active low) with a 10 nF capacitor to ground. A pushbutton to ground may be added to permit a manual reset.

A general problem with modern, low-power microcontrollers—not just the MSP430—is that they seem to be painfully sensitive to noise on their reset inputs. Nagy [4] has some good advice on this. I used a demonstration board for another processor where there was enough coupling from a square wave on the analog input to reset the MCU.

This is a classic example where any problem will probably not arise during early testing. The connections to the debugger usually include the $\overline{\text{RST}}$/NMI pin and the output impedance of the interface pod is often low enough to suppress any noise picked up on the board. The sensitivity to noise appears only when the debugger is disconnected.

Poor-Quality Power Supply

The brownout detector is intended to bring the device into operation safely from a well-behaved power supply. It is not sufficient for poor-quality supply, particularly if the voltage V_{CC} rises very slowly to its full value.

To take an example, the operation of a F2013 is specified down to $V_{CC} = 1.8$ V. The brownout detector therefore holds the chip in its reset state until V_{CC} rises through a value just below this threshold, waits for a delay of up to 2 ms, and releases the device to run.

However, the F2013 needs $V_{CC} > 2.2\,V$ for programming flash memory and $V_{CC} > 3.3\,V$ for its maximum speed of 16 MHz. There is no guarantee that V_{CC} meets either of these criteria when the brownout detector releases the chip, only that it is at least 1.8 V. The speed must not exceed 6 MHz at this lower limit and flash memory cannot be programmed.

A supply voltage supervisor should be used if there is any danger that the supply voltage may fall too low for normal operation, but remain too high to cause a brownout reset. The device could be run at minimum speed until the supervisor reports that the voltage is high enough for full speed and programming the flash memory. The larger MSP430s contain an internal SVS, some voltage regulators provide a "power good" signal, and a wide range of dedicated supervisor ICs is available.

5.7.4 Starting up in Special Modes

The above description applies to a device that executes a user's program after it has started up. This is the usual behavior for a finished product but the debugger or bootstrap loader should take control of the device instead during development. This is normally handled by the debugging software so a user need not be concerned about the details, but they are in the application notes *Programming a Flash-Based MSP430 Using the JTAG Interface* (slaa149) and *Features of the MSP430 Bootstrap Loader* (slaa089) if you are interested.

The MSP430 examines the voltages on its $\overline{\text{RST}}$/NMI pin and either TEST or TCK as $\overline{\text{RST}}$/NMI goes high to release the device from reset. A particular sequence of edges launches the bootstrap loader (BSL) rather than taking the user's reset vector. This works in almost all variants of the MSP430 but not in the F2013 because it has no bootstrap loader.

JTAG can take control of a larger device at any time through the four dedicated pins mentioned in the section "Access to the Microcontroller for Programming and Debugging" on page 57. If necessary, a debugger can also drive $\overline{\text{RST}}$/NMI low to force a reset if the pin has not been reconfigured to provide an interrupt instead. In smaller devices, where the JTAG pins can also be used for conventional input and output, the TEST pin must be pulled high to enable the JTAG functions. Finally, in devices with the two-wire Spy-Bi-Wire interface, the SBWTCK pin is shared with TEST and again activates JTAG if it is pulled high. The other Spy-Bi-Wire pin, SBWTDIO, is shared with $\overline{\text{RST}}$/NMI.

Checklist—Have You ...

❏ Initialized the stack pointer if you are using assembly language?

❏ Connected the $\overline{\text{RST}}$/NMI pin correctly?

❏ Disabled the watchdog timer or provided code to service it regularly?

5.8 Clock System

All microcontrollers contain a clock module to drive the CPU and peripherals. The conflicting requirements for clocks in high-performance, low-power microcontrollers were described in the section "Clock Generator" on page 33. Figure 5.8 shows a simplified diagram of the Basic Clock Module+ (BCM+) for the MSP430F2xx family. I concentrate on this because it is the most recent design; I mention the extra features of the MSP430x4xx later. The details vary between devices, even in the same family, and the second crystal oscillator XT2 is often omitted. Recall that the clock module provides three outputs:

- **Master clock, MCLK** is used by the CPU and a few peripherals.

- **Sub-system master clock, SMCLK** is distributed to peripherals.

- **Auxiliary clock, ACLK** is also distributed to peripherals.

Most peripherals can choose either SMCLK, which is often the same as MCLK and in the megahertz range, or ACLK, which is typically much slower and usually 32 KHz. A few peripherals, such as analog-to-digital converters, can also use MCLK and some, such as

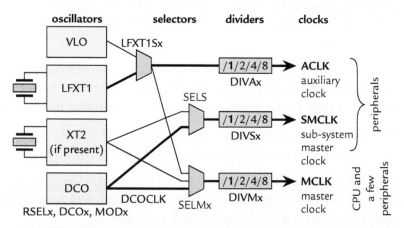

Figure 5.8: Simplified block diagram of the clock module of the MSP430F2xx family, showing some of the more important bits in the peripheral registers that control its operation. Heavy lines indicate the default configuration.

timers, have their own clock inputs. The frequencies of all three clocks can be divided in the BCM+ as shown in Figure 5.8. For example, you might wish to run the CPU at 8 MHz for rapid execution of code and therefore choose $f_{\text{MCLK}} = f_{\text{DCOCLK}} = 8\,\text{MHz}$. On the other hand, it may be more convenient if the peripherals run from a slower clock, in which case you might configure the divider for SMCLK with DIVSx to give $f_{\text{SMCLK}} = f_{\text{DCOCLK}}/8 = 1\,\text{MHz}$. Most
peripherals have their own dividers for their clock sources, which gives yet more control.

Up to four sources are available for the clock, depending on the family and variant:

Low- or high-frequency crystal oscillator, LFXT1: Available in all devices. It is usually used with a low-frequency watch crystal (32 KHz) but can also run with a high-frequency crystal (typically a few MHz) in most devices. An external clock signal can be used instead of a crystal if it is important to synchronize the MSP430 with other devices in the system.

High-frequency crystal oscillator, XT2: Similar to LFXT1 except that it is restricted to high frequencies. It is available in only a few devices and LFXT1 (or VLO) is used instead if XT2 is missing.

Internal very low-power, low-frequency oscillator, VLO: Available in only the more recent MSP430F2xx devices. It provides an alternative to LFXT1 when the accuracy of a crystal is not needed.

Digitally controlled oscillator, DCO: Available in all devices and one of the highlights of the MSP430. It is basically a highly controllable *RC* oscillator that starts in less than 1 μs in newer devices.

It is not quite true that any of the clocks can come from any of the sources, but the selection can seem bewildering. Fortunately the default configuration described in the section "Clock Generator" on page 33 is a good starting point. As a reminder, this is as follows:

- ACLK comes from a low-frequency crystal oscillator at 32 KHz. There is no choice in almost all devices, the exceptions being those with a VLO.

- Both MCLK and SMCLK are supplied by the DCO with a frequency of around 1 MHz. This is stabilized by the FLL where present. You may wish to raise this frequency provided that V_{CC} is high enough to support it.

Most applications do not need MCLK to be highly accurate, so there is rarely a need for a high-frequency crystal. A possible exception is those that use fast, asynchronous communication, where the frequency must remain within a few percent of its nominal value.

We look at the characteristics of the sources in more detail. The Basic Clock Module+ is controlled by four registers, DCOCTL and BCSCTL1–3. In addition there are bits in the special function registers IFG1 and IE2 for reporting faults with the oscillators. Some of the clock signals can be brought out to pins if needed to supply external components or for testing (I use this later for Figure 5.10). This typically needs the pin to be configured for output using P1DIR and the signal switched to the clock from the usual digital port with P1SEL.

5.8.1 Crystal Oscillators, LFXT1 and XT2

Crystals are used when an accurate, stable frequency is needed:

- *Accurate* means that the frequency is close to what it says on the package, typically within 1 part in 10^5.

- *Stable* means that does not change significantly with time or temperature.

Crystals are cut from carefully grown, high-quality quartz with specific orientations to give them high stability. Traditional crystals oscillated at frequencies of a few MHz but most small microcontrollers use low-frequency watch crystals with a frequency of 32 KHz. These are machined into complicated tuning fork shapes to give the low frequency. A disadvantage is that their frequency is more sensitive to temperature than high-frequency crystals, but they are designed to be most stable near 25°C. A change of 10°C in the temperature causes the frequency to fall by about 4 parts per million (ppm). Detailed specifications are given by the manufacturers, such as Micro Crystal [58].

Example 5.8

An accuracy of 1 part in 10^5 for a crystal sounds impressive but is it good enough for a real-time clock? What is the error per day or per year? How does it compare with John Harrison's H-4 chronometer, which accumulated an error of 5 s over 81 days when it was taken on its first sea voyage for testing in 1761–1762?

Oscillator circuits are integrated into the MSP430 and the crystal should be connected to pins XIN and XOUT. The circuit is known as a *Pierce oscillator* and also requires a capacitance to ground from each pin. The value of the capacitance depends on the crystal and is typically around 10 pF for a watch crystal. Part of this is inevitably contributed by stray capacitance of the PCB tracks to the crystal, which must be kept as short as possible. External capacitors are needed with many microcontrollers but they are integrated into the MSP430 for low-frequency crystals. The value is selected with the XCAPx bits in the BCSCTL3 register of the F20xx. The older Basic Clock Module (BCM, without the +) in the MSP430x1xx has a fixed capacitance of 12 pF per pin. For a highly accurate frequency, the drive current of the LFXT1 should also match the specification of the crystal. This sounds rather specialized but systems with the MSP430 may run for 10 years on a single battery and extreme accuracy and stability are needed if the clock is not to drift over this lifetime. The application note *MSP430 LFXT1 Oscillator Accuracy* (slaa225) describes some of the issues.

The oscillator is designed to run at low power and this renders it susceptible to electromagnetic interference. This means that the printed circuit board must be laid out carefully. There is detailed advice in the application note *MSP430 32 KHz Crystal Oscillators* (slaa322). Having said that, the crystal on the Olimex 1121STK is perched up in the air over the MCU itself, which seems to violate the rules but nevertheless works reliably.

It is also possible to use high-frequency crystals (above 400 KHz) with LFXT1 in most devices and some have a second oscillator XT2, which works only at high frequencies. External capacitors must be used with high-frequency crystals and the module must be configured for a suitable range of frequencies. The module also accepts an external clock signal on XIN.

The stability of crystals is reflected electrically in their high Q factor. This means that they oscillate for a long time after being excited, like the ringing tone from a wine glass after it has been tapped. The disadvantage of this is that the oscillator takes an equally long time to reach a stable state, typically around 10^5 cycles. An oscillator based on a watch crystal therefore takes nearly a second to start and is almost always left running continuously. It might seem better that a 10 MHz crystal starts in "only" 10 ms but the CPU could have used around 10^5 cycles from the DCO in this time, which is enough to complete many tasks and return to a low-power mode. A software delay loop can be included if it is important to wait for the crystal to stabilize. This can be similar to those in the section "Automatic Control Flashing Light by Software Delay" on page 91.

The auxiliary clock ACLK can be derived only from LFXT1 in most devices so ACLK will not be available if there is no crystal. Beware if you are using a TI development kit because these often come without crystals installed. The exceptions are most devices in the MSP430F2xx family, which can take ACLK from the VLO instead.

5.8.2 Internal Low-Power, Low-Frequency Oscillator, VLO

The VLO is an internal *RC* oscillator that runs at around 12 KHz and can be used instead of LFXT1 in some newer devices. It saves the cost and space required for a crystal and reduces the current drawn. The data sheet for the F2013 shows that LFXT1 draws about 0.8 μA, which falls to 0.5 μA with the VLO. (Both are impressively small currents.)

Of course this comes at a cost: accuracy and stability. The same data sheet quotes a range of frequency for f_{VLO} from 4 to 20 KHz. This looks terrible at first sight but closer reading shows that it covers the full operating range of the device in voltage and temperature. A variation of 10°C changes the frequency by about 5% instead of 5 ppm for a crystal. Clearly you would not use the VLO for serious timing. On the other hand, its purpose is often to wake the device periodically to check whether any inputs have changed, and accuracy is not important.

ACLK is taken from LFXT1 by default even where the VLO is present. This means that current is wasted in LFXT1 and that pins P2.6 and P2.7 in the F20xx are configured for a crystal. It is usually a good idea to reconfigure the BCM+ to use the VLO and redirect port 2 for digital input/output. This needs a few lines of code:

```
BCSCTL3 = LFXT1S_2;      // Select ACLK from VLO (no crystal)
P2SEL = 0;               // Digital i/o rather than crystal
P2DIR = 0;               // Input direction
P2REN = BIT6|BIT7;       // Pull Rs on unused pins (6 and 7)
```

This port is unusual because it is not entirely an input by default; P2.7 is initially the output that drives the crystal and P2DIR.7 is set. Finally I enabled the pull resistors to provide a well-defined voltage on the pins and avoid floating inputs.

5.8.3 Digitally Controlled Oscillator, DCO

One of the aims of the original design of the MSP430 was that it should be able to start rapidly at full speed from a low-power mode, without waiting a long time for the clock to settle. Early versions of the DCO started in 6 μs, which has been reduced to 1–2 μs in the MSP430F2xx family. There are no erratic pulses: The output from the DCO starts cleanly after this delay. The stability and accuracy also improved significantly since the early days of the MSP430 and calibration values are now stored for a set of frequencies, giving an accuracy of 1–2%.

The frequency can be controlled through sets of bits in the module's registers at three levels. The numbers are taken from the data sheet for the F2013. The first two levels set the DCO to a constant frequency:

RSELx: Selects one of 16 coarse ranges of frequency. The frequencies in each range are larger than those in the one below by a factor of 1.3–1.4. The overall range is about 0.09–20 MHz.

DCOx: Selects one of eight steps within each range. Each step increases the frequency by about 8%, giving a factor of 1.7 from bottom to top of the range. Thus the ranges overlap slightly.

Figure 5.9 shows the frequency of the DCO in the low part of its range as a function of RSELx and DCOx. There is roughly a constant *ratio* between values so a logarithmic scale would be needed to show the full range clearly.

TI does not provide much information about the innards of the DCO. It appears to be based on a type of *RC* oscillator whose frequency is programmed by a current, which is selected by RSEL. This feeds a ring counter whose period is adjusted with DCO. An external resistor R_{osc} can be connected in some devices to regulate the current instead of RSEL. This could improve the stability of the frequency in older devices but is less useful in newer ones. It could also be used to control the frequency by an external analog signal.

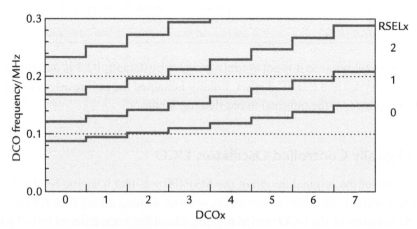

Figure 5.9: Frequency of digitally controlled oscillator in the F20xx. There are eight steps selected with DCOx in the lowest few ranges, which are chosen with RSELx.

Finer control of the *average* frequency is obtained by modulating the frequency of the oscillator between the selected value of DCO and the next step up (DCO + 1). This needs DCO < 7 of course. Each period of 32 clock cycles contains MOD cycles with the higher frequency given by (DCO + 1) and (32 − MOD) cycles with the lower frequency given by DCO. The average period over these 32 cycles is therefore

$$T_{\text{average}} = \frac{\text{MOD} \times T_{\text{DCO}+1} + (32 - \text{MOD}) \times T_{\text{DCO}}}{32}. \tag{5.1}$$

The frequencies are given by $f = 1/T$ so the average frequency is

$$f_{\text{average}} = \frac{32 f_{\text{DCO}} f_{\text{DCO}+1}}{\text{MOD} \times f_{\text{DCO}} + (32 - \text{MOD}) \times f_{\text{DCO}+1}}. \tag{5.2}$$

The DCO does not simply produce MOD pulses of one frequency followed by (32 − MOD) of the other, but mixes them thoroughly. For example, setting MOD = 16 gives an equal number of pulses of the two frequencies and they alternate: Each period given by DCO is followed by one given by (DCO + 1). This is shown on the oscilloscope in Figure 5.10. The top trace shows the clean square wave seen in a single sweep. There is a longer period (DCO = 0) in the center with shorter periods (DCO = 1) on either side. The lower trace shows repeated sweeps with persistence, as would be seen on an analog oscilloscope. The clock now appears to have jitter on the falling edges because of the two different periods, but the positive edges, which are used for triggering, all overlap. A truly periodic clock is found over the full period of 32 pulses (or fewer in special cases, such as that shown here). The values of RSEL, DCO, and MOD can be changed at any time to alter the frequency. Modulation can be turned off by setting MOD = 0 if a constant period is more important than an accurate, average frequency.

The modulator serves little purpose if the DCO runs freely without calibration. There is nearly a factor of 2 between minimum and maximum frequencies given in the data sheet for given values of RSEL and DCO, although this covers the full range of temperature and supply voltage. Therefore it is pointless to specify the frequency better than can be done with RSEL and DCO alone. A possible advantage of modulation is that there is less electromagnetic interference (EMI) from the clock, because the energy is spread over a wider range of frequencies, but this is not a large effect with periods that are only 8% apart. A method to reduce EMI further is described in the application note *Spread-Spectrum Clock Source Using an MSP430* (slaa291). Modulation is also used to generate accurate baud rates for asynchronous communication as we shall see in the section "Setting the Baud Rate with the USCI_A" on page 581.

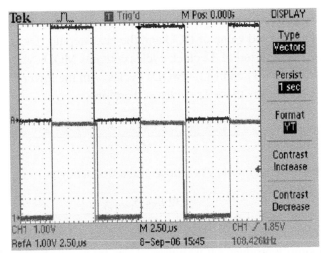

Figure 5.10: Oscilloscope traces of the clock from the DCO in a F2013 to show the effect of modulation. The basic settings are RSEL = 0 and DCO = 0, which give the lowest frequency of about 105 KHz. The modulator is set to MOD = 16, which gives alternating cycles of DCO = 1 with DCO = 0.

Calibration values for certain frequencies are programmed into the MSP430F2xx family and can be used where better accuracy is needed. The values are simply copied into the registers for the clock module:

```
    BCSCTL1 = CALBC1_1MHZ;   // Set range
    DCOCTL = CALDCO_1MHZ;    // Set DCO step and modulation
// May need to reconfigure other bits of BCSCTL1!
```

The write to BCSCTL1 affects the whole byte, not just the RSELx bits, and clears the other bits. It may therefore be necessary to reconfigure the other functions controlled by this register (XT2OFF, XTS, and DIVAx) but the defaults are usually satisfactory. The calibration is typically accurate within about 2% over the full range of operating conditions and to 0.2% at room temperature with a 3 V supply. This is still not as good as a crystal but is impressive compared with a "plain vanilla" *RC* oscillator, whose frequency might vary by $\pm 25\%$.

The F20xx has calibrated frequencies of 1, 8, 12, and 16 MHz. The values are stored in segment A of the information memory, which is locked by default against programming and erasing. You should not be able to overwrite these values by accident when downloading your program unless you override the settings in the debugger.

5.8.4 Control of the Clock Module through the Status Register

The clock module is controlled by 4 bits in the status register as well as its own peripheral registers. This is because of the intimate connection between clocks and low-power modes, which is discussed fully in the section "Low-Power Modes of Operation" on page 198. It is rarely necessary to alter these bits directly because there are intrinsic functions or predefined constants for each low-power mode.

All bits are clear in the full-power, active mode. This is the main effect of setting each bit in the MSP430F2xx:

- **CPUOFF** disables MCLK, which stops the CPU and any peripherals that use MCLK.

- **SCG1** disables SMCLK and peripherals that use it.

- **SCG0** disables the DC generator for the DCO (disables the FLL in the MSP430x4xx family).

- **OSCOFF** disables VLO and LFXT1.

It is not as straightforward as this because the effects of the different bits interact with each other. For example, setting only SCG0 and SCG1 does not stop the DCO if it supplies MCLK, because that would paralyze the processor. The DCO stops only if the source of MCLK is also switched to LFXT1 or VLO. This is illustrated by the code example `msp430x20x3_1_vlo`. Here are the relevant two lines. You might worry that the first line would stop the CPU but it does not; the DCO remains active while MCLK needs it.

```
bis.w   #SCG1+SCG0,SR            ; Stop DCO
bis.b   #SELM_3+DIVM_3,&BCSCTL2  ; MCLK = LFXT1/8
```

Detailed logical diagrams for the on–off controls are found in the family user's guides.

5.8.5 Oscillator Faults

A failure of a clock, MCLK in particular, is crippling. The clock module therefore detects and recovers from the most likely fault, a failure of an oscillator that relies on a vulnerable external crystal. Each oscillator has a flag that is raised to indicate a fault, which also sets the OFIFG bit in the interrupt flag register IFG1. This in turn requests a nonmaskable interrupt if it has been enabled. It also switches MCLK to the DCO if it was not already being used, which ensures that the CPU remains active. The user's software can then take appropriate action.

The flag LFXT1OF in BCSCTL3 is set if a fault is detected in LFXT1. A device with a VLO can switch to this instead of LFXT1 but otherwise there is no other source for ACLK. It might be possible to reconfigure peripherals to use SMCLK from the DCO rather than ACLK. Alternatively, the CPU can repeatedly poll LFXT1OF, which is cleared by the hardware if the oscillator recovers. The OFIFG flag is not cleared automatically; this must be done in software. The high-frequency oscillator XT2 has similar protection but a weakness of the MSP430x1xx family is that it cannot detect a failure of LFXT1 with a low-frequency crystal.

OFIFG is also set by a power-on reset, which ensures that MCLK is initially taken from the DCO. Again the flag must be cleared by the user. This should be done repeatedly in a loop because the flag will be set again immediately until the crystal oscillator has reached a stable state. This also means that unused oscillators must not be enabled, or they will appear to have a permanent fault and OFIFG will never clear.

TI code examples demonstrate various configurations of the clocks. For example, `msp430x20x3_lpm3_vlo` shows how to use the VLO instead of the default LFXT1 in the F20xx and `msp430x20x3_LFxtal_nmi` shows how to handle an oscillator fault. There are similar examples for other devices.

5.8.6   Frequency-Locked Loop, FLL+

The MSP430x4xx family has the more sophisticated FLL+ clock module. Much of this, such as LFXT1 and XT2, is similar to the MSP430F2xx but the registers and bits have different names. For example, the load capacitance for a low-frequency crystal is controlled by the XCAPxPF bits in the FLL_CTL0 register. There are no dividers for the internal clocks but the external signal from ACLK can be divided.

The main difference is of course the *frequency-locked loop*. This is hardware that aims to lock the frequency of the DCO to that of LFXT1. The DCO in the FLL+ has only five ranges but each covers a factor of about 10 in frequency and is divided into 29 taps. The modulator works in the same way as that in the BCM+. The name makes the FLL sound complicated but its basic mode of operation is simple. It relies on a feedback loop shown in Figure 5.11:

- The range of the DCO is set with the FN_x bits and modulation may be suppressed by setting SCFQ_M. Its output is at a frequency f_{DCO}.

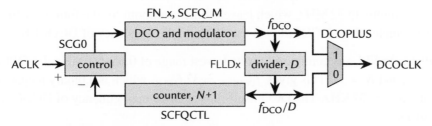

Figure 5.11: Simplified block diagram of the frequency-locked loop in the FLL+ module and the principal bits that configure it.

- This is divided by a factor D specified with the FLLDx bits. This gives a frequency of f_{DCO}/D.

- The divided signal is fed into a counter with a period of $(N+1)$, where N is stored in the lower 7 bits of SCFQCTL.

- The counter overflows at a frequency of $f_{DCO}/[D(N+1)]$, which is compared with the frequency of ACLK.

- The controller adjusts f_{DCO} one step up or down with the aim of bringing these frequencies together.

Thus the frequency of the DCO itself is given by $f_{DCO} = D(N+1)f_{ACLK}$ when the loop has locked but this is not necessarily the frequency of the output. DCOCLK can be taken either before or after the divider according to the setting of the DCOPLUS bit. When this bit is clear, the divided output is taken and

$$f_{DCOCLK} = (N+1)f_{ACLK} \qquad (\text{DCOPLUS} = 0). \qquad (5.3)$$

This does not depend on D. When DCOPLUS is set,

$$f_{DCOCLK} = D(N+1)f_{ACLK} \qquad (\text{DCOPLUS} = 1). \qquad (5.4)$$

The nomenclature is a little confusing because the divider apparently increases the frequency of the clock, which is the opposite of the usual case. Do not forget the $(N+1)$ in the multiplier—it is not just N.

There are 7 bits for N in SCFQCTL so the maximum value of $(N+1)$ is 128 and the maximum value of f_{DCOCLK} is 4 MHz (binary megahertz to be precise) if $f_{ACLK} = 32$ KHz and DCOPLUS = 0. This is well below the maximum frequency at which the CPU can run, which is 16 MHz in newer devices. DCOPLUS must be set for higher frequencies. It raises the maximum to 32 MHz, which leaves some headroom for the future. Listing 10.2 shows an example where DCOPLUS is used to raise the frequency of DCOCLK.

After a PUC, the FLL+ is configured for its lowest range of 0.65–6.1 MHz, DCOPLUS is clear, $D = 2$, and $N = 31$. This gives $f_{DCOCLK} = 32 f_{ACLK}$, which is a binary megahertz (2^{20} Hz) if $f_{ACLK} = 32$ KHz. The DCO itself runs at twice the frequency of DCOCLK because of D.

It takes some time for the FLL to lock and a software delay loop can be used to wait for this (and for the crystal to stabilize). The FLL starts at the bottom of its range after a PUC. It may have to reach the top of the range for the desired frequency, which requires up to $32 \times 28 \approx 900$ steps. This is the number of modulator steps times the number of usable taps—the highest tap is not useful because it cannot be modulated. Each step takes one cycle of ACLK, which corresponds to $(N + 1)$ or $D(N + 1)$ cycles of MCLK. This tells us the length of the delay loop needed. A simple loop takes three cycles of MCLK per iteration, as we found in Listing 4.12, so the stabilization loop needs about $300(N + 1)$ or $300D(N + 1)$ iterations. This is about $10,000 \approx 0x2700$ iterations using the default settings. You can therefore use the few lines of C that follow to check that the FLL has locked to the default frequency. I also configured the capacitors to suit the TI MSP430FG4618/F2013 Experimenter's Board.

```
volatile uint16_t i;                      // Loop counter to stabilize FLL+
FLL_CTL0 = XCAP14PF;                       // 14pF load caps; 1MHz default
do {                                       // Wait until FLL has locked
    for (i = 0x2700; i > 0;  --i) {   // One loop should be enough
    }                                      // Delay for FLL+ to lock
    IFG1_bit.OFIFG = 0;                    // Attempt to clear osc fault flag
} while (IFG1_bit.OFIFG != 0);             // Repeat if not yet clear
```

The program attempts to clear the oscillator fault flag OFIFG after the delay and checks that this was successful. If not, the delay loop is repeated. This is simple but the program will never leave the loop if OFIFG cannot be cleared. This could arise if ACLK fails or the FLL has been incorrectly configured so that the desired frequency lies outside the range of the DCO. The DCO error flag DCOF is set if the DCO's frequency tap is in either its bottom or its top position and this in turn sets OFIFG. DCOF clears when the tap is moved from its extremes.

An FLL is much simpler than its analog equivalent, the phase-locked loop (PLL), and is far quicker to come into lock. Having said that, there appears to be no way of telling whether the FLL has locked—only if it fails so badly that it moves to its top or bottom tap, which sets DCOF.

The FLL is disabled by setting the SCG0 bit in the status register, SR. This occurs when the processor is put into low-power modes LPM3 or LPM4. The DCO restarts from its previous configuration in a few microseconds when an interrupt wakes the processor again. It should still be close to its previously locked frequency unless the temperature or supply voltage has changed significantly while the device was asleep, because these have a large effect on f_{DCO}. However, the FLL is *not* automatically reenabled when the device wakes because the SCG0 bit is not cleared when an interrupt is accepted, unlike the rest of SR. The FLL must be enabled by clearing SCG0 from software if it is wanted during an ISR. This is unlikely in practice because interrupt service routines are usually kept short. Lengthy processing should be carried out in the main function and this is explained in the section "Returning from a Low-Power Mode to the Main Function" on page 203.

The FLL+ module can detect faults with the crystal oscillators in the same way as the MSP430F2xx but a failure of LFXT1 also affects the FLL. The FLL continues to try to lock to LFXT1 by lowering the frequency of the DCO until it hits the lower limit. This sets the flag DCOF to indicate an error. The DCO continues to function, however, producing the lowest frequency in its range, so MCLK remains operational.

A frequency-locked loop can be written in software to provide the same function as the FLL module in devices that lack the hardware. The principle is described in the section "Measurement of Frequency: Comparison of SMCLK and ACLK" on page 312. Either a crystal or an accurate external frequency, such as the mains at 50 or 60 Hz, can be used as a reference. See the application note *Controlling the DCO Frequency of the MSP430x11x* (slaa074).

Checklist—Have You ...

❏ Checked that V_{CC} is high enough to support your clock frequency?

❏ Activated the VLO if you want ACLK in an MSP430F2xx without a crystal?

❏ Selected the appropriate load capacitors for a watch crystal?

❏ Remembered that the frequency of the FLL depends on $N + 1$ in the multiplier, not just N? Also that the "divider" *increases* the clock frequency in this case.

❏ Allowed time for the crystal and FLL to stabilize if the frequency of the clock must be accurate?

CHAPTER 6

Functions, Interrupts, and Low-Power Modes

A well-structured program should be divided into separate modules—functions in C or subroutines in assembly language. There is nothing special about this in embedded systems. If you wish to write subroutines in assembly language, you need to know how arguments are passed, local variables allocated, and results returned. This is occasionally useful for efficiency but the main reason for studying these details is to assist the debugging of C programs. It is particularly important to understand the central role of the stack.

Interrupts are a major feature of most embedded software. They are vaguely like functions that are called by hardware rather than software. The distinction sounds trivial but it makes them much harder to handle because the processor must be able to return correctly to its activity before it was interrupted. The application note *MSP430 Software Coding Techniques* (slaa294) describes the overall structure of a typical, interrupt-driven program for the MSP430 and describes a range of techniques to ensure that programs are robust and can easily be debugged.

The final topic in this chapter is the range of low-power modes of operation. They are described here because the MSP430 needs an interrupt to wake it from a low-power mode. In fact we see that no extra effort is usually needed to handle low-power modes in interrupts: The MSP430 automatically goes to active mode when an interrupt is requested, services the interrupt, and resumes its low-power mode afterward.

6.1   Functions and Subroutines

It is good practice to break programs into short functions or subroutines, for reasons that are explained in any textbook on programming: It makes programs easier to write and more reliable to test and maintain. Functions are obviously useful for code that is called from more than one place but should be used much more widely, to encapsulate every distinct function. They hide the detailed implementation of an activity from the high-level, strategic level of the software. Functions can readily be reused and incorporated into libraries, provided that their documentation is clear.

I find that many students are reluctant to break programs into functions. Frankly I think that the real reason is laziness but "efficiency" is usually offered as an excuse. The cost of calling a function is tiny. Looking back at Listing 4.14, the statement `call #DelayTenths` takes five cycles and the `ret` requires a further two. This is usually negligible. There is a further overhead in setting up the arguments for a more complicated function, retrieving the return value, and saving registers, which is covered in the next few sections, but the time is still unlikely to be significant. Against this, functions make better use of RAM because local variables need exist only while the function is being called. The space can be released when the function has finished and is available for the rest of the program. In any case, the argument of efficiency is largely irrelevant when set against the advantages of testing and reliability. What use is an "efficient" program that does not work correctly?

There is not much else to say about writing functions in C other than a reminder about systematic documentation. Use prototypes to ensure that functions are called with appropriate arguments and put them in a header file so that they are consistent across a project. Subroutines in assembly language are much more tricky because arguments, return values, and local variables must be handled explicitly. We explore this next.

6.2   What Happens when a Subroutine Is Called?

The basic operation of subroutines in assembly language was explained in the section "Automatic Control: Use of Subroutines" on page 99. To recap briefly, the `call` instruction first causes the return address, which is the current value in the PC, to be pushed on to the stack. The address of the subroutine is then loaded into the PC and execution continues from there. At the end of the subroutine the `ret` instruction pops the return address off the stack into the PC so that execution resumes with the instruction following the call of the subroutine. The return operation is so straightforward that `ret` is emulated with a `mov` instruction.

Nothing else is done automatically when a subroutine is called. This means that the subroutine inherits the existing contents of the CPU registers, including the status register. The behavior is different for interrupts, as we see shortly. Therefore a subroutine and its calling routine must agree on whether the contents of these registers should be preserved or may be overwritten. This must be made clear in the documentation. For example, the comments at the start of the subroutine in Listing 4.14 describe the use of R4 and R12.

It is always wise to follow a convention for the use of registers and this becomes essential when assembly code is mixed with C. EW430 uses the following calling convention:

- The scratch registers R12 to R15 are used for parameter passing and hence are not normally preserved across the call.

- The other general-purpose registers, R4 to R11, are used mainly for register variables and temporary results and must be preserved across a call. This means that you must save the contents of any register that you wish to use and restore its original contents at the end.

The subroutine in Listing 4.14 infringes the second item because it uses R4 but does not preserve its value when the subroutine was called. I omitted this for simplicity as there was no value to save in that program.

If you are writing in assembly language, remember to initialize the stack pointer before calling any subroutines. It is easy to forget this before calling an `initialize` subroutine to configure the MCU. The other point that is only too easy to forget in assembly language is the # sign before the name of the subroutine, as in `call #DelayTenths`.

The stack grows each time a function calls another function within it; in other words, the functions are nested. There is therefore a danger of running out of space if the nesting becomes too deep. Simulators monitor the stack and offer a warning when space is running low. Another way of monitoring the stack is to fill RAM initially with a distinctive pattern, as was done in Listing 5.1. There is no problem when a second function is called after another has returned because the space on the stack used by the first function should have been released.

6.3   Storage for Local Variables

Most functions need local variables and there are three ways in which space for these can be allocated:

- CPU registers are simple and fast. One was used in the delay loop in Listing 5.1. The convention is that registers R4–R11 may be used and their values should be preserved. This is done by pushing their values on to the stack.

- A second approach is to use a fixed location in RAM, as in the delay loop in Listing 4.13. There are two serious disadvantages to this approach. The first is that the space in RAM is reserved permanently, even when the function is not being called, which is wasteful. Second, the function is not *reentrant*. This means that the function cannot call a second copy of itself, either directly or with nested functions between. Both copies would try to use the same variable and interfere with each other, unless the intention was for them to share the variable. This would obviously wreck a software delay loop, for instance.

- The third approach is to allocate variables on the stack and is generally used when a program has run out of CPU registers. This can be slow in older designs of processors, but the MSP430 can address variables on the stack with its general indexed and indirect modes. Of course it is still faster to use registers in the CPU when these are available.

Listing 6.1 shows a corrected version of the delay subroutine from Listing 4.14, which now saves and restores R4 before using it as a loop counter within the subroutine. The operation of the stack when the subroutine is called is illustrated in Figure 6.1.

Listing 6.1: Subroutine from `substk0.s43`, which now saves and restores R4 correctly.

```
; Subroutine to give delay of R12*0.1s
; Parameter is passed in R12 and destroyed
; R4 used for loop counter, stacked and restored
; -----------------------------------------------------------------
DelayTenths:
      push.w  R4                    ; Stack R4: will be overwritten
      jmp     LoopTest              ; Start with test in case R12 = 0
OuterLoop:
      mov.w   #DELAYLOOPS,R4        ; Initialize loop counter
DelayLoop:                          ; [clock cycles in brackets]
      dec.w   R4                    ; Decrement loop counter   [1]
      jnz     DelayLoop             ; Repeat loop if not zero [2]
      dec.w   R12                   ; Decrement number of 0.1s delays
LoopTest:
      cmp.w   #0,R12                ; Finished number of 0.1s delays?
      jnz     OuterLoop             ; No: go around delay loop again
      pop.w   R4                    ; Yes: restore R4 before returning
      ret                          ; Return to caller
```

Figure 6.1: **Operation of the stack in the MSP430F1121A for the subroutine in Listing 6.1. The stack is shown in words, all values are in hexadecimal, and the top of RAM is at 0x02FF in this device. Register R4 held 0xA5A5 on entry to the subroutine.**

Example 6.1

Try the program in Listing 6.1 in the simulator and watch the operation of the stack. Open windows for the registers, stack, and RAM (memory). Use the contextual menu to view the stack in 2x Units (words), which is clearer. The stack window clears to show that it is empty when SP has been initialized (there may be complaints before this that the stack pointer is outside the stack range). After the `call` instruction, the return address appears, located at the highest word in RAM. This is joined by the value in R4 after the `push.w R4` instruction and the return address is now shown at location +2, which is its address relative to the stack pointer. The window shows the top of the stack at the top of the list, which means that addresses increase down the window. Unfortunately this is the opposite of the convention used in TI's documents.

The contents of the stack can also be viewed at the highest addresses in the RAM window. At the end of the subroutine the saved value of R4 and the return address are popped in turn from the stack, but the RAM window shows that the values remain in memory. Only SP changes and no longer points to these locations.

The bar in the upper right corner of the Stack window shows how much of the space reserved for the stack is being used. It is barely visible in this program because the stack is so small.

The stack is used for local variables when the registers have been exhausted. This does not happen often in the MSP430, which has eight registers available by convention, so the example in Listing 6.2 is contrived. It is yet another software delay with three nested loops this time. The outer loop is the number of 0.1 s delays, which is passed as a parameter in R12 as before and this register is again used for its counter. Two loops are nested within this, both of which use variables on the stack to show its operation.

Listing 6.2: Subroutine from `substk1.s43`, whose delay loop uses two local variables on the stack.

```
; Subroutine to give delay of R12*0.1s
; Parameter is passed in R12 and destroyed
; Space for two loop counters is created on stack, after which
;   0(SP) is innermost (little) loop, 2(SP) is big loop counter
;-----------------------------------------------------------------
; Iterations of delay loop for about 0.1s (6 cycles/iteration):
BIGLOOPS      EQU      130
LITTLELOOPS   EQU      100
;-----------------------------------------------------------------
DelayTenths:
    sub.w   #4,SP                 ; Allocate 2 words (4 bytes) on stack
    jmp     LoopTest              ; Start with test in case R12 = 0
OuterLoop:
    mov.w   #BIGLOOPS,2(SP)       ; Initialize big loop counter
BigLoop:
    mov.w   #LITTLELOOPS,0(SP)    ; Initialize little loop counter
LittleLoop:                       ; [clock cycles in brackets]
    dec.w   0(SP)                 ; Decrement little loop counter [4]
    jnz     LittleLoop            ; Repeat loop if not zero         [2]
    dec.w   2(SP)                 ; Decrement big loop counter      [4]
    jnz     BigLoop               ; Repeat loop if not zero         [2]
    dec.w   R12                   ; Decrement number of 0.1s delays
LoopTest:
    cmp.w   #0,R12                ; Finished number of 0.1s delays?
    jnz     OuterLoop             ; No: go around delay loop again
    add.w   #4,SP                 ; Yes: finished, release space on stack
    ret                           ; Return to caller
```

Space for the variables is reserved with the instruction `sub.w #4,SP`. Remember that the stack grows downward, so this subtraction increases the size of the stack by 4 bytes or two words. I chose to use the word at the top of the stack for the innermost (little) loop counter. Its address is `0(SP)`. The second word, located at `2(SP)`, is used for the big loop counter. The allocation of the stack is shown in Figure 6.2. It is much clearer and safer to define symbols instead of the explicit addresses, indexed on SP, but I wanted to show precisely what is happening.

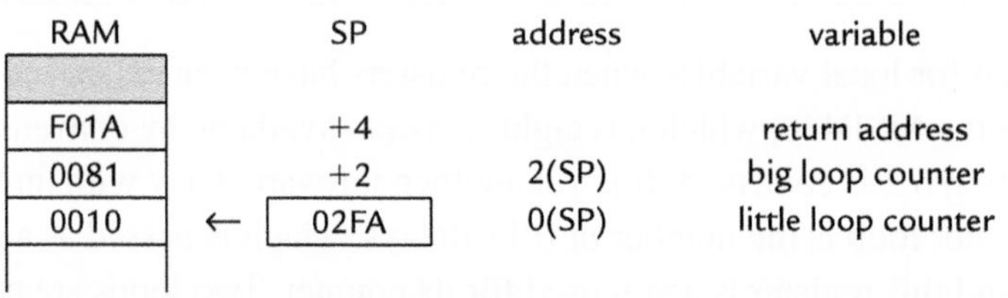

Figure 6.2: Stack in the MSP430F1121A for the subroutine in Listing 6.2, showing the return address and two local variables.

It is *painfully* easy to forget the final instruction in the subroutine, add.w #4, SP. This releases the space on the stack that was reserved for the variables and leaves SP pointing to the return address. It is an instant disaster if this is forgotten.

Example 6.2

Try the program in Listing 6.2 in the simulator and again watch the operation of the stack.

Example 6.3

What is the "instant disaster" if the space for the variables is not released before the subroutine returns?

Example 6.4

Look back at the first software delay loop in Listing 4.10, which was written in C. Check that there is no optimization (Project Options > C/C++ compiler > Optimizations tab) and simulate this program. You should see that the compiler allocates the LoopCtr variable on the stack and refers to it with address 0x0(SP). A CPU register is used instead if the optimization is increased.

6.4 Passing Parameters to a Subroutine and Returning a Result

The final issue concerning subroutines in assembly language is how to pass arguments to them and return values to the calling routine. Registers R12–R15 are reserved for this purpose by convention and are used first. Further arguments are pushed on to the stack, which is also used to return a large value, typically a structure (which must be very large if it does not fit into four registers). Listing 6.3 is the software delay loop yet again, but this time with the length of the delay passed on the stack rather than in a register. Space for this argument could have been created explicitly with sub.w #2, SP, followed by an assignment to this location, but push.w #5 does the same job more simply.

Listing 6.3: Subroutine and calling code from substk2.s43, where the stack is used for passing the parameter and for local variables.

```
InfLoop:                        ; Loop forever
    push.w   #5                 ; Push delay parameter on to stack
    call     #DelayTenths       ; Call subroutine: don't forget the #!
    incd.w   SP                 ; Release space used for parameter
    xor.b    #LED1,&P2OUT       ; Toggle LED
```

```
    jmp      InfLoop                  ; Back around infinite loop
; ------------------------------------------------------------------
; Subroutine to give delay of n*0.1s
; Parameter n is passed on stack
; Space for two loop counters created on stack. After this:
;    0(SP) is innermost (little) loop counter
;    2(SP) is big loop counter
;    4(SP) is return address
;    6(SP) is parameter n passed on stack
; ------------------------------------------------------------------
; Iterations of delay loop for about 0.1s (6 cycles/iteration):
BIGLOOPS     EQU      130
LITTLELOOPS  EQU      100
; ------------------------------------------------------------------
DelayTenths:
    sub.w    #4,SP                    ; Allocate 2 words (4 bytes) on stack
    jmp      LoopTest                 ; Start with test in case R12 = 0
OuterLoop:
    mov.w    #BIGLOOPS,2(SP)          ; Initialize big loop counter
BigLoop:
    mov.w    #LITTLELOOPS,0(SP)       ; Initialize little loop counter
                                      ; [clock cycles in brackets]
LittleLoop:
    dec.w    0(SP)                    ; Decrement little loop counter [4]
    jnz      LittleLoop               ; Repeat loop if not zero        [2]
    dec.w    2(SP)                    ; Decrement big loop counter     [4]
    jnz      BigLoop                  ; Repeat loop if not zero        [4]
    dec.w    6(SP)                    ; Decrement number of 0.1s delays
LoopTest:
    cmp.w    #0,6(SP)                 ; Finished number of 0.1s delays?
    jnz      OuterLoop                ; No: go around delay loop again
    add.w    #4,SP                    ; Yes: finished, release space on stack
    ret                              ; Return to caller
```

It is now very important to keep track of the order in which variables are stored on the stack. Figure 6.3 shows the structure, which is called the *stack frame* for the subroutine. The arguments are pushed *before* the subroutine is called and are therefore above the return address, while the local variables are below the return address as before (which means that they are closer to the "top" of the stack). A complete stack frame also includes

RAM	SP	address	variable
0005	+6	6(SP)	parameter passed
F01A	+4		return address
0081	+2	2(SP)	big loop counter
0010 ←	02FA	0(SP)	little loop counter

Figure 6.3: Stack in the MSP430F1121A for the subroutine in Listing 6.2, showing the parameter passed, return address and two local variables.

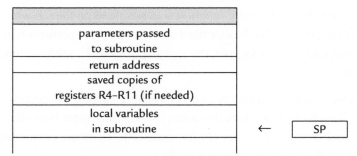

Figure 6.4: Complete stack frame for a subroutine with parameters passed, return address, saved copies of registers, and local variables.

copies of the registers R4–R11 to preserve their values for the calling routine if the subroutine wishes to use the registers itself. This is shown in Figure 6.4.

A critical feature is again the release of space on the stack after the subroutine has returned and the argument is no longer needed. This is done with incd.w SP, which is emulated by add.w #2, SP. It will not cause an immediate crash if this is omitted but the program will eventually fail.

Arguments are always passed to functions by value in C. You must pass the address of a variable in the calling function by using the & operator if you want a function to modify that variable. The same rules apply when subroutines are called in assembly language.

Example 6.5

Try the program in Listing 6.3 in the simulator and again watch the operation of the stack.

Example 6.6

How does the program fail if the space for the arguments is not released after the subroutine returns?

6.5   Mixing C and Assembly Language

Most programs are now written in C but it is occasionally worthwhile to write parts of the code in assembly language. This can be done in several ways, described in the reference guide for the EW430 C compiler and in the application note *Mixing C and Assembler with the MSP430* (slaa140). It is not complicated in principle but there are plenty of pitfalls.

Check first to see whether an intrinsic function is available to do the job without leaving C. Many of these are declared in the header file `intrinsics.h` to perform functions that are not possible in standard C. For example, the `__swap_bytes()` intrinsic function calls the `swpb` instruction.

Another possibility, when only a line or two of assembly language is needed, is to use inline assembly. This looks like a function `asm()` whose argument is the line(s) of assembly code, such as `asm("mov.b &P1IN,&dest")`. There are some dangers with this approach, set out in the compiler reference guide.

The third method is to write a complete subroutine in assembly language and call it from C. Obviously it is essential to get the calling convention correct. This is defined in the reference guides but there is an easier way. Write as much of the subroutine as possible in C and compile it. Copy the resulting assembly code and use it as the shell of your subroutine. This can be done from the disassembly window in the Kickstart version of EW430; you have to pay to get a real output file with the assembly code. Do not change the level of optimization afterward because this may change the code that calls the subroutine. An example will be given in "Conversion from Binary to BCD in Assembly Language" on page 272.

6.6 Interrupts

Interrupts were introduced in the section "Exceptions: Interrupts and Resets" on page 36. They are like functions but with the critical distinction that they are requested by hardware at unpredictable times rather than called by software in an orderly manner. (Well, a periodic interrupt should be highly predictable in real time, but this is not apparent to the CPU.) Interrupts are commonly used for a range of applications:

- Urgent tasks that must be executed promptly at higher priority than the main code. However, it is even faster to execute a task directly by hardware if this is possible.

- Infrequent tasks, such as handling slow input from humans. This saves the overhead of regular polling.

- Waking the CPU from sleep. This is particularly important in the MSP430, which typically spends much of its time in a low-power mode and can be awakened only by an interrupt.

- Calls to an operating system. These are often processed through a trap or software interrupt instruction but the MSP430 does not have one. A substitute is for software to set an unused interrupt flag for one of the peripherals, such as port P1 or P2.

The code to handle an interrupt is called an *interrupt handler* or *interrupt service routine* (ISR). It looks superficially like a function but there are a few crucial modifications. The feature that interrupts arise at unpredictable times means that an ISR must carry out its action and clean up thoroughly so that the main code can be resumed without error—it should not be able to tell that an interrupt occurred.

Interrupts can be requested by most peripheral modules and some in the core of the MCU, such as the clock generator. Each interrupt has a flag, which is raised (set) when the condition for the interrupt occurs. For example, Timer_A sets the TAIFG flag in the TACTL register when the counter TAR returns to 0. We polled this flag to pace the loop in the section "Automatic Control: Flashing a Light by Polling Timer_A" on page 105. Each flag has a corresponding enable bit, TAIE in this case. Setting this bit allows the module to request interrupts.

Most interrupts are *maskable*, which means that they are effective only if the general interrupt enable (GIE) bit is set in the status register (SR). They are ignored if GIE is clear. Therefore both the enable bit in the module and GIE must be set for interrupts to be generated. The (non)maskable interrupts cannot be suppressed by clearing GIE. The reason for the parentheses around (non) is that these interrupts also require bits to be set in special function or peripheral registers to enable them. Thus even the (non)maskable interrupts can be disabled and indeed all are disabled by default. I drop the parentheses around *non* for clarity.

The MSP430 uses *vectored* interrupts, which means that the address of each ISR—its vector—is stored in a *vector table* at a defined address in memory. In most cases each vector is associated with a unique interrupt but some sources share a vector. The ISR itself must locate the source of interrupts that share vectors. For example, TAIFG shares a vector with the capture/compare interrupts for all channels of Timer_A other than 0. Channel 0 has its own interrupt flag TACCR0 CCIFG and separate vector.

Each interrupt vector has a distinct priority, which is used to select which vector is taken if more than one interrupt is active when the vector is fetched. The priorities are fixed in hardware and cannot be changed by the user. They are given simply by the address of the vector: A higher address means a higher priority. The reset vector has address 0xFFFE, which gives it the top priority, followed by 0xFFFC for the single nonmaskable interrupt vector. The vectors for the maskable interrupts depend on the peripherals in a particular device and are listed in a table of *Interrupt Vector Addresses* in the data sheet. Taking the

F2013 as an example, the vector for TACCR0 CCIFG has an address 0xFFF2 and therefore has a higher priority than the shared vector for TAIFG and TACCR1 CCIFG, whose address is 0xFFF0. Five other vectors are used in this device to give a total of 9, although space for up to 32 is reserved (16 in older devices).

Interrupts must be handled in such a way that the code that was interrupted can be resumed without error. This means in particular that the values in the CPU registers must be restored. The hardware can take two extreme approaches to this:

- Copies of all the registers are saved on the stack automatically as part of the process for entering an interrupt. This is done in the Freescale HCS08, for example, which is a CISC and has only a few registers. The disadvantage is the time required, which means that the response to an interrupt is delayed. An alternative is to switch to a second set of registers, which is done in the Z80 and descendants.

- The opposite approach is for the hardware to save only the absolute minimum, which is the return address in the PC as in a subroutine. This is much faster but it is up to the user to save and restore values of the critical registers, notably the status register. The Microchip PIC16 takes this approach, consistent with its minimalist philosophy.

The MSP430 is close to the second extreme but stacks both the return address and the status register. The SR gets this privileged treatment because it controls the low-power modes and the MCU must return to full power while it processes the interrupt. This is explored further in the section "Low-Power Modes of Operation" on page 198. The other registers must be saved on the stack and restored if their contents are modified in the ISR. Instructions have been added to the MSP430X to push and pop multiple registers, which makes this process faster.

6.7   What Happens when an Interrupt Is Requested?

A lengthy chain of operations lies between the cause of a maskable interrupt and the start of its ISR. It starts when a flag bit is set in the module when the condition for an interrupt occurs. For example, TAIFG is set when the counter TAR returns to 0. This is passed to the logic that controls interrupts if the corresponding enable bit is also set, TAIE in this case. The request for an interrupt is finally passed to the CPU if the GIE bit is set. Hardware then performs the following steps to launch the ISR:

1. Any currently executing instruction is completed if the CPU was active when the interrupt was requested. MCLK is started if the CPU was off.

2. The PC, which points to the next instruction, is pushed onto the stack.

3. The SR is pushed onto the stack.

4. The interrupt with the highest priority is selected if multiple interrupts are waiting for service.

5. The interrupt request flag is cleared automatically for vectors that have a single source. Flags remain set for servicing by software if the vector has multiple sources, which applies to the example of TAIFG.

6. The SR is cleared, which has two effects. First, further maskable interrupts are disabled because the GIE bit is cleared; nonmaskable interrupts remain active. Second, it terminates any low-power mode, as explained in the section "Low-Power Modes of Operation" on page 198. (The SCG0 bit is not cleared in the MSP430x4xx family, which means that the frequency-locked loop is not automatically reactivated; see "Frequency-Locked Loop, FLL+" on page 172.)

7. The interrupt vector is loaded into the PC and the CPU starts to execute the interrupt service routine at that address.

This sequence takes six clock cycles in the MSP430 before the ISR commences. The stack at this point is shown in Figure 6.5. The position of SR on the stack is important if the low-power mode of operation needs to be changed.

The delay between an interrupt being requested and the start of the ISR is called the *latency*. If the CPU is already running it is given by the time to execute the current instruction, which might only just have started when the interrupt was requested, plus the six cycles needed to execute the launch sequence. This should be calculated for the slowest instruction to get the worst case. Format I instructions take up to 6 clock cycles so the overall latency is 12 cycles. The time required to start MCLK replaces the duration of the

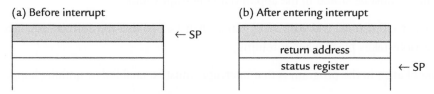

Figure 6.5: Stack before and after entering an interrupt service routine. The return address (PC) and status register (SR) have been saved, with SR on the top of the stack.

current instruction if the device was in a low-power mode. The delay varies on each occasion because the interrupt may be requested at different points during an instruction, whose length may also differ. Thus there is no fixed interval between the request of an interrupt and the start of its ISR. Use the hardware of a timer to read an input or change an output at a precise time. Figure 6.6 shows an example of this and there are many more in Chapter 8.

An interrupt service routine must always finish with the special *return from interrupt* instruction `reti`, which has the following actions:

1. The SR pops from the stack. All previous settings of GIE and the mode control bits are now in effect, regardless of the settings used during the interrupt service routine. In particular, this reenables maskable interrupts and restores the previous low-power mode of operation if there was one.

2. The PC pops from the stack and execution resumes at the point where it was interrupted. Alternatively, the CPU stops and the device reverts to its low-power mode before the interrupt.

This takes a further five cycles in the MSP430. The stack is restored to its state before the interrupt was accepted.

6.8   Interrupt Service Routines

The framework of an interrupt service routine is more straightforward in assembly language so I'll describe this before explaining how they are implemented in C.

6.8.1   Interrupt Service Routines in Assembly Language

An ISR looks almost identical to a subroutine but with two distinctions:

- The address of the subroutine, for which we can use its name (a label on its first line), must be stored in the appropriate interrupt vector.

- The routine must end with `reti` rather than `ret` so that the correct sequence of actions takes place when it returns.

The other change in the program is that interrupts must be enabled or nothing happens.

I use the same old example, to toggle the LEDs on the Olimex 1121STK. The program is shown in Listing 6.4, adapted from the C program in Listing 4.17. The timer runs in Up mode with a period set by the value in TACCR0.

We have a choice of two interrupts. The programs in Chapter 4 polled the TAIFG flag, which is set when the counter TAR returns to 0. This flag can also generate an interrupt but it has the minor inconvenience that its vector is shared with other sources and is therefore slightly trickier to handle. It is more straightforward to use the TACCR0 CCIFG flag, which is set when TAR counts up to the value in TACCR0, because it has its own interrupt vector. This interrupt is enabled by setting the CCIE bit in the control register for channel 0, TACCTL0. It is safer to do this before starting the timer. Timer_A now requests interrupts but these are not accepted unless the GIE bit in the status register is also set. The main routine has nothing more to do after this. It can sit back and let the interrupts do all the work. Here I left the CPU in an infinite, empty loop but it would be better to put it into a low-power mode, which is done in the section "Low-Power Modes of Operation" on page 198.

Listing 6.4: Program `timrint1.s43` in assembly language to toggle LEDs using interrupts generated by timer_A in up mode.

```
; timrint1.s43 - toggles LEDs with period of about 1s
; TACCR0 interrupts from timer A with period of about 0.5s
; Timer clock is SMCLK divided by 8, up mode, period 50000
; Olimex 1121STK, LED1,2 active low on P2.3,4
; J H Davies, 2006-09-20; IAR Kickstart version 3.41A
; ---------------------------------------------------------------------
#include  <msp430x11x1.h>              ; Header file for this device
; ---------------------------------------------------------------------
; Pins for LED on port 2
LED1        EQU     BIT3
LED2        EQU     BIT4
; ---------------------------------------------------------------------
        RSEG    CSTACK                  ; Create stack (in RAM)
; ---------------------------------------------------------------------
        RSEG    CODE                    ; Program goes in code memory
Reset:                                  ; Execution starts here
    mov.w   #SFE(CSTACK),SP             ; Initialize stack pointer
main:                                   ; Equivalent to start of main() in C
    mov.w   #WDTPW|WDTHOLD,&WDTCTL      ; Stop watchdog timer
    mov.b   #LED2,&P2OUT                ; Preload LED1 on, LED2 off
    bis.b   #LED1|LED2,&P2DIR           ; Set pins with LED1,2 to output

    mov.w   #49999,&TACCR0              ; Period for up mode
    mov.w   #CCIE,&TACCTL0              ; Enable interrupts on Compare 0
    mov.w   #MC_1|ID_3|TASSEL_2|TACLR,&TACTL    ; Set up Timer A
; Up mode, divide clock by 8, clock from SMCLK, clear TAR
    bis.w   #GIE,SR                     ; Enable interrupts (just TACCR0)
    jmp     $                           ; Loop forever; interrupts do all
; ---------------------------------------------------------------------
; Interrupt service routine for TACCR0, called when TAR = TACCR0
; No need to acknowledge interrupt explicitly - done automatically
TA0_ISR:                                ; ISR for TACCR0 CCIFG
    xor.b   #LED1|LED2,&P2OUT           ; Toggle LEDs
```

```
        reti                                ; That's all: return from interrupt
;------------------------------------------------------------------------
        COMMON   INTVEC                     ; Segment for vectors (in Flash)
        ORG      TIMERA0_VECTOR
        DW       TA0_ISR                    ; ISR for TA0 interrupt
        ORG      RESET_VECTOR
        DW       Reset                      ; Address to start execution
        END
```

The interrupt service routine TA0_ISR looks much like a subroutine and starts with the usual label to give it a name. There is no significance to the name—I chose something obvious but you can call it anything that you want. In many processors the first task is to "acknowledge" the interrupt, usually by clearing the associated flag in the peripheral register. If this is not done, the interrupt is reasserted continually and the program repeats the ISR until an interrupt of higher priority is called. This step is not needed for most interrupts in the MSP430 because the flag is cleared automatically while the ISR is launched. Although this is fast and convenient it can sometimes be a nuisance when debugging because there is no pause between interrupts.

The only task in this ISR is to toggle the LEDs, which is done in the usual way, before the routine returns with reti. No registers are used so there is no need to save and restore them.

The final part of the program is to associate the name of the ISR, TA0_ISR here, with the vector for the TACCR0 CCIFG interrupt. This means that the address of the routine—its vector—is stored in the corresponding entry of the vector table. The absolute addresses are listed in the data sheet and can be used with an ORG directive to locate the vector in absolute assembly:

```
        ORG  0xFFF2                         ; Address of TACCR0 CCIFG vector in table
        DW   TA0_ISR                        ; Vector: address of TACCR0 CCIFG ISR
```

The absolute addresses are *not* defined in the header files. As usual I prefer to use relocatable assembly. The vectors are stored in the segment INTVEC (what a surprise) and the addresses *within this segment* are defined in the header file, as used in the preceding listing. The reset vector can be defined either in the same way as the interrupt vectors, which I do here, or in a segment of its own, as done in Listing 4.4. The effect is the same. I introduce the INTVEC segment with COMMON rather than RSEG so that more than one file can store interrupt vectors in the segment. This is unnecessary in a project with only a single file but there is no harm in doing things properly.

Example 6.7

Try this in the simulator. Interrupts introduce several new aspects. The first is that EW430 simulates only the core of the MSP430 and not the peripherals. A full chip simulator would also model the timer and gives interrupts automatically after the correct number of clock cycles. Here we must mimic the interrupts by hand. Open the window Simulator > Forced Interrupts, which brings up a list of possible interrupts for the selected device, and select TIMERA0_VECTOR.

Step through the code in the usual way. It is a good idea to open the CPU Registers window and expand SR. You see the GIE bit become set immediately before the loop. This makes interrupts possible but nothing happens if the program is stepped further; it remains stuck in the loop and only the cycle counter changes.

Now click the Trigger button in the Forced Interrupt window and step again. The simulator jumps to the first instruction in the ISR, which toggles the LEDs. Control returns to the infinite loop after the `reti` instruction and remains there until the next interrupt is triggered. You can also see the stack grow by two words on entry to the ISR and empty again after the `reti`.

Example 6.8

Try the emulator next. There is no need for forced interrupts and the window is not displayed. Step through the code to the infinite loop again. Note that the GIE button in the Debug toolbar becomes "pressed" when the bit is set. The counter TAR in Timer_A increases every time a single step is taken. I found that it rose by about 49 (decimal) each step, which is due to the number of clock cycles needed to read the changed registers through JTAG to the debugger. The debugger halts MCLK and SMCLK on this device (F1121A) between steps.

It is clearly going to take a long time to reach the value of 49,999 in TACCR0 that triggers the interrupt. Use Edit > Toggle Breakpoint, to put a breakpoint on the first instruction in `TA0_ISR ()`, and let the program Go. It will stop very shortly on the breakpoint. The GIE button is now "released" to show that the bit is clear, disabling further maskable interrupts. Choosing Go again toggles the LEDs and the program runs for a further 0.5 s until the next interrupt. The GIE button can be used to clear the bit and disable interrupts, which stops further toggling of the LEDs.

6.8.2   Interrupt Service Routines in C

An interrupt service routine cannot be written entirely in standard C because there is no way to identify it as an ISR rather than an ordinary function. In fact it would appear that the function was dead code, meaning that it could never be called, so the compiler would optimize it away. Some extensions are therefore needed and these inevitably differ between compilers. The usage for EW430 is given in Listing 6.5. Two additions are needed to the ISR. First is the #pragma line, which associates the function with a particular interrupt vector. The second is the _ _interrupt keyword at the beginning of the line that names the function. This ensures that the address of the function is stored in the vector and that the function ends with reti rather than ret. Again there is no significance to the name of the function; it is the name of the vector that matters.

An intrinsic function is needed to set the GIE bit and turn on interrupts. This is called _ _enable_interrupt(), singular rather than plural. It is declared in intrinsics.h, which must be included.

Listing 6.5: Program timintC1.c to toggle LEDs using interrupts generated by channel 0 of Timer_A in up mode.

```
// timintC1.c - toggles LEDs with period of about 1.0s
// Toggle LEDs in ISR using interrupts from timer A CCR0
//   in Up mode with period of about 0.5s
// Timer clock is SMCLK divided by 8, up mode, period 50000
// Olimex 1121STK, LED1,2 active low on P2.3,4
// J H Davies, 2006-10-11; IAR Kickstart version 3.41A
// -------------------------------------------------------------------
#include <io430x11x1.h>                  // Specific device
#include <intrinsics.h>                  // Intrinsic functions
// -------------------------------------------------------------------
// Pins for LEDs
#define  LED1     BIT3
#define  LED2     BIT4
// -------------------------------------------------------------------
void main (void)
{
    WDTCTL = WDTPW|WDTHOLD;              // Stop watchdog timer
    P2OUT = ~LED1;                       // Preload LED1 on, LED2 off
    P2DIR = LED1|LED2;                   // Set pins with LED1,2 to output
    TACCR0 = 49999;                      // Upper limit of count for TAR
    TACCTL0 = CCIE;                      // Enable interrupts on Compare 0
    TACTL = MC_1|ID_3|TASSEL_2|TACLR;    // Set up and start Timer A
// "Up to CCR0" mode, divide clock by 8, clock from SMCLK, clear timer
    __enable_interrupt();                // Enable interrupts (intrinsic)
    for (;;) {                           // Loop forever doing nothing
    }                                    // Interrupts do the work
}
// -------------------------------------------------------------------
// Interrupt service routine for Timer A channel 0
```

```
//------------------------------------------------------------
#pragma vector = TIMERA0_VECTOR
__interrupt void TA0_ISR (void)
{
        P2OUT ^= LED1|LED2;              // Toggle LEDs
}
```

Example 6.9

This program can be simulated and emulated in much the same way as the version in assembly language. About the only difference is that the stack has an extra, permanent entry, which is because the main() function is called from a "shell" in case the user's program should exit.

Additional work is needed to handle interrupts that have multiple sources. For example, the counter TAR of Timer_A and all its channels except 0 share a single interrupt vector. Similarly, all 8 bits of each input/output port share a common vector. The ISR must first determine the source of its interrupt in these cases. Timer_A has a register TAIV to identify the origin rapidly. This must be read to acknowledge the interrupt, even if only one source has been enabled and there is no other need to read TAIV. Alternatively, the flag associated with the specific source can be cleared directly, as explained in "Interrupts from Timer_A" on page 296. There seem to be "gotchas" like this associated with interrupts on all processors.

The status register is cleared during the sequence of actions at the entry to an ISR. This includes GIE = 0, which disables further maskable interrupts. This prevents nested interrupts, where one ISR is interrupted by another. The problem with nested interrupts is that the stack grows by at least two words for each ISR. This can cause the stack to run out of memory—overflow—if there is a continuous stream of interrupts, probably because something is out of control. Nested interrupts are for experts only. On the other hand, clearing GIE does not affect nonmaskable interrupts, so it is possible for one of these to arise during an ISR.

6.8.3   Nonmaskable Interrupts

There are a few small differences in the handling of nonmaskable interrupts compared with the maskable type. All the sources share a single vector, which has the highest address and therefore the highest priority except for the reset vector. Three modules can request a nonmaskable interrupt:

- Oscillator fault, OFIFG. This was described in the section "Oscillator Faults" on page 171. Remember that this flag is set by a power-up clear as a warning that the oscillator may not yet have stabilized. This interrupt should therefore not be enabled until the oscillator is running correctly.

- Access violation to flash memory, ACCVIFG.

- An active edge on the external $\overline{\text{RST}}$/NMI pin if it has been configured for interrupts rather than reset.

The function of the $\overline{\text{RST}}$/NMI pin is configured in the control register for the watchdog timer module, WDTCTL. By default it is an active low, reset input $\overline{\text{RST}}$. This can be switched to an NMI interrupt input with the WDTNMI bit. It then generates an interrupt if it receives either a positive or negative edge, depending on the value of the WDTNMIES edge select bit. The interrupt is enabled by the NMIIE bit in the special function register IE1 and the corresponding flag is NMIFG in IFG1. An external pull-up or pull-down may be required. The $\overline{\text{RST}}$/NMI pin is also used for the two-wire Spy-Bi-Wire JTAG interface in the MSP430F2xx, which may limit its usefulness for external interrupts. (External *maskable* interrupts can be generated by ports P1 and P2.)

The shared vector means that the ISR must first identify the source of the interrupt. This is explained clearly in the family user's guides.

One of the steps in starting a maskable ISR is to clear the status register, which disables other maskable interrupts. This has no effect on nonmaskable interrupts so the hardware clears all the enable flags in the modules to prevent further interrupts during the ISR. The desired nonmaskable interrupts must therefore be reenabled by the ISR as its last instruction. One of TI's standard code examples shows how to handle NMI.

6.9   Issues Associated with Interrupts

Interrupts are fun to program, if occasionally frustrating to debug. They can also give rise to problems that are very hard to locate. It is a good idea to follow a few simple rules to avoid some of these difficulties.

Keep Interrupt Service Routines Short

Other interrupts, which may be urgent, are disabled during an ISR unless you choose to allow nested interrupts. If lengthy processing is necessary, do the minimum in the ISR and send a message to the main function, perhaps by setting a flag. The main function can then do the time-consuming work between interrupts. Alternatively, use the ISR simply to wake the processor and hand control immediately to the main function. This is explained later in "Returning from a Low-Power Mode to the Main Function" on page 203.

Configure Interrupts Carefully

Make sure that unwanted interrupts are disabled. This should be the case by default and a problem is likely to arise only if an interrupt has been used at one time but not required later.

Make sure that the module is fully configured before interrupts are enabled. Sometimes the flag can become set before interrupts are wanted so it should be cleared before the enable bit is set. Do not set the GIE bit until you are ready to handle interrupts.

Define All Interrupt Vectors

It is good practice for safety to fill *all* the defined entries of the vector table, even for interrupts that are not used. A simple approach for assembly language is to define a single ISR for unused interrupts that traps the processor in an infinite loop. The address of this routine can then be stored in all the unused interrupt vectors. Use the debugger to trace back if one of these interrupts arises. It is a little clumsier in C because a separate function must be provided for each interrupt, but the body can again be an empty, infinite loop.

An infinite loop is helpful for debugging but not for a production version of a program. It would probably be better to force the device to reset itself, perhaps by writing an illegal value to the watchdog control register WDTCTL. Nagy [4] has further advice.

The Shared Data Problem

This is one of the classic issues with interrupts and multitasking systems and is explained thoroughly by Simon [25]. It arises when variables are used both in the main function and in ISRs. Here is a trivial example from a real-time clock. Suppose that a variable `MinsOfDay` is updated once per minute in an ISR. The main function drives separate displays for the minutes and hours, with a function for each:

```
DisplayMinutes(MinsOfDay);
DisplayHours(MinsOfDay);
```

(I assume that `MinsOfDay` has been declared `volatile` or there is even worse trouble.) The shared data problem arises if the ISR updates `MinsOfDay` between the two function calls. Suppose that the time was 01:59 when `DisplayMinutes` was called but has been updated to 02:00 by the time that `DisplayHours` is called. The displayed time will be 02:59. Oops!

Example 6.10

What would happen if the functions were called in the reverse sequence? What if they were called around midnight, assuming a 24-hour clock?

This is typical of a shared data problem. It does not occur often, even in theory. In practice it is seen only at inconvenient times, usually during a demonstration (such as when your project is being assessed in a university). There are various solutions. In this trivial case, we could take a local copy of `MinsOfDay` and use the copy as the argument to both functions. More generally it may be necessary to disable interrupts during critical sections that are susceptible to this problem.

Checklist—Have You ...

❏ Enabled interrupts both in their module and generally (GIE bit)?

❏ Provided an interrupt service routine for all enabled interrupts?

❏ Included code to acknowledge interrupts that share a vector (such as TAIFG or the input ports) even if only one source is active?

❏ Ended interrupt service routines in assembly language with `reti` and initialized the stack?

6.10 Low-Power Modes of Operation

The MSP430 was designed from the outset for low power and this is reflected in a range of low-power modes of operation. There are five in all but two are rarely employed in current devices (they were useful with earlier versions of the DCO). The most important modes are summarized here with typical currents I for the F2013 taken from its data sheet. These are for $V_{CC} = 3$ V, the DCO running at 1 MHz, and LFXT1 at 32 KHz from a crystal. The current in active mode rises with frequency to about 4 mA at 16 MHz, although this requires a higher supply voltage.

Active mode: CPU, all clocks, and enabled modules are active, $I \approx 300\,\mu\text{A}$. The MSP430 starts up in this mode, which must be used when the CPU is required. An interrupt automatically switches the device to active mode. The current can be reduced by running the MSP430 at the lowest supply voltage consistent with the frequency of MCLK; V_{CC} can be lowered to 1.8 V for $f_{DCO} = 1$ MHz, giving $I \approx 200\,\mu A$.

LPM0: CPU and MCLK are disabled, SMCLK and ACLK remain active, $I \approx 85\,\mu A$. This is used when the CPU is not required but some modules require a fast clock from SMCLK and the DCO.

LPM3: CPU, MCLK, SMCLK, and DCO are disabled; only ACLK remains active; $I \approx 1\,\mu A$. This is the standard low-power mode when the device must wake itself at regular intervals and therefore needs a (slow) clock. It is also required if the MSP430 must maintain a real-time clock. The current can be reduced to about 0.5 μA by using the VLO instead of an external crystal in a MSP430F2xx if f_{ACLK} need not be accurate.

LPM4: CPU and all clocks are disabled, $I \approx 0.1\,\mu A$. The device can be wakened only by an external signal. This is also called *RAM retention mode*.

It is common to describe a device as *sleeping* when it is in a low-power mode, although this is a little vague in the MSP430 with its range of modes. Similarly, *waking* means that the device returns to active mode. Many processors are put into low-power modes by issuing special `sleep` or `stop` instructions but these are not required with the MSP430. Instead its operation is controlled through the four bits: SCG0, SCG1, CPUOFF, and OSCOFF in the status register. All these are clear in active mode and particular combinations are set for each low-power mode. For example, setting only CPUOFF disables the CPU and MCLK, putting the MSP430 into LPM0.

The location in SR is vital because it means that any low-power mode is suspended automatically when an interrupt is accepted. Recall from the section "What Happens when an Interrupt Is Requested?" on page 188 that the current SR is stacked and the SR is cleared for the interrupt service routine, including the low-power bits (except SCG0 in some families). The low-power mode is resumed at the end of the ISR when SR is restored from the stack. All this is automatic.

The usual instructions in assembly language can be used to modify the bits that control the low-power modes. Their values for each operating mode are given in the user's guides but rarely is there any need to use these because symbols such as `LPM0` are defined in the header file for each mode. Thus the single instruction

```
bis.w    #LPM3,SR               ; enter LPM3 but can we wake again?
```

puts the device into LPM3. There is one catch, however. The MSP430 can be awakened only by an interrupt so these must be enabled by setting the GIE bit in SR if it has not been done already. The following line is therefore much safer:

```
    bis.w    #GIE|LPM3,SR          ; enter LPM3 with interrupts enabled
```

This cannot be done in standard C, of course, and an intrinsic function is required. The function that follows is taken from `intrinsics.h` and sets both the low-power and GIE bits:

```
    _low_power_mode_3();          // enter LPM3 with interrupts enabled
```

This simply calls another intrinsic function, `__bis_SR_register()`, with the same set of bits as the assembly code. Different versions of the intrinsic functions are found in EW430 and another possibility is `_BIS_SR(GIE|LPM3_bits)`, where GIE must be included. The names inevitably differ in other development environments.

6.10.1   Waking from a Low-Power Mode

An interrupt is needed to awaken the MSP430. The processor handles an interrupt from a low-power mode in almost the same way as in active mode. The only difference is that MCLK must first be started so that the CPU can handle the interrupt; this replaces the first step when the CPU is active, which is to complete the current instruction. MCLK is started automatically by the hardware for servicing interrupts and requires no intervention from the programmer. Remember that the status register is cleared when an interrupt is accepted, which puts the processor into active mode. Similarly, MCLK is automatically turned off at the end of the ISR if the MSP430 returns to a low-power mode when the status register is restored.

Thus interrupts from low-power modes are written in *exactly the same way* as those from active mode. This has the attractive feature that low-power modes fit naturally into the structure of many programs. In simple cases the main function configures the peripherals, enables interrupts, puts the MSP430 into a low-power mode, and plays no further role. Interrupts wake the device so that it can perform each task, after which it returns to its low-power mode until the next interrupt. We shall see many examples of this.

A minor drawback is that it is a little harder to put the processor into a low-power mode with one instruction and have it execute the next instruction after awakening. I explain how to do this in the section "Returning from a Low-Power Mode to the Main Function" on page 203.

Listing 6.6 shows the tiny changes to Listing 6.5 needed to put the device into low-power mode 0 between interrupts. The power cannot be reduced further in this program because the timer runs from SMCLK. The only difference from before is that the infinite loop in the main function now contains _ _low_power_mode_0() instead of being empty. In principle this could have been executed just once before the loop, which could have been left empty. Putting it in the loop sends the processor back into a low-power mode if it ever returns to the main function, although something fairly drastic must have gone wrong for this to happen.

Listing 6.6: Part of program `timintC2.c` to toggle LEDs using interrupts from Timer_A. The device enters low-power mode 0 between interrupts.

```
        __enable_interrupt();                   // Enable interrupts (intrinsic)
        for (;;) {                              // Loop forever doing nothing
                __low_power_mode_0();           // Enter low power mode LPM0
        }                                       // Interrupts do the work
}
//-------------------------------------------------------------------------
// Interrupt service routine for Timer A channel 0
// Processor returns to LPM0 automatically after ISR
//-------------------------------------------------------------------------
#pragma vector = TIMERA0_VECTOR
__interrupt void TA0_ISR (void)
{
        P2OUT ^= LED1|LED2;                     // Toggle LEDs
}
```

The line with _ _enable_interrupt() is unnecessary because interrupts are enabled when the MSP430 is put into the low-power mode with _ _low_power_mode_0().

The ISR is identical to the previous version, as pointed out earlier: *No changes are needed* to handle the wake-up from low power on entry and the return to low power at the end of the ISR. This is a great advantage of the way in which the MSP430 controls its mode of operation and makes it very easy to use the low-power modes. The DCO continues to run and provide SMCLK in LPM0, which means that restarting MCLK is simply a matter of enabling a gate; there is no need to wait for an oscillator to start in this case.

Example 6.11

Try this listing in the simulator. It is very similar to the version with an infinite loop except that the highlighted line in the source window goes dim to show the low-power mode. Interrupts can be forced in the same way as before.

The best way of exploring the operation with the emulator is to put a breakpoint in the ISR. If you single-step out of the ISR, you will see the line with _ _low_power_mode_0() highlighted dimly to show that the MSP430 is in a low-power mode.

The program in assembly language is again almost identical to Listing 6.4. The empty loop jmp $ is replaced by the three lines in Listing 6.7.

Listing 6.7: A few lines from program `timrint2.s43` that toggles LEDs using interrupts from Timer_A. The device enters low-power mode 0 between interrupts.

```
InfLoop:
    bis.w    #LPM0|GIE,SR        ; Enable ints and low power mode 0
    jmp      InfLoop             ; Infinite loop (should never happen)
```

We could save more power by using low-power mode 3 rather than mode 0. Two small changes are needed. The TASSEL bits in TACTL must be adjusted so that the timer runs from ACLK rather than SMCLK. Then the MCU can be put into LPM3 instead of LPM0 between interrupts. No changes are needed to the ISR itself.

Example 6.12

Try LPM3 instead of LPM0 using either C or assembly language. You could again set a breakpoint in the ISR and try single-stepping but this time you will never reach the line in the main function with _ _low_power_mode_3(). This is because ACLK runs continuously, unless you have a recent device with clock control and change this behavior, so there is always an interrupt waiting for service.

The only clock that runs in LPM3 is ACLK. This means that the DCO must be started to provide MCLK before the ISR can be executed. Obviously there is a greater delay than in starting from LPM0, when DCO is already running, but it was explained in the section "Digitally Controlled Oscillator, DCO" on page 167 that the MSP430 was designed for this sort of activity and DCO starts rapidly. This is shown by the data from an oscilloscope plotted in Figure 6.6 for the F2013. The top trace is the OUT0 signal taken directly from Timer_A. Its edge triggers the interrupt. Below this is SMCLK, which is taken from the DCO with the maximum calibrated frequency of 16 MHz. The bottom trace is the output on P1.0, which is toggled in the ISR.

There is a delay of 0.36 μs between the OUT0 signal from the timer and the first edge of SMCLK, which is the time needed to start the DCO. Sixteen clock cycles occur before the

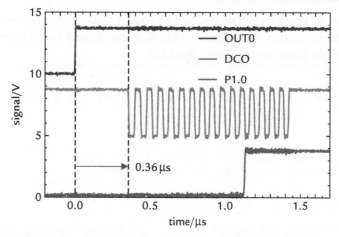

Figure 6.6: Execution of an interrupt from Timer_A that awakens an F2013 from low-power mode 3. The traces were taken from an eZ430–F2013. Top to bottom: output direct from timer, which triggers the interrupt; SMCLK, which is supplied by the DCO calibrated to 16 MHz; and signal to toggle LEDs, generated by software in the ISR. The upper 2 traces have been shifted up the voltage axis for clarity.

device returns to LPM3, assuming that each cycle has a falling edge followed by a rising edge. The expected number of cycles is six to accept the interrupt, four for xor.b, and five to return from the interrupt. This comes to 15 and I am not sure where the extra cycle comes from. Note that P1.0 changes right at the end of its instruction, which follows from the sequence of steps described in the section "Instruction Timing" on page 141.

A good point for future reference is the delay of about 1.1 μs between the direct output from the timer and the pin controlled by software in the ISR. It is best to use the timer directly to drive outputs that need precise timing. This is described in the section "Timer_A" on page 287.

6.10.2 Returning from a Low-Power Mode to the Main Function

Sometimes it is not appropriate to carry out all actions in an ISR and it is better to return to the main function in active mode. To be more precise, this means returning to the function that put the device into the low-power mode. In this case we must clear all the low-power bits in the saved value of SR on the stack before it is restored at the end of the interrupt service routine. This sounds alarming but is actually straightforward, particularly in C because an intrinsic function _ _low_power_mode_off_on_exit() is available to do the job. Listing 6.8 shows how this is used to flash the LEDs yet again.

Listing 6.8: Part of program `timainC1.c` to toggle LEDs using interrupts from Timer_A. The device enters low-power mode 0, is awakened by an interrupt and returns to the main function to toggle the LED.

```
    for (;;) {                         // Loop forever
        __low_power_mode_3();          // Enter low power mode LPM3
// Wait here, pace loop until timer expires and ISR restores Active Mode
        P2OUT ^= LED1|LED2;            // Toggle LEDs
    }
}
//-------------------------------------------------------------------
// Interrupt service routine for Timer A channel 0
// Processor remains in Active Mode after ISR
//-------------------------------------------------------------------
#pragma vector = TIMERA0_VECTOR
__interrupt void TA0_ISR (void)
{
    __low_power_mode_off_on_exit();    // Restore Active Mode on return
}
```

The main function is now like the paced loop of Listing 4.16 except that the processor goes to sleep and waits to be awakened by an interrupt from the timer instead of polling the overflow flag. In fact `__low_power_mode_3()` behaves just like a simple delay function. The ISR is trivial and contains only one line for the intrinsic function to clear the stacked low-power mode. The more general function `__bic_SR_register_on_exit()` can be used to clear selected bits in SR if finer control is required. It is also called `_BIC_SR_IRQ()`.

Assembly language requires a mildly tricky line of assembly language to clear the bits set for the low-power mode. The state of the stack is shown in Figure 6.5. Recall that SP points to the most recently added word, which is the pushed value of SR. The simplest way to address this would be `@SP` but this is not permitted for a destination so the indexed mode is used with an offset of 0. This is shown in Listing 6.9. It would be safer to use `bic.w #LPM4,0(SP)` to ensure that *all* low-power bits are cleared.

Listing 6.9: Part of program `timain1.s43` to toggle LEDs using interrupts from Timer_A. The device enters low-power mode 0, is awakened by an interrupt and returns to the main function to toggle the LED.

```
InfLoop:
    bis.w    #LPM3|GIE,SR             ; Enable interrupts and enter LPM3
    nop                               ; Helps debugging
; Wait here, pace loop until timer expires and ISR restores Active Mode
    xor.b    #LED1|LED2,&P2OUT        ; Toggle LEDs
    jmp      InfLoop                  ; Infinite loop
;-------------------------------------------------------------------
```

```
; Interrupt service routine for TACCR0, called when TAR = TACCR0
; No need to acknowledge interrupt explicitly - done automatically
TA0_ISR:                          ; ISR for TACCR0 CCIFG
    bic.w    #LPM3,0(SP)          ; Delete LPM3 on exit: Active Mode
    reti                          ; That's all: return from interrupt
```

Example 6.13

The example in the text is obviously contrived. The Morse code exercise in the section "Morse Code on the Buzzer" on page 110 provides a slightly less artificial opportunity because there are tasks to be handled on two time scales. The piezoelectric buzzer needs to be serviced most frequently, at about 1 KHz, whereas the durations of the dots, dashes, and spaces are multiples of 0.1 s. Here is a possible structure for the program:

- Set up the timer to produce interrupts at 1 KHz. Use ACLK if available because this allows LPM3 to be used between interrupts (use SMCLK and LPM0 otherwise).

- Toggle the piezoelectric buzzer in the ISR and update a tick counter to measure 0.1 s. Use the method described in this section to return from the interrupt to active mode after 0.1 s has passed, otherwise return normally to the previous low-power mode.

- Write a loop in the main function to turn the LED and piezoelectric buzzer on or off according to the current character. It then puts the processor into LPM3, during which the timer toggles the piezo regularly. Control returns to the main function after 0.1 s and the outputs are updated. The _ _low_power_mode_3() call behaves effectively as a delay 0.1 s function.

- Use the lowest power mode LPM4 after the message because there is nothing more for the processor to do.

Debugging is sometimes easier in EW430 if a nop instruction is placed after the entry to low-power mode as in Listing 6.9.

Checklist—Have You …

❏ Checked that there is something to awaken the MSP430 from a low-power mode—are interrupts enabled and is there a clock running in LPM3, for instance?

CHAPTER 7

Digital Input, Output, and Displays

A microcontroller interacts in many ways with the system in which it is embedded. It may receive inputs from a human user through switches, for example. These are digital in the sense that they are either on or off, high or low. Similar signals arise from some sensors, such as detectors for the water level or door lock in a washing machine. Going in the opposite direction, the microcontroller turns external devices on or off. These might be indicators, such as simple light-emitting diodes (LEDs) or more complicated seven-segment displays. The MSP430 can supply these directly if they work from the same voltage and draw a sufficiently small current. Heavier loads require transistors or integrated circuits to drive them.

The early sections of this chapter concern digital inputs and outputs like these. Although they are digital in the sense that they should take only two levels, it is not possible even here to ignore the uncomfortable feature that the real world is analog. For example, you might think that a push button gives the simplest input of all. Unfortunately its signal is spoiled by *bounce* caused by the mechanical motion of the contacts as they are brought together and released. This can make it appear that the button was pressed several times when the user operated it only once. The effects of bounce must be removed by hardware or software so that the user sees the expected result.

Liquid crystal displays (LCDs) are the second major topic of this chapter because they are widely used with the MSP430. LCDs are a natural choice for low-power systems and the MSP430x4xx family can drive segmented LCDs directly. Plenty of demonstration boards have LCDs and it is much more satisfying to work with a system that provides informative, numerical output instead of a couple of LEDs.

More complicated forms of input and output are treated in later chapters. Chapter 9 is concerned with the conversion of "real" analog inputs into digital values that the microcontroller can process. Special interfaces, such as SPI or I²C, are used to communicate between the microcontroller and other digital components or systems. These are covered in Chapter 10.

7.1   Digital Input and Output: Parallel Ports

The most straightforward form of input and output is through the digital input/output ports using binary values (low or high, corresponding to 0 or 1). We already used these for driving LEDs and reading switches. In this section we look at their wider capabilities.

There are 10–80 input/output pins on different devices in the current portfolio of MSP430s; the F20xx has one complete 8-pin port and 2 pins on a second port, while the largest devices have ten full ports. Almost all pins can be used either for digital input/output or for other functions and their operation must be configured when the device starts up. This can be tricky. For example, pin P1.0 on the F2013 can be a digital input, digital output, input TACLK, output ACLK, or analog input A0+. This is a choice of five functions and therefore needs at least 3 bits for selection. It was hard to puzzle this out for older devices but newer data sheets have an admirably clear table for each pin in the section *Application Information*. There is also a schematic drawing of the circuit associated with the pin. For example, the function of P1.0 depends on P1DIR, P1SEL, and the analog enable register SD16AE. This pin is a digital input by default after reset, which is true for most pins but not all.

A convenient feature of all peripherals in the MSP430 is that they are implemented in much the same way in all devices and families. For example, ports P1 and P2 have interrupts in all cases, from the 14-pin F20xx to the 100-pin FG4618. Up to eight registers are associated with the digital input/output functions for each pin. Here are the registers for port P1 on a MSP430F2xx, which has the maximum number. Each pin can be configured and controlled individually; thus some pins can be digital inputs, some outputs, some used for analog functions, and so on.

Port P1 input, P1IN: reading returns the logical values on the inputs if they are configured for digital input/output. This register is read-only and volatile. It does not need to be initialized because its contents are determined by the external signals.

Port P1 output, P1OUT: writing sends the value to be driven to each pin if it is configured as a digital output. If the pin is not currently an output, the value is stored in a buffer and appears on the pin if it is later switched to be an output. This register is not initialized and you should therefore write to P1OUT *before* configuring the pin for output.

Port P1 direction, P1DIR: clearing a bit to 0 configures a pin as an input, which is the default in most cases. Writing a 1 switches the pin to become an output. This is for digital input and output; the register works differently if other functions are selected using P1SEL.

Port P1 resistor enable, P1REN: setting a bit to 1 activates a pull-up or pull-down resistor on a pin. Pull-ups are often used to connect a switch to an input as in the section "Read Input from a Switch" on page 80. The resistors are inactive by default (0). When the resistor is enabled (1), the corresponding bit of the P1OUT register selects whether the resistor pulls the input up to V_{CC} (1) or down to V_{SS} (0).

Port P1 selection, P1SEL: selects either digital input/output (0, default) or an alternative function (1). Further registers may be needed to choose the particular function.

Port P1 interrupt enable, P1IE: enables interrupts when the value on an input pin changes. This feature is activated by setting appropriate bits of P1IE to 1. Interrupts are off (0) by default. The whole port shares a single interrupt vector although pins can be enabled individually.

Port P1 interrupt edge select, P1IES: can generate interrupts either on a positive edge (0), when the input goes from low to high, or on a negative edge from high to low (1). It is not possible to select interrupts on both edges simultaneously but this is not a problem because the direction can be reversed after each transition. Care is needed if the direction is changed while interrupts are enabled because a spurious interrupt may be generated. This register is not initialized and should therefore be set up before interrupts are enabled.

Port P1 interrupt flag, P1IFG: a bit is set when the selected transition has been detected on the input. In addition, an interrupt is requested if it has been enabled. These bits can also be set by software, which provides a mechanism for generating a software interrupt (SWI).

In some cases the configuration of a pin selected by these registers can be overruled by another function. For example, P1.0 in the F2013 can also be used as input A0+ to the analog-to-digital converter (SD16_A). This module includes an analog input enable register SD16AE. Selecting channel 0 with this register connects P1.0 to the SD16_A, regardless of the settings of P1SEL and P1DIR. This is made clear in the *Application Information* but needs careful reading.

Other ports are similar, although most have fewer registers: Ports other than P1 and P2 have only the four registers PnIN, PnOUT, PnDIR, and PnSEL in the MSP430x1xx and MSP430x4xx families. Here are a few points to watch about the input/output ports:

- Interrupts are available only on ports P1 and P2 so the PnIE, PnIES, and PnIFG registers are provided for only these two.

- Pull resistors and the PnREN register are provided only in the MSP430F2xx family and newer MSP430x4xx devices. *Do not activate pull/resistors unless you are using a pin for digital input.* Selecting a pull-up or pull-down on an output pin removes the full output drive and gives only a feeble current through the pull-up to resistor instead.

- Pins P2.6 and P2.7 on many devices in the MSP430F2xx family are exceptions to the general rule and are *not* digital inputs by default. These pins can also be used for a crystal, which is their default configuration. You should reconfigure these pins if the internal VLO is used instead of a crystal.

- There is no port 0 (P0) on modern devices. It was present in some members of MSP430x3xx family and differed from the other ports in several ways, notably the handling of interrupts.

Example 7.1

Write code to configure the pins of a F2013 as follows:

- P1.0 and P1.1 are inputs A0+ and A0− to the SD16_A analog-to-digital converter (ADC).

- P1.2 is input CCI1A to Timer_A.

- P1.3 is connected to the voltage reference VREF of SD16_A.

- P1.4 is a digital output, initially driven low.

- P1.5 is output TA0 from Timer_A.

- P1.6 and P1.7 are not used; leave them unconfigured for now.

- P2.6 and P2.7 are digital inputs with pull-up resistors; ACLK should be derived from VLO.

The registers associated with input/output ports on many other microcontrollers are a little peculiar because they are not simple memories. Many designs, including the Microchip PIC16 and the Freescale HCS08, use the *same* register for input and output through a port. In this case, reading the register usually returns the values on the pins, while writing to it drives the values onto pins if they are configured as outputs. This has the curious effect that writing a value to a port and immediately reading it back does not return the same value for bits that are configured as input: The write operation puts values into the output buffer but the read operation gives the values on the input pins due to the surrounding circuitry. Great care must be used when handling these schizophrenic registers. This is not an issue with the MSP430 because it has separate input and output registers. You would expect P1IN.x = P1OUT.x when pin x is an output, which should be true provided that the output is not overloaded.

The digital input/output ports are sometimes called *parallel ports* because all eight pins can be used simultaneously to read or write a complete byte. However, they are nothing like the parallel port found on the back of old PCs. One byte is the largest unit that can be handled in most MSP430 devices because of the way in which the addresses of the registers are arranged. The exception is the FG6461X, where the registers for ports P7–P10 are laid out in the memory map so that they may be accessed individually as bytes or in pairs as words. Thus the bytewide ports P7 and P8 can also be handled as the wordwide port PA. For example, the (byte) register P7IN has address 0x0038 and P8IN is at 0x039, so the 2 bytes can be treated as the word at address 0x0038, which is the (word) register PAIN. Similarly, port PB is equivalent to ports P9 and P10.

7.1.1 Circuit of an Input/Output Pin

It is a lot easier to understand the peculiarities of input and output if you have a rough idea of the circuit. Figure 7.1(a) shows a basic CMOS inverter. This is much simpler than a real input/output pin but contains enough to explain the main features. The inverter itself requires only a complementary pair of metal–oxide–silicon field-effect transistors (MOSFETs). Almost all very large-scale integrated (VLSI) circuits are based on MOSFETs. They come in two "polarities," n-channel and p-channel, both of which are used in complementary metal–oxide–silicon (CMOS) technology.

A highly simplified cross-section of an n-channel MOSFET with its symbol appears in Figure 7.1(b). The basis of its operation is that a positive voltage on the *gate* creates a channel that permits current to flow from the *drain* to the *source*. The *drain* is defined to have a positive voltage relative to the source although the structure of the transistor itself is

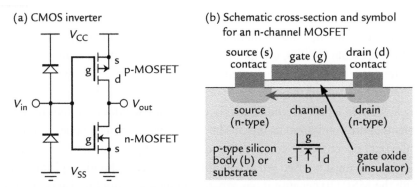

Figure 7.1: (a) Circuit of a simple CMOS inverter including the input protection diodes. (b) Schematic cross-section of an n-channel MOSFET.

often symmetric. More precisely, the channel exists when the voltage between the gate and source, V_{gs}, exceeds a critical value called the *threshold voltage*, V_t. In most MOSFETs $V_t > 0$ so current cannot flow between the source and drain when there is no difference in voltage between the gate and source. This, called an *enhancement-mode device*, is indicated by the broken line in the symbol for the devices in the figure.

We normally think of a MOSFET as a three-terminal device but in practice there is a fourth connection to the *body* or substrate. The reason is that the source–body and drain–body junctions act like n–p diodes, which must be kept reverse biased for correct operation of the device. The body is typically connected to V_{SS} in an integrated circuit, which is the most negative point. In a discrete device, the body is connected to the source, as shown in the symbol. A diode remains between the source and drain, which comes into play with inductive loads as we see in the section "Driving Heavier Loads" on page 247.

The signs of the voltages and current are reversed in a p-channel MOSFET. The drain is negative with respect to the source and the gate–source voltage must be made more negative than the threshold voltage to turn on the channel.

One of the most significant features of a MOSFET is that the gate is separated from the channel by a thin layer of silicon dioxide, which is an insulator. Thus there is no direct electrical connection between the source and channel. Instead, the gate, oxide, and channel form a capacitor. This is reflected in the symbol by a gap between the gate and channel. No current flows into a capacitor when the voltage across it is constant, which is a key factor in the low power consumption of CMOS circuits. Thus the MSP430 can retain the contents of its registers in LPM4 while drawing less than 1 μA. On the other hand, current must flow to charge and discharge the gate–channel capacitance when the transistors change state and this accounts for most of the supply current in CMOS.

This gate oxide is only a few nanometers (10^{-9} m) thick in a modern transistor. Silicon dioxide is an excellent insulator but even this breaks down if the electric field across it becomes too large. It is easy for a person to acquire large voltages through static electricity—walking across a nylon carpet in dry weather generates enough charge to cause a spark when you next touch a grounded conducting object. CMOS devices must therefore be handled only at workstations that are protected against static electricity by grounding and an appropriate choice of materials. Integrated circuits themselves are also protected by connecting diodes between inputs and the supply rails, as shown in Figure 7.1(a). Recall that current flows only in the direction shown by the arrow symbol for the diode. These are reverse-biased in normal operation, where $V_{SS} < V_{in} < V_{CC}$. They turn on to protect the circuit when the input strays by more than about 0.3 V outside the supply rails, $V_{in} < V_{SS} - 0.3$ V or $V_{in} > V_{CC} + 0.3$ V. The magnitude of the current through these diodes should not exceed 2 mA.

The input protection diodes can cause a puzzling side effect. Suppose that a logical high input is applied to a circuit whose power supply is not connected. Current flows through the protection diode from the input to V_{CC}, from where it supplies the rest of the circuit. Thus the circuit works almost normally, despite having no apparent source of power.

After that diversion, we can explain the operation of a CMOS inverter with the aid of Figure 7.2. A model inverter can be made with a pair of controlled switches, one between the output and V_{SS} to pull the output down to logic 0 and the other between the output and V_{CC} to provide a logic 1. Ideally one switch should always be closed and one open.

If the input is a logical 1, the output should be a logical 0, which needs the switch to V_{SS} closed and that to V_{CC} open, as in Figure 7.2(a). The lower switch should therefore close when the input is relatively positive (near V_{CC}) and the upper switch should be open under the same conditions, passing no current. Everything is reversed when the input is a logical 0, near V_{SS}.

This can be achieved by using an n-channel MOSFET (n-MOSFET for short) for the lower switch and a p-MOSFET for the upper switch, as shown in Figure 7.2(b). A channel is created in the n-MOSFET, allowing conduction in the same way as a closed switch, when its gate is driven positive by a high input or logical 1. When the input falls to a logic 0, so $V_{in} = V_{SS}$, there is no difference in potential between the gate and source of the MOSFET. Thus $V_{gs} = 0$, the channel vanishes, and no current flows. This is just like an open switch.

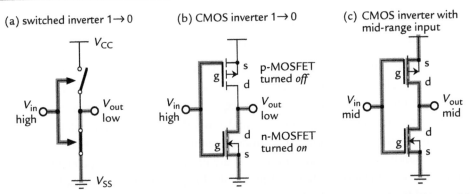

Figure 7.2: Operation of a CMOS inverter. (a) Model inverter using switches with a logical input of 1 and output of 0. (b) Corresponding operation of MOSFETs. (c) Operation when the input lies near the middle of the supply voltages V_{SS} and V_{CC} rather than close to either extreme. Both MOSFETs conduct and a large current flows from V_{CC} to V_{SS}.

The operation of the p-MOSFET is just a little harder to understand. A logical 1 on the input gives $V_{in} = V_{CC}$. This means that the gate and source of the transistor are at the same potential and it switches off. A logical 0 on the input pulls the gate down to V_{SS}. Now the gate–source voltage is $V_{gs} = V_g - V_s = V_{SS} - V_{CC}$, which is strongly negative and the transistor turns on. No voltages in the circuit are absolutely negative, taking V_{SS} as 0, but the gate of the p-MOSFET is negative with respect to its source. The output is pulled to V_{SS} by the p-MOSFET. No current flows from V_{CC} to V_{SS} because one transistor is on but the other is off.

Example 7.2

How does the CMOS inverter operate when its input is low, $V_{in} = V_{SS}$?

Everything is therefore well defined when the input is a clear logical 0 or 1. What happens when the input makes a transition between these levels? For some time V_{in} is near the midpoint of V_{SS} and V_{CC} and *both* transistors are turned on. This allows a *shoot-through current* to flow from V_{CC} to V_{SS}, shown in Figure 7.2(c). This is one of the contributions to the current drawn by a CMOS circuit whenever it changes state. Normally it flows for only a very short time during a transition of the clock but this may not be true if the inverter is connected to an input pin with a slowly changing signal, which leads naturally into the next section.

7.1.2 Configuration of Unused Pins

Not all of the input/output pins are used in most applications. *Unused pins must never be left unconnected in their default state as inputs.* This follows a general rule that inputs to CMOS must never be left unconnected or "floating." A surprising number of problems can be caused by floating inputs. The most trivial is that the input circuit draws an excessive current from the power supply. This is because the input is likely to float to the midpoint of V_{SS} and V_{CC}, turning on both MOSFETs and leading to the situation shown in Figure 7.2(c). The shoot-through current may exceed 40 μA, a huge waste by the standards of the MSP430.

Old CMOS circuits, such as the 74HC family, are amazingly sensitive to floating inputs. They may oscillate or refuse to work at all if an input is floating, even if the input belongs to an unused gate or flip-flop. I have seen this happen many times when students have not taken heed of the rule about floating inputs. Missing decoupling (bypass) capacitors can cause similar problems. Floating inputs are also susceptible to noise and to static electricity if the product is handled, which may lead to permanent damage.

There are three ways of avoiding these problems:

1. Wire the unused pins externally to a well-defined voltage, either V_{SS} or V_{CC}, and configure them as inputs. The danger with this is that you might damage the MCU if the pins are accidentally configured as outputs.

2. Leave the pins unconnected externally but connect them internally to either V_{SS} or V_{CC} by using the pull-down or pull-up resistors. They are again configured as inputs. I prefer this approach but it is restricted to the MSP430F2xx family because the others lack internal pull resistors.

3. Leave the pins unconnected and configure them as outputs. The value in the output register does not matter. This is perhaps the most robust solution and is recommended for MSP430 devices that lack internal pull resistors. I am less keen on this approach for experimental systems because it is easy to short-circuit pins with a test probe.

There is a helpful list of recommended connections for unused pins at the end of the chapter on *System Resets, Interrupts, and Operating Modes* in the family user's guides.

Example 7.3

Complete Example 7.1 by configuring the unused pins P1.6 and P1.7. They have no external connections.

In some manufacturers' devices the unconnected bits of a port may be present on the chip itself but are not bonded to pins. In this case it is important to configure *all* bits of the port correctly, including those that are not connected to the outside world. This issue arises when the same integrated circuit (the same "silicon") may be offered in packages with different numbers of pins. I believe that this does not apply to the MSP430. However, a few input/output registers contain bits without corresponding pins. For example, the F1121A has all 8 bits of P2IFG implemented on silicon but there are pins for only P2.0–P2.5. The bits for the missing pins 6 and 7 can be used for software interrupts.

Checklist—Have You…

❏ Configured all unused input/output pins so that they do not float?

7.2   Digital Inputs

Digital inputs to the MSP430 are typically connected to digital outputs from other circuits or to components such as switches. We already used the digital inputs many times for straightforward connections to a push button using the standard circuit shown in Figure 4.4. The programs in Chapter 4 used polling but this is wasteful for inputs that change on a human timescale—very slowly by electronic standards. Interrupts may be more efficient. A different approach is also needed when a large number of inputs must be read.

7.2.1   Interrupts on Digital Inputs

Ports P1 and P2 can request an interrupt when the value on an input changes. This is one of the few interrupts that remains active in LPM4 and is therefore useful to wake the CPU in portable equipment that lies idle for a long time.

Interrupts for port P1 are controlled by the registers P1IE and P1IES, mentioned previously, and similarly for port P2. There is a single vector for each port, so the user must check P1IFG to determine the bit that caused the interrupt. This bit must be cleared explicitly; it does not happen automatically as with interrupts that have a single source.

The direction of the transition that causes the interrupt can be changed in P1IES at any time by the program. This is useful if both edges of a pulse need to be detected, for example, when a button is pressed and released. There is a danger that spurious interrupts may be generated, so it is a good idea to disable this interrupt, change P1IES, and clear any spurious flags in P1IFG before reenabling the interrupt. In fact, to practice this should not be a problem if the direction is changed in the most obvious way. For instance, the port may wait for a low-to-high transition while the input is low. An interrupt is requested when the input goes high. The sensitivity is then changed to high-to-low to detect the next edge.

The use of interrupts is illustrated in Listing 7.1, which is perhaps the ultimate development of the programs to light an LED when a button is pressed. The device spends most of its time in LPM4, waiting for an interrupt on pin P2.1. Both the LED and the direction of the transition for an interrupt are toggled in the ISR. Any pending requests for an interrupt are cleared by a loop before returning to LPM4. I included this code as an example of how to avoid spurious interrupts. Unfortunately it is not a particularly good idea here because it loses the second edge of short pulses. These arise from a problem that is described in the section "Switch Debounce" on page 225.

Listing 7.1: Program `butled4.c` in C to light LED1 when button B1 is pressed using interrupts and low-power mode 4.

```
// butled4.c - press button B1 to light LED1
// Responds to interrupts on input pin, LPM4 between interrupts
// Olimex 1121STK board, LED1 active low on P2.3,
//    button B1 active low on P2.1
// J H Davies, 2006-11-18; IAR Kickstart version 3.41A
//-------------------------------------------------------------------
#include <io430x11x1.h>            // Specific device
#include <intrinsics.h>           // Intrinsic functions
//-------------------------------------------------------------------
void main (void)
{
    WDTCTL = WDTPW | WDTHOLD;      // Stop watchdog timer
    P2OUT_bit.P2OUT_3 = 1;         // Preload LED1 off (active low!)
    P2DIR_bit.P2DIR_3 = 1;         // Set pin with LED1 to output
    P2IE_bit.P2IE_1 = 1;           // Enable interrupts on edge
    P2IES_bit.P2IES_1 = 1;         // Sensitive to negative edge (H->L)
    do {
        P2IFG = 0;                 // Clear any pending interrupts...
    } while (P2IFG != 0);          // ...until none remain
    for (;;) {                     // Loop forever (should not need)
        __low_power_mode_4();      // LPM4 with int'pts, all clocks off
    }                              //   (RAM retention mode)
}
//-------------------------------------------------------------------
```

```
// Interrupt service routine for port 2 inputs
// Only one bit is active so no need to check which
// Toggle LED, toggle edge sensitivity, clear any pending interrupts
// Device returns to low power mode automatically after ISR
//------------------------------------------------------------------
#pragma vector = PORT2_VECTOR
__interrupt void PORT2_ISR (void)
{
    P2OUT_bit.P2OUT_3 ^= 1;          // Toggle LED
    P2IES_bit.P2IES_1 ^= 1;          // Toggle edge sensitivity
    do {
        P2IFG = 0;                   // Clear any pending interrupts...
    } while (P2IFG != 0);            // ...until none remain
}
```

Example 7.4

Take out the loop with `P2IFG = 0` from the ISR in Listing 7.1 and clear the register only once. Does this lead to any spurious interrupts?

Example 7.5

The Olimex 1121STK board has two buttons and two LEDs, so it can be used as a quiz detector. It should light the LED for the first button that is pressed and ignore the second button. Leave the LED on for about 1 s, then put the device back to sleep and wait for the next answer. Ignore switch bounce.

You could sound the buzzer as well to show that a button has been pressed. How can you prevent a player cheating by pressing the button while the LED is still on for the previous question?

7.2.2   Multiplexed Inputs: Scanning a Matrix Keypad

Many products require numerical input and provide a keypad for the user. These often have 12 keys, like a telephone, or more. An individual connection for each switch would use an exorbitant number of pins so they are usually arranged as a matrix instead. Only seven pins are needed for a 12-key pad, as shown in Figure 7.3, or eight pins for 16 keys. As usual this economy comes at a price. The matrix must be scanned, which is more complicated than reading individual inputs. Moreover, the reading may become ambiguous if more than one key is depressed.

There are, as usual, many ways of dealing with a keypad. Here is a straightforward approach, although it needs refinement in practice. I do not worry about debouncing at this stage and assume that no more than one switch is closed.

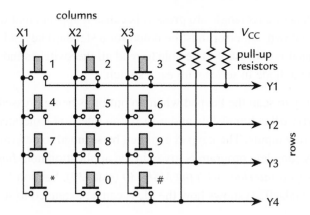

Figure 7.3: Connection of a 12-button telephone keypad as a 3 × 4 matrix.

Connect the rows Y1–Y4 as inputs to the microcontroller while the columns X1–X3 are driven by outputs. (It could equally well be the other way around.) Pull-up resistors are required on the inputs. These could be internal for the MSP430F2xx family but must be provided externally for other devices.

1. Drive X1 low and the other columns X2 and X3 high. This makes the switches in column X1 active and the corresponding Y input goes low if a button is pressed. Thus we can detect the state of switches 1, 4, 7, or *. The switches in the other columns have no effect because both of their terminals are at V_{CC}.

2. Drive X2 low and the other columns high to read the switches in column X2.

3. Repeat this for column X3.

This process can be repeated as often as required.

A problem with this simple method arises if two buttons, such as 1 and 2, are pressed, which short-circuits the column drives X1 and X2. This damages the output of the microcontroller if they are connected directly. Resistors should therefore be connected between the pins of the microcontroller and the columns of the keypad. Diodes could be used instead. Another possibility for the MSP430F2xx family is to use the internal pull-ups instead of full-strength outputs for columns that are not being addressed. Please remember to disable the pull-up before trying to use the pin for "real" output.

Usually only one key should be pressed at a time on the pad. In fact there should be no problem identifying two keys held down simultaneously. Difficulties arise when three buttons on the corners of a rectangle are pressed because it appears that the button on the fourth corner is also down. An error must be noted for a standard keypad in this case. In other applications it is necessary to be able to read all the switches independently and the solution is to put a diode in series with each switch.

It is a waste of energy to scan the keypad when no button is being pressed. In this case it is more efficient to drive *all* columns low and wait for an interrupt generated by a falling edge on any of the row inputs. The keypad can then be scanned to determine which key has been pressed. A complete program is given in the application note *Implementing an Ultralow-Power Keypad Interface with the MSP430* (slaa139). Section 5.5.5 of *Application Reports* (slaa024) shows how the ideas can be extended to scan different types of input.

Example 7.6

The keypad in Figure 7.3 is attached to port 1 of a F20xx with Y1, Y2, Y3, Y4, X1, X2, and X3 connected to P1.0–P1.6. No other components are provided, which means that you should use the internal pull-ups. Write a function to scan the keypad and return the value as a signed, 16-bit integer, using 10 for * and 11 for #. The return value should be −1 if no keys are pressed and −2 if there is an error, such as more than one key pressed. Leave the keypad with the columns pulled low so that interrupts can be detected when a button is pressed.

Example 7.7

Another way of reading the keypad [26] uses two steps. It is assumed that only a single key is pressed.

- Configure the rows Y1–Y4 as inputs with pull-ups and drive all columns X1–X3 low. Read Y1–Y4 to find the row that contains the button pressed.

- Configure the columns X1–X3 as inputs with pull-ups and drive all rows Y1–Y4 low. Read X1–X3 to find the column that contains the button pressed.

The active key can now be located and its value can conveniently be stored in a two-dimensional array. Try this. It works best with the internal pull-ups in the MSP430F2xx. What happens if more than one button is pressed?

7.2.3   Analog Aspects of "Digital" Inputs

It is usually safe to assume that signals inside the microcontroller are straightforward logical zeros and ones although there are a few exceptions to this comfortable situation, such as the output of a comparator (see the section "Comparator_A" on page 371). Outside the microcontroller there is no escape from the fact that the real world is analog. This raises many issues, of which the most basic is the question, What analog voltages correspond to the digital values 0 and 1?

Take $V_{SS} = 0$ for clarity. The precise input voltages V_{in} that correspond to logical 0 and 1 depend on the technology but typical values for CMOS are

- Inputs of 0 to 0.3 V_{CC} give logic 0.

- Inputs of 0.7 V_{CC} to V_{CC} give logic 1.

These are symmetric. The output voltage V_{out} for CMOS is typically below 0.1 V_{CC} for a logical 0 and above 0.9 V_{CC} for a logical 1. This is again symmetric and the large gap between the ranges for logical 0 and 1 means that CMOS is relatively insensitive to noise.

The logical value is undefined for an input that lies in the transition region between the two ranges, which is typically 0.3 V_{CC} to 0.7 V_{CC} for CMOS. Inputs in this range also cause an excessive current to flow through the input buffer, as shown in Figure 7.2(c). The apparent logical value as seen from inside the microcontroller may also oscillate wildly between 0 and 1. The voltage on inputs should therefore pass rapidly through the transition zone. This is particularly important for inputs that generate interrupts or provide clocks (to a timer, for instance).

An excellent feature of the MSP430 is that its inputs are provided with *Schmitt triggers*, which eliminate many of these problems. This includes the standard port inputs, which also protects their interrupts. Other digital inputs, such as the external clock to the timer, also pass through the Schmitt triggers. The details are shown in *Port pin schematics* in the data sheets.

Figure 7.4 shows the output voltage as a function of the input voltage for a conventional buffer and a Schmitt trigger. The major difference is that a Schmitt trigger displays *hysteresis*. This is most easily explained by looking at the response to a slow triangular wave on the input in Figure 7.4(c).

The input and output are initially at 0 voltage and the input voltage rises gradually. The output remains safely in the range for logical 0 until the input passes through the upper,

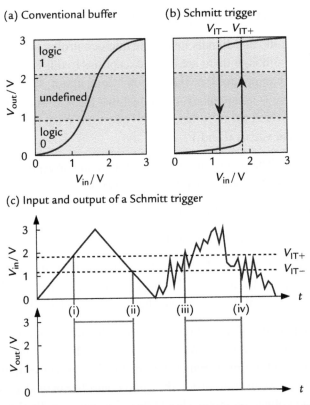

Figure 7.4: Transfer characteristic (output voltage as a function of input voltage) for (a) conventional buffer and (b) Schmitt trigger. The Schmitt trigger shows hysteresis and never gives an output with an undefined logic value. (c) Input to and output from a Schmitt trigger as a function of time. The trigger turns a slowly varying input into sharp transitions and eliminates noise.

positive-going threshold voltage V_{IT+} at (i). At this point the output jumps abruptly to a value that is well inside the range for a logical 1. It stays here while the input continues to rise and after it has started to fall again. The output remains at a logical 1 even after the input has fallen through V_{IT+} and does not change until the input crosses the lower threshold voltage V_{IT-} at (ii). At this point the output jumps to a logical 0 and remains here as the input falls further.

The second half of the plot shows the effect of (rather fanciful) noise on the input. The output jumps from 0 to 1 when the input first goes above V_{IT+} at (iii). The noise pulls the input back down below this threshold but the output remains cleanly at 1. Similarly, on the downward half of the wave, the output falls from 1 to 0 at point (iv) when the input first falls below V_{IT-} and is not affected by fluctuations that take the input back above this threshold. Thus a Schmitt trigger has two desirable effects:

- It turns slowly varying inputs, which might cause problems while they pass slowly through the undefined range of input voltages, into abrupt, clean logical transitions.

- It eliminates the effect of noise on the input, provided that it is not large enough to span the gap between the upward and downward thresholds.

Schmitt triggers have many other applications. A simple oscillator can be made by adding a resistor and capacitor, for instance. The principle is described in the section "Comparator_A" on page 371.

Another analog aspect of the inputs is that a small current flows into or out of the pin. Ideally this would be 0 because the gates of MOSFETs act like capacitors but in reality the capacitors leak slightly, as do the input protection diodes. Of course the circuit associated with each pin is much more complicated than a simple inverter too. Having said all that, the pins of the MPS430 have a low input leakage current, below ±50 nA. This is roughly equivalent to a resistance of 50 MΩ and means that leakage should rarely cause a serious drop in voltage across a resistor in series with the input. The leakage current would not drop more than ±0.1 V across a 2 MΩ resistor, for example. This allows large pull-up resistors to be used to save current, although the input may then become sensitive to noise. The low leakage current is also important if the input is used to detect the voltage on a small capacitor (a few pF). This arises in touch sensors, which are also mentioned in the section "Capacitative Touch Sensing with Comparator_A" on page 391. Finally, low leakage contributes to a long battery life, particularly in devices with over 100 pins.

Before leaving this topic, let us look more closely at the way in which thresholds and other parameters are specified in the data sheet. Table 7.1 shows a small extract for the F20xx. There is a minimum and maximum for each of the two threshold voltages. When is each important? Suppose that the input is initially at V_{SS}:

- The system should not respond to noise on the input, so any fluctuations in voltage must be kept below the threshold V_{IT+}. In this case we should choose the *minimum* value of 1.35 V to ensure that the input is never triggered.

- On the other hand, we want to guarantee that the microcontroller responds when the desired signal appears on the input. We must ensure that the input goes above the *maximum* value of 2.25 V to be certain that the Schmitt input is triggered.

Table 7.1:   Selected electrical characteristics of Schmitt trigger inputs with $V_{CC} = 3\,V$ adapted from the data sheet for the F20xx.

	Parameter	Minimum	Typical	Maximum	Unit
V_{IT+}	Positive-going input threshold voltage	1.35		2.25	V
V_{IT-}	Negative-going input threshold voltage	0.75		1.65	V
V_{hys}	Input voltage hysteresis, $V_{IT+} - V_{IT-}$	0.3		1.0	V
R_{Pull}	Pull-up/pull-down resistor	20	35	50	kΩ
C_I	Input capacitance		5		pF
$T_{(int)}$	External interrupt timing	20			ns
I_{lkg}	High-impedance leakage current			±50	nA

The same arguments apply to the negative-going threshold V_{IT-}. There is no point in specifying typical figures for these thresholds because we should always design to the appropriate extreme. I describe this shortly in "Debouncing in Hardware" on page 229.

Look next at the hysteresis, defined by $V_{hys} = V_{IT+} - V_{IT-}$. There are two opposite cases to consider:

- The input is most sensitive to noise when the voltage has just gone through one of the thresholds. Suppose that it has just risen through V_{IT+}. Noise must be kept below the *minimum* hysteresis of 0.3 V to ensure that a negative fluctuation does not trigger an unwanted downward transition by pulling the input down through V_{IT-}. Thus the minimum hysteresis sets the noise margin of the input. This does not add anything new because it is the same as the minimum difference between V_{IT+} and V_{IT-}.

- The peak-to-peak magnitude of an input must exceed the *maximum* value of 1.0 V to ensure that both its positive and negative edges trigger the input. In practice this specification is less demanding than the individual figures for V_{IT+} and V_{IT-}, which require a difference of 1.5 V. (Effectively this tells us that an input with the minimum value of V_{IT-} does not have the maximum value of V_{IT+}.)

The other specifications are more straightforward. For example, the single figures given for $T_{(int)}$ and I_{lkg} are both the worst cases.

Example 7.8

What problems may be caused by noise of 1 V peak-to-peak in an input signal?

7.3 Switch Debounce

We have just seen that there is no escape from analog electronics when dealing with signals outside the microcontroller. Unfortunately a switch also involves mechanical engineering and we cannot avoid a problem called *switch bounce*. The problem is that the voltage across a real switch in the usual circuit with a pull-up in Figure 4.4 does not go instantly and cleanly between V_{SS} and V_{CC} when the button is pressed and released. Instead it may make multiple transitions, partial transitions with pauses on the way, or just change slowly on the timescale of a microcontroller. A good system should accommodate these deficiencies rather than act as if the button has been pressed several times when the user pressed it only once. You have probably used products where this has not been done—it is annoying at best.

7.3.1 What Does Switch Bounce Look Like?

Older textbooks tell you that bounce is worse when a switch is closed than when it is opened and may last for around 50 ms. This may be true for toggle switches, the sort with an arm that sticks out, which you click up and down. These are expensive and now rarely used. Most switches are push buttons, mounted directly on a PCB. In fact the PCB forms part of the switch itself in products such as remote controls. Ganssle wrote a comprehensive guide to debouncing [17] which I heartily recommend. This section is heavily based on this reference. He shows results from a range of different types of switch, whose characteristics vary wildly.

I made some measurements of the behavior of a push button on a simple, homemade demonstration board for the MSP430F20xx. Figure 7.5 shows oscilloscope traces from a few of my experiments. The input was the usual active low circuit, a push button between the pin and ground, using the internal pull-up. The output was an LED driven active high and the program was a simple loop that copied the input port to the output port. Each loop took roughly 0.4 μs and the latency, the time between reading from the input and writing to the output, varied from 0.2 to 0.6 μs depending on when the input changed in relation to the sequence of instructions in the loop. Note that the scale of the time base on the different traces varies over a wide range but all times are shown in microseconds.

The first two traces are for pressing the button. Most transitions are clean and effectively instantaneous as in Figure 7.5(a). I found no multiple transitions at all. Figure 7.5(b) shows the worst behavior that I saw, with several steps on the input between V_{CC} and V_{SS}. It looks bad but lasted for only about 0.5 μs and caused no false changes in the output. Pressing the button gave perfect behavior for practical purposes.

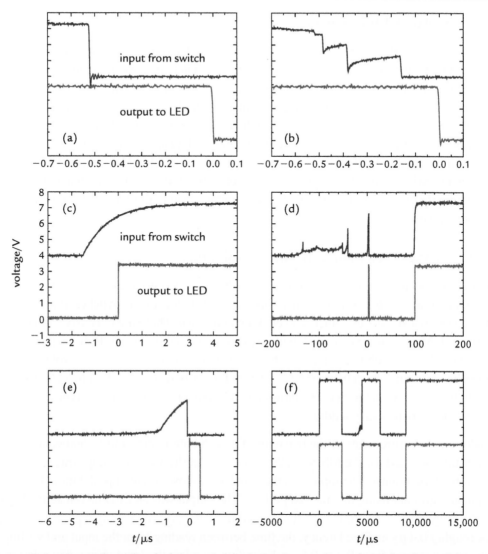

Figure 7.5: Examples of switch bounce measured on a demonstration board for an MSP430. The program is a simple loop that copies the input from an active low push button to an active high LED. Traces (a) and (b) are for pressing the button while (c)–(f) are for releasing it. All times are in microseconds but note the different scales for the time base. The origin of time is at the first transition on the output.

Releasing the button was quite different. Figure 7.5(c) shows a clean transition. This is slow compared with the response when the button is pressed and looks like a classic, exponential *RC* charging curve. The reason is that the input pin is not connected directly to V_{CC} but only through the weak internal pull-up. This has a resistance of around 35 kΩ and current must flow though this to charge the capacitance associated with the pin. I explain about *RC* charge and discharge curves in the next section. For now, the time-constant τ is about 1 μs in Figure 7.5(c) and $\tau = RC$ so the capacitance $C \approx 30$ pF. This is a lot larger than the value of 5 pF quoted for an input pin in the data sheet. The demonstration board was made on stripboard, which certainly contributes a few more pF of stray capacitance. Where has the rest come from? The answer is the oscilloscope, whose probe presents a capacitance of about 16 pF. Never forget that measuring equipment disturbs the circuit under test. The transition therefore would be about twice as fast if the probe were removed, and faster still with better layout of the circuit.

Although it is a smooth curve, this slow *RC* charging curve could cause problems with a microcontroller that did not have Schmitt triggers on its input. The signal spends around 1 μs in the range of voltages that give undefined outputs, which could cause multiple transitions in the logical value seen from inside the microcontroller. This is one reason why pull-up resistors should not be made too high in value. There is no problem with the MSP430.

Unfortunately many transitions were far from clean when the button was released. The range of behavior was so wide that it is impossible to pick out a typical trace but Figure 7.5(d) shows the sort of result that causes the microcontroller to see multiple transitions. The input fluctuates away from V_{SS} for about 100 μs without triggering the input of the microcontroller before a more determined but brief attempt to reach V_{CC} causes a short pulse on the output at 0 time. The signal from the switch then returns to V_{SS} for another 100 μs before the final transition to V_{CC}. The F20xx has MCLK running at 16 MHz and can therefore carry out several hundred instructions between the two transitions that trigger its input. This bounce is fast on a mechanical timescale but slow to the microcontroller.

Even a short glitch on the input can trigger the Schmitt input. Figure 7.5(e) shows the beginning of a usual *RC* charging curve, but it is interrupted as the switch closes again and the voltage returns to V_{SS} after about 1.5 μs. This may be brief but the input rose above 2 V and crossed the threshold of the Schmitt trigger. The result was a brief pulse on the output.

In contrast, Figure 7.5(f) shows the longest bounce that I saw. There was an apparently clean transition at 0 time but the switch closed again after about 2000 μs, opened again, and repeated this before finally opening after 9000 μs. Thus the switch bounced for nearly 10 ms.

The conclusion is that this particular switch does not bounce when pressed but may bounce for up to 10 ms when released. Unfortunately there is no guarantee that another push button of the same type will behave in the same way, let alone other switches. One of Ganssle's push buttons bounced for over 150 ms when released, for example. I think that the results in Figure 7.5 are fairly typical but my experience of using large numbers of demonstration boards in classes shows that a few switches display much worse behavior. Some never seem to close properly at all; instead they just bounce continuously while pressed, but fortunately these are rare.

The difference between pressing and releasing the button must arise from its internal construction. I took one apart and found that the active element is a thin piece of metal shaped into a shallow dome. The button presses the center of the dome, which pops between two shapes. This is the same as the safety lid on a glass jar of food after the vacuum has been released. Other switches are constructed differently. I also dismantled an old TV remote control. The moving part of the switch was a molded, rubbery sheet, presumably of a conductive material, which pressed against two contacts on a PCB when operated. I did not measure its electrical characteristics, which may be quite different from the push button with the metal dome.

A defect of my experiment is that the output is taken from a polling loop. Some pulses as long as 0.4 μs (400 ns) do not cause an output simply because they are not present at the critical time when the input register is read. In contrast, an interrupt can be triggered by a pulse of only 20 ns or less on the input. My oscilloscope is not fast enough to see this either.

The last trace in Figure 7.5 makes it clear that a good system must filter its input in some way to avoid the effects of switch bounce. This can be done either in external hardware or software.

Example 7.9

The simplest way of showing the effect of switch bounce is with a program to count the number of times the button is pressed, set as Example 4.9. Try this if you have not done it before. It would be interesting to look for any difference in behavior between polling the input and using interrupts.

If you are lucky enough to have a digital oscilloscope available, try recording the voltage on the input to the microcontroller. Do your results resemble Figure 7.5? In particular, what is the longest duration of the bounce?

7.3.2 Debouncing in Hardware

Debouncing was traditionally done using hardware. The most effective solution is to use a two-way toggle switch (single pole, double throw, or SPDT) and a set–reset (*SR*) flip-flop in the circuit of Figure 7.6(a). See Horowitz and Hill [42] for the details; alternatively, look in any book on elementary digital electronics because it is a classic application of an *SR* flip-flop. This gives a clean output, provided that the contacts of the switch never bounce all the way back to their original position. (I used to take this for granted but Figure 7.5(f) made me worry. Of course that is for a push button, not a toggle switch.) Special ICs were formerly made for this application but are now far more expensive than an MSP430. It would be possible to use two inputs to a microcontroller to emulate a flip-flop but toggle switches are now rarely used and the method cannot be applied to a single push button. Despite these drawbacks it is the most robust solution because all other approaches require an assumption about the duration of bounces.

Example 7.10

It is tempting to use an MSP430 with a pair of push buttons for "on" and "off" to emulate a two-way toggle switch and flip-flop. The circuit in Figure 7.6(a) uses a standard *SR*

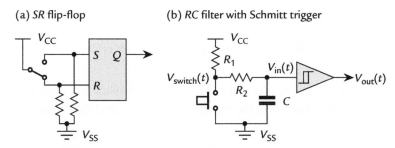

(a) *SR* flip-flop   (b) *RC* filter with Schmitt trigger

Figure 7.6: Two classic circuits for debouncing a switch in hardware: (a) digital approach using a *SR* flip-flop and a SPDT switch; (b) analog approach using a *RC* filter. This *must* be followed by a Schmitt trigger as shown, not just a simple CMOS input.

flip-flop, in which case the rocker of the switch must be connected to V_{CC} and the inputs need pull-down resistors. Most switches are connected to ground instead with pull-up resistors. In this case you must emulate an $\bar{S}\bar{R}$ flip-flop rather than an SR, but it is no more complicated. Try this if your demonstration board has two buttons.

Now for a problem, which does not arise with a toggle switch but is introduced with two buttons. What happens if an uncooperative user presses *both* buttons? Is it possible to handle this robustly, given that they may be pressed at the same time and bounce simultaneously?

The flip-flop cannot be used to debounce a simple push button (single pole, single throw, or SPST switch) so the classic solution comes from analog electronics: an RC filter. The circuit is shown in Figure 7.6(b). In a nutshell, the resistors and capacitor slow down the signal seen by the microcontroller and smooth out any rapid changes caused by bouncing. An unavoidable side effect is that the output is slowly varying and *must* be connected to a Schmitt trigger, not a standard logic input. Otherwise, the slow passage of the input signal through the undefined range of voltages gives more trouble than the original bounce. Of course this is not a problem with the MSP430.

I have to admit that the detailed analysis that follows may not appeal to readers with a background in computer science rather than electronic engineering. You may prefer to skip the rest of this section. However, I think that it is a lot clearer to see the problem in terms of voltages as a function of time rather than successive values in a shift register.

First, Figure 7.7(a) provides a reminder of the behavior of an RC circuit, a capacitor C in series with a resistor R. This has a time-constant $\tau = RC$. Suppose that the capacitor is initially uncharged and the circuit is connected at $t = 0$ across a battery of voltage V_{CC}. The equation for the voltage $V_C(t)$ on the capacitor as it charges is

$$V_C(t) = V_{charge}(t) = V_{CC}[1 - \exp(-t/\tau)].^*$$ (7.1)

Similarly, if the capacitor is first charged to V_{CC}, then the resistor and capacitor are connected in a short circuit at $t = 0$, the capacitor discharges and the voltage is given by

$$V_C(t) = V_{discharge}(t) = V_{CC}\exp(-t/\tau).$$ (7.2)

---

* The notation $\exp(x)$ is equivalent to e^x.

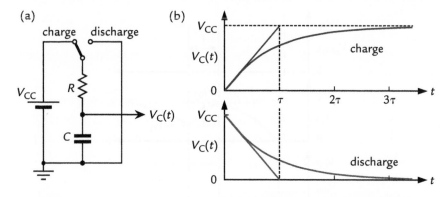

Figure 7.7: Charge and discharge curves for a simple *RC* circuit.

See the sketches in Figure 7.7(b). These equations are needed throughout Chapter 9 and in several other places, such as the behavior of the I²C bus.

Now we can explain how the circuit in Figure 7.6(b) behaves with an ideal switch. Suppose that the switch has been open a long time so that the capacitor is fully charged to V_{CC}. Take $V_{SS} = 0$ to make the equations clearer. When the switch is pressed, the capacitor discharges through R_2 according to equation (7.2) with a time-constant $\tau_2 = R_2C$. When the switch is released after being held down for a long time, the capacitor recharges through R_1 and R_2 in series according to equation (7.1) with a time-constant $\tau_1 = (R_1 + R_2)C$.

Thus the *RC* circuit smooths out the changes in voltage from the switch. This reduces the influence of short transitions, which is how it suppresses the effect of bounce. We need to design the circuit to function correctly with the worst case input, but what should this be? Looking back at the extract from the data sheet in Table 7.1, the most worrying figure is the minimum hysteresis V_{hys} of 0.3 V. Assume for simplicity that the thresholds are symmetric between V_{SS} and V_{CC}, which gives $V_{IT+} = 1.65$ V and $V_{IT-} = 1.35$ V. Suppose that the button has been held down for a long time and is then released, because we know that this is when bounce is a problem. Figure 7.8 shows a simplified version of the experimental trace in Figure 7.5(f). This is the sequence of events.

1. The switch is opened at t_1 and the capacitor begins to charge through R_1 and R_2 with time-constant τ_1.

2. At t_2 the input to the Schmitt trigger crosses the positive-going threshold voltage V_{IT+} and the output goes from a logical 0 to 1. Immediately afterward, the switch bounces open and the capacitor begins to discharge through R_2 with time-constant τ_2.

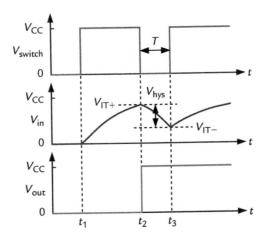

Figure 7.8: Voltages across the switch, input to the Schmitt trigger and output of the Schmitt trigger for a worst case bounce using the circuit shown in Figure 7.6(b).

3. The switch bounces closed again at time t_3, just before the voltage reaches the negative-going threshold voltage V_{IT-}. Thus a false transition is just avoided. The output voltage subsequently rises again as the capacitor recharges.

This shows the worst case, meaning that the switch recloses for the maximum time $T = t_3 - t_2$ that we consider likely. The debounce circuit fails if the switch stays closed again for longer than this.

The critical part of this episode is from t_2 to t_3, while the capacitor discharges with time-constant τ_2. Equation (7.3) gives the voltage at these two times:

$$V_C(t_2) = V_{CC} \exp(-t_2/\tau_2) = V_{IT+} \quad \text{and} \quad V_C(t_3) = V_{CC} \exp(-t_3/\tau_2) = V_{IT-}. \quad (7.3)$$

Dividing the two equations shows that

$$\frac{V_{IT+}}{V_{IT-}} = \exp\left(\frac{t_3 - t_2}{\tau_2}\right) = \exp\left(\frac{T}{\tau_2}\right). \quad (7.4)$$

Finally, we can take natural logarithms to find

$$T = \tau_2 \log_e\left(\frac{V_{IT+}}{V_{IT-}}\right). \quad (7.5)$$

We estimated values for the threshold voltages earlier, which give $T \approx 0.2\tau_2$. What value should we choose for T, the maximum length of pulse that we expect? The experiments showed durations of 3 ms and it is best to play it safe, so take $T = 5$ ms. Therefore $\tau_2 = 5T \approx 25$ ms. This is set by $R_2 C$. Capacitors of 0.1 μF are readily available, being a favorite value for decoupling. This requires $R_2 = 250$ kΩ, which can be rounded up to the standard value of 270 kΩ.

This fixes R_2 and C but not yet R_1. A similar analysis can be done for an unwanted high pulse rather than the low pulse that we have just studied. The result for the time-constant $\tau_1 = (R_1 + R_2)C$ is the same because the threshold voltages are symmetric. Thus we can make R_1 as small as we wish as far as the debouncing is concerned. We do not want to waste current, on the other hand, so something like 100 kΩ is reasonable.

A circuit without R_2 is often used, which is just the usual pull-up resistor with an added capacitor to ground. It has a couple of disadvantages. The first is that the circuit does not provide protection against low pulses due to noise or bounce when the input is high. This could be a problem with the sort of signal shown in Figure 7.8. The time-constant must be large enough that the rising voltage does not cross the threshold $V_{\text{IT+}}$ until all bouncing has ceased. This might give a slow response. The other problem is that the capacitor is short-circuited when the button is pressed, which produces a short pulse of current as it is discharged. The resistance and inductance of the connections produce a corresponding pulse of voltage, which can interfere with other parts of the circuit.

Example 7.11

How would the traces for $V_{\text{in}}(t)$ and $V_{\text{out}}(t)$ in Figure 7.8 change if R_2 were omitted, assuming the same signal $V_{\text{switch}}(t)$?

Whew! This seems like a lot of effort to calculate the values of two resistors and a capacitor. Perhaps you think that it would be easier to perform debouncing in software? It saves a few components too. We look at this next.

7.3.3   Debouncing in Software

In practice debouncing is usually carried out inside the microcontroller. A wide variety of algorithms is in use, which range from trivial approaches to sophisticated digital equivalents of the *RC* filter described in the previous section.

The simplest method is to detect a transition from the switch, wait for a fixed delay, and test the input again. If the input remains the same, it is accepted as a valid new value, otherwise it is assumed to be an error and ignored. The delay should be longer than the expected duration of any bounce and 10 ms is a common choice. This would just be satisfactory for the switch that I tested, provided that it never behaves worse than the trace in Figure 7.5(f). The delay is best implemented with a timer and the MSP430 can be returned to a low-power mode if there is nothing else to do. This method is unsuitable for switches that may bounce longer, such as toggle switches rather than push buttons, because the maximum foreseeable delay is likely to be so long that the response is unacceptably sluggish.

More versatile algorithms are typically based on counters or shift registers. These are updated at regular intervals of around 5 ms, usually triggered by interrupts from a timer. Here is an example from Ganssle [17] that uses a counter. The routine stores the current debounced state of the switch so that it can be compared with the raw input from the switch itself.

- No action is needed if the raw input matches the debounced state. If they differ, a down counter is started.

- The counter is decremented every time the routine is called, provided that the raw input still differs from the debounced state. The counter is reinitialized if the raw input has reverted to the debounced state.

- If the counter reaches 0, the raw input is accepted as the new debounced state. A flag can be set to indicate that the input has changed in the same way as the P2IFG flags are set by hardware.

The initial value for the counter can be different for pressing and releasing the key. This is useful because it is often desirable to respond more quickly to the press than the release. It is also convenient for modern push buttons, which bounce less when pressed than when released.

Another way of implementing this idea is to use a shift register, which was mentioned in the section "Recording the State of a Push Button" on page 151. Again the input from the switch is read regularly and rotated into the most significant bit of a register. Here is an outline of the algorithm in pseudocode, assuming as usual that the switch is active low:

```
rotate raw input into msb of shift register
if (debounced state == released) {
    if (shift register <= threshold for press) {
        debounced state = pressed
        set flag to indicate H->L change
    }
```

```
    } else {
        if (shift register >= threshold for release) {
            debounced state = released
            set flag to indicate L->H change
        }
    }
```

Suppose that the button has been up for a long time, which fills the shift register with ones. When the button is pressed, the next call of the routine shifts a 0 into the msb of the register. This lowers (roughly by half) the value stored in the register if it is regarded as an unsigned number. If the button is still pressed at the next call, another 0 is shifted in and the value is further reduced. The value falls below the threshold after enough zeros occupy the more significant bits and the press is accepted. On the other hand, a shorter run of zeros does not reduce the value sufficiently and does not trigger a change in the debounced state. The opposite happens when the button is released and bits of 1 are rotated into the register.

Again the number of sequential zeros and ones for pressing and releasing the button can be made different. Suppose that the register holds a single byte. We might consider two successive low inputs as sufficient to confirm that the button has been pressed, in which case 00111111b = 0x3F is the threshold for pressing. (This gives a fast response but might be a little lax.) However, we might wish to be more careful when the button is released and wait for six successive ones, which needs a threshold of 11111100b = 0xFC.

Listing 7.2 shows a program to light an LED when a button is pressed (yet again—sorry) including a shift register to debounce the raw input. The main loop of the program puts the processor into LPM3 with Timer_A running in Up mode from ACLK. This generates interrupts every 5 ms. The interrupt service routine samples the raw input, feeds it into the shift register, and compares the value with the thresholds suggested earlier. When a threshold is satisfied, the new debounced state is stored and an intrinsic function is called to terminate the low-power mode. This causes the processor to return to the main function, where the actions are carried out in response to the change in debounced input. Here we just turn the LED on or off.

Listing 7.2: Program `debtim1.c` to light an LED when a button is pressed, using a shift register to debounce the input.

```
// debtim1.c - press button B1 to light LED1 with debounce
// Samples input at 5 ms intervals set by Timer_A, 32KHz ACLK
// Shift reg for debounce, different thresholds for press and release
// Olimex 1121STK board, LED1 active low on P2.3,
//   button B1 active low on P2.1
// J H Davies, 2007-04-10; IAR Kickstart version 3.42A
```

```
//------------------------------------------------------------------
#include <io430x11x1.h>              // Specific device
#include <intrinsics.h>             // Intrinsic functions
#include <stdint.h>                 // Standard integer types

union {                            // Debounced state of P2IN
    unsigned char DebP2IN;          // Complete byte
    struct {
        unsigned char DebP2IN_0 : 1;
        unsigned char DebP2IN_1 : 1;
        unsigned char DebP2IN_2 : 1;
        unsigned char DebP2IN_3 : 1;
        unsigned char DebP2IN_4 : 1;
        unsigned char DebP2IN_5 : 1;
        unsigned char DebP2IN_6 : 1;
        unsigned char DebP2IN_7 : 1;
    } DebP2IN_bit;                  // Individual bits
};
#define RAWB1    P2IN_bit.P2IN_1
#define DEBB1    DebP2IN_bit.DebP2IN_1
#define LED1     P2OUT_bit.P2OUT_3

void main (void)
{
    WDTCTL = WDTPW | WDTHOLD;        // Stop watchdog timer
    P2OUT_bit.P2OUT_3 = 1;          // Preload LED1 off (active low)
    P2DIR_bit.P2DIR_3 = 1;          // Set pin with LED1 to output
    DebP2IN = 0xFF;                 // Initial debounced state of port
    TACCR0 = 160;                   // 160 counts at 32KHz = 5ms
    TACCTL0 = CCIE;                 // Enable interrupts on Compare 0
    TACTL = MC_1|TASSEL_1|TACLR;    // Set up and start Timer A
// "Up to CCR0" mode, no clock division, clock from ACLK, clear timer
    for (;;) {                      // Loop forever
        __low_power_mode_3();       // Enter LPM3, only ACLK active
// Return to main function when a debounced transition has occurred
        LED1 = DEBB1;               // Update LED1 from debounced button
    }
}
//------------------------------------------------------------------
// Interrupt service routine for Timer A chan 0; no need to acknowledge
// Device returns to LPM3 automatically after ISR unless input changes
// PRESS_THRESHOLD = 0x3F = 0b00111111, needs 2 successive 0s (enough?)
// RELEASE_THRESHOLD = 0xFC = 0b11111100, 6 successive 1s (too many?)
//------------------------------------------------------------------
#define PRESS_THRESHOLD    0x3F
#define RELEASE_THRESHOLD  0xFC

#pragma vector = TIMERA0_VECTOR
__interrupt void TA0_ISR (void)
{
    static uint8_t P21ShiftReg = 0xFF;  // Shift reg for history of P2.1

    P21ShiftReg >>= 1;              // Update history in shift register
    if (RAWB1 == 1) {               // Insert latest input from B1
        P21ShiftReg |= BIT7;        // Set msb if input high
    }
```

```
    if (DEBB1 == 0) {
// Current debounced value low, looking for input to go high (release)
        if (P21ShiftReg >= RELEASE_THRESHOLD) { // button released
            DEBB1 = 1;                  // New debounced state high
            __low_power_mode_off_on_exit(); // Wake main routine
        }
    } else {
// Current debounced value high, looking for input to go low (press)
        if (P21ShiftReg <= PRESS_THRESHOLD) {    // button pressed
            DEBB1 = 0;                  // New debounced state low
            __low_power_mode_off_on_exit(); // Wake main routine
        }
    }
}
```

The byte that holds the debounced value is defined as a union in the same way as the hardware register P2IN in the header file io430x11x1.h. This allows it to be addressed as a complete byte or as individual bits. Here we process only 1 bit but all bits could be used to make a virtual, debounced input register. For clarity I defined some compact names for the individual bits used for the input and output. The shift register P21ShiftReg must be declared static so that it retains its value between interrupts. It is also unsigned, which means that the shift-right is logical and clears the msb. This bit is then set if the input is high. These operations are more efficient in assembly language.

This program reacts to a change in the debounced value by leaving its low-power mode and returning to the main function. Another approach is to generate a software interrupt. This would be like the interrupts flagged in P1IFG but due to the debounced input rather than the raw input. A neat trick in the F1121A is to use the flags for the unbonded pins P2.6 and P2.7.

Example 7.12

Modify the program in Listing 7.2 so that the LED(s) on your demonstration board count the number of apparent presses. Does it debounce the input successfully? You might like to experiment with the criteria for accepting a press and release.

Example 7.13

Sometimes it is necessary to detect only when the switch is pressed but not when it is released. Does this lead to a simpler algorithm? Ensure that short glitches do not trigger a false transition.

This approach sounds quite different to the hardware with an *RC* filter but is in fact closely related. The value in the shift register is analogous to the voltage stored on the capacitor. The shift to the right is like the decay of the stored charge through the resistance and the bit rotated into the msb mimics the current through the switch. Therefore the shift register can be viewed as a simple digital equivalent of an analog *RC* filter. The two thresholds for pressing and releasing the button are something like the threshold voltages of a Schmitt trigger too. Mazzocca [24] analyzes the relation thoroughly and gives a more complete algorithm.

Often several switches are to be debounced and Ganssle [17] gives an algorithm for handling them in parallel. The external inputs are sampled simultaneously and stored together in a byte (or larger unit if necessary). The bytes from a number of successive samples are kept in a queue. The stable states of each switch are found from these as follows:

- Take the bitwise AND of all the bytes. This leaves bits of 1 in only those positions where there was a 1 in every byte. The input is therefore stable at 1 for these bits.

- Take the bitwise OR of all the bytes. This leaves bits of 0 in only those positions where there was a 0 in every byte. The input is therefore stable at 0 for these bits.

- In other cases, the input has changed during the duration of the queue and the previously stored value should be retained.

This could be useful in debouncing a keypad, for instance.

A disadvantage of the counter and shift register is that they need regular polling of the input, which does not suit low-power applications. A solution is to put the MSP430 into LPM4 with interrupts enabled on inputs from switches. Polling is started after an interrupt and continues until the new input is accepted or it is clear that the change was due to a short pulse of interference that should be ignored.

7.4   Digital Outputs

The standard circuits for connecting an LED to a pin of a microcontroller are shown in Figure 4.3 and are repeated in Figure 7.9, which includes the transistors inside the MSP430. Always include current-limiting resistors in series with the LEDs. Remember also that LED stands for light-emitting *diode* and that a diode passes current in only one direction, shown by the arrow in the symbol. This refers to conventional current, which flows from positive to negative. No light is produced if the LED is connected backward.

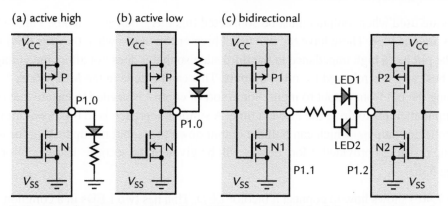

Figure 7.9: Standard connection of an LED to pin P1.0 in (a) active high and (b) active low configurations. (c) Connection of a bidirectional pair of LEDs between a pair of pins, P1.1 and P1.2.

In the active high circuit (a) the LED lights when the p-MOSFET is switched on and the n-MOSFET is off. Current flows from V_{CC} through the p-MOSFET, out of the pin and through the LED to V_{SS}. The pin therefore acts as a *source* of current. The opposite is true in the active low circuit (b). This time current flows from V_{CC} through the LED, into the pin and through the n-MOSFET to V_{SS}. The pin is said to be a current *sink*. In general n-MOSFETs have better performance than p-MOSFETs and this is why many older ICs were better at sinking current than sourcing current. LEDs were therefore usually connected active low. Most modern microcontrollers are designed so that the performance of the output is more or less symmetric.

An important parameter is the maximum recommended current in or out of the port pins. The data sheets are surprisingly reticent: No limiting currents are specified at the time of writing. The section on *Electrical Characteristics* includes plots of the output voltage as a function of current that go up to ±40 mA. This would be a startlingly high current for any microcontroller, let alone a low-power device. I presume that it is measured for very short pulses to avoid destructive overheating. In contrast, the product information center recommends that the current should be limited to 4 or 5 mA per pin and 25 mA per port. Consult them if your application approaches these bounds or see the section "Driving Heavier Loads" on page 247 for circuits that allow the MSP430 to switch heavier loads.

There is usually no problem with connecting a few inputs to a single output. On the other hand, you should *never* connect two ordinary outputs together because this causes contention if they attempt to produce different outputs, which may damage them. Special circuits are used where outputs must be connected together, such as on a bus. Three-state outputs are one type. These have the two usual high or low states when they are driving the bus. The pin has a high impedance in the third state so that it does not affect the voltage on the bus, which is released for other outputs. This can be done in the MSP430 by switching the pin from output to input. Some sort of control is needed to ensure that only one output attempts to drive the bus at a time. A simpler approach is to use open-drain or open-collector outputs, which can pull the output down to V_{SS} but not up to V_{CC}, for which a pull-up resistor is provided. More details will be given in the section "Hardware for I²C" on page 535.

Figure 7.9(c) shows how to connect a bicolor LED. This has two LEDs in a common package, connected so that one color lights when the current flows in a particular direction and the other color lights for current in the opposite direction. The package must be connected between a pair of pins, which act as a simple H-bridge (see the section "Driving Heavier Loads" on page 247). Suppose that pin P1.1 is driven high and P1.2 low. Current flows from V_{CC} through P1, LED1, and N2 to V_{SS}. Similarly, LED2 is lit by driving P1.1 low and P1.2 high. Neither LED is active if both pins are driven high or both low. The same technique can be used for other loads that need both directions of current.

7.4.1   Multiplexed Displays

The idea behind the bidirectional output can be carried further as a way of multiplexing LEDs, as shown in Figure 7.10. In general, $N(N-1)$ LEDs can be driven from N pins by extending this circuit. It relies on the one-way characteristic of a diode and its nonlinear relation between current and voltage. A single LED is selected by driving one pin high,

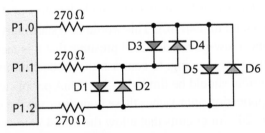

Figure 7.10: Circuit to select one LED by multiplexing. Up to $N(N-1)$ LEDs can be driven from N pins of the microcontroller.

one low, and configuring all other pins as inputs. For example, suppose that P1.0 is high, P1.2 is low, and P1.1 is an input and therefore effectively disconnected. Current flows through D5 and two of the series resistors, which limit the current in the usual way. A parallel path lies through D3 and D1 but this has two LEDs in series. Each receives only half the voltage, so very little current flows. The remaining LEDs are reverse biased. We could light D6 instead by driving P1.0 low and P1.2 high, and similarly for the other LEDs.

Example 7.14

It feels uncomfortable to rely on the absence of any significant current flowing through two forward-biased diodes in series, such as D3 and D1 in the preceding text. Test this. Suppose that a single LEDs passes 4 mA when 1.8 V is applied across it. Assume that the current is given by the ideal diode equation for forward bias, $I(V) = I_S \exp(V/V_T)$, where the thermal voltage $V_T \approx 26\,\mathrm{meV}$ at room temperature. Calculate the scale current I_S and hence the current that flows when $V = 1.5$ V. The result may surprise you (in other words, it surprised me).

Only one LED can be addressed at a time, which may seem a serious disadvantage. This is resolved by repeatedly addressing each LED in turn. The eye does not detect that the LEDs are flashing rather than continuously illuminated, provided the frequency is above 100 Hz or so. In fact LEDs are often more efficient when operated in this way, in the sense that they need a smaller average current to produce the same apparent brightness. As usual this feature comes at a price: The current must be higher during each pulse, and this may exceed the limit of the MSP430's pins.

A more conventional form of multiplexing is often used with seven-segment LED displays. These tend to consume a large current, which clashes with the low-power ethos of the MSP430, but are simpler than liquid crystal displays. The layout and circuit of a single digit are shown in Figure 7.11(a). Note that there are usually eight segments, despite the name; the eighth segment is a decimal point. In the circuit shown, the cathodes of all the LEDs are connected to give a *common cathode* display. (The cathode is the negative terminal when the diode is forward biased, shown by the bar on the symbol.) Common anode displays are equally common. The usual resistors should be connected in series with each segment to limit the current. A higher value may be needed for the decimal point because it draws a lower current. Sometimes the segments comprise

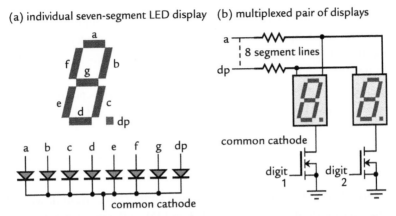

Figure 7.11: (a) Physical layout and circuit of a single, common-cathode, seven-segment LED display. The segments are labeled in a standard way and *dp* stands for decimal point. (b) Multiplexed connection of two digits. The eight segment lines and the gates of the n-MOSFETs, one per display, are connected to pins of the microcontroller.

two or more LEDs in *series*, but this requires a higher voltage than an MSP430 can provide.

Suppose that two digits are needed. This would require 16 pins of the microcontroller if the displays were connected separately, which is excessive. They are usually multiplexed instead, as shown in Figure 7.11(b). These are again common-cathode devices. The corresponding segment pins of the displays are connected in parallel and the common cathodes are used to select a particular digit. In principle each cathode could be connected to a pin of the microcontroller but the current would exceed the rating of the MSP430 so I use an n-MOSFET as a switch instead; this is described fully in the section "Driving Heavier Loads" on page 247.

The two digits are driven alternately. To select digit 1, the gate of its FET is driven to V_{CC}, which turns it on, while the gate of the FET for digit 2 is driven to V_{SS} to turn it off. Individual segments of digit 1 can then be lit by bringing the corresponding pins high or turned off by pulling them low. Both displays feel the voltages on the segment lines but only digit 1 is able to respond to them. The voltages on the FETs are then reversed to make digit 2 active and the segment lines are changed to give the desired pattern. The details of the software, such as using lookup tables to get the correct patterns on the segment lines for the digits, are very similar to LCD displays and I cover this in the section "Simple Applications of the LCD" on page 264.

Example 7.15

How might it be possible to drive two seven-segment displays without using transistors, assuming that the MSP430 can provide enough current for each segment? Hint: Try something similar to the circuit in Figure 7.10.

A large, seven-segment display needs more current than the MSP430 can provide itself. This is not a problem because plenty of special ICs are available to drive LEDs. The LEDs are usually driven from constant-current sources to give better control of illumination than a simple resistor. Many have serial interfaces such as SPI or I²C (see Chapter 10), which save pins on the microcontroller.

7.5   Interface between 3 V and 5 V Systems

It is often necessary to connect a low-voltage microcontroller, such as the MSP430, to devices that work at 5 V or higher voltages. Electronic systems in cars are designed for 12 V supplies, for instance. There are still many peripherals, such as motor drivers, designed to work from 5 V logic. Some have "TTL compatible" inputs and it may be possible to drive them directly from a 3 V system but others require a level translator of some sort. The opposite problem also arises, where an MSP430 must receive input from a system that runs at a higher voltage. The application note *Interfacing the 3 V MSP430 to 5 V Circuits* (slaa148) covers both directions.

Figure 7.12 illustrates the problem by showing the ranges of voltage recognized by the inputs and produced by the outputs of traditional, 5 V, bipolar transistor–transistor logic (TTL), CMOS at 5 V, and the MSP430 at 3 V and 2 V. This is the notation:

- V_{OL} is the maximum low output voltage for a logic 0.
- V_{OH} is the minimum high output voltage for a logic 1.
- V_{IL} is the maximum high input voltage for a logic 0.
- V_{IH} is the minimum high input voltage for a logic 1.

All of these represent the worst cases. For example, V_{IL} is the minimum value of the threshold voltage V_{IT-} for the Schmitt trigger inputs of the MSP430. The following criteria

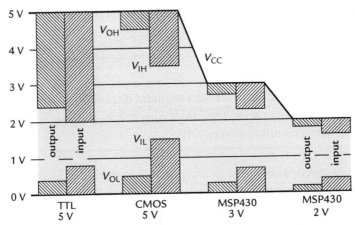

Figure 7.12: Ranges of input and output voltages specified for various families of logic.

should be obeyed for two devices to communicate reliably. I have assumed that the ground supplies are connected together.

- The minimum high output voltage V_{OH} of the sender must be *above* the minimum high input voltage V_{IH} of the receiver.

- The maximum low output voltage V_{OL} of the sender must be *below* the maximum low input voltage V_{IL} of the receiver.

- The worst case output voltages of the sender must not damage the receiver. This means that the voltage never goes below $V_{SS} - 0.3\,V$ nor above $V_{CC} + 0.3\,V$.

The two devices can be connected directly if these criteria are satisfied. If not, an interface of some sort will be needed. A wide range of special integrated circuits is available for this application. They are effectively buffers with two power supplies, one to define the input voltage and one for the output. Most are unidirectional but some are bidirectional, which means that they are able to detect the driver and receiver automatically; this is particularly useful on buses such as I²C, where data travel in both directions. Full details of TI's products are given in its *Logic Selection Guide* (sdyu001) and a subset is highlighted in *Analog and Logic for Low-Power Processors* (slyt052).

It is sometimes possible to avoid the expense of dedicated level translators and I now describe a few other ways of handling input and output with higher voltages.

7.5.1   Input from a Higher Voltage

An obvious problem with connecting higher voltages to an input is the presence of the input protection diodes, seen in Figure 7.1(a). These start to conduct if the voltage on an input goes outside the supply rails by more than about 0.3 V. It is therefore not possible to connect an input of a MSP430 directly to an output that may go up to 5 V or above $V_{CC} + 0.3$ V in general. Some microcontrollers have "5 V tolerant" inputs, whose protection circuits are modified to permit higher voltages (within limits). These are not currently available on the MSP430. Instead the simplest approach is to use the potential divider drawn in Figure 7.13. This divides the voltage by a factor of

$$\frac{V_{MSP430}}{V_{external}} = \frac{R_2}{R_1 + R_2}. \tag{7.6}$$

The values of the resistors should be small enough that the MSP430 does not load the circuit and upset the division. That just means small compared with 50 MΩ, which is unlikely to be an issue. On the other hand, the values should be kept reasonably large because the current through the resistors is wasted.

The resistors should be chosen so that the voltage seen by the MSP430 obeys the three previous criteria. This is trivial for connecting a 5 V CMOS output to the MSP430 but proves more awkward for TTL output. It is unlikely that you will use TTL itself but plenty of sensors still produce the same output voltages so it is worth a closer look. Suppose that the TTL output comes from a 5 V system and the MSP430 runs at 2 V:

- The minimum high output voltage V_{OH} is 2.4 V for TTL, which must be above the minimum high input voltage V_{IH} of 1.5 V for the MSP430. A direct connection is acceptable. If a potential divider is used, it must keep the voltage above 1.5 V. This needs $R_2/(R_1 + R_2) > 1.5/2.4 = 0.63$.

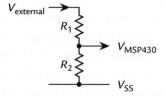

Figure 7.13: Potential divider to reduce an external voltage $V_{external}$ before applying it to the input of an MSP430.

- The maximum low output voltage $V_{OL} = 0.4$ V for TTL must be below the maximum low input voltage $V_{IL} = 0.5$ V for the MSP430. This is already obeyed so no division is required. Therefore we could connect the two components directly as far as the two logic levels are concerned.

- The maximum output voltage could be 5.0 V. This is well above the voltage at which the protection diode starts to conduct and a potential divider is therefore essential. It should preferably give an output below $V_{CC} = 2.0$ V for the MSP430, which needs $R_2/(R_1 + R_2) < 2.0/5.0 = 0.40$.

Oops! There is a problem: We need $R_2/(R_1 + R_2) > 0.63$ for the logical high voltages but $R_2/(R_1 + R_2) < 0.40$ to avoid possible damage to the input. These are incompatible.

Choose $R_2/(R_1 + R_2)$ to be just above 0.63 to satisfy the logic levels. Suitable standard values are $R_1 = 100$ kΩ and $R_2 = 180$ kΩ, which give $R_2/(R_1 + R_2) = 0.64$. A high TTL output of 2.4 V gives 1.54 V input to the MSP430, just above the threshold of 1.5 V. Unfortunately the maximum output of 5.0 V gives 3.2 V so the input protection diodes conduct. However, only a small current flows. A rough estimate is $(5.0\,V - 2.0\,V)/100\,k\Omega = 30$ μA. This is well below the permitted limit of ± 2 mA and should therefore be acceptable. Larger values could be chosen for the resistors to reduce the current further.

Example 7.16

Design a potential divider for connecting a 5 V CMOS output to the input of a MSP430 running at 3 V.

7.5.2   Output to a Higher Voltage

A glance at Figure 7.12 shows that it is not possible for a 3 V MSP430 to drive 5 V CMOS directly. The input voltage needs to be above $V_{IH} \approx 3.5$ V for a high input to the CMOS but this is impossible because it is above V_{CC} for the MSP430. However, a direct connection to the input of 5 V TTL would be fine. This is convenient because the interface between TTL and CMOS at 5 V is a problem that was solved long ago. The high output voltage of TTL, V_{OH}, is far too low for the input threshold V_{IH} of normal CMOS. Special families of CMOS were therefore introduced with input thresholds adapted to suit TTL. For example, the HC family has the thresholds shown as CMOS in Figure 7.12. The HCT family is similar but has the same input thresholds as TTL. Therefore a 3 V MSP430 can drive 5 V HCT logic directly. This is the simplest approach to connecting 5 V logic input to the outputs of a MSP430 with a 3 V supply. Unfortunately you are (just) out of luck if your MSP430 runs from 2 V.

The inputs of many other components that run from 5 V can be connected directly for the same reason. Look for TTL-compatible inputs. If these are not available, an HCT gate may be the simplest solution for a level translator. If this is unsatisfactory, try one of the methods for powering heavier loads described in the next section.

7.6   Driving Heavier Loads

Often it is necessary to drive a load that requires more current than can be provided by the MSP430 itself, or which needs a higher voltage. Heavier loads are always switched either fully on or off where possible, rather than varying the current through them continuously. The main reason is for efficiency. Very little power is dissipated in the "switch," which is usually a transistor of some sort, when it is fully on or off. The load is switched on and off periodically to control its average power, as explained in the section "Output in the Up Mode: Edge-Aligned Pulse-Width Modulation" on page 330. In contrast, suppose that the load runs at half its maximum voltage. Equal voltages are dropped across the switch and the load, which therefore dissipate the same power. This is wasteful of energy and requires precautions to prevent the transistor overheating.

Several ways of switching a load are shown in Figure 7.14. The load is switched on when the pin of the microcontroller is driven high in the first three. The traditional method is to use an npn bipolar transistor in Figure 7.14(a), where the transistor is driven by pin P1.0 of the MSP430. The load is connected between its supply voltage and the collector (c) of the transistor. The transistor controls the current from the load to ground and is therefore called a *low-side* switch or driver. The load and microcontroller are connected to the same ground V_{SS} but the supply voltage to the load, V_{supply}, is often higher than V_{CC} for the microcontroller.

I now give a brief explanation of how this circuit works—skip the next couple of paragraphs if bipolar transistors are unfamiliar. The transistor is cut off if the output from the microcontroller is low, in which case no current flows through the load. Raising the output voltage from the microcontroller causes current to flow into the base of the transistor, which in turn causes a larger current to flow into the collector and turns the load on. The aim is to drive the transistor into *saturation*. This causes the voltage at the collector to fall to its minimum value, called the saturation voltage $V_{ce,sat}$. A typical value is around 0.2 V. This gives the maximum voltage across the load and minimizes the power dissipated in the transistor.

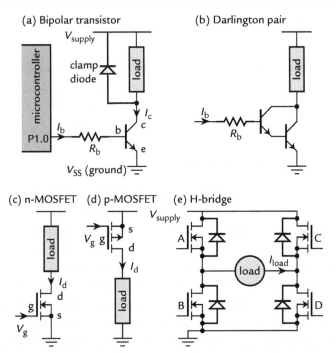

Figure 7.14: Circuits to drive a heavier load than the MSP430 can supply directly using (a) npn bipolar transistor, (b) Darlington pair, (c) n-MOSFET, (d) p-MOSFET, and (e) H-bridge. A clamp diode to suppress the back-emf from an inductive load is shown in (a) and should be used in *all* cases for inductive loads.

Two steps are needed to design this circuit. First, choose a transistor that can safely handle the supply voltage and current through the load. This is not always straightforward because many loads do not obey Ohm's law. A classic (if old-fashioned) example is an incandescent bulb. The rated voltage and current apply to normal operating conditions, when the filament is hot. However, the resistance is very low when the bulb is turned on and the filament is cold. The initial current is therefore much higher than the rated value and the transistor must be able to handle this safely. The same is true for a load with significant capacitance.

The second step is to calculate a suitable value for the resistor R_b in series with the base (b). The currents into the collector and base are related by $I_c = \beta I_b$. This looks simple but the gain β (also called h_{FE}) varies with current and falls when the transistor is saturated. A typical value is around 50 for a small transistor in saturation. The voltage on the base is $V_{be} \approx 0.7\,\text{V}$, almost independent of the current, and the voltage at the other end of the resistor is the output voltage of the microcontroller. The value of $R_b = (V_{CC} - V_{be})/I_b$ then follows from Ohm's law.

Do not worry if that explanation was gobbledegook because manufacturers provide ready-made devices for this application with the resistors built in. They are called *digital transistors* or *resistor-equipped transistors*. Nothing is digital about the devices themselves; it is just the intended application. Follow this checklist to select a suitable device:

1. Ensure that the digital transistor is suitable for a 3 V system: It should turn on fully with 3 V applied to the base. Older devices may have been designed for 5 V systems. It is important that the *input on* voltage be below 3 V.

2. Check that the collector–emitter breakdown voltage is large enough. It should exceed V_{supply}, allowing a margin for any transients.

3. Choose a device whose rated collector current is sufficient for the load.

4. Divide the collector current by the *minimum* value of the current gain β or h_{FE} to find the base current. Check that this will not overload the MSP430.

A problem with a single transistor is that the base current may be too large. For example, $I_c = 500$ mA requires $I_b \approx 10$ mA if $\beta \approx 50$, which exceeds the recommended limit of the MSP430. The classic solution is to use a *Darlington pair* of transistors as in Figure 7.14(b). The effective value of β is squared, greatly reducing the input current. A snag is that the collectors are pulled down only to about 0.9 V. This would be a serious loss of voltage in a 3 V system but may be acceptable if V_{supply} is larger. An alternative is to use a more modern type of bipolar transistor, such as the BISS range from NXP (formerly Philips). These are designed for higher current in a given package, better gain β in saturation, and lower $V_{ce,sat}$. A useful side effect is that they dissipate less power. Having said that, field-effect transistors may be a better choice and are described shortly.

Darlington pairs were widely used in the past and are readily available in packages that include the base resistors and often provide further components to improve the performance. Arrays of Darlington drivers were popular because several outputs are often needed. Many stepper motors have four windings, for instance, and eight drivers are needed for a seven-segment LED display. The ULN2003 is a well-known array of seven Darlington pairs rated at 50 V and 500 mA per load. It also includes clamp diodes, whose purpose is explained shortly.

It is sometimes necessary to control loads that work directly from the mains or line voltage, 110 or 230 V AC. Special semiconductor or electromechanical switches are made for this and include vital isolation for safety. Mechanical relays are old-fashioned but remain widely used. They include an electromagnet that operates one or more switches when it is energized. The contacts may close, open, or change over when the relay is active. In most relays the contacts return to their inactive state when the current through the coil is removed but there are also latching relays. These stay in their new state after the current has been removed until the coil is energized again, which makes them attractive for battery-powered systems. The coil of the electromagnet in a relay is rated for a particular voltage. A few are available for 3 V but require a current of around 50 mA, far too high for an MSP430 to provide directly. Most relays are intended for higher voltages, often 5 V, 12 V, or more. Even a small relay consumes a power of around 0.1–0.2 W so they are not low-power components. They also bring another problem, which is dealing with the inductance of the coil.

There is no particular difficulty in driving resistive loads, once you have chosen a suitable transistor for the voltage and current. Unfortunately many loads are *inductive* and require more care. Motors and relays are the most common examples. An inductor attempts to keep a constant current flowing through it and generates a voltage called a *back-emf* to oppose any change. This causes trouble when an inductive load is turned off: The voltage across the load reverses sign compared with its normal value in an attempt to keep the current flowing. The back-emf will damage the electronics unless precautions are taken. In Figure 7.14(a), for example, the voltage at the collector of the transistor is raised above V_{supply} by the back-emf of an inductive load. This may exceed the transistor's rating and destroy it. The solution is to connect a *clamp*, *freewheeling*, or *catch* diode across the load. Under normal conditions this is reverse biased and passes very little current but it becomes forward biased when a back-emf is generated and safely short-circuits the current across the load. These diodes are built into arrays like the ULN2003 and must be used with all inductive loads.

The modern alternative to bipolar transistors are MOSFETs, larger versions of the devices from which the MSP430 is built. Their operation is explained in the section "Circuit of an Input/Output Pin" on page 211. Figure 7.14(c) shows an n-channel MOSFET used as a low-side driver. This is clearly analogous to the circuit in Figure 7.14(a) with an npn bipolar transistor. The vital improvement is that no current flows into the gate in a steady state. This makes MOSFETs much easier to drive (unless switching is so fast that their capacitance becomes an issue). No resistor is needed.

MOSFETs with appropriate threshold voltages for controlling loads from microcontrollers are marketed as *digital FETs*. I recommend that you use one of these if you need to drive a heavy load from the MSP430. They are very simple to use and their characteristics are better than bipolar devices in most respects. The checklist for selecting a device is similar to that for a bipolar transistor except that there is no need to worry about the gate current. Make sure that the device is designed for your value of V_{CC}, such as 3 V rather than 5 V. Also check the resistance of the channel when the MOSFET is turned on, to be sure that it does not drop too much voltage and dissipate excessive power.

Integrated arrays of MOSFETs are available but, like most modern integrated circuits, tend to be more specialized than the ULN2003. An example is the TPL9201, which is aimed at domestic appliances. This contains eight low-side n-MOSFET drivers, particularly intended for relays and motors. They do not have individual analog inputs as in the ULN2003 but are controlled digitally through a serial peripheral interface (SPI). The device also includes a 5 V regulator and a zero-crossing detector so that loads can be switched synchronously with the AC mains or line. Other ICs are available to drive DC motors, stepper motors, and indeed any load in common use. The application note *MSP430 Stepper Motor Controller* (slaa223) describes the use of the UC3717, for instance.

The mirror image of these circuits can be used if the load must be connected between the switch and ground. This, called a *high-side driver*, is shown for a p-MOSFET in Figure 7.14(d). The load is turned on when the pin goes low. This simple circuit can be used only if the load and microcontroller operate from the same voltage, $V_{supply} = V_{CC}$. A level shifter is needed to drive the gate of the MOSFET if the voltages are different.

All these circuits switch the current through the load on and off. In some cases it may be necessary to reverse the direction of the current through the load as well. Figure 7.9(c) shows how this can be done using the port pins to drive a bicolor LED directly. The most common heavier load that needs this approach is a motor whose direction needs to be reversible. The standard circuit used to accomplish this is the *full bridge* or *H-bridge* shown in Figure 7.14(e). Suppose that transistors A and D are on while B and C are off. Current flows from V_{supply} through A, the load, and D to ground, so $I_{load} > 0$. Now change the inputs so that B and C are on with A and D off. Current can again flow from V_{supply} to ground but now the direction through the load has been reversed so that $I_{load} < 0$. Effectively the H-bridge turns the DC supply into AC—the opposite of a bridge rectifier.

The high-side switches A and C could be p-MOSFETs as in the bridge shown in Figure 7.9(c) but they are usually n-MOSFETs because of their lower resistance when switched on. Level-shifting circuits are needed to drive their gates because the voltage needs to be around V_{supply}, which is usually larger than the microcontroller can provide. A wide range of integrated circuits is available for H-bridges. They include all the components for lower powers, but a separate driver chip and discrete MOSFETs are used for high powers. Motors are inductive and protection against the back-emf is essential.

Wilmshurst [26] has a good chapter on interfacing to external devices and, as always, an abundance of material is found in Horowitz and Hill [42].

Checklist—Have You...

❑ Included diodes to protect against the back-emf caused by inductive loads?

❑ Checked that your transistor can handle the worst case current drawn by the load?

❑ Inserted a resistor in series with the base if you use a bipolar transistor?

❑ Put resistors in series with all LEDs (unless you are using a dedicated driver IC)?

7.7   Liquid Crystal Displays

A liquid crystal display (LCD) uses much less power than LEDs and is therefore a natural companion for the MSP430. An LCD does not emit light itself but controls the intensity of reflected or transmitted light. It therefore works well in strong ambient lighting but a backlight must be provided for a display to be used in dark surroundings, which erodes the advantage of an LCD. The backlight is usually provided by LEDs so it might require less power to use LEDs for the display itself in these circumstances.

LCDs fall into three classes. The display itself is often called the *glass* but many displays are packaged into modules that include the supporting electronics.

> **Segmented LCDs:** the simplest and can be driven directly by the MSP430x4xx family. These displays include the familiar seven-segment numerical displays found in watches, meters, and many other applications. Further segments can be added to each character to allow alphanumeric display. Individual segments can have any shape desired and most displays include special symbols. I use the SoftBaugh SBLCDA4 as an example because it is fitted to several demonstration boards. It can display 7½ digits (the ½ is a 1 rather than a complete digit), a progress bar, envelope, antenna, battery, and many other symbols.

Character-based LCDs: have a dot-matrix display, often with 1–4 rows of 8–72 characters. They can typically show a set of around 256 characters drawn from the ASCII characters, arrows, and a selection of other symbols. They are addressed by sending a byte to define each character. Further control bytes are used to clear the display, return the cursor to home and so on. These displays are usually incorporated into modules with a "Hitachi" interface. In its full form this uses eight wires to transfer a complete byte and a further three wires for control. This is extravagant for a device that is driven at only a few kilohertz so the data can be sent in two nibbles instead of a single byte, requiring only seven pins. A few displays are available with SPI or serial interfaces, which reduces the number of precious pins needed. The modules are considerably more expensive than the LCD glass alone. Several demonstration boards have this type of LCD connected to a MSP430x1xx device.

Fully graphical LCDs: found on every mobile (cell) phone, to say nothing of games and computers. There is a simple example on the Olimex MSP430-169LCD demonstration board, which has a F169 and a Nokia 3310 black and white display with 84×48 pixels. The interface to the microcontroller needs only a few pins but the MSP430 has to send the value of each pixel to the display and must therefore generate the shape of every character.

Figure 7.15(a) shows the basic construction of a reflective LCD. Two glass plates carry transparent electrodes on their opposing faces and there is a mirror below the lower plate. The gap between is filled with a liquid crystal. Incident light is reflected and the display appears clear when no bias is applied to the electrodes. A sufficiently large bias changes the optical properties of the liquid crystal so that reflected light is no longer transmitted through the upper glass. The segment now appears dark, as in Figure 7.15(b). Electrically the display is similar to a capacitor, albeit rather lossy.

A complication is that LCDs must be driven with AC, not DC. A steady voltage of only a few tens of millivolts leads to electrolysis of the liquid crystal, which eventually destroys the display. The two electrodes of a segment are therefore driven with square waves in antiphase to produce an alternating voltage with zero mean. This is the same technique that was used for the piezoelectric sounder in Figure 4.9. The frequency is low, typically around 100 Hz, but must not be close to multiples of the AC mains (line) frequency (50 or 60 Hz). The output of many lights fluctuates at twice the frequency of the mains and the LCD appears to flicker if it is updated at a similar rate.

Almost all displays have more than one segment. The simplest approach is to drive these individually as shown in Figure 7.15(c). There is a common backplane for all segments,

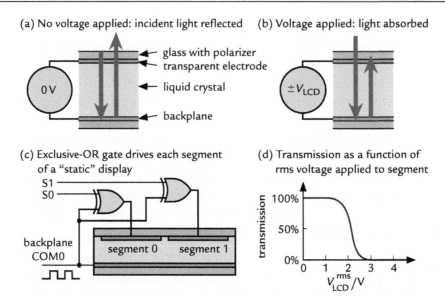

(a) No voltage applied: incident light reflected   (b) Voltage applied: light absorbed

glass with polarizer
transparent electrode
liquid crystal
backplane

(c) Exclusive-OR gate drives each segment
of a "static" display

(d) Transmission as a function of
rms voltage applied to segment

Figure 7.15: Structure and operation of a reflective LCD display. (a) Light is reflected when no bias is applied. (b) An AC bias between the electrodes changes the optical properties so that light is no longer reflected. (c) Method of driving each segment of a "static" LCD. (d) Transmission as a function of the rms voltage across a segment.

called COM0 here, and each segment on the front has a separate connection. A square wave provides a clock to bias the display. This signal is applied directly to the backplane and through an exclusive-OR gate with a control signal to each segment:

- If the control signal is low the exclusive-OR gate transmits the clock unchanged, the same voltage is applied to the front and back electrodes, there is no potential difference, and the segment remains clear.

- A high control signal causes the exclusive-OR to invert the clock so that an alternating bias is applied to the segment, which turns dark.

The exclusive-OR gates could be real devices but it is straightforward to implement this inside the MCU by toggling the outputs periodically.

This type of display is called *static* despite the requirement for an AC drive. It is simple but suffers from the obvious disadvantage of needing a large number of pins, one per segment plus the backplane. Most displays are therefore multiplexed to fewer pins, like displays with LEDs. Unfortunately it is much more tricky to multiplex LCDs because of the AC drive. The method of two-way multiplexing is shown in Figure 7.16. Each segment

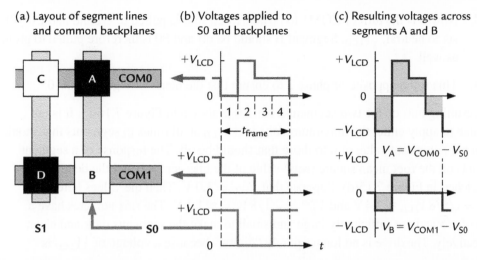

Figure 7.16: LCD with two-way multiplexing. (a) Two segment lines, S0 and S1, are shown with the two common backplanes COM0 and COM1. The waveforms in (b) are applied to the backplanes and segment line S0. Segments A and B feel the differences in voltage shown in (c), which turn A on (dark) and leave B off (clear).

line Sn now controls two individual segments, one on each of the two backplanes COM0 and COM1. The segments in the diagram are laid out in two lines for simplicity; the geometry would be more complicated in a real display but the connections follow the same pattern.

The sketches in Figure 7.16(b) show the waveforms applied to the backplanes and S0. Segment A is driven by S0 and COM0 and is on (dark), while B is driven by the same segment line S0 but the other backplane COM1 and is off (clear). Each period of the waveforms, called a *frame*, is divided into four phases:

1. The segments on COM0 are addressed in the first phase by pulling COM0 to ground (0 V). Segment A should be on and S0 is therefore driven to its maximum value, V_{LCD}. The bias across the segment is $V_{\text{A}} = V_{\text{COM0}} - V_{\text{S0}} = -V_{\text{LCD}}$. The segments on COM1 should be inactive during this phase and it is therefore put at a "neutral" voltage of $\frac{1}{2}V_{\text{LCD}}$. This gives $V_{\text{B}} = V_{\text{COM1}} - V_{\text{S0}} = -\frac{1}{2}V_{\text{LCD}}$.

2. The voltages in the second phase are the opposite of those in the first to ensure a pure AC signal with zero mean. This time COM0 is driven to V_{LCD} and S0 is pulled to ground to give $V_{\text{A}} = +V_{\text{LCD}}$. The backplane that is not being addressed, COM1, remains at its neutral voltage of $\frac{1}{2}V_{\text{LCD}}$ so that $V_{\text{B}} = +\frac{1}{2}V_{\text{LCD}}$.

3. Now it is the turn of COM1 to be addressed so it is pulled to ground and COM0 is set to neutral, $\frac{1}{2}V_{LCD}$. Segment B should be off and S0 is therefore pulled to ground as well.

4. This is the opposite of phase 3 to ensure that the mean voltage remains 0.

The resulting bias on the two segments A and B is shown in Figure 7.16(c). It is not possible to apply either the maximum voltage $\pm V_{LCD}$ at all times to segments that should be on nor a constant value of 0 to those that should be off. The response of a segment depends on the root mean square (rms) value of the bias across it and the dependence is sketched in Figure 7.15(d). Suppose that $V_{LCD} = 3.0$ V. Then the values here are $V_A^{rms} = \sqrt{5/8}\,V_{LCD} \approx 2.4$ V and $V_B^{rms} = \sqrt{1/8}\,V_{LCD} \approx 1.1$ V. The rms voltages have a ratio of $\sqrt{5}$ and are sufficiently large and small to make the segments dark and clear, respectively. The drive is no longer purely "digital" because a voltage of $\frac{1}{2}V_{LCD}$ is needed.

Example 7.17

What waveform should be applied to segment line S1 in Figure 7.16 so that C is clear and D is dark?

This scheme can be extended to greater multiplexing, such as the four-way multiplexing used in the SBLCDA4. However, greater contrast is obtained if two intermediate voltages of $\frac{1}{3}V_{LCD}$ and $\frac{2}{3}V_{LCD}$ are used instead of one. This makes the waveforms more complicated; see the *MSP430x4xx Family User's Guide*. Fortunately the hardware in the LCD driver generates the voltages automatically for static displays or those with two, three, or four-way multiplexing so there is no need to plow through the details here.

7.8   Driving an LCD from an MSP430x4xx

All devices in the MSP430x4xx family contain an LCD controller and newer variants have an enhanced version called the LCD_A. I use the TI MSP430FG4618/F2013 Experimenter's Board as an example for the rest of this chapter. The board provides a four-way multiplexed SoftBaugh SBLCDA4 display driven by an FG4618, which has the LCD_A controller. The board is described in the application note *MSP430FG4618/F2013 Experimenter's Board* (slaa213). To use the display, the program writes the desired pattern of segments to a buffer and the hardware does the rest. As usual the LCD controller must first be configured. It may also need some external components as well as the display.

7.8.1 Hardware to Drive the LCD

We saw in the previous section that the LCD needs a clock and that multiplexed displays require intermediate bias voltages to generate suitable waveforms. These are provided in different ways in the two controllers.

Clock and Bias for the LCD Controller

The older LCD controller does not contain a clock generator itself and obtains its clock from the BTCNT1 counter in the Basic Timer1 module. This is described in the section "Basic Timer1" on page 281 and divides ACLK by 32, 64, 128, or 256. The frequency of the clock for the LCD, f_{LCD}, should be chosen in the same way as for the LCD_A controller, explained later.

An external chain of resistors sets the levels of the intermediate bias voltages used to drive the segments and backplanes. Figure 7.17 shows the circuit for one-third bias, which is usually used with four-way multiplexing. The three resistors R have the same value, typically $100\,k\Omega$–$1\,M\Omega$. They are supplied from R33, which is connected internally to the analog supply at AV_{CC} when the module is enabled. A variable resistor R_x can be connected between R03 and ground (V_{SS}). Increasing its value decreases the voltages developed across the other resistors, which reduces the amplitude of the waveforms applied to the display. This acts as a contrast control. It can be omitted, in which case R03 is connected directly to ground.

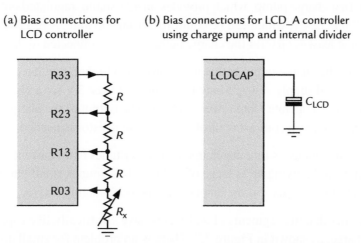

(a) Bias connections for LCD controller

(b) Bias connections for LCD_A controller using charge pump and internal divider

Figure 7.17: Circuit to provide the bias for (a) LCD controller and (b) LCD_A controller, assuming that the internal divider and charge pump are selected.

Clock and Bias for the LCD_A Controller

The newer LCD_A controller contains many of the functions that required other modules or components in the older LCD controller. It has its own prescaler to derive a clock from ACLK and does not need the Basic Timer1. The frequency can be divided by powers of 2 from 32 to 512.

A typical LCD needs to be refreshed at 30 Hz or faster to avoid flicker. Higher frequencies give a clearer display but consume more current. Figure 7.16 shows that each frame of the LCD clock contains two cycles for each common backplane, so a four-way multiplexed display needs eight clock cycles per frame. Thus the frequency f_{LCD} must be at least 240 Hz. Assuming that ACLK is at the usual 32 KHz, it should be divided by 32 K/240 = 136 or less, so a factor of 128 would probably be chosen.

There is an internal chain of resistors in LCD_A, which means that no external components are needed other than the display itself. Curiously, an external resistor chain can be used to reduce the current required. A variable resistor can be attached as a contrast control but there is a better way because the LCD_A offers three choices for the voltage to drive the display:

- Internal AV_{CC}, as in the LCD controller.

- An external voltage, which may be used with either the internal or an external divider.

- An internal charge pump, which provides an adjustable, regulated output in the range 2.60–3.44 V, which can be controlled from software. A reservoir capacitor C_{LCD} of at least 4.7 μF for the charge pump *must* be connected to the LCDCAP pin.

The advantage of the charge pump is that it can provide a higher, regulated voltage to the LCD than the general supply AV_{CC}. Many LCDs need at least 3 V for a clear display but the MSP430 itself can operate from lower voltages. The output of the charge pump can also be adjusted to act as a contrast control. No variable resistor is needed.

An inevitable disadvantage is that the charge pump consumes extra current. It draws short pulses of around 2 mA, giving an average of 2–4 μA. This sounds small but the rest of the module requires only 4–5 μA and the display may be permanently active.

I mentioned earlier that the segments of an LCD behave electrically like capacitors. These take time to charge, as shown in Figure 7.7. There is no problem for small displays but larger ones, those with digits more than about 20 mm high, may have so high a capacitance that they do not charge fully. Thus the voltage does not reach its intended value and gives poor contrast. Remedies are suggested in the application note *Driving Large LCDs with the LCD Peripheral of the MSP430* (slaa272). Another application note also concerns LCDs, *Using Two MSP430F4xx Devices to Drive Additional LCD Segments* (slaa293).

7.8.2 Software to Drive the LCD

The most tricky aspect of setting up a display in a new system is working out the relation between segments on the display and individual bits in the memory of the controller. The data sheet for the display gives the mapping between segments and pins, both for the segment pins and for the common backplanes. This is illustrated in Figure 7.18 for the SBLCDA4, where I take the least significant (right-hand) digit as an example. The four segments A, B, C, and D are connected to pin P14 and segments E, F, and G with the decimal point DP are connected to pin P13. The backplanes COM0–COM3 select which of the four segments is addressed by each pin during the phases of the LCD waveforms.

The next step is to trace the wiring of each pin on the display to the corresponding segment pin on the MSP430. Here we find that P14 and P13 are connected to S4 and S5, respectively. The remaining segment pins of the display are connected to S6–S25.[1] Pins S0–S3 of the FG4618 have reduced functionality and are therefore not used in this design.

Finally, the display buffer occupies registers LCDM1–LCDM20 (there is no LCDM0). Each LCDMn holds 1 byte, which is split into two nibbles for the 4 bits associated with

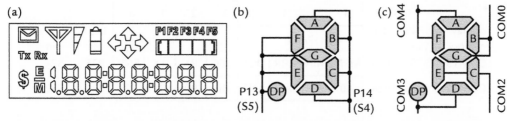

Figure 7.18: SoftBaugh SBLCDA4 display showing (a) all segments and detail of its least significant digit (1), with (b) the two segment lines and (c) the four common backplanes. Source: www.softbaum.com.

---

[1] Unfortunately the connections are labeled S0–S21 on some versions of the schematic diagram, not S4–S25. Yes, it caught me out.

each segment pin. For example, the lower nibble of LCDM1 holds the bits for S0 and its upper nibble holds S1. The least significant bit is for COM0 and the most significant bit of each nibble is for COM3. Twenty registers allow up to 160 segments to be driven.

We can now complete the mapping from segments of the display to bits in memory. Part of this map for the MSP430FG4618/F2013 Experimenter's Board is shown in Figure 7.19. Registers LCDM1 and LCDM2 are omitted because the corresponding segment pins S0–S3 are not used. The SBLCDA4 has 88 segments and therefore needs 22 segment pins and 11 display registers. The two segment pins for each digit are mapped into two nibbles of the same byte, which makes it very easy to address the display. For example, LCDM3 contains the bits for all segments of digit 1 and its associated decimal point. Four-way multiplexing is most convenient.

There is one more issue about the connections, which is to configure the pins for the correct functions. This is done in a slightly inconsistent way:

- The segment pins are configured using the registers LCDAPCTL0 and LCDAPCTL1 in the LCD_A module. Setting each bit assigns a set of four pins to the LCD. For example, LCDS0, the lsb of LCDAPCTL0, assigns S0–S3. The groups of four reduce the number of bits needed but mean that a few more pins may have to be assigned to the LCD than are needed. These settings generally override those set by registers in the port module.

display memory	MSP430 pin	LCD pin	COM: 3	2	1	0	3	2	1	0	LCD pin	MSP430 pin
			S_{n+1}				S_n					
LCDM13	S25	P26	MEM	MIN	ERR	DOL	8BC	RX	TX	ENV	P25	S24
LCDM12	S23	P24	A0	A1	A2	ANT	BB	B0	B1	BT	P23	S22
⋮	⋮	⋮									⋮	⋮
LCDM4	S7	P11	DP2	2E	2G	2F	2D	2C	2B	2A	P12	S6
LCDM3	S5	P13	DP1	1E	1G	1F	1D	1C	1B	1A	P14	S4
		Bit:	7	6	5	4	3	2	1	0		

Figure 7.19: Mapping of segments from the pins of the SBLCDA4 display to the segment pins and display memories of the FG4618 on the TI MSP430FG4618/F2013 Experimenter's Board.

- The backplanes are configured using the usual registers in the port module. The COM0 pin needs no configuration because it has no other function in the FG4618 but COM1–COM3 are shared with P5.2–P5.4. The corresponding bits in P5SEL must be set to enable the backplanes. It is easy to forget this because the registers are not in the LCD_A module. The result is a ghostly display where the segments initially darken when they are turned on but rapidly fade.

You will want to test the LCD after digesting all this. Listing 7.3 shows a program to do this. It starts with a blank display, then activates each segment in turn, starting with the lsb of LCDM3. After they have all been turned on, it turns them off in the same sequence. (The pattern of bits is the same as in a Johnson counter, a shift register with its output returned to the input through an inverter.) There is a delay between changing each segment, which is provided by Timer_A while the device is put into LPM3. It seemed easier to write the loop in the main function and return the MSP430 to active mode after each interrupt, as in the section "Returning from a Low-Power Mode to the Main Function" on page 203. The alternative would be to put all the work in the interrupt service routine.

Listing 7.3: Program `test1cd1.c` to test the LCD on the TI MSP430FG4618/F2013 Experimenter's Board by turning on all segments in turn, then turning them off again.

```
// testlcd1.c - test SBLCD4 display on TI Experimenter's Board
// PCB tracks S0-S21 connected to MSP430 pins S4-S25 - beware
// Default clocks; ACLK from 32KHz crystal; minimal configuration
// Turns on all LCD segments from least significant bit and segment,
//   then turns them off in the same sequence - a Johnson counter
// Flashes LEDs 1 and 2 on P2.2 and P2.1 to confirm operation of timer
// Timer clock 32KHz ACLK, no division, up mode, period 0x1000 = 1/8 s
// J H Davies, 2006-11-22; IAR Kickstart version 3.41A
//--------------------------------------------------------------------
#include <io430xG46x.h>              // Specific device
#include <intrinsics.h>             // Intrinsic functions
#include <stdint.h>                 // Integers of defined sizes
//--------------------------------------------------------------------
#define LCDMEMS 11                  // LCD memories used (3-13)
// Pointer to LCD memory used: allows use of array LCDMem[]
uint8_t * const LCDMem = (uint8_t *) &LCDM3;
//--------------------------------------------------------------------
// Function prototypes
void InitLCD (void);
//--------------------------------------------------------------------
void main (void)
{
    int i, j;                       // Loop counters
    enum {up, down} direction;      // Direction of Johnson counter

    WDTCTL = WDTPW|WDTHOLD;          // Stop watchdog timer
```

```
    P2OUT = BIT2;                          // LED1 (P2.2) on, LED2 (P2.1) off
    P2DIR = BIT1|BIT2;                     // Set pins with LEDs to output
    InitLCD();                            // Initialize SBLCDA4
    TACCR0 = 0x1000;                      // Upper limit of count for TAR
    TACCTL0 = CCIE;                       // Enable Compare 0 interrupts
    TACTL = MC_1|TASSEL_1|TACLR;          // Set up and start Timer A
// "Up to CCR0" mode, no clock division, clock from ACLK, clear timer
    direction = up;                       // Start Johnson counter up
    for (;;) {                            // Loop forever
        for (i = 0; i < LCDMEMS; ++i) {   // Count through LCD memories
            for (j = 0; j < 8; ++j) {     // Count through bits
                __low_power_mode_3();     // LPM3 provides delay
                if (direction == up) {    // Shift left, insert 1 on right
                    LCDMem[i] = (LCDMem[i] << 1) | BIT0;
                } else {                  // Shift left, insert 0 (default)
                    LCDMem[i] = (LCDMem[i] << 1);
                }
            }
            __low_power_mode_3();         // extra delays after each character
            __low_power_mode_3();
        }
        if (direction == up) {            // Reverse direction after all chars
            direction = down;
        } else {
            direction = up;
        }
    }
}
//-------------------------------------------------------------------------
// Initialize SBLCDA4
//-------------------------------------------------------------------------
void InitLCD (void)
{
    int i;
    for(i = 0; i < LCDMEMS; ++i) {        // Clear LCD memory used
        LCDMem[i] = 0;
    }
    P5SEL = BIT4 | BIT3 | BIT2;           // Select COM[3:1] function
    LCDAPCTL0 = LCDS4|LCDS8|LCDS12|LCDS16|LCDS20|LCDS24;
                                          // Enable LCD segs 4-27 (4-25 used)
    LCDAVCTL0 = 0;                        // No charge pump, everything internal
    LCDACTL = LCDFREQ_128 | LCD4MUX | LCDSON | LCDON;
}                                         // ACLK/128, 4mux, segments on, LCD_A on
//-------------------------------------------------------------------------
// Interrupt service routine for Timer A channel 0
// Processor returns to "calling routine" in active mode
//-------------------------------------------------------------------------
#pragma vector = TIMERA0_VECTOR
__interrupt void TIMERA0_ISR (void)
{
    P2OUT ^= BIT1|BIT2;                   // Toggle LEDs
    __low_power_mode_off_on_exit();       // Restore Active Mode on return
}
```

A few points about this program are worth mentioning:

1. The buffer memory is defined as separate bytes in io430xG46x.h but it is much more convenient to address them as an array. I therefore defined a pointer that can be treated as an array LCDMem[LCDMEMS]. Its base address is set to LCDM3 because the two lowest registers are not used.

2. I use an enumerated type for the direction for clarity.

3. To save paper I have not configured the device fully: neither the oscillators nor the ports. Sorry, this is poor practice but keeps the printout concise.

4. Several steps are needed to configure LCD_A, most of which have been covered already.

   - The display buffer is first cleared to ensure that the display will be blank when turned on. This is much easier with the array.

   - The shared pins of port 5 are assigned to the backplanes COM1–COM3; this is not needed for COM0, which is a dedicated pin.

   - The segment pins are assigned next. Further pins can be assigned in LCDAPCTL1 for a larger display.

   - Register LCDAVCTL0 controls the options for powering and biasing the display. I left it with the default values so that the charge pump is not used and the LCD is supplied directly from internal power, the bias chain is internal, there is no variable resistor to adjust the contrast, and one-third bias is used. If the charge pump is activated, LCDAVCTL1 is used to adjust its output voltage.

   - Finally, LCDACTL configures the overall functions of LCD_A. The clock is divided by 128 from ACLK (this would need Basic Timer1 in a device with the LCD controller rather than LCD_A), four-way multiplexing is selected and LCDON turns the module on. The behavior of the LCDSON bit is a little strange. Clearing this bit causes the display to be blanked but keeps the module running, which is useful for flashing the display. It is cleared on reset but appears to be set automatically when the module is turned on. It may therefore not be necessary to set LCDSON explicitly but I have played safe and included it.

5. The main task of the interrupt service routine for the timer is to restore active mode on return. I have also toggled a pair of LEDs. This makes the board look

pretty. More seriously, it shows that Timer_A and interrupts are working. This eliminates some possible causes if the display does not function.

6. The main loop acts like a shift register on buffer memory. Bits are set successively, starting with the lsb, in the "up" direction and cleared going "down." This is done by a left-shift of the register that contains the current transition between clear and set bits. A 1 is introduced in the lsb if the direction is up. The loop is paced by placing the device in low-power mode 3, which is terminated by Timer_A. There is an extra delay when the direction changes.

Example 7.18

Rewrite the program to test the display with a different pattern. Set only bit 0 in *all* bytes of the display buffer, then pause. Next, clear bit 0, set bit 1, and pause again. Repeat from bit 0 after bit 7 has been set and cleared. It would be better to increase the delay.

Example 7.19

Turn on all the segments as in Listing 7.3 once and then flash the display on and off a few times using the LCDSON bit. This sort of display is often included as part of the startup code to reassure the user that the display is working.

Checklist—Have You...

❏ Connected a capacitor to LCDCAP if you are using the charge pump to supply an LCD?

❏ Routed pins to the LCD for both the segments and backplanes?

7.9   Simple Applications of the LCD

There is one more step to make the LCD useful, which is to display meaningful characters rather than patterns. This requires appropriate patterns of segments to be defined for the digits 0–9 and other symbols of interest. The extra hexadecimal digits A–F are useful during development, for instance. Here are a few simple examples.

7.9.1   A Very Simple Clock

A good starting place is the program for a clock in Listing 7.4. It displays hours, minutes, and seconds in a 24-hour format. There is no way for the user to set the time, which might be a small impediment in a commercial product, but you can use the debugger.

Listing 7.4: Program `sclock1.c` to display a simple, real-time clock on the LCD.

```c
// sclock1.c - simple clock using SBLCD4 on TI Experimenter's Board
// Default clocks; ACLK from 32KHz crystal; minimal configuration
// Timer clock 32KHz ACLK, no division, up mode, period 0x8000 = 1s
// J H Davies, 2007-04-23; IAR Kickstart version 3.42A
//-------------------------------------------------------------------
#include <io430xG46x.h>                  // Specific device
#include <intrinsics.h>                  // Intrinsic functions
#include <stdint.h>                      // Integers of defined sizes
//-------------------------------------------------------------------
#define LCDMEMS 11                       // LCD memories used (3-13)
// Pointer to LCD memory used: allows use of array LCDMem[]
uint8_t * const LCDMem = (uint8_t *) &LCDM3;
// LCD segment definitions (SoftBaugh SBLCDA4)
#define SEG_A    BIT0                    //    AAAA
#define SEG_B    BIT1                    //   F    B
#define SEG_C    BIT2                    //   F    B
#define SEG_D    BIT3                    //    GGGG
#define SEG_E    BIT6                    //   E    C
#define SEG_F    BIT4                    //   E    C
#define SEG_G    BIT5                    //    DDDD
#define SEG_H    BIT7                    // colon, point etc
// Patterns for hexadecimal characters
const uint8_t LCDHexChar[] = {
    SEG_A | SEG_B | SEG_C | SEG_D | SEG_E | SEG_F,          // '0'
    SEG_B | SEG_C,                                          // '1'
    SEG_A | SEG_B | SEG_D | SEG_E | SEG_G,                  // '2'
    SEG_A | SEG_B | SEG_C | SEG_D | SEG_G,                  // '3'
    SEG_B | SEG_C | SEG_F | SEG_G,                          // '4'
    SEG_A | SEG_C | SEG_D | SEG_F | SEG_G,                  // '5'
    SEG_A | SEG_C | SEG_D | SEG_E | SEG_F | SEG_G,          // '6'
    SEG_A | SEG_B | SEG_C,                                  // '7'
    SEG_A | SEG_B | SEG_C | SEG_D | SEG_E | SEG_F | SEG_G,  // '8'
    SEG_A | SEG_B | SEG_C | SEG_D | SEG_F | SEG_G,          // '9'
    SEG_A | SEG_B | SEG_C | SEG_E | SEG_F | SEG_G,          // 'A'
    SEG_C | SEG_D | SEG_E | SEG_F | SEG_G,                  // 'b'
    SEG_A | SEG_D | SEG_E | SEG_F,                          // 'C'
    SEG_B | SEG_C | SEG_D | SEG_E | SEG_G,                  // 'd'
    SEG_A | SEG_D | SEG_E | SEG_F | SEG_G,                  // 'E'
    SEG_A | SEG_E | SEG_F | SEG_G,                          // 'F'
};
// Patterns for AM ('A'), PM ('P') and blank
const uint8_t LCDAMChar = SEG_A | SEG_B | SEG_C | SEG_E | SEG_F | SEG_G;
const uint8_t LCDPMChar = SEG_A | SEG_B | SEG_E | SEG_F | SEG_G;
const uint8_t LCDBlankChar = 0;
//-------------------------------------------------------------------
// Function prototypes
void InitLCD (void);
//-------------------------------------------------------------------
void main (void)
{
    WDTCTL = WDTPW|WDTHOLD;              // Stop watchdog timer
    InitLCD();                          // Initialize SBLCDA4
    TACCR0 = 0x8000;                    // Upper limit of count for TAR
```

```
    TACCTL0 = CCIE;                        // Enable Compare 0 interrupts
    TACTL = MC_1|TASSEL_1|TACLR;           // Set up and start Timer A
// "Up to CCR0" mode, no clock division, clock from ACLK, clear timer
    for (;;) {                             // Loop forever
        __low_power_mode_3();              // ACLK continues to run
    }
}
//-----------------------------------------------------------------------
// Initialize SBLCDA4
//-----------------------------------------------------------------------
void InitLCD (void)
{
    int i;
    for(i = 0; i < LCDMEMS; ++i) {         // Clear LCD memories used
        LCDMem[i] = 0;
    }
    P5SEL = BIT4|BIT3|BIT2;                // Select COM[3:1] function
    LCDAPCTL0 = LCDS4|LCDS8|LCDS12|LCDS16|LCDS20|LCDS24;
                                 // Enable LCD segs 4-27 (4-25 used)
    LCDAVCTL0 = 0;               // No charge pump, everything internal
    LCDACTL = LCDFREQ_128 | LCD4MUX | LCDSON | LCDON;
                                 // ACLK/128, 4mux, segments on, LCD_A on
}
//-----------------------------------------------------------------------
// Interrupt service routine for Timer A channel 0
// Time stored as seconds, minutes and hours in BCD format
// Need "unsigned short" to suit intrinsic function __bcd_add_short()
//-----------------------------------------------------------------------
#pragma vector = TIMERA0_VECTOR
__interrupt void TIMERA0_ISR (void)
{
    static unsigned short seconds = 0x59;   // initial time in BCD
    static unsigned short minutes = 0x59;
    static unsigned short hours = 0x23;
// Update time
    seconds = __bcd_add_short(seconds, 0x01);   // ++seconds
    if (seconds >= 0x60) {
        seconds = 0;                             // start new minute
        minutes = __bcd_add_short(minutes, 0x01);
        if (minutes >= 0x60) {
            minutes = 0;                         // start new hour
            hours = __bcd_add_short(hours, 0x01);
            if (hours >= 0x24) {
                hours = 0;                       // start new day
            }
        }
    }
// Update all digits of display; SEG_H is colon separator
    LCDMem[1] = LCDHexChar[seconds & 0x0F];
    LCDMem[2] = LCDHexChar[(seconds >> 4) & 0x0F] | SEG_H;
    LCDMem[3] = LCDHexChar[minutes & 0x0F];
    LCDMem[4] = LCDHexChar[(minutes >> 4) & 0x0F] | SEG_H;
    LCDMem[5] = LCDHexChar[hours & 0x0F];
    LCDMem[6] = LCDHexChar[(hours >> 4) & 0x0F];
}
```

The definitions start with the mapping from segments to bits in each buffer. These are assembled into the patterns for hexadecimal characters in the array LCDHexChar[]. A few extra characters are defined for convenience.

In this case I chose to embed all the action in the interrupt service routine. Again I used Timer_A although there might be better choices—wait for Chapter 8. The time is stored with hours, minutes, and seconds in separate variables. They are defined slightly wastefully as unsigned short to be compatible with the intrinsic functions for BCD arithmetic. Note that they must be static so that they retain their values between invocations of the ISR. The time will never advance if you forget this. They are initialized to their maximum values so that the display will update to midnight when the ISR is first called.

This is a good example of the use of binary-coded decimal (BCD) because it avoids any need to decode the values before displaying them. Each decimal digit is held in a separate nibble (4 bits). The variables are incremented decimally by calling the intrinsic function _ _bcd_add_short(), which in turn uses the MSP430's dadd instruction. This ensures that 0x09 is incremented in BCD to 0x10, rather than to 0x0A as in a straightforward binary number. The two digits in each variable must then be displayed separately. Masking with 0x0F picks out the lowest nibble and the shift operator ≫ 4 moves the second nibble into the lowest nibble so that it can be selected.

Example 7.20

Try this in the debugger. Add the variables of the clock to a Watch window so that you can change them to check the program. Use hexadecimal format to match the storage of the time as BCD.

Example 7.21

Modify the clock so that it does the minimum work to update the stored time and display. The displayed value of minutes and hours should be updated only when they change, not every second. Unfortunately this makes the program awkward to debug. If you change the value of hours, for instance, it will not become apparent until the minutes next roll over.

Example 7.22

Modify the clock so that it no longer displays seconds, but just hours and minutes. Flash the separating colon with a period of 1 s to show that the clock is alive.

Example 7.23

Rewrite the clock for a 12 hour display. Use digit 0 to show a trailing A or P for AM and PM. A leading zero is usually suppressed to show 1:23 rather than 01:23 although this is a matter of taste. If I remember correctly, the display of time during AM or PM should run from 12:00 to 12:59, then 1:00 to 11:59. (Sorry—the 24 hour clock is much more straightforward.)

The application note *Implementing a Real-Time Clock on the MSP430* (slaa076a) addresses important practical issues, such as the accuracy and power consumption of the clock. The basic timer has been extended to include a real-time clock module in MSP430FG4618, which makes time-keeping much easier. See the section "Real-Time Clock" on page 283.

7.9.2   Hexadecimal Display

It is useful to be able to display hexadecimal words (16 bits) when debugging a program. This requires four digits of the display, which leaves room to show an *h* or *H* on the right as a reminder that the value is hexadecimal. You could use the $ sign or add a leading zero as well if you wished. A straightforward function is shown in Listing 7.5. Successive nibbles are moved into the least significant position so that each digit can be displayed from right to left. The unused positions of the seven-digit display are cleared.

Listing 7.5: Function from `disphex1.c` to display a hexadecimal word on the LCD.

```
// Display word in hexadecimal, 4 digits followed by 'h' (or 'H')
//-----------------------------------------------------------------
#define LCDDIGITS    7                 // Number of digits in display
void DisplayHex (uint16_t HexValue)
{
    uint8_t i;                         // Index for LCD array

    LCDMem[0] = LCDhChar;              // 'h' for hexadecimal on right
    for (i = 1; i <= 4; ++i) {         // Display 4 hex digits
        LCDMem[i] = LCDHexChar[HexValue & 0x000F];
        HexValue >>= 4;               // Move next nibble into position
    }
    while (i < LCDDIGITS) {            // Clear more significant digits
        LCDMem[i++] = LCDBlankChar;   //   of numerical display
    }
}
```

Example 7.24

Extend this function to display a 32-bit hexadecimal value. This sounds trivial but the SBLCDA4 display has only 7½ digits. You will need to think of a good way of handling values over 0x1FFFFFFF.

7.9.3 Decimal Display

It takes more effort to display a value as a decimal number rather than hexadecimal because it must first be converted from binary to BCD so that each decimal digit can be shown separately. Suppose that we are dealing with signed integers (16 bits). Their values run from −32,768 to 32,767, which require five digits and a minus sign to display. Leading zeros should be suppressed for clarity so that 1 is shown as a single digit, not as 00001. You could call library functions such as sprintf() for complicated formatting but these take up a lot of memory that would be wasted for a simple display.

The first step in Listing 7.6 is to store the sign of the number and convert the magnitude to BCD. I describe the function UIntToBCD() that does the conversion shortly. The binary value fits into a word (16 bits) but 32 bits are needed for the result because the BCD value may require five decimal digits. The display is loaded in a slightly different way because the number of nonzero digits is not known and the loop therefore runs until the remaining value is zero. There is no danger of exceeding the number of digits on the display so I have not checked. The do-while loop ensures that at least one digit is always written so that a value of 0 does not vanish completely. A minus sign is displayed if needed and unused digits are again cleared.

Listing 7.6: Function from dispint1.c to display a signed, decimal word on the LCD.

```
// Display signed, 16-bit integer (int16_t)
// Strip sign, convert unsigned value to BCD and display
// Leading zeros suppressed; BCD value does not exceed 5 digits
// -------------------------------------------------------------
#define LCDDIGITS   7                  // Number of digits in display
void DisplayInt (int16_t IntValue)
{
    uint8_t i;                         // Index for LCD array
    uint32_t BCDValue;                 // Value converted bin to BCD
    enum {plus, minus} sign;

    if (IntValue >= 0) {               // Keep track of sign
        sign = plus;
    } else {
```

```
        sign = minus;
        IntValue = -IntValue;                  // Conversion needs IntValue>=0
    }
    BCDValue = UIntToBCD (IntValue);    // Convert binary to BCD
    i = 0;                              // Index for LCD memories
    do {                               // Store pattern for next digit
        LCDMem[i++] = LCDHexChar[BCDValue & 0x000F];
        BCDValue >>= 4;                 // Move next nibble down
    } while (BCDValue > 0);             // (Always display first digit)
    if (sign == minus) {
        LCDMem[i++] = LCDMinusChar;     // Prepend minus sign
    }
    while (i < LCDDIGITS) {             // Clear more significant digits
        LCDMem[i++] = LCDBlankChar;     //   of numerical display
    }
}
```

7.9.4   Conversion from Binary to Binary-Coded Decimal

The hardest part of the process for displaying a decimal value is the conversion from binary to decimal. There are many approaches. The most straightforward way is to use division and multiplication by 10 to extract each decimal digit. The method relies on the property that integer division rounds down:

$$\text{tens} = \text{value}/10$$

$$\text{units} = \text{value} - 10 \times \text{tens}$$

This extracts the units and the process can be repeated on the tens to produce the digits in ascending significance. Unfortunately this approach is slow on a processor that lacks hardware for division and multiplication. There are more economical methods that rely on basic, binary operations, such as the "shift and add 3" algorithm. However, a better approach is to use the instruction for decimal addition built into the MSP430.

Suppose that you want to find the decimal value of a binary number by hand. Take the 4-bit number *DCBA* for simplicity. Its value is

$$\text{value} = 8D + 4C + 2B + A$$
$$\equiv 2^3 D + 2^2 C + 2B + A.$$

This requires a lot of multiplications by 2, which can be reduced by looking for common factors. One factor of 2 is common to D, C, and B and another factor of 2 is shared by D and C. Thus the value can be evaluated more efficiently as

$$\text{value} = 2(2(2D + C) + B) + A.$$

This is known as *Horner's algorithm* and can be written as a sequence of operations, which lends itself to a loop that works from the inside out:

$$\text{value} = D$$
$$\text{value} = 2 \times \text{value} + C$$
$$\text{value} = 2 \times \text{value} + B$$
$$\text{value} = 2 \times \text{value} + A.$$

All bits can be treated identically by replacing the first line with these two:

$$\text{value} = 0$$
$$\text{value} = 2 \times \text{value} + D$$

The values are calculated and displayed decimally if you perform the operations on a pocket calculator. We can achieve the same result by using the decimal addition instruction dadd in the MSP430. The result then is binary-coded decimal rather than the usual binary value. Multiplication by 2 just requires the addition of the value to itself. We need to use the intrinsic function _ _bcd_add_long() to call dadd for the decimal addition of two 32-bit numbers in C.

Listing 7.7 shows the function UIntToBCD(). I put it in a separate file because the next section shows how to use assembly language for this function instead. It is simple to use more than one source file in a project: Just add the extra files to the project and choose Make to compile and link all of them. The main file needs a prototype for UIntToBCD(), which is used to check that the call of the function matches its definition.

Listing 7.7: Function in C to convert an unsigned 16-bit number to binary-coded decimal.

```
// unittobcd.c: Convert unsigned, 16-bit integer to BCD
// Needs 5 nibbles for converted result, so 32-bit value
// Based on Application Reports, section 5.5.3, Lutz Bierl
// Loop works from msb to lsb of input using Horner's algorithm
// In each interation, the converted value is doubled decimally
//   to allow for the factor of 2 going from bit to bit
// It is also incremented if the current bit of the input is set
//---------------------------------------------------------------
#include <io430xG46x.h>            // Specific device (for BITF!)
#include <intrinsics.h>            // Intrinsic functions
#include <stdint.h>                // Integers of defined sizes
//---------------------------------------------------------------
```

```
uint32_t UIntToBCD (uint16_t UIntValue)
{
    uint32_t converted;
    uint8_t i;

    converted = 0;
    for (i = 0; i < 16; ++i) {
        converted = __bcd_add_long (converted, converted);
        if ((UIntValue & BITF) != 0) {
            converted = __bcd_add_long (converted, 1);
        }
        UIntValue <<= 1;
    }
    return converted;
}
```

Inside the loop the current value is first doubled, then the current bit is tested and 1 is added if the bit is set. The binary input value is shifted one place in each iteration so that it can be tested against a fixed mask. The mask could be shifted instead but this version is closer to one in assembly language that is described next.

7.9.5   Conversion from Binary to BCD in Assembly Language

The function in Listing 7.7 relies heavily on an intrinsic function, which provides access to a native instruction not available directly from C. Why not write the complete function in assembly language instead? It is not difficult and this conversion provides a convenient example, although it would be pointless to go to this trouble for a function that is not called often.

I mentioned the general approach briefly in the section "Mixing C and Assembly Language" on page 185. The vital feature is to ensure that the assembly language reads the parameters from, and returns its value to, the places where the C compiler expects. *This depends on the conventions used by each compiler*, which means that a program with mixed C and assembly language lacks portability and may not work if you change development environment. Be careful. This is a good reason for avoiding the technique wherever possible. Here we pass a single 16-bit value to the function, which returns a single 32-bit value. The details of the interface between C and assembly language for EW430 are given in the chapter "Assembler Language Interface" of the *Compiler Reference Guide*:

- A single 16-bit parameter is passed in register R12.

- A single 32-bit value is returned in registers R13:R12.

- Values in registers R4–R11 must be saved by the function.

The calculations here are simple enough that we do not have to worry about the last point; the function uses only R12–R15, whose contents need not be preserved.

Overall, Listing 7.8 looks very similar to subroutines in assembly language that we have written before. The only new feature is the directive PUBLIC UIntToBCD. This tells the assembler to make the function available outside the file (export it) so that it can be called by the function in C. There is no need for a header file because we do not use any special function registers. Nor do any of the interrupt or reset vectors appear, because all that is handled by the functions in C. On the other hand, RSEG is still needed to ensure that the code is stored in the correct part of memory. Extra memory for local variables can be allocated on the stack or in RAM as in the section "Storage for Local Variables" on page 179.

Listing 7.8: Function in assembly language called from C to convert an unsigned 16-bit number to binary-coded decimal.

```
; Assembly routine to convert 16-bit unsigned binary value to BCD
; Taken from Application Reports, section 5.5.3
; Called from C as: uint32_t UIntToBCD (uint16_t UIntValue);
; 16-bit parameter "UIntValue" passed in R12
; 32-bit value "UIntToBCD" returned in R13:R12
; Uses only the scratch registers R12-R15 so no stacking necessary
; ----------------------------------------------------------------
        PUBLIC  UIntToBCD       ; Export symbol outside this file
        RSEG    CODE            ; Essential as usual
UIntToBCD:                      ; Name of function as usual
    mov.w   R12,R14         ; Move input and leave R12 free for result
    clr.w   R12             ; Clear registers for result
    clr.w   R13
    mov.w   #0x0010,R15     ; Initialize loop counter to number of bits
LoopStart:
    rla.w   R14             ; Shift msb of input into carry bit
    dadd.w  R12,R12         ; R13:R12 = 2*R13:R12 + carry bit DECIMALLY
    dadd.w  R13,R13         ;    lsword then msword
    dec.w   R15             ; Decrement loop counter
    jnz     LoopStart       ; Repeat if nonzero
    reta                    ; Return instruction in MSP430X
;   ret                     ; Usual MSP430 return instruction
    END
```

The algorithm is taken from Section 5.5.3 of *Application Reports* (slaa024). The principle is the same as the function in C but is tuned to exploit the details of assembly language. In particular, dadd is really "decimal add *with carry*." The first step is to copy the input value from R12 into R14, which leaves R12 and R13 free to accumulate the result. The only other register needed is R15 for the loop counter. In each iteration, the most significant remaining bit of input is rotated from R14 into the carry bit, where it is ready to be added to the less significant word of the result, R12. The single instruction `dadd.w R12,R12` doubles R12 and adds the carry bit decimally. The carry bit is set if the result overflows and is added to the more significant word R13 when that is doubled in the next line. It is a remarkably compact code. The subroutine ends with `reta` rather than the usual `ret` because the FG4618 has the extended MSP430X CPU (see the section "Architecture of the MSP430X" on page 601).

Is this worth the trouble? I found that `UIntToBCD()` took 302 cycles in C and 110 cycles in assembly language. This large difference arises mainly because the intrinsic function `__bcd_add_long ()` is more general than we need for this application and a lot of time is wasted in copying the arguments and return value.

Example 7.25

If you have a suitable board, investigate how the clarity of the LCD depends on the voltage from the charge pump. Use the LCD to show the voltage produced by the pump, which can be changed by pushing a button or at regular intervals. Compare the display driven directly from AV_{CC}.

CHAPTER 8

Timers

Most modern microcontrollers provide a range of timers and the MSP430 is no exception. All devices contain two types of timer and some have five. Each type of timer module works in essentially the same way in all devices. Timer_A is identical in almost all MSP430s, for instance, except that a few have a different number of capture/compare channels.

Watchdog timer: Included in all devices (newer ones have the enhanced watchdog timer+). Its main function is to protect the system against malfunctions but it can instead be used as an interval timer if this protection is not needed.

Basic timer1: Present in the MSP430x4xx family only. It provides the clock for the LCD and acts as an interval timer. Newer devices have the LCD_A controller, which contains its own clock generator and frees the basic timer from this task.

Real-time clock: In which the basic timer has been extended to provide a real-time clock in the most recent MSP430x4xx devices.

Timer_A: Provided in all devices. It typically has three channels and is much more versatile than the simpler timers just listed. Timer_A can handle external inputs and outputs directly to measure frequency, time-stamp inputs, and drive outputs at precisely specified times, either once or periodically. There are internal connections to other modules so that it can measure the duration of a signal from the comparator, for instance. It can also generate interrupts. We used a few of its capabilities in earlier chapters and most of this chapter is devoted to Timer_A.

Timer_B: Included in larger devices of all families. It is similar to Timer_A with some extensions that make it more suitable for driving outputs such as pulse-width modulation. Against this, it lacks a feature of sampling inputs in Timer_A that is useful in communication.

The simplest way to generating a delay is to use a software loop as in the section "Automatic Control Flashing Light by Software Delay" on page 91. This should be avoided wherever possible in favor of one of the hardware timers because the timers are more precise and leave the CPU free for more productive activities. Alternatively, the device can be put into a low-power mode if there is nothing else to be done. We already saw that the MSP430 spends much of its time asleep in many applications and is awakened periodically by a timer.

8.1 Watchdog Timer

The main purpose of the watchdog timer is to protect the system against failure of the software, such as the program becoming trapped in an unintended, infinite loop. Left to itself, the watchdog counts up and resets the MSP430 when it reaches its limit. The code must therefore keep clearing the counter before the limit is reached to prevent a reset.

The operation of the watchdog is controlled by the 16-bit register WDTCTL. It is guarded against accidental writes by requiring the password WDTPW = 0x5A in the upper byte. A reset will occur if a value with an incorrect password is written to WDTCTL. This can be done deliberately if you need to reset the chip from software. Reading WDTCTL returns 0x69 in the upper byte, so reading WDTCTL and writing the value back violates the password and causes a reset.

The lower byte of WDTCTL contains the bits that control the operation of the watchdog timer, shown in Figure 8.1. The $\overline{\text{RST}}$/NMI pin is also configured using this register, which you must not forget when servicing the watchdog—we see why shortly. This pin is described in the section "Nonmaskable Interrupts" on page 195. Most bits are reset to 0 after a power-on reset (POR) but are unaffected by a power-up clear (PUC). This distinction is important in handling resets caused by the watchdog. The exception is the

7	6	5	4	3	2	1	0
WDT-HOLD	WDT-NMIES	WDTNMI	WDT-TMSEL	WDT-CNTCL	WDTSSEL	WDTISx	
rw–(0)	rw–(0)	rw–(0)	rw–(0)	r0(w)	rw–(0)	rw–(0)	rw–(0)

Figure 8.1: The lower byte of the watchdog timer control register WDTCTL.

WDTCNTCL bit, labeled r0(w). This means that the bit always reads as 0 but a 1 can be written to stimulate some action, clearing the counter in this case.

The watchdog counter is a 16-bit register WDTCNT, which is not visible to the user. It is clocked from either SMCLK (default) or ACLK, according to the WDTSSEL bit. The reset output can be selected from bits 6, 9, 13, or 15 of the counter. Thus the period is $2^6 = 64$, 512, 8192, or 32,768 (default) times the period of the clock. This is controlled by the WDTISx bits in WDTCTL. The intervals are roughly 2, 16, 250, and 1000 ms if the watchdog runs from ACLK at 32 KHz.

The watchdog is always active after the MSP430 has been reset. By default the clock is SMCLK, which is in turn derived from the DCO at about 1 MHz. The default period of the watchdog is the maximum value of 32,768 counts, which is therefore around 32 ms. You must clear, stop, or reconfigure the watchdog before this time has elapsed. In almost all programs in this book, I take the simplest approach of stopping the watchdog, which means setting the WDTHOLD bit. This goes back to the first program to light LEDs, Listing 4.2.

If the watchdog is left running, the counter must be repeatedly cleared to prevent it counting up as far as its limit. This is done by setting the WDTCNTCL bit in WDTCTL. The task is often called *petting*, *feeding*, or *kicking the dog*, depending on your attitude toward canines. The bit automatically clears again after WDTCNT has been reset.

The MSP430 is reset if the watchdog counter reaches its limit. Recall from the section "Resets" on page 157 that there are two levels of reset. The watchdog causes a power-up clear, which is the less drastic form. Most registers are reset to default values but some retain their contents, which is vital so that the source of the reset can be determined. The watchdog timer sets the WDTIFG flag in the special function register IFG1. This is cleared by a power-on reset but its value is preserved during a PUC. Thus a program can check this bit to find out whether a reset arose from the watchdog.

Listing 8.1 shows a trivial program to demonstrate the watchdog. I selected the clock from ACLK (WDTSSEL = 1) and the longest period (WDTISx = 00), which gives 1 s with a 32 KHz crystal for ACLK. It is wise to restart any timer whenever its configuration is changed so I also cleared the counter by setting the WDTCNTCL bit. LED1 shows the state of button B1 and LED2 shows WDTIFG. The watchdog is serviced by rewriting the configuration value in a loop while button B1 is held down. If the button is left up for more than 1 s the watchdog times out, raises the flag WDTIFG, and resets the device with a PUC. This is shown by LED2 lighting.

Listing 8.1: Program `wdtest1.c` to demonstrate the watchdog timer.

```
// wdtest1.c - trival program to demonstrate watchdog timer
// Olimex 1121STK board, 32KHz ACLK
// J H Davies, 2007-05-10; IAR Kickstart version 3.42A
//-------------------------------------------------------------
#include <io430x11x1.h>                      // Specific device
//-------------------------------------------------------------
// Pins for LEDs and button
#define LED1    P2OUT_bit.P2OUT_3
#define LED2    P2OUT_bit.P2OUT_4
#define B1      P2IN_bit.P2IN_1
// Watchdog config: active, ACLK/32768 -> 1s interval; clear counter
#define WDTCONFIG   (WDTCNTCL|WDTSSEL)
// Include settings for _RST/NMI pin here as well
//-------------------------------------------------------------
void main (void)
{
    WDTCTL = WDTPW | WDTCONFIG;              // Configure and clear watchdog
    P2DIR = BIT3 | BIT4;                     // Set pins with LEDs to output
    P2OUT = BIT3 | BIT4;                     // LEDs off (active low)
    for (;;) {                               // Loop forever
        LED2 = ~IFG1_bit.WDTIFG;             // LED2 shows state of WDTIFG
        if (B1 == 1) {                       // Button up
            LED1 = 1;                        // LED1 off
        } else {                             // Button down
            WDTCTL = WDTPW | WDTCONFIG;      // Feed/pet/kick/clear watchdog
            LED1 = 0;                        // LED1 on
        }
    }
}
```

I have to admit with embarrassment that this program gave me a little trouble. In my first
version I serviced the watchdog with the following line of C. The obvious idea was to clear
the counter and leave everything else as it was:

```
    WDTCTL = WDTPW | WDTCNTCL;        // Feed/pet/kick/clear watchdog
```

In my defense, this is given as an example in the family user's guide. Distressingly, the
MSP430 seemed to reset as soon as the button was released, even if this was less than 1 s
after the power was applied. The reason is that this line overwrites the complete register and
clears all the bits other than WDTCNTCL. In particular, the clock reverts to SMCLK, which
reduces the period to 32 ms instead of 1 s. It also wipes out the settings for the $\overline{\text{RST}}$/NMI
pin if I changed them. The correct approach is to rewrite the *complete* configuration
to WDTCTL as in the listing. Do not try to set only the WDTCNTCL bit either, as follows:

```
    WDTCTL_bit.WDTCNTCL = 1;          // Crashes
```

This causes a reset because it violates the password protection. Maybe the watchdog is not quite so simple after all.

Example 8.1

Experiment with Listing 8.1, which you may need to adapt to your hardware. You might wish to try some of the stupid mistakes too—it is a good way of remembering them. You should find that LED2 remains lit permanently after a PUC because the WDTIFG bit needs to be cleared in software and I deliberately omitted this. It is cleared only after a POR, which can be requested by the Reset command in the debugger.

Listing 8.2 shows a traditional example of how the watchdog is used to protect a program built around a paced loop. The general idea was explained in the section "Returning from a Low-Power Mode to the Main Function" on page 203. A timer is configured to generate interrupts every 10 ms. Its interrupt service routine restores active mode, which causes the program to resume the main loop. This services the watchdog before carrying out its tasks. Both the watchdog and the other timer run from ACLK and therefore continue to operate in low-power mode 3.

Listing 8.2: Outline of a traditional paced loop protected by the watchdog.

```
// Configure a timer for interrupts every 10ms and return to active mode
// Configure watchdog for 16ms period from 32\,KHz ACLK
#define WDTCONFIG   (WDTCNTCL | WDTSSEL | WDTIS1)
    WDTCTL = WDTPW | WDTCONFIG;            // Configure and clear watchdog
    for (;;) {
        __low_power_mode_3();             // Enter low power mode LPM3
// Wait here, pace loop until timer expires and ISR restores Active Mode
        WDTCTL = WDTPW | WDTCONFIG;       // Service watchdog
// Tasks of paced loop
    }
```

What should we choose for the period of the watchdog? Clearly it should be greater than 10 ms. I chose the next period up, 16 ms, which is given by selecting ACLK with a period of 512 and setting WDTISx = 10. This may be a bit short. It does not matter too much if one pass through the paced loop is slower than 10 ms provided that the next is shorter, so that the program never lags by more than 10 ms and misses an interrupt. Unfortunately the next interval up is 250 ms, which is too long.

It is important that the watchdog is serviced from the highest level of the program. This should never be done in an interrupt service routine. In this example it would be tempting to reset the watchdog in the ISR for the timer but this would lose a great deal of protection.

Suppose that the program became stuck in one of the tasks of the paced loop. Interrupts would still be generated periodically and the watchdog would continue to be serviced correctly if this were done in the ISR. On the other hand, the watchdog would expire and cause a reset with the structure in Listing 8.2.

It is possible that the watchdog may time out during the initialization of a program, which is carried out by the startup code before the user's `main()` function is called. This would happen if it took longer than 32 ms to initialize the RAM, which is possible if a large number of global or static variables are used. In EW430 you can supply a function `__low_level_init()`, which is called before the RAM is initialized. The watchdog can be stopped or reconfigured here.

Watchdogs vary considerably from one type of microcontroller to another. Some have a set of passwords that must be used in a prescribed order, rather than a single value; a reset occurs if a password is used out of sequence. Windowed watchdogs must be serviced only during a particular part of their period, such as the last quarter; clearing the watchdog earlier than this causes a reset. Some have their own built-in oscillator, which protects them from failure of the main clocks. Many watchdogs are controlled by write-once registers, which means that their configuration cannot be changed after an initial value has been written. This would be a problem in the MSP430, where the watchdog may need to be reconfigured for low-power modes.

The reset caused by the watchdog can be a nuisance during development because a PUC destroys much of the evidence that could help you to detect a problem that caused the watchdog to time out. A solution might be to generate an interrupt rather than a reset by using the watchdog as an interval timer, which is described shortly. The interrupt service routine could copy critical data to somewhere safe, signal a problem by lighting an LED, or simply cause execution to stop on a breakpoint. Nagy [4] has further suggestions.

8.1.1   Failsafe Clock Source for Watchdog Timer+

Newer devices, including the MSP430F2xx family and recent members of the MSP430x4xx, have the enhanced watchdog timer+ (WDT+). This includes fail-safe logic to preserve the watchdog's clock. Suppose that the watchdog is configured to use ACLK and the program enters low-power mode 4 to wait for an external interrupt, as in Listing 7.1. The old watchdog (WDT) stops during LPM4 and resumes counting when the device is awakened. In contrast, WDT+ does not let the device enter LPM4 because that would disable its clock. Therefore it is not possible to use LPM4 with WDT+ active; the watchdog must first be stopped by setting WDTHOLD. Similarly, it is not possible to use LPM3 if WDT+ is active and gets its clock from SMCLK. If its clock fails, WDT+ switches from ACLK or SMCLK to MCLK and takes this from the DCO if an external crystal fails. The watchdog interval may change dramatically but there must be serious problems elsewhere if this happens.

8.1.2   Watchdog as an Interval Timer

The watchdog can be used as an interval timer if its protective function is not desired. Set the WDTTMSEL bit in WDTCTL for interval timer mode. The periods are the same as before and again WDTIFG is set when the timer reaches its limit, but no reset occurs. The counter rolls over and restarts from 0. An interrupt is requested if the WDTIE bit in the special function register IE1 is set. This interrupt is maskable and as usual takes effect only if GIE is also set. The watchdog timer has its own interrupt vector, which is fairly high in priority but not at the top. It is not the same as the reset vector, which is taken if the counter times out in watchdog mode. The WDTIFG flag is automatically cleared when the interrupt is serviced. It can be polled if interrupts are not used.

Many applications need a periodic "tick," for which the watchdog timer could be used in interval mode. The disadvantage is the limited selection of periods, but 1 s is convenient for a clock. Some of the previous examples that used Timer_A could be rewritten for the watchdog instead and its use is illustrated in the standard sets of code examples from TI.

Example 8.2

Rewrite the simple clock in Listing 7.4 to use the watchdog as a 1 s interval timer instead of Timer_A.

Checklist—Have You . . .

❏ Stopped the watchdog or ensured that it is serviced frequently enough?

❏ Set the WDTIE bit in the special function register IE1 if you want interrupts? (It is easy to forget this because it is not in WDTCTL.)

8.2   Basic Timer1

Basic Timer1 is present in all MSP430xF4xx devices and is perhaps the most straight-forward module in the MSP430, as you might guess from its name. It provides the clock

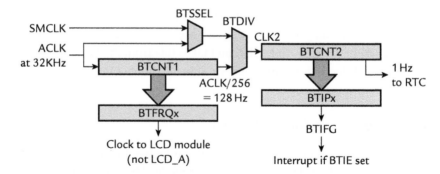

Figure 8.2: Simplified block diagram of Basic Timer1.

7	6	5	4	3	2	1	0
BTSSEL	BTHOLD	BTDIV	BTFRFQx		BTIPx		

Figure 8.3: The Basic Timer1 control register BTCTL.

for the LCD module (but not LCD_A) and generates periodic interrupts. A slightly simplified block diagram is shown in Figure 8.2. Newer devices also contain a real-time clock driven by a signal at 1 Hz from Basic Timer1.

The register BTCTL shown in Figure 8.3 controls most of the functions of Basic Timer1 but there are also bits in the special function registers IFG2 and IE2 for interrupts. An unusual feature of this module is that BTCTL is not initialized by a reset: This must be done by the user. It is not even specified whether the timer is running or not. The counters are not initialized either and this should be done before the timer is started or the first interval will be incorrect.

There are two 8-bit counters in Basic Timer1, which can either be used independently or cascaded for longer intervals:

BTCNT1: Takes its input from ACLK and provides the clock for the LCD module at frequency f_{LCD}. The two BTFRFQx bits select the value of f_{LCD}, which can vary from $f_{ACLK}/256$ to $f_{ACLK}/32$ in powers of 2. It is assumed that $f_{ACLK} = 32\,\text{KHz}$. This gives f_{LCD} from 128 Hz to 1 KHz, which should be suitable for the LCD. The calculation of f_{LCD} is explained in the section "Driving an LCD from an MSP430x4xx" on page 256. The LCD_A controller does not need a clock from BTCNT1, so the counter's only function in this case is as a prescaler for BTCNT2.

BTCNT2: Can be used independently of BTCNT1, in which case the BTSSEL bit selects the clock from ACLK or SMCLK. For longer intervals, BTCNT2 can be clocked from the output of BTCNT1 at a frequency of $f_{ACLK}/256$. Set the BTDIV bit to cascade the counters in this way. Setting the BTHOLD bit stops BTCNT2, but stops BTCNT1 only if BTDIV is also set.

BTCNT2 provides no output signals. Instead it raises the BTIFG flag at a frequency determined by the BTIPx bits. The range goes from $f_{CLK2}/256$ to $f_{CLK2}/2$, where f_{CLK2} is the frequency of the clock input to BTCNT2. With the counters cascaded this gives a period from about 16 ms to 2 s.

The BTIFG flag is in the IFG2 register. An interrupt also is requested if the BTIE bit is set in IE2. The interrupt is maskable so GIE must also be set for the interrupt to be accepted. The BTIFG flag is cleared automatically when the interrupt is serviced. Alternatively the flag can be polled, in which case it must be cleared in software.

We used Timer_A for many applications in Chapters 4 and 6. It would have been better to have used Basic Timer1 for most of these examples, except that it is restricted to the MSP430xF4xx family. The first instance was to flash a light by polling the TAIFG flag in the section "Automatic Control: Flashing a Light by Polling Timer_A" on page 105. This could be done in exactly the same way with Basic Timer1 and the BTIFG flag. The only difference is that a restricted range of intervals is available, unlike Timer_A in Up mode. Similarly, Listings 6.4 and 6.5 show how to do the same task using interrupts. Basic Timer1 could again be used instead. This is a convenient way of generating a *real-time interrupt* (RTI), which can be used as a "heartbeat" to wake a program periodically. Again this was done using Timer_A in Listing 6.8 but Basic Timer1 is often a better choice.

Example 8.3

This is not very exciting but completes the set of possible timers: Rewrite the simple clock in Listing 7.4 to use Basic Timer1 as a 1 s interval timer instead of Timer_A.

8.2.1 Real-Time Clock

A Real-Time Clock (RTC) module has been added to recent devices in the MSP430xFxx family. It counts seconds, minutes, hours, days, months, and years. Alternatively it can be used as a straightforward counter, which I describe briefly at the end of this section, but for now I assume that it is configured in calendar mode by setting RTCMODEx = 11 in the control register RTCCTL. The bits are shown in Figure 8.4. This register is initialized after

7	6	5	4	3	2	1	0
RTCBCD	RTCHOLD	RTCMODEx		RTCTEVx		RTCIE	RTCFG

Figure 8.4: The Real-Time Clock control register RTCCTL.

a power-on reset, unlike BTCTL, and the RTCHOLD bit is set so that the clock does not run by default.

The current time and date are held in a set of registers that contain the following bytes:

- Second (RTCSEC).

- Minute (RTCMIN).

- Hour (RTCHOUR), which runs from 0–23 (24-hour format).

- Day of week (RTCDOW), which runs from 0–6.

- Day of month (RTCDAY).

- Month (RTCMON).

- Year (RTCYEARL), assuming BCD format.

- Century (RTCYEARH), assuming BCD format.

The registers are arranged in pairs that can also be accessed as words. For example, RTCYEAR = RTCYEARH:RTCYEARL and RTCTIM0 = RTCMIN:RTCSEC. Their values can be stored either as normal binary numbers or as BCD. The latter is more convenient for driving a display and is selected by setting the RTCBCD bit in RTCCTL. The registers that hold the date and time are initialized when calendar mode is selected but the user will obviously need to store the current values if "real" time is to be real.

The module automatically accounts for the different number of days in the months and allows for leap years during the current century. In principle it could compute the day of week from the rest of the date but it does not: RTCDOW is effectively an independent 0–6 counter, incremented daily. The user must initialize this appropriately and decide which day is the start of the week. There is no provision for a 12-hour clock with an AM/PM flag so this would have to be done in software.

The clock needs a 1 Hz input, which it takes from Basic Timer1. The RTC module therefore takes control over Basic Timer1 and cascades the counters as if BTDIV = 1. It remains possible to take interrupts from BTCNT2 so the BTIPx bits are still useful.

The Real-Time Clock has an interrupt flag RTCFG and corresponding enable bit RTCIE in RTCCTL. The flag is set every minute, every hour, daily at midnight, or daily at noon depending on the RTCTEVx bits. The interrupt vector is shared with Basic Timer1. It is maskable and can be used in two ways:

- Interrupts are generated by the Real-Time Clock module if RTCIE is set. Both the BTIFG and RTCFG flags are set at an interrupt and cleared automatically when it is serviced. The interval is determined by RTCTEVx.

- Interrupts come from Basic Timer1 as described earlier if RTCIE is clear. The interval is determined by BTIPx. The Real-Time Clock sets its RTCFG flag according to RTCTEVx but this does *not* request an interrupt. A program can poll the flag to check whether an interval of time has elapsed and must clear RTCFG in software.

There is no alarm clock—an interrupt requested at a specified time and date—which seems a curious omission. This must be implemented in software if desired. "**Gotcha!**" The flag is called RTCFG, not RTCIFG as you might expect.

Listing 8.3 shows the relevant functions from a program to run a clock on the TI MSP430FG4618/F2013 Experimenter's Board. The display shows hours and minutes, separated by a colon that flashes at 1 Hz. An interrupt from Basic Timer1 every 0.5s toggles the colon, polls RTCFG and updates the digits if a minute has passed. The flag is cleared by software. The current version of the header file does not define bitfields for BTCTL and RTCCTL so I have had to use masks instead.

Listing 8.3: Part of `rtclock3.c` to configure the Real-Time Clock and handle interrupts from Basic Timer1.

```
void InitRTC (void)
{
// BCD mode, stop RTC, calendar mode (clock from BTCNT2.6), 1min flag
//    (interrupt remains under control of basic timer, not RTC)
// Clears RTC registers and configures basic timer appropriately
    RTCCTL = RTCBCD | RTCHOLD | RTCMODE_3 | RTCTEV_0;
// Initialize time and date
//  RTCTIM0 = 0x0000;                    // Minutes and seconds
    RTCHOUR = 0x17;                      // After-tea relaxation
    RTCDOW = 0;                          // Day of week - Sunday = 0?
    RTCDATE = 0x0513;                    // Month and day
    RTCYEAR = 0x2007;                    // Year (4 digits)
// Set basic timer for 0.5s interrupts (RTC overrides SSEL, HOLD, DIV)
    BTCTL = BT_fCLK2_DIV64;              // Divide frequency by 64
    BTCNT1 = 0;                          // Clear counters
    BTCNT2 = 0;
```

```
        IE2_bit.BTIE = 1;                   // Enable basic timer interrupts
        RTCCTL &= ~RTCHOLD;                 // Start RTC - No bitfield!
        RTCCTL |= RTCFG;                    // Set minute flag and...
        IFG2_bit.BTIFG = 1;                 // ...request an interrupt to
}                                           //   start display
//------------------------------------------------------------------------
// Interrupt service routine for basic timer; flag cleared automatically
// Time stored in RTC as seconds, minutes, and hours in BCD format
// WARNING: potential problem of accessing RTC registers asynchronously
//------------------------------------------------------------------------
#pragma vector = BASICTIMER_VECTOR
__interrupt void BASICTIMER_ISR (void)
{
        LCDMem[2] ^= SEG_H;                 // Toggle colon separator
        if ((RTCCTL & RTCFG) != 0) {        // Time needs to be updated?
            RTCCTL &= ~RTCFG;               // Clear flag manually
            LCDMem[1] = LCDHexChar[RTCMIN & 0x0F];
            LCDMem[2] = LCDHexChar[(RTCMIN >> 4) & 0x0F];
            LCDMem[3] = LCDHexChar[RTCHOUR & 0x0F];
            LCDMem[4] = LCDHexChar[(RTCHOUR >> 4) & 0x0F];
        }
}
```

There are warnings in the user's guide about accessing the registers of the Real-Time Clock. The problem is that changes in these registers are triggered by ACLK but reading or writing from a program is synchronized to MCLK. It is therefore possible that you might try to read the registers while they are being updated, in which case the result would be corrupted. This is a classic example of the issue described in the section "The Shared Data Problem" on page 197. It should not be a problem in Listing 8.3 because the interrupt is requested just after the clock has been updated, at which point the values are stable for nearly 1 s. The safest, general approach is to stop the clock while a reading is taken but this could lose 1 s if a tick is missed during the read. Alternatively the registers can be read several times and a majority vote taken, which should discard any erroneous values. Another solution is to copy the values from the module into a set of "shadow" registers in RAM during interrupts from the Real-Time Clock. The shadow registers can be read safely from the main program *provided that interrupts are disabled during the read*. This will not cause any time to be lost because the Real-Time Clock is updated by hardware.

The clock is stopped while values are written to its registers, which could again cause the loss of 1 s. I stopped the Real-Time Clock with the RTCHOLD bit in Listing 8.3 and started it only when all registers had been initialized.

Example 8.4

The simplest program for a clock uses the Real-Time Clock to display only minutes and seconds without the flashing colon. Try this. Use interrupts from the RTC rather than Basic Timer1.

Example 8.5

Extend this program to display the date (month and day) when a button is pressed down. Button S1 is connected to P1.0 on the TI MSP430FG4618/F2013 Experimenter's Board. This is much easier with the Real-Time Clock module. Without it, we would have to service interrupts regularly to update the time even while the date was displayed.

The Real-Time Clock can instead be used as a straightforward, 32-bit counter if preferred. This is chosen with the RTCMODEx bits in RTCCTL, which also select the source of the clock. Interrupts can be taken from bits 8, 16, 24, or 32 of the counter. The maximum interval is 2^{32} s, which is over 100 years.

8.3 Timer_A

This is the most versatile, general-purpose timer in the MSP430 and is included in all devices. It was introduced in the section "Automatic Control: Flashing a Light by Polling Timer_A" on page 105 and its general features will be familiar if you have used a general-purpose timer in any other modern microcontroller. There are two main parts to the hardware:

Timer block: The core, based on the 16-bit register TAR. There is a choice of sources for the clock, whose frequency can be divided down (prescaled). The timer block has no output but a flag TAIFG is raised when the counter returns to 0.

Capture/compare channels: In which most events occur, each of which is based on a register TACCRn. They all work in the same way with the important exception of TACCR0. Each channel can

- **Capture** an input, which means recording the "time" (the value in TAR) at which the input changes in TACCRn; the input can be either external or internal from another peripheral or software.

- **Compare** the current value of TAR with the value stored in TACCRn and update an output when they match; the output can again be either external or internal.

- **Request an interrupt** by setting its flag TACCRn CCIFG on either of these events; this can be done even if no output signal is produced.

- **Sample** an input at a compare event; this special feature is particularly useful if Timer_A is used for serial communication in a device that lacks a dedicated interface.

Timer_A is modular and the number of capture/compare channels varies between devices. Most have three channels but the smallest members of the MSP430F2xx family have only two and some earlier devices had more. The number of channels is sometimes appended to the name as in Timer_A3. Capture/compare channel 0 is special in two ways. Its register TACCR0 is taken over for the modulus value in Up and Up/Down modes, so that it is no longer available for its usual functions. It also has its own interrupt vector with a higher priority than the other interrupts from Timer_A, which all share a common vector. Therefore channel 0 should be chosen for the most urgent tasks if it is free.

It is important to realize that all channels within Timer_A share the same timer block: There is only one TAR. This ensures that actions performed by the different channels are precisely synchronized. The drawback is that they all work at the same fundamental frequency. Outputs must be supervised by software rather than driven purely by hardware if you need them to change at different frequencies or not to be periodic at all. A few devices have two Timer_A modules, which operate with independent time bases.

A general principle is to use the *hardware* of Timer_A for the part of an event that needs precise timing and to reserve *software* for the less critical parts. This means that signals to be timed should be connected directly to capture inputs so that there is no delay. Outputs should be driven directly from the timer so that they change as soon as a compare event happens. Of course this works only if suitable pins are available or there is an internal connection to the peripheral concerned. Software can then respond to the event—calculate the duration of an input or set up the next output—but the delay required for this does not compromise the timing of the signals.

Timer_A is well documented. There are extensive application notes, some of which I mention later, and over 160 pages in *Application Reports* (slaa024). Many of TI's code examples illustrate its applications.

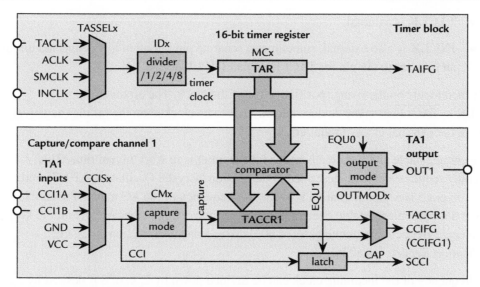

Figure 8.5: Simplified block diagram of Timer_A showing the timer block and capture/compare channel 1. The circles show external signals that may be brought out to pins of the device.

Figure 8.5 shows a simplified block diagram of the timer block and a typical capture/compare channel. Many of the internal signals are omitted for clarity and it emphasizes different features from Figure 4.7. The internal compare signal EQU0 from channel 0 is particularly important because it plays an important role in the outputs from the other channels and controls the timer itself in Up and Up/Down modes.

8.3.1 Timer Block

This contains the 16-bit timer register TAR, which is central to the operation of the timer. It is controlled by the register TACTL shown in Figure 4.8.

Remember that a timer is really no more than a counter and has no direct concept of time (the Real-Time Clock is an exception). It is the programmer's job to establish a relation between the value in the counter and real time. This depends essentially on the frequency of the clock for the timer. It can be chosen from four sources by using the TASSELx bits:

- **SMCLK** is internal and usually fast (megahertz).

- **ACLK** is internal and usually slow, typically 32 KHz from a watch crystal but may be taken from the VLO in the MSP430F2xx family.

- **TACLK** is external.

- **INCLK** is also external, sometimes a separate pin but often it is connected through an inverter to the pin for TACLK so that INCLK = $\overline{\text{TACLK}}$.

TAR increments on the rising (positive) edge of the clock. The arrangement INCLK = $\overline{\text{TACLK}}$ in many devices allows TAR to be clocked on the falling (negative) edge of the external clock if required.

An accurate, stable clock source is needed if the timer is to work in real time. This generally requires a crystal, described in the section "Crystal Oscillators, LFXT1 and XT2" on page 165. An alternative is to use the frequency of the AC mains if available. This is not particularly stable over short times but power companies usually aim to keep the average frequency accurate at 50 or 60 Hz over a day. See Section 6.3.8.7 of *Application Reports* (slaa024).

The frequency of the incoming clock can be divided down by 2, 4, or 8 if desired by configuring the IDx bits. A slower clock reduces the resolution of the timer so that events are timed less precisely. Against this, it increases the natural period of the timer before it overflows, rolls over, and restarts from 0. It is not difficult to count the number of overflows for intervals longer than the period but of course it is easier to avoid it. The trade-off between resolution and period by choosing different clocks is shown in Table 8.1. The period of the timer can range from 4 ms with the fastest SMCLK to 16s with a 32 KHz ACLK and maximum division (even longer with the 12 KHz VLO). I assume that SMCLK runs at the same frequency as MCLK but it can also be divided in the MSP430x1xx and MSP430F2xx families, which allows longer periods from SMCLK.

Table 8.1:   Resolution and period of Timer_A in the Continuous mode with different clocks and input dividers. Values have been rounded for clarity.

Input clock		Timer clock		Range of timer	
Source	Frequency	Divider	Resolution	Frequency	Period
SMCLK	16 MHz	1	$\frac{1}{16}$ μs	240 Hz	4 ms
SMCLK	1 MHz	1	1 μs	15 Hz	66 ms
SMCLK	1 MHz	8	8 μs	2 Hz	0.5 s
ACLK	32 KHz	1	31 μs	$\frac{1}{2}$ Hz	2 s
ACLK	32 KHz	8	240 μs	$\frac{1}{16}$ Hz	16 s

These periods all apply to the Continuous mode, in which TAR cycles through its full range. The timer has four modes of operation, selected with the MCx bits:

Stop (MC = 0): The timer is halted. All registers, including TAR, retain their values so that the timer can be restarted later where it left off.

Continuous (2): The counter runs freely through its full range from 0x0000 to 0xFFFF, at which point it overflows and rolls over back to 0. The period is $2^{16} = 65,536$ counts. This mode is most convenient for capturing inputs and is also used when channels provide outputs with different frequencies or that are not periodic at all.

Up (1): The counter counts from 0 up to the value in TACCR0, the capture/compare register for channel 0. It returns to 0 on the next clock transition. The period is (TACCR0 + 1) counts. For example, if TACCR0 = 4, the sequence of counts is 0, 1, 2, 3, 4, 0, 1, ... with period 5. Up mode is usually used when all channels provide outputs at the same frequency, often for pulse-width modulation.

Up/Down (3): The counter counts from 0 up to TACCR0, then down again to 0 and repeats. The period is $2 \times \text{TACCR0}$ counts. For example, if TACCR0 = 3, the sequence of counts is 0, 1, 2, 3, 2, 1, 0, 1, ... with period 6. This is a specialized mode, typically used for centered pulse-width modulation.

Precise control of the period of the timer is available in Up and Up/Down modes. The disadvantage is the loss of channel 0, whose register TACCR0 is taken over to hold the upper limit of the count—the modulus. Most modern timers include a modulus register in the timer block and Timer_A seems a bit primitive in this respect.

The count in TAR and the divider can be cleared by writing a 1 to the TACLR bit in TACTL. This also resets the direction of counting in Up/Down mode. It is a good idea to do this whenever the timer is configured to ensure that the first period will be correct. The TACLR bit automatically clears itself after use.

The flag TAIFG in TACTL is set when the timer *counts* to 0 and a maskable interrupt is requested if the TAIE bit is set. We used this several times already. The family user's guide always shows *count* in italics to emphasize that actions such as setting TAIFG occur only as a result of normal counting. They do not occur if the appropriate value is written directly to a register. For example, setting TACLR clears TAR but does not set TAIFG.

You are strongly advised to stop the timer before modifying its operation (except the interrupt enables, interrupt flags, and TACLR) to avoid possible errors. Often the timer clock is not synchronous with the CPU clock, such as when the timer is supplied from ACLK. In this case it is best to stop the timer before reading the value of TAR. Alternatively, the timer may be read multiple times while operating and a majority vote taken in software to determine the correct reading. This is not an issue when reading values from the TACCRn registers after a capture because they should be stable.

An advantage of the 16-bit architecture of the MSP430 is that registers of the timer can be read in a single instruction. It can be awkward to read 16-bit timers in an 8-bit processor because the time may change between the two read instructions.

8.3.2   Capture/Compare Channels

Timer_A has three channels in most MSP430s although channel 0 is lost to many applications because its register TACCR0 is needed to set the limit of counting in Up and Up/Down modes, as we have just seen. Each channel is controlled by a register TACCTLn, shown in Figure 8.6.

The central feature of each channel is its capture/compare register TACCRn. In Capture mode this stores the "time"—the value in TAR—at which an event occurs on the input; in Compare mode it specifies the time at which the output should next be changed and an interrupt requested. The mode is selected with the CAP bit. This is cleared by default so that the channel is in Compare mode. Any mixture of capture and compare channels can be used and the mode can be switched freely from one to the other.

Remember to configure pins if you want to connect them to the timer. Otherwise the timer may produce a signal but it does not appear outside the chip. Typically this needs PnSEL.x $= 1$ to select the module rather than digital input/output and the appropriate value in PnDIR.x for input or output. Check the *Application Information* tables in the data sheet.

15	14	13	12	11	10	9	8
CMx		CCISx		SCS	SCCI		CAP

7	6	5	4	3	2	1	0
OUTMODx			CCIE	CCI	OUT	COV	CCIFG

Figure 8.6: The Timer_A capture/compare control register TACCTLn.

Capture Mode Hardware

An *event* can be a rising edge, falling edge, or either edge on the input according to the capture mode bits CMx. The CCISx bits in TACCTLn select the input to be captured. Two of these, CCInA and CCInB, come from outside the timer module. They have many possible connections, set out in the section on Timer_A in the data sheet for the particular device. Often CCInA is connected to external pin TAn while CCInB is connected internally to another module. Here are the internal connections for the F20x1 as an example:

- CCI0B is connected to ACLK, which allows the frequency of SMCLK to be compared with ACLK. This enables a frequency-locked loop to be completed in software to synchronize the two clocks.

- CCI1B comes from CAOUT, the output of the comparator. This allows precise timing of a measurement without the overhead and delay that would arise if software were needed to trigger a capture when CAOUT changed. This provides an economical form of analog-to-digital conversion and is described in the section "Comparator_A" on page 371.

The internal connections are examples of the "system on chip" (SoC) theme of the MSP430, which enables peripherals to work together effectively. It avoids the delays that would be introduced by software and saves power because the CPU need not be restarted simply to notify one module about an event in another.

The other two inputs that can be selected by CCISx are constants, GND and VCC. This looks pointless but their purpose is to allow captures from software. To do this, set CMx = 11 to capture both edges and set CCIS1 = 1 to select the pair of internal inputs. Toggling CCIS0 then inverts the apparent input and thus generates an event that can be captured.

The state of the selected input can always be read in the CCI bit of TACCTLn. This input can change at any time and it is possible for a hardware "race" to occur if the input changes at the same time as the timer clock. The capture hardware therefore includes a synchronizer, which is enabled by setting the SCS bit. Capture events are then recorded on the next falling edge of the timer clock that follows the trigger, midway between increments of TAR. This bit should normally be set for safety; in fact I do not know when the synchronizer should *not* be used.

When a capture occurs, the current value of TAR is copied into TACCRn and the channel flag TACCRn CCIFG (CCIFGn for short) is set. As usual, a maskable interrupt is requested if the corresponding enable bit CCIE in TACCTLn is set. It is important to react rapidly to a capture because most applications need to compute the difference in time between successive captures. The first value must be stored away so that TACCRn is ready for the next capture. The capture overflow bit COV is set if another capture occurs before TACCRn has been read following the previous event. This warns the program that an event has been missed. Some timers include an extra buffer to allow recovery from a single missed event but Timer_A does not.

Compare Mode Hardware

The purpose of Compare mode is to produce an output and interrupt at the time stored in TACCRn. Several actions are triggered when TAR counts to the value in TACCRn:

- The internal signal EQUn is set.

- This in turn raises the CCIFGn flag and requests an interrupt if enabled.

- The output signal OUTn is changed according to the mode set by the OUTMODx bits in TACCTLn.

- The input signal to the capture hardware, CCI, is latched into the SCCI bit. I prefer to regard this as a separate *Sample* mode of Timer_A because it is an input but controlled by the hardware usually used for output.

Often there is no need for an external output and a channel is used purely to provide a flag or interrupt, either periodically or after a single delay. We have done this several times already and will see more examples later. In some devices internal SoC connections enable CCIFGn to trigger other modules directly. These include the digital-to-analog converter (DAC, described in the section "Digital-to-Analog Conversion" on page 485) and direct memory access (DMA) controller.

The output modes are listed in Table 8.2. They are more complicated than you might expect because changes in *all* channels can be triggered by EQU0 as well as EQUn. I illustrate the operation in later sections but here is a quick summary of their typical uses:

Output (mode 0): In which output is controlled directly by the OUT bit in TACCTLn; TAR has no influence. It is as if the pin is used for normal, digital output but operated via Timer_A. This mode is used to set up the initial state of the output before compare events take over. For example, the first period of pulse-width modulation (PWM) may be erratic if this is not done, but it often does not matter.

Table 8.2: Output modes for capture/compare channel *n* (not all are applicable to TACCR0) and the counter mode in which each is most useful. The precise point at which the 'TACCR0' actions occur is explained in the text.

Mode	Actions at TACCRn/"TACCR0"	Most useful in counter mode
0	Output	(Output is controlled by OUT bit)
1	Set	Continuous
2	Toggle/Reset	Up/Down
3	Set/Reset	Up
4	Toggle	(Doubles period)
5	Reset	Continuous
6	Toggle/Set	Up/Down
7	Reset/Set	Up

Toggle (mode 4): Provides a simple way of switching a load on and off for equal times (50% duty cycle) and can be used even with channel 0 in Up and Up/Down modes. The disadvantage is that the load is toggled only once per cycle of the timer and its frequency is therefore halved (period doubled).

Set (mode 1) and Reset (mode 5): Typically used for single changes in the output, usually in the Continuous mode.

Reset/Set (mode 7) and Set/Reset (mode 3): Typically used for periodic, edge-aligned PWM in Up mode of the counter. In this case the first action takes place when TAR matches TACCRn and the second action occurs when TAR returns to 0, one count after the match to TACCR0. (This is not made entirely clear in the user's guides.)

Toggle/Reset (mode 2) and Toggle/Set (mode 6): Typically used for center-aligned PWM in Up/Down mode. The first action takes place when TAR matches TACCRn and the second occurs when TAR matches TACCR0. The second action is needed only once to fix the sign of the waveform; all subsequent action is done by toggling at matches of TACCRn.

Please note that these are the *typical*, most straightforward uses of the different output modes in the different modes of the counter. In general you can use any output mode in any counter mode but the results may be confusing. Be warned that the EQU0 signal *always* influences modes 2, 3, 6, and 7. This applies even in the Continuous mode of the counter, when TACCR0 does not determine the maximum count. These four modes are useless with channel 0 because the two actions occur simultaneously.

A weakness in the design of the output unit can cause glitches in the output when the output mode is changed. Details are given in the user's guides. The most common changes in practice are between modes 1 and 5, set and reset, which are safe in both directions. If there is any doubt, use mode 7 as an intermediate step to avoid glitches.

The TACCTLn registers are cleared by a power-on reset (they are not affected by a power-up clear). This means that the default state of a channel is in the Compare mode with the output determined by the OUT bit, which is clear. Thus an output pin immediately is driven low when it is connected to Timer_A. This is incorrect if the output should be high in its idle state, which often applies to communications. The OUT bit should be set before configuring the pin in this case.

8.3.3   Interrupts from Timer_A

Interrupts can be generated by the timer block itself (flag TAIFG) and each capture/compare channel (flag TACCRn CCIFG or CCIFGn for short). TACCR0 is privileged and has its own interrupt vector, TIMERA0_VECTOR. Its priority is higher than the other vector, TIMERA1_VECTOR, which is shared by the remaining capture/compare channels and the timer block. The CCIFG0 flag is cleared automatically when its interrupt is serviced but this does not happen for the other interrupts because the interrupt service routine (ISR) must first determine the source of the interrupt. The obvious way of doing this is to poll TAIFG and all the CCIFGn flags to locate which is active, clear the flag, and service the interrupt. Unfortunately this is slow, which is particularly undesirable in an ISR. The MSP430 therefore provides an *interrupt vector register* TAIV to identify the source of the interrupt rapidly.

When one or more of the shared and enabled interrupt flags is set, TAIV is loaded with the value corresponding to the source with the highest priority. The possible values in TAIV for Timer_A3 are listed in Table 8.3 with their priorities and you see from the missing

Table 8.3:   Interrupt vector register TAIV for Timer_A3.

TAIV contents	Source	Flag	Priority
0x0000	No interrupt pending		
0x0002	Capture/compare channel 1	CCIFG1	Highest
0x0004	Capture/compare channel 2	CCIFG2	
0x0006	—		↑
0x0008	—		
0x000A	Timer overflow	TAIFG	Lowest

entries that this was designed with five channels (0–4) in mind. An interrupt service routine can read TAIV to locate the source immediately.

Any access to TAIV automatically resets both TAIV and the corresponding flag. If another interrupt is pending, TAIV is reloaded with the value for the source with the highest priority and another interrupt is requested as soon as the current one is serviced. This approach is similar to the handling of other interrupts in the MSP430 and saves the program having to acknowledge interrupts by clearing the flag, which is required by most processors. A side effect is that it makes debugging "interesting."[1] Similar vectors are provided for other modules with shared interrupts.

The values in TAIV are deliberately chosen to be even numbers because this allows a *jump table* to be used for selecting the appropriate action. The idea is to add TAIV to the program counter PC. The add.w instruction is followed by a sequence of jmp instructions, one for each possible value of TAIV. Each jmp occupies 2 bytes, hence the multiples of two in TAIV. Alternatively, reti can be used if no action is required and again takes 2 bytes of memory.

This sounds complicated but is straightforward in practice, as shown in Listing 8.4. This flashes the LED (yet again—sorry) in an eZ430–F2013. The LED is connected to P1.0, which cannot be routed to the timer, so the action must be performed by software. I configured the timer for the Continuous mode and enabled interrupts on TAIFG when TAR returns to 0. The LED is turned on when this interrupt is serviced. It is turned off again in another ISR when TAR matches TACCR1. This is equivalent to the Reset/Set mode (7) of the output unit. The two interrupts share a vector and must therefore be decoded with TAIV.

Timer_A runs in the Continuous mode from SMCLK divided by 8. This is a frequency of about 125 KHz and the period of TAR is about ½ s from Table 8.1. The value of 0x2000 in TACCR1 causes the interrupt to occur one eighth of the way through the cycle of TAR. The LED is therefore illuminated for about $\frac{1}{16}$ s and dark for $\frac{7}{16}$ s in each cycle.

Listing 8.4: Part of program `tmrints1.s43` to show use of TAIV in assembly language.

```
; Set up timer channel 1
    mov.w   #0x2000,TACCR1      ; Full range/8 (short flash of LED)
    mov.w   #CCIE,TACCTL1       ; Interrupts on TACCR1 compare
; Start timer: SMCLK, prescale/8, continuous mode, interrupts, clear TAR
```

---

[1] This applies to smaller devices, such as the F20xx. Larger devices, such as the FG416x, have a more powerful emulation module and TAIV seems to be better protected against the debugger.

```
        mov.w      #TASSEL_2|ID_3|MC_2|TACLR|TAIE,&TACTL
InfLoop:
        bis.w      #LPM0|GIE,SR            ; LPM0 with interrupts (need SMCLK)
        jmp        InfLoop                ; Infinite loop
; ------------------------------------------------------------------
; Interrupt service routine for TACCR1.CCIFG, called when TAR -> TACCR1
;   and for TAIFG, called when TAR -> 0; share vector
; Interrupt flag cleared automatically with access to TAIV
; ------------------------------------------------------------------
TIMERA1_ISR:                              ; ISR for TACCR1 CCIFG and TAIFG
        add.w      &TAIV,PC               ; Add offset to PC for jump table
        jmp        EndTA1_ISR             ; Vector 0: No interrupt pending
        jmp        CCIFG1_ISR             ; Vector 2: CCIFG1
        jmp        FalseInterrupt         ; Vector 4: Not possible (CCIFG2)
        jmp        FalseInterrupt         ; Vector 6: Not possible (CCIFG3)
        jmp        FalseInterrupt         ; Vector 8: Not possible (CCIFG4)
;       jmp        TAIFG_ISR              ; Vector A: TAIFG, last value possible
TAIFG_ISR:                                ; Start of PWM cycle
        bis.b      #BIT0,&P1OUT           ; Light LED
        jmp        EndTA1_ISR
CCIFG1_ISR:                               ; End of duty cycle
        bic.b      #BIT0,&P1OUT           ; Extinguish LED
        jmp        EndTA1_ISR
FalseInterrupt:
        jmp        $                      ; Disaster. Loop here forever
EndTA1_ISR:
        reti
```

There are six possible values in TAIV from 0 to 0x000A and I provide a jump for all of them except the last. Three should never occur with Timer_A2 but I trap them in an infinite loop in case I do something stupid while playing around with the debugger. The entry for a 0 value, jmp EndTA1_ISR, could be replaced by reti but I prefer functions to have single entry and exit points; admittedly it is just fussy in this trivial example. No jump is required for the highest value, which corresponds to TAIFG here, so there is a commented line as a reminder. There is no need to clear any of the interrupt flags because it is all done automatically when TAIV is read.

Example 8.6

Try Listing 8.4 in the debugger. The program should run happily without modification but you may need to adapt it for another demonstration board.

Now try the interesting aspects of debugging. Open a Register window for Timer_A2 and set a breakpoint on the line add.w &TAIV,PC in the ISR. The debugger now shows the value in TAIV on entry to the ISR, which alternates between 0x0002 and 0x000A. Unfortunately nothing else happens in the ISR and the LED never changes. Why not?

Take out this breakpoint and set breakpoints on the lines with bis.b and bic.b instead. What happens now?

Listing 8.5 shows the corresponding ISR in C. It uses a switch as you might expect. A standard switch with a few cases would normally be compiled into a sequence of comparisons but the intrinsic function __even_in_range(TAIV, 10) tells the compiler that TAIV can take only even values up to 10 and encourages it to implement a more efficient jump table instead. A possible disadvantage is that no checks are made that the value is either even or within the range but this should not be an issue with hardware. I use symbolic constants from the header file instead of the numerical values for the cases and could write TAIV_TAIFG instead of 10 in the switch itself.

Listing 8.5: Interrupt service routine from `tmrints2.c` to show use of TAIV in C.

```
#pragma vector = TIMERA1_VECTOR
__interrupt void TIMERA1_ISR (void) // ISR for TACCR1 CCIFG and TAIFG
{
//  switch (TAIV) {                  // Standard switch
    switch (__even_in_range(TAIV, 10)) {     // Faster intrinsic fn
    case 0:                          // No interrupt pending
        break;                       // No action
    case TAIV_CCIFG1:                // Vector 2: CCIFG1
        P1OUT_bit.P1OUT_0 = 0;       // End of duty cycle: Turn off LED
        break;
    case TAIV_TAIFG:                 // Vector A: TAIFG, last value possible
        P1OUT_bit.P1OUT_0 = 1;       // Start of PWM cycle: Turn on LED
        break;
    default:                         // Should not be possible
        for (;;) {                   // Disaster. Loop here forever
        }
    }
}
```

Example 8.7

Look at Listing 8.5 in the debugger. The disassembly code should look like a slightly more economical version of Listing 8.4. Try setting breakpoints or single-stepping through the ISR. You may find that debugging in C is even more confusing.

There is no need to do any decoding if only one source for TIMERA1_VECTOR is active. On the other hand, you must remember to acknowledge the interrupt yourself, either by clearing the flag or by reading TAIV. The interrupt is reasserted immediately if you forget, which it is easy to do.

Here is another mistake to avoid:

```
if (TAIV == 2) {      // WRONG way to decode shared interrupt vector
    // Actions for channel 1
} else if (TAIV == 10) {
    // Actions for TAIFG
}
```

The problem is that TAIV is reset after the first comparison and does not have the same value in the second test. If you are lucky the compiler will save you by taking a copy of TAIV. Use a switch instead.

The interrupt for TACCR0 is faster to service because it has a unique vector and decoding is not required. It also has higher priority and should therefore be chosen for the most urgent task in the Continuous mode. (Channel 0 has a special role in the Up and Up/Down modes so there is no choice there.)

Checklist—Have You . . .

❏ Designed your system so that Timer_A can drive outputs directly for precise timing?

❏ Routed pins to Timer_A if you want it to handle external signals directly?

❏ Cleared the interrupt flag, either explicitly or by reading TAIV (except for channel 0)?

8.4   Measurement in the Capture Mode

The Capture mode is used to take a time stamp of an event: to note the time at which it occurred. A measurement typically requires two or more captures, although a trivial exception to this rule is given in an example. The timer can be used in two opposite ways illustrated in Figure 8.7:

(a)   In most cases the timer clock is either ACLK or SMCLK, whose frequency is known, and the unknown signal is applied to the capture input. To measure the length of a single pulse, we should capture both edges and subtract the captured times. This gives the duration of the pulse in units of the timer clock's period. For a periodic signal we might capture only the rising edges (or falling if preferred) and the difference gives the period directly. The period of the timer clock should be much less than the duration of the signal to give good resolution, not as in Figure 8.7.

(b) The opposite approach is used to measure a signal with a high frequency. The signal is used as the timer clock and the captured events are typically edges of ACLK, whose frequency is known. The difference between that and the captured value gives the number of cycles of the signal in one cycle of ACLK. This gives the frequency rather than the period.

The first method is much more common.

The timer usually runs in the Continuous mode for captures because this makes it easy to calculate differences of times when TAR has rolled over between them. It is not too difficult to handle this in the Up mode but the Up/Down is definitely tricky.

Here are a few examples of the use of the Capture mode:

- Many speed sensors produce a given number of pulses per revolution of a shaft—often just one for a bicycle wheel but many more in machinery. The signal is usually slow by electronic standards and the Capture mode is used to measure the period between pulses to determine the speed. Rotary knobs on the front panel of equipment can be read in the same way.

- Some sensors encode their outputs as a frequency, length of a pulse, or the duty cycle of a square wave, the fraction of the time during which the signal is high. The

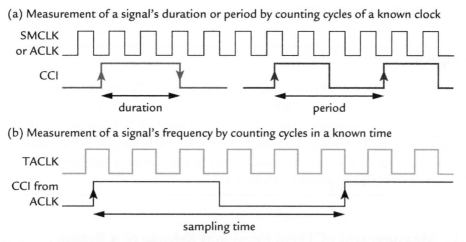

(a) Measurement of a signal's duration or period by counting cycles of a known clock

SMCLK or ACLK

CCI

duration   period

(b) Measurement of a signal's frequency by counting cycles in a known time

TACLK

CCI from ACLK

sampling time

Figure 8.7: Two ways in which the Capture mode is used to time a signal. (a) The duration or period of the signal CCI is measured by counting the number of cycles of a known clock. (b) The frequency of the external signal TACLK is measured by using it as the clock and counting the number of cycles during a known interval.

reason is that only one wire is needed to transmit the output and that time is easy to measure accurately. For example, the Maxim MAX6675–6677 are sensors whose output is a frequency, period, or delay proportional to temperature.

- The delay between transmission and reception of an ultrasonic pulse is measured in the range finder described in application note *Ultrasonic Distance Measurement with the MSP430* (slaa136).

- In communications, the Capture mode is used to detect and time stamp the start of data received, which typically begins with a falling edge on the input.

- Some forms of digital communication encode 0 and 1 as pulses of different length, sometimes with further lengths to denote the start and end of a packet. The Capture mode can decode these signals.

- A frequency-locked loop can be emulated by using the Capture mode to compare the frequencies of SMCLK and ACLK.

- Many MSP430 devices contain an analog comparator, which can be used to detect when an external signal passes through particular levels. The output of the comparator can be routed internally to a Capture channel and the combination acts as an analog-to-digital converter. This is explained in the section "Comparator_A" on page 371. It is also possible to use the Schmitt trigger inputs as less precise comparators.

Example 8.8

A simple example that requires only single captures is to use Timer_A as a random number generator. Try this for simplified "dice" with four values on the Olimex 1121STK. One of the buttons, B2, is conveniently connected to capture input CCI1A and you can display 2 bits of the captured value in TACCR1 on the LEDs. Ignore bounce. A fancier version with the usual range of 1–6 (or more) and a better display could be built on the TI MSP430FG4618/F2013 Experimenter's Board.

8.4.1   Measurement of Time: Press and Release of a Button

As a simple example, let us measure the time between each press or release of button S2 on the TI MSP430FG4618/F2013 Experimenter's Board. (I chose this button because the board does a somersault if you press S1 too forcefully.) Listing 8.6 shows a simple program. Button S2 is connected to P1.1, active low with a pull-up resistor. It can be routed to compare input CCI0B so we must use capture/compare channel 0. This makes programming easy because of its dedicated interrupt vector but its higher priority and faster processing are wasted on such a slow signal and you would usually choose another channel. I put all the code for the display into a separate file LCDutils.c with prototypes in LCDutils.h. There are a few extra functions whose purpose should be obvious from their names. Look at the Web site for the details.

Listing 8.6: Program press1.c to display the time between presses and releases of button S2 on the TI MSP430FG4618/F2013 Experimenter's Board. The configuration of ports P2–P10 is not shown.

```
// press1.c - Time interval between events on S2 = CCI0B using Timer_A
// TI Experimenter's Board with F4619, default clocks
// J H Davies, 2007-06-05; IAR Kickstart version 3.42A
//-----------------------------------------------------------------------
#include <io430xG46x.h>                 // Specific device
#include <intrinsics.h>                 // Intrinsic functions
#include <stdint.h>                     // Integers of defined sizes
#include "LCDutils.h"                   // SBLCDA4 utility functions
void PortsInit (void);                  // Set up input-output ports
//-----------------------------------------------------------------------
void main (void)
{
    WDTCTL = WDTPW | WDTHOLD;           // Stop watchdog timer
    FLL_CTL0 = XCAP14PF;                // Configure load caps (14pF)
    PortsInit();                        // Initialize ports
    LCDInit();                          // Initialize SBLCDA4
    DisplayLine();                      // Display ------- on LCD
// Capture either edge of CCI0B, synchronized, interrupts enabled
    TACCTL0 = CM_3 | CCIS_1 | SCS | CAP | CCIE;
// Start timer: ACLK, no prescale, continuous mode, no ints, clear
    TACTL = TASSEL_1 | ID_0 | MC_2 | TACLR;
    for (;;) {                          // Loop forever with interrupts
        __low_power_mode_3();           // Only ACLK continues to run
    }
}
//-----------------------------------------------------------------------
// Interrupt service routine for TACCR0.CCIFG, called on capture
//-----------------------------------------------------------------------
#pragma vector = TIMERA0_VECTOR
__interrupt void TIMERA0_ISR (void)     // Flag cleared automatically
{
    static uint16_t LastTime = 0;       // Last time captured

    DisplayUint(TACCR0 - LastTime);     // Display interval (counts)
    LastTime = TACCR0;                  // Save time for next capture
}
//-----------------------------------------------------------------------
// Initialize ports for board: outputs driven low by default
// Many of these will be overwritten by LCD initialization later
//-----------------------------------------------------------------------
```

```
void PortsInit (void)
{
// Port 1: 0 = SW1, 1 = SW2, others used for Chipcon (not placed)
    P1OUT = 0;
    P1DIR = 0xFF & ~(BIT0|BIT1);        // P1.0,1 input, others output
    P1SEL = BIT1;                       // P1.1 = S2 to Timer_A CCI0B
```

Timer_A runs in the Continuous mode with ACLK. Why did I choose this clock? Look back at Table 8.1 to see the most straightforward options. I wanted to get as fine a resolution as possible to detect switch bounce. This requires a fast clock. On the other hand, the period of the timer should preferably be longer than a reasonable press of the button, which is around 0.5 s. Thus ACLK without division looked like the best choice because it gives a period of 2 s, although SMCLK/8 might also work for users with nimble fingers. Another point is that ACLK is always accurate if it comes from a crystal but SMCLK is less accurate unless a frequency-locked loop is used. This is not an issue in the MSP430x4xx family.

Capture/compare channel 0 is configured to capture and request interrupts on both rising and falling edges, corresponding to releases and presses of the button. The input is selected from CCI0B and captures are synchronized to the timer clock by setting SCS. Remember that configuring TACCTL0 is not sufficient by itself to connect the pin to CCI0B. You must also redirect the pin from digital input/output to Timer_A by setting the corresponding bit in P1SEL.

All the action takes place in the ISR for TACCR0 CCIFG, which is straightforward. The time since the last edge is computed by subtracting the previous captured time from the latest value in TACCR0. The difference is then displayed on the LCD. Finally, the latest value is saved in a static variable LastTime to be ready for the next capture. Recall that this must be declared static so that it retains its value between calls of the function. If you forget this, you get a new variable at each interrupt and the memory of the last time is lost.

An obvious concern is, What happens if TAR rolls over and restarts from 0 during the measurement? Then we would have TACCR0 < LastTime and it looks as though TACCR0 − LastTime will be negative. In fact this is no problem at all—declare the variables as *unsigned* integers and the difficulty takes care of itself. To illustrate this, suppose that the counter has 4 bits, LastTime = 0b0111 and TACCR0 = 0b0010. Binary, 4-bit subtraction gives $0010 − 0111 = 1011$ and this sets the borrow bit to show that 1 needs to be borrowed from the next more significant column (which does not exist). Effectively the sum is $10010 − 00111$, which is just what we want. Take a look in Maxfield and Brown's book [37] if you need any help with binary arithmetic.

Example 8.9

Rewrite the program so that it detects only the time spent *down* after the button is pressed, not the time spent up as well.

A more serious issue arises if more than 2^{16} cycles are counted, which exceeds the range of TAR. This can be resolved by counting the number of overflows by using interrupts on TAIFG. It is easier to choose the timer clock so that the number of pulses stays below 2^{16} but this can be done only if the input is known to have a maximum duration. Humans are not so cooperative.

Example 8.10

Extend the program to measure longer durations by counting the number of times that TAR rolls over. This part is not difficult but the routines to display values on the LCD need to be expanded for larger integers, which is harder. You may wish to look at the functions provided in LCDutils.c.

It is worth a brief review of the advantages of capturing the input in hardware. A capture/compare channel copies the value of TAR into TACCRn as soon as a transition is detected on the input. The only delays are the inevitable propagation delay in the electronics and synchronization with the timer clock (which can be deselected, although this should be done with care). Subsequent processing, such as the DisplayUint() function in the interrupt service routine, does not upset the value stored. Even the interrupt latency has no effect, nor the time required to wake the processor if it were in a low-power state. These delay the processing but the captured value itself is accurate.

Suppose that we could not use a capture/compare channel to record the time of the transition. The alternative would be to request an interrupt when the input changes and record the value of TAR during the ISR. This obviously suffers from the interrupt latency and the time to restart MCLK from a low-power mode. A further delay occurs if another interrupt is being serviced at the time and we have to wait for it to complete. The capture/compare hardware of Timer_A avoids all these problems.

Example 8.11

Suppose that the designers of the MSP430FG4618/F2013 Experimenter's Board had been less cooperative and that the push buttons were not connected directly to capture inputs of

Timer_A. In this case the captures would have to be mediated by software. This could be done by setting the CCIS1 bit in TACCTL0, which connects the capture input to either V_{CC} or ground, depending on the state of CCIS0. Set up port 1 to generate interrupts when S2 is pressed or released and copy its new state to TACCTL0.CCIS0 for capture. The rest of the program needs few changes. Measure only the time spent down.

8.4.2   Measurement of Time: Reaction Timer

Now for a slightly more complicated example: a reaction timer on the TI MSP430FG4618/F2013 Experimenter's Board. Here are instructions for the user:

1. Press and release button S2 to start a new trial.

2. LED4 lights after a random delay between 1 and 2 s.

3. Press button S1 as soon as you see LED4 light. The LED goes out and your reaction time in milliseconds is displayed on the LCD.

We use Timer_A to record the reaction time and the first question, as usual, is which clock to use. The time is displayed in milliseconds so a 1 KHz clock is most convenient. Unfortunately the slowest internal clock is ACLK and the maximum divider is 8, which gives a minimum frequency of 4 KHz. It will prove to be convenient to use the undivided ACLK. The only disadvantage is that this limits the slowest reaction that can easily be timed to 2 s, which is not a problem in most cases. This is how the FG4618 is configured:

*   Interrupts are enabled for the start button S2, connected to P1.1.

*   The timing button S1, connected to P1.0, is routed to capture input CCI0A of Timer_A.

*   Capture/compare channel 0 is used for the captures because S1 provides the input CCI0A.

*   Timer_A runs from ACLK with no division, giving a period of 2 s.

We now look at the basic actions needed and the different ways in which they can be implemented.

Wait for Button S2 to Be Pressed

A new trial starts when S2 is pressed, which generates a falling edge on the input (active low). In principle we could poll the input but that would be a ridiculous waste of power and an interrupt is the correct approach. The ISR disables further interrupts from this source, which conveniently means that there are no problems from any subsequent transitions caused by switch bounce. There is no need to detect the rising edge when S2 is released.

The only real decision is the selection of low-power mode while waiting for the interrupt: LPM3 or LPM4? I use LPM3 for simplicity because this leaves ACLK running for the LCD. The display has to be shut down if LPM4 is used. Probably the best solution would be to stay in LPM3 for a reasonable time, until the user appears to have lost interest, before entering LPM4. Remember to reactivate the LCD_A module after wakening from LPM4. (An alternative approach would be to reset the MSP430 by writing an illegal value to the watchdog control register WDTCTL.)

Random Delay before Starting the Reaction Timer

There should be a random delay after pressing S2 before the LED lights to test the reaction time so that the user cannot predict when to press S1. I experimented and found that a delay of between 1 and 2 s seem about right. It is awkward if the delay is too short and boring if it goes on for too long.

The simplest way of getting a random delay is to use a free-running timer with a period of 2 s, the maximum duration of the random delay. By good fortune this is just what Timer_A does with an undivided ACLK. Button S2 can be pressed at any time during the period of the timer, so the delay until the next overflow is random between zero and the full period. This is not quite what we want, though, because short delays are undesirable. These can be avoided by testing the value in TAR. If it is in the upper half of its range, which means that the timer overflows in less than 1 s, a value of 0x8000 is subtracted from TAR to extend the delay by 1 s. This does not destroy the randomness. I stop the timer while TAR is tested and adjusted to avoid possible problems. The interrupt on overflow (TAIFG) is enabled to trigger the next step.

Start the Reaction Timer

Timer_A generates a TAIFG interrupt when it overflows after the random delay. We now need to turn on the LED to start the measurement and record the starting time. However, the second step is unnecessary because TAR has just overflowed so the starting time is 0. Easy!

Ideally the LED should be lighted precisely when TAIFG is set, which means that it should be done in hardware. This could in principle be done by driving the LED from Timer_A and using a Compare event but no LEDs are connected in this way on the MSP430FG4618/F2013 Experimenter's Board. In any case, a few microseconds are insignificant on the human timescale. Instead, the LED is turned on during the ISR. Further interrupts from TAIFG are disabled and those from TACCR0 are enabled instead to be ready for the user to press the button.

Measure, Compute, and Display the Reaction Time

A capture occurs when the user presses S1 and the reaction time in counts of ACLK is given simply by the value in TACCR0 because the measurement started at 0. The ISR also turns the LED off and enables interrupts on the start button S2 so that the system is ready for the next trial.

Counts of ACLK are not meaningful to most users so the time is rescaled to milliseconds. The formula is

$$\text{time (ms)} = \frac{1000 \times \text{time (counts)}}{f_{\text{ACLK}}} \tag{8.1}$$

with $f_{\text{ACLK}} = 32\,\text{KHz} = 32,768\,\text{Hz}$. It is tempting to "simplify" the numbers to give

$$\text{time (ms)} = \frac{\text{time (counts)}}{32,768} \tag{8.2}$$

but this requires floating-point division, which is very slow indeed. I therefore do the calculation more economically in two steps. The first is multiplication by 1000, which needs a 32-bit variable to hold the result. The second step is division by $32\,\text{K} = 2^{15}$. Do not try to do the operations in the opposite order because the division would destroy the data. The division implicitly rounds the value down, which is the most favorable outcome for the contestant. I wrote the division as a shift because I expected this to be more efficient. It was a waste of time because the compiler implements it as a single left-shift instead, which is a much better way. It did the same when I wrote the division instead. There is no point in doing trivial optimizations yourself with modern compilers.

How Should the Program Be Organized?

Clearly the action needs to be triggered by interrupts, but does this mean that the code should all be placed in interrupt service routines? Listing 8.7 shows the program written in this way, with the main loop reduced to putting the processor into a low-power mode.

Listing 8.7: Program `react1.c` based on interrupt service routines to display the user's reaction time on the TI MSP430FG4618/F2013 Experimenter's Board. The configuration of ports P2–P10 is not shown.

```c
// react1.c - Reaction timer using Timer_A, display in ms
// Press S2, wait for LED4, press S1 as quickly as possible
// Timer_A used for random delay and reaction time
// Actions take place in ISRs
// TI Experimenter's Board with F4619, default clocks
// J H Davies, 2007-06-11; IAR Kickstart version 3.42A
//-------------------------------------------------------------------
#include <io430xG46x.h>          // Specific device
#include <intrinsics.h>          // Intrinsic functions
#include <stdint.h>              // Integers of defined sizes
#include "LCDutils.h"            // SBLCDA4 utility functions
#define LED       P5OUT_bit.P5OUT_1   // Pin for LED
void PortsInit (void);           // Set up input-output ports
//-------------------------------------------------------------------
void main (void)
{
    WDTCTL = WDTPW | WDTHOLD;     // Stop watchdog timer
    FLL_CTL0 = XCAP14PF;          // Configure load caps (14pF)
    PortsInit();                  // Initialize ports
    LCDInit();                    // Initialize SBLCDA4
    DisplayHello();               // Display HELLO on LCD
// Capture falling edge of CCI0A, synchronized, no ints yet
    TACCTL0 = CM_2 | CCIS_0 | SCS | CAP;
// Start timer: ACLK, no prescale, continuous mode, no ints yet, clear
    TACTL = TASSEL_1 | ID_0 | MC_2 | TACLR;
    P1IFG = 0;                    // Clear any pending interrupts
    P1IE = BIT1;                  // Enable interrupts on P1.1 = B2
    for (;;) {                    // Loop forever with interrupts
        __low_power_mode_3();     // Only ACLK continues to run
    }
}
//-------------------------------------------------------------------
// Interrupt service routine for port 1; need to clear flag
// Enable TAIFG interrupts to give 1-2s delay before reaction test
//-------------------------------------------------------------------
#pragma vector = PORT1_VECTOR
__interrupt void PORT1_ISR (void)    // Flag NOT cleared automatically
{
    P1IFG = 0;                    // Acknowledge interrupt
    P1IE = 0;                     // Disable further P1 interrupts
    TACTL_bit.TAMC = 0;           // Hold timer
    if (TAR > 0x8000) {           // Will delay be < 1s?
        TAR -= 0x8000;            // Yes: add 1s so 1s < delay < 2s
    }
    TACTL_bit.TAIFG = 0;          // Clear any pending interrupt
    TACTL |= MC_2 | TAIE;         // Restart timer with interrupts
}
//-------------------------------------------------------------------
// Interrupt service routine for TAIFG, called after random interval
// Check source to clear flag and "service" erroneous interrupts
//-------------------------------------------------------------------
```

```
#pragma vector = TIMERA1_VECTOR
__interrupt void TIMERA1_ISR (void)   // Flag NOT cleared automatically
{
      if (TAIV == TAIV_TAIFG) {        // Vector 10: TAIFG
          LED = 1;                     // Turn on LED to start trial
          TACTL_bit.TAIE = 0;          // Disable further TAIFG interrupts
          TACCTL0_bit.CCIFG = 0;       // Clear any pending interrupt
          TACCTL0_bit.CCIE = 1;        // Enable TACCR0 capture interrupts
      }
}
//-----------------------------------------------------------------
// Interrupt service routine for TACCR0.CCIFG, called on capture
//-----------------------------------------------------------------
#pragma vector = TIMERA0_VECTOR
__interrupt void TIMERA0_ISR (void)   // Flag cleared automatically
{
      uint32_t ReactionTime;           // Computed reaction time, 32 bits

      TACCTL0_bit.CCIE = 0;            // Disable further CCIFG0 interrupts
      ReactionTime = TACCR0;           // Starting time was 0 counts
      ReactionTime *= 1000;            // Convert seconds to ms
      ReactionTime >>= 15;             // Economical /= 32K (f_ACLK)
      DisplayUint ((uint16_t) ReactionTime);  // Truncate and display
      LED = 0;                         // Turn off LED
      P1IFG = 0;                       // Clear any pending interrupts
      P1IE = BIT1;                     // Reenable interrupts on P1.1 = B2
}
//-----------------------------------------------------------------
// Initialize ports for board: pins are outputs driven low by default
// Many of these will be overwritten by LCD initialization later
//-----------------------------------------------------------------
void PortsInit (void)
{
// Port 1: 0 = SW1, 1 = SW2, others used for Chipcon (not placed)
      P1OUT = 0;
      P1DIR = 0xFF & ~(BIT0|BIT1);     // P1.0,1 input, others output
      P1SEL = BIT0;                    // P1.0 = S1 to Timer_A CCI0A
      P1IES = BIT1;                    // Interrupts on falling edge
```

The alternative is to put the action in the main loop, in which case the interrupt service routines simply clear their flags where necessary and return the processor to active mode on exit. This is shown in Listing 8.8. Which style do you prefer? I normally prefer to put the actions in ISRs but here the overall strategy is much easier to follow when it is in the main loop. To be honest this is an untypical example because the interrupts occur in a fixed sequence rather than arising randomly, which is often the case.

Listing 8.8: Program react2.c concentrated on the main loop to display the user's reaction time on the TI MSP430FG4618/F2013 Experimenter's Board. The configuration of the ports is not shown.

```
// react2.c - Reaction timer using Timer_A, display in ms
// Press S2, wait for LED4, press S1 as quickly as possible
```

```
// Timer_A used for random delay and reaction time
// Actions take place in main loop; minimal ISRs return to active mode
// TI Experimenter's Board with F4619, default clocks
// J H Davies, 2007-06-11; IAR Kickstart version 3.42A
//--------------------------------------------------------------------
#include <io430xG46x.h>            // Specific device
#include <intrinsics.h>            // Intrinsic functions
#include <stdint.h>                // Integers of defined sizes
#include "LCDutils.h"              // SBLCDA4 utility functions
#define LED      P5OUT_bit.P5OUT_1 // Pin for LED
void PortsInit (void);             // Set up input/output ports
//--------------------------------------------------------------------
void main (void)
{
    uint32_t ReactionTime;         // Computed reaction time

    WDTCTL = WDTPW | WDTHOLD;       // Stop watchdog timer
    FLL_CTL0 = XCAP14PF;            // Configure load caps (14pF)
    PortsInit();                    // Initialize ports
    LCDInit();                      // Initialize SBLCDA4
    DisplayHello();                 // Display HELLO on LCD
// Capture falling edge of CCI0A, synchronized, no ints yet
    TACCTL0 = CM_2 | CCIS_0 | SCS | CAP;
// Start timer: ACLK, no prescale, continuous mode, no ints yet, clear
    TACTL = TASSEL_1 | ID_0 | MC_2 | TACLR;
    for (;;) {                      // Loop forever with interrupts
        P1IFG = 0;                  // Clear any pending interrupts
        P1IE = BIT1;                // Enable interrupts on P1.1 = B2
        __low_power_mode_3();       // Wait for button to be pressed
        P1IE = 0;                   // Disable further P1 interrupts
        TACTL_bit.TAMC = 0;         // Hold timer
        if (TAR > 0x8000) {         // Will delay be < 1s?
            TAR -= 0x8000;          // Yes: add 1s so 1s < delay < 2s
        }
        TACTL_bit.TAIFG = 0;        // Clear any pending interrupt
        TACTL |= MC_2 | TAIE;       // Restart timer with interrupts
        __low_power_mode_3();       // Wait for random delay
        LED = 1;                    // Turn on LED to start trial
        TACTL_bit.TAIE = 0;         // Disable further TAIFG interrupts
        TACCTL0_bit.CCIFG = 0;      // Clear any pending interrupt
        TACCTL0_bit.CCIE = 1;       // Enable TACCR0 capture interrupts
        __low_power_mode_3();       // Wait for user to press button
        TACCTL0_bit.CCIE = 0;       // Disable further interrupts
        ReactionTime = TACCR0;      // Starting time was 0 counts
        ReactionTime *= 1000;       // Convert seconds to ms
        ReactionTime /= 0x8000;     // Divide by f_ACLK
        DisplayUint ((uint16_t) ReactionTime); // Truncate and display
        LED = 0;                    // Turn off LED
    }                               // Finished - go and start again
}
//--------------------------------------------------------------------
// Interrupt service routine for port 1; need to clear flag
//--------------------------------------------------------------------
#pragma vector = PORT1_VECTOR
__interrupt void PORT1_ISR (void)   // Flag NOT cleared automatically
{
```

```
    P1IFG = 0;                          // Acknowledge interrupt
    __low_power_mode_off_on_exit();
}
//---------------------------------------------------------------
// Interrupt service routine for TAIFG, called after random interval
//---------------------------------------------------------------
#pragma vector = TIMERA1_VECTOR
__interrupt void TIMERA1_ISR (void) // Flag NOT cleared automatically
{
    TACTL_bit.TAIFG = 0;                // Acknowledge interrupt
    __low_power_mode_off_on_exit();
}
//---------------------------------------------------------------
// Interrupt service routine for TACCR0.CCIFG, called on capture
//---------------------------------------------------------------
#pragma vector = TIMERA0_VECTOR
__interrupt void TIMERA0_ISR (void) // Flag cleared automatically
{
    __low_power_mode_off_on_exit();
}
```

There is no safeguard in either version of this program against overflows of TAR if the user waits for more than 2 s before pressing the button. This is not hard to add because it is signaled by an overflow of TAR. Of course there should be no overflow if users genuinely try to measure their reaction time. Mine was about 0.16 s.

Example 8.12

Extend either version of the program so that it stops the trial after 2 s and displays "Error" on the LCD using the DisplayErr() function.

Example 8.13

Develop your program further so that it enters LPM4 if there is no activity for some time, which can be monitored by counting overflows of TAR. Remember to disable the LCD before entering LPM4 and to restart it afterward (or reset the MSP430).

8.4.3   Measurement of Frequency: Comparison of SMCLK and ACLK

The general principle for measuring a frequency f is to count the number of cycles N in a known interval of time T, whence $f = N/T$. Look at Figure 8.7(b). If $T = 1$ s then N gives the frequency in hertz directly. Alternatively, if $T = 1$ ms then N gives the frequency in kilohertz.

The signal under test is used as the timer clock and an external signal should be applied to the TACLK pin. Remember to set the PnSEL.x bit to connect this pin to the timer and select TACLK or INCLK using the TASSELx bits. In the example that follows I use an internal signal, SMCLK.

We need to count the number of cycles of the test signal in a known time. This requires an accurate reference, which is provided by ACLK and the usual watch crystal. An obvious way of counting cycles is to start the timer at the beginning of a cycle of ACLK and stop it after a fixed number of cycles of ACLK have passed. The problem is that this requires software, which introduces errors in the timing. A far better approach is to do the measurement entirely in hardware. This is straightforward because ACLK is connected internally to one of the capture inputs of Timer_A.

The channel is configured to capture the value of TAR and request an interrupt at a rising edge of ACLK. This value N_1 is stored during the ISR. The next rising edge of ACLK stimulates another capture and interrupt. The difference between the new and old values of TAR, $N_2 - N_1$, gives the number of cycles of the test signal in one cycle of ACLK. Thus the measured frequency is $f = (N_2 - N_1) f_{\text{ACLK}}$.

The result is restricted to multiples of f_{ACLK} if only a single cycle of ACLK is used, which gives a rather coarse set of values. Finer resolution is obtained by counting for more cycles of ACLK. For example, we might use 32 cycles. Then $f = (N_2 - N_1) f_{\text{ACLK}}/32$. Now $f_{\text{ACLK}} = 32\,\text{KHz}$ so $f = (N_2 - N_1)\,\text{KHz}$, which is convenient. Note the capital K because this is a binary kilo, meaning $2^{10} = 1024$, rather than the decimal kilo of 1000 as in KHz. There is no problem if TAR overflows once during the measurement, as discussed already, and an overflow counter can be added if more than 2^{16} cycles need to be recorded.

As an example I chose to use SMCLK as the "unknown" signal and display its frequency on the LCD of the TI MSP430FG4618/F2013 Experimenter's Board. The quotation marks are there because f_{SMCLK} should be a precise multiple of f_{ACLK} in a MSP430x4xx due to the frequency-locked loop (FLL+). Turning this around, it gives us a chance to explore the operation of the FLL+ so you might want to look back at the section "Frequency-Locked Loop, FLL+" on page 172. Listing 8.9 shows the relevant parts of the program:

Listing 8.9: Program `clkcmp1.c` to display the frequency of SMCLK on the LCD of a TI MSP430FG4618/F2013 Experimenter's Board.

```
// clkcmp1.c - Show ratio of SMCLK : ACLK on display using Timer_A
// Compared every 0.5s, triggered by Basic Timer; flashes LED as signal
// TI Experimenter's Board with F4619, fairly complete initialization
// J H Davies, 2007-06-03; IAR Kickstart version 3.42A
```

```c
//-------------------------------------------------------------
#include <io430xG46x.h>              // Specific device
#include <intrinsics.h>              // Intrinsic functions
#include <stdint.h>                  // Integers of defined sizes
#include "LCDutils.h"                // SBLCDA4 utility functions
//-------------------------------------------------------------
void PortsInit (void);
void TimersInit (void);
//-------------------------------------------------------------
void main (void)
{
    volatile uint16_t i;             // Counter for FLL+ delay

    WDTCTL = WDTPW | WDTHOLD;         // Stop watchdog timer
// Clock default: MCLK = 32 x ACLK = 1MHz, FLL+ operating
    FLL_CTL0 = XCAP14PF;             // Configure load caps (14pF)
    do {
        IFG1_bit.OFIFG = 0;          // Try to clear OSCFault flag
        for (i = 50000; i > 0; --i) {   // Delay for flag to reset
        }                            //   if not yet stable
    } while (IFG1_bit.OFIFG != 0);   // OSCFault flag still set?
    PortsInit();                     // Initialize ports
    LCDInit();                       // Initialize SBLCDA4
    TimersInit();                    // Initialize Timer_A and BT
    for (;;) {                       // Loop forever with interrupts
        __low_power_mode_0();        // SMCLK and DCO continue to run
    }                                //    and FLL+ remains active
}
//-------------------------------------------------------------
// Initialize Timer_A for capture from ACLK, BT for 0.5s interrupts
//-------------------------------------------------------------
void TimersInit (void)
{
// Initialize basic timer: ACLK, cascaded, 0.5s interrupts, stopped
    BTCTL = BTHOLD | BT_ADLY_500;
    BTCNT1 = BTCNT2 = 0;             // Clear counters
    IE2_bit.BTIE = 1;                // Enable basic timer interrupts
    BTCTL &= ~BTHOLD;                // Start basic timer running
// Capture on CCI2B, rising edge, sync; do not yet enable interrupts
    TACCTL2 = CCIS_1 | CM_1 | SCS | CAP;
// Start timer: SMCLK, no prescale, continuous mode, no ints, clear
    TACTL = TASSEL_2 | ID_0 | MC_2 | TACLR;
}
//-------------------------------------------------------------
// Interrupt service routine for basic timer: start captures on CCI2B
// Flag cleared automatically
//-------------------------------------------------------------
#pragma vector = BASICTIMER_VECTOR
__interrupt void BASICTIMER_ISR (void)
{
    TACCTL2_bit.COV = 0;             // Clear previous overruns
    TACCTL2_bit.CCIFG = 0;           // Clear pending interrupt
    TACCTL2_bit.CCIE = 1;            // Enable interrupts on capture
}
//-------------------------------------------------------------
// Interrupt service routine for Timer_A, called on capture
```

```
// Read TAIV to clear flag and "service" erroneous interrupts
//-------------------------------------------------------------------
#define NCAPTURES   32              // Number of captures to accumulate
#pragma vector = TIMERA1_VECTOR
__interrupt void TIMERA1_ISR (void)  // ISR for CCIFGn and TAIFG
{
    static uint8_t Captures = 0;     // Number of captures accumulated
    static uint16_t StartTime;       // Time at start of sampling

    if (TAIV == TAIV_CCIFG2) {       // CCIFG2 vector (4)
        switch (Captures) {
        case 0:                      // Starting new sequence of captures
            StartTime = TACCR2;      // Starting time
            Captures = NCAPTURES;    // Initialize down counter
            P5OUT_bit.P5OUT_1 = 1;   // Light LED to mark start captures
            break;
        case 1:                      // Final capture of sequence
            DisplayUint (TACCR2 - StartTime);   // Display result
            Captures = 0;            // Finished
            TACCTL2_bit.CCIE = 0;    // Disable further interrupts
            P5OUT_bit.P5OUT_1 = 0;   // Turn off LED to mark end captures
            break;
        default:                     // Sequence of captures continues
            --Captures;
            break;
        }
    }
}
```

- I have been less lazy than useful and configured the clocks better. A loop tests the fault flag OFIFG and allows the oscillators to stabilize. Thus the frequencies should be stable before the measurements start and the displayed value should be reliable.

- Basic Timer1 is configured for a real-time interrupt every 0.5s to stimulate the measurements.

- Capture/compare channel 2 is used for the measurements because ACLK is connected to CCI2B. It is configured for rising edges and synchronized captures.

- The timer block takes its clock from SMCLK without division and runs in the Continuous mode.

- The ISR for Basic Timer1 enables interrupts on TACCR2 to start a new measurement. It first clears the COV and CCIFG bits, which are set by previous capture events that were ignored while interrupts were disabled.

- Most of the work takes place in the ISR for TIMERA1_VECTOR. Only one interrupt is enabled but I checked TAIV in case the program is later extended or if

something terrible goes wrong. The action depends on the value of the variable
Captures, which is tested in a switch.

— Captures is 0 at the start of a new measurement. The captured value of TAR
  in TACCR2 is stored in StartTime and Captures is initialized to the
  number of captures required for the measurement, 32 here. I also lit an LED
  to show that a measurement is in progress.

— The value of Captures is decremented in subsequent captures; there is no
  need to store the value of TACCR2 again.

— The final capture is taken when Captures is 1. The difference between the
  most recent capture, in TACCR2, and the stored value gives the frequency
  in binary kilohertz as explained already. This is displayed as an unsigned
  integer on the LCD. Further interrupts are disabled and the LED is turned
  off to show that the measurement is finished.

• The MSP430 returns to LPM0 until the next interrupt from Basic Timer1.

I chose low-power mode 0 so that the DCO and FLL+ run continuously. Usually we put
the device into LPM3 because only ACLK is needed until the next measurement. The
problem with this is that the FLL+ is not automatically reengaged when the MSP430
enters active mode after an interrupt.

The default configuration for MCLK in the FG4618 is that the FLL+ has a multiplier of 32,
in which case $f_{\text{MCLK}} = 32 f_{\text{ACLK}} = 1024\,\text{KHz}$ ($= 1{,}048{,}576\,\text{Hz}$, not that the frequency is
this accurate). The display should therefore show 1024 when the program runs.

However, it does not. I found 1376 typically, although there were fluctuations. The reason
is straightforward: Only 32 cycles of MCLK are available to service each interrupt before
the next capture occurs, which is not long enough. How many cycles are required? The
FG4618 has the MSP430X CPU, which needs fewer clock cycles to service interrupts than
the original MSP430. It takes five cycles to call the ISR and three to return from it, plus
one to synchronize the capture with MCLK. This represents 9 cycles of overhead, leaving
only 23. Typically the CPU must process the decisions and decrement Captures. The
complete interrupt requires 42 clock cycles according to the debugger, too many.

There are several possible solutions to this problem. The code could be streamlined,
perhaps by removing the check on TAIV. Perhaps the best approach would be to divide
ACLK so that captures occurred less frequently. This can be done in the MSP430x1xx and

MSP430F2xx families but is not possible in the clock module for the MSP430x4xx (it can divide the external signal ACLK/n but not the internal clock). I doubled the frequency of the FLL+ instead by raising the multiplier in SCFQCTL from 31 to 63. The display then showed 2048 as expected.

Example 8.14

Experiment with Listing 8.9 in the debugger. Does the unmodified version fail as advertised? Do you see the COV bit set to show that a capture has been missed?

Double the frequency of the DCO and confirm that the program now works correctly. Try higher frequencies as well but remember that 8 MHz needs $V_{CC} = 3.6$ V.

Example 8.15

Use the debugger to disengage the frequency-locked loop by setting the SCG0 bit in the status register. The DCO now runs freely at its last setting. Resume the program. Has the frequency changed?

Test the sensitivity to temperature by pressing your finger on the FG4618. Does the frequency change? Reengage the FLL+ and confirm that the frequency no longer changes if you put your finger on the chip again.

Example 8.16

It seems a bit extravagant to use two timers for this measurement. Why not use TAIFG to start measurements instead of the basic timer? Try this. Use a switch on TAIV.

Example 8.17

Save power by entering LPM3 between measurements. What low-power mode should be used between captures during a measurement?

A practical use of this specific example is to emulate a frequency-locked loop in software for devices that lack the hardware. Algorithms are suggested in the application note *Controlling the DCO Frequency of the MSP430x11x* (slaa074). It also shows how the 50 or 60 Hz supply can be used as a reference instead of ACLK.

8.5 Output in the Continuous Mode

In the Continuous mode, TAR counts from 0 up to 0xFFFF and returns to 0 on the next clock transition, setting TAIFG as it does so. The only control over the period of TAR in real time is through the choice of clock, which is so coarse that it will rarely give the desired value. Therefore the duration of output signals must almost always be controlled by software rather than the period of the timer. This contrasts with the Up and Up/Down modes, where outputs can usually be generated by hardware alone. The Continuous mode is typically used in the following circumstances:

- All channels are needed for output, including channel 0.

- Outputs must be driven at different, unrelated frequencies.

- Single delays are required rather than periodic signals.

- Some channels are used for capture and some for compare events.

The Continuous mode is convenient when times have to be calculated because the 16-bit range of TAR matches the size of a simple unsigned integer variable, which means that there is no need to worry about overflows: The arithmetic and the counter overflow in the same way. This is the same as in the Capture mode.

A fundamental limitation of Continuous mode is that the time for the next Compare event must be updated in software, usually during an interrupt service routine. Intervals therefore cannot be too short or there is insufficient time for the next event to be set up. This is the same issue that hampered frequent captures in the section "Measurement of Frequency: Comparison of SMCLK and ACLK" on page 312.

The application note *Implementing IrDA with the MSP430* (slaa202) shows how both TACCR1 and TACCR0 can be used to control OUT1 in the Continuous mode. This is tricky. There are several standard applications of the Continuous mode to digital communications and I describe an example in the section "A Software UART Using Timer_A" on page 590. Meanwhile, here are some more straightforward illustrations of its use.

8.5.1 Generation of Independent, Periodic Signals

An example that uses the Continuous mode to produce several independent, period signals is shown in Listing 8.10. It runs on the Olimex 1121STK, which has LEDs that can be connected to outputs TA1 and TA2 of Timer_A3. There are four outputs with different frequencies, one from each channel and one from TAR itself:

- LED1 is driven from TA1 and toggled every 0.25 s so that it flashes with a 50% duty cycle (0.25 s on, 0.25 s off). The overall period is 0.5 s.

- LED2 is driven from TA2 at roughly 100 Hz with a 10% duty cycle. The eye cannot resolve the flashes and the LED simply appears dim.

- The piezo sounder is driven at nearly 440 Hz, the frequency of the standard note A used for tuning western music. It is not connected directly to Timer_A so its outputs are toggled in the interrupt service routine for channel 0.

- The sounder is toggled on and off every 2 s by enabling and disabling its interrupts in the ISR for TAIFG.

Listing 8.10: Program `conta11.c` to illustrate periodic outputs from Timer_A in the Continuous mode.

```
// conta11.c - All channels in operation using Timer_A continuous mode
// LED1 = TA1 in toggle mode, 50% duty cycle, period 0.5s
// LED2 = TA2 using set and reset modes, 10% duty, period roughly 10ms
// Piezo toggled by TA0 interrupts at close to 440Hz (standard 'A'),
//  turned on and off every 2s using TAIFG interrupts
// Olimex 1121STK, LED1 on P2.3 = TA1, LED2 on P2.4 = TA2, active low;
//  piezo between P2.0 and P2.5, Timer_A from ACLK = 32KHz
// J H Davies, 2007-06-19; IAR Kickstart version 3.42A
// ---------------------------------------------------------------
#include <io430x11x1.h>          // Specific device
#include <intrinsics.h>          // Intrinsic functions
// ---------------------------------------------------------------
// Delays in cycles of timer clock = ACLK = 32KHz = 0x8000Hz
#define LED1half    0x2000       // (1/4)s half-period for toggling
#define LED2duty    30           // Duration of "on" duty cycle
#define LED2period  300          // Total period
#define PIEZhalf    37           // Toggle piezo to give f = 440Hz nearly
// ---------------------------------------------------------------
void main (void)
{
    WDTCTL = WDTPW | WDTHOLD;     // Stop watchdog timer
// Channel 0 of timer, interrupts only to drive piezo
    TACCR0 = PIEZhalf;           // "Compare" value for first interrupt
    TACCTL0 = CCIE;              // Enable interrupts only (no output)
// Channel 1 of timer, drive LED1 (P2.3) in toggle mode with interrupts
    TACCR1 = LED1half;          // Compare value for first interrupt
    TACCTL1 = OUTMOD_4 | CCIE;   // Toggle mode with interrupts
// Channel 2 of timer, drive LED2 (P2.4) with set, reset and interrupts
    TACCR2 = LED2duty;           // Compare value for first interrupt
    TACCTL2 = OUTMOD_1 | CCIE;   // "Set" to turn LED off at next match
// Configure ports 1 and 2; redirect P2.3,4 to Timer_A
```

```
    P1OUT = BIT0 | BIT1;              // Output, high for Freq pin and TXD
    P1DIR = BIT0 | BIT1;
    P2SEL = BIT3 | BIT4;              // Re-route P2.3 to TA1, P2.4 to TA2
    P2OUT = BIT0 | BIT3 | BIT4;       // LEDs off (active low); prepare piezo
    P2DIR = BIT0 | BIT3 | BIT4 | BIT5;  // Piezo and LED outputs
// Start timer from ACLK, Continuous mode, clear, TAIFG interrupts
    TACTL = TASSEL_1 | ID_0 | MC_2 | TACLR | TAIE;
    for (;;) {                        // Loop forever
        __low_power_mode_3();         // LPM3, all action in interrupts
    }
}
//-----------------------------------------------------------------
// Interrupt service routine for TACCR0.CCIFG, called on compare
//-----------------------------------------------------------------
#pragma vector = TIMERA0_VECTOR
\__interrupt void TIMERA0_ISR (void) // Flag cleared
automatically {
    P2OUT ^= (BIT0|BIT5);            // Toggle piezo sounder
    TACCR0 += PIEZhalf;              // Calculate next compare value
}
//-----------------------------------------------------------------
// Interrupt service routine for TACCRn.CCIFG (n > 0) and TAIFG
//-----------------------------------------------------------------
#pragma vector = TIMERA1_VECTOR
__interrupt void TIMERA1_ISR (void) // ISR for TACCRn CCIFG and TAIFG
{
    switch (__even_in_range(TAIV, 10)) {     // Acknowledges interrupt
    case 0:                          // No interrupt pending
        break;                       // No action
    case TAIV_CCIFG1:                // Vector 2: CCIFG1
        TACCR1 += LED1half;          // Calculate next compare value
        break;                       // (LED1 toggled automatically)
    case TAIV_CCIFG2:                // Vector 4: CCIFG2
        if (TACCTL2 & OUTMOD_4) {    // Was last action Reset (on)?
            TACCR2 += LED2duty;      // Duration of On part of cycle
            TACCTL2 = OUTMOD_1 | CCIE;  // "Set" turns LED off next
        } else {                     // Last action was Set (off)
            TACCR2 += (LED2period - LED2duty);  // Duration of Off part
            TACCTL2 = OUTMOD_5 | CCIE;  // "Reset" turns LED on next
        }
        break;
    case TAIV_TAIFG:                 // Vector A: TAIFG
        TACCTL0_bit.CCIE ^= 1;       // Toggle CCIFG0 ints (piezo on/off)
        break;
    default:                         // Should not be possible
        for (;;) {                   // Disaster! Loop here for ever
        }
    }
}
```

The first job is, as always, to choose an appropriate clock for Timer_A. All the outputs are slow, which indicates ACLK. It could be divided, which would save a little power, but this would spoil the accuracy of the A note. If this were not important it would make sense to use ACLK/8.

Next we need to work out the durations of the output signals. LED1 should be toggled every 0.25 s. There are 32 K = 0x8000 cycles of ACLK per second so the interval is 0x2000 counts. Next, the frequency for LED2 should be around 100 Hz, which is a period of around 328 counts. I rounded this to 300 for convenience. The LED should be lit for 10% of the time or 30 counts. Finally, we would like a note of 440 Hz from the piezo sounder, which means that it should be toggled at 880 Hz. This is 37.24 counts, which I round to 37. The note therefore is a little sharp at 443 Hz but I certainly will not notice the error. (The next highest note is A♯ at 466 Hz.) We see how to do better in the section "Generation of a Precise Frequency" on page 326.

After stopping the watchdog, the channels of the timer are configured. All are done differently—this example was devised to show most of the likely modes. In all cases, the time at which the first event should occur is loaded into the capture/compare register TACCRn:

- Channel 0 produces only interrupts because the piezo sounder cannot be driven directly.

- Channel 1 is set up to toggle the output because it is on and off for equal times.

- Channel 2 is more complicated because the LED is on and off for different times. The first event is configured as a Set, which turns LED2 off because it is active low.

The ports are configured next. Do not forget to use P2SEL to connect the external pins to the outputs of Timer_A. The timer itself is started last of all and the main routine enters LPM3 so that all subsequent action takes place in the interrupt service routines.

- Channel 0 is simple because the vector is not shared. The ISR toggles the bits for the piezo sounder to produce a square wave, as shown in Figure 4.9. The time of the next interrupt is computed by adding the delay PIEZhalf to the current value.

- The other interrupts share a vector and are decoded using TAIV as usual. The output for channel 1 is toggled automatically by the hardware as soon as Timer_A detects the match between TAR and TACCR1: It happens perfectly on time without intervention from software. An interrupt is requested at the same time and the ISR needs only to update TACCR1 for the next change in output. These are equally spaced so we just add LED1half.

- It is more complicated to handle channel 2 because it is on and off for different times. The ISR first tests the OUTMOD2 bit to distinguish between Set (mode 1) and Reset (mode 5). It is slightly confusing because there is no individual definition for this bit in the io430x11x1.h header file so I use the constant OUTMOD_4 instead, which has the same value.

— If this bit is *set*, the channel is in mode 5 and the event that triggered this interrupt was a Reset of the output, which turned on the LED. We therefore increase TACCR2 by the time that the LED spends on, given by LED2duty. The LED should be turned off at the end of this time so the channel is therefore reconfigured to Set (mode 1).

— The opposite happens if the bit is *clear*. The output has just been Set to turn off the LED and this should last for the difference between the total period and the duty time. The LED should be turned on at the next event, which requires Reset (mode 5).

• Finally, TAIFG is easy to handle: It simply toggles the bit that enables interrupts on channel 0 to turn the piezo sounder on and off. There is silence if the outputs are not toggled; we could instead disable the outputs.

There is a subtle point about the value in TACCR0. This will fall out of synchronization with TAR if interrupts stop and the first Compare event might therefore be incorrect when interrupts resume. Fortunately it is not a problem here because we both stop and restart the interrupts when TAR returns to 0.

There is an important distinction between channels 1 and 2 on the one hand and channel 0 and TAIFG on the other. The outputs TA1 and TA2 are driven directly and therefore change at *precise* times regulated by the hardware of Timer_A: There is no delay while MCLK is restarted and the ISR is called. This is the same feature as for capture events. The processing of the interrupt takes place a little later but this has no effect on the timing provided that the time of the next compare event in TACCRn has been updated before it arrives. Thus the delay between two events cannot be too short. In contrast, the piezo sounder is driven from software and its timing is therefore not precise. The same applies to any signals controlled by TAIFG.

Example 8.18

Try this in the debugger. Single-stepping is instructive. You will find that the LEDs turn on after the lines with P2SEL and P2DIR have been executed. Setting P2OUT has no effect because the pins have been redirected to Timer_A and the initial state of the output units is low. You will also find that LED1 flashes slowly even when the program is stopped but the other outputs halt. Why is this?

Example 8.19

Increase the duty cycle of the PWM on TA2 by 10% every time TAR overflows. Start again from 10% after 90% because special treatment is needed to get a duty cycle close to 0 or 100% in the Continuous mode. (You might wish to disable the piezo sounder if it drives you mad.)

8.5.2 A Single Pulse or Delay with a Precise Duration

It is often necessary to change an output once after a precise delay, to trigger another module such as the analog-to-digital converter or to request an interrupt. The Continuous mode is most convenient for this. Listing 8.11 shows a rather daft application: a doorbell that lights an LED for precisely 0.5 s on the Olimex 1121STK. I chose this illustration because you can easily see what is happening. Usually the delay would be much shorter and you would need an oscilloscope rather than your eyes. (The "bell" is only an LED because I could not tolerate the noise from the piezo sounder any more.) More precisely, this is the sequence of events:

1. The processor waits in LPM4 until wakened by a falling edge when the button is pressed. Timer_A is stopped during the wait (in any case it would not run because ACLK is disabled).

2. Channel 1 of Timer_A is set up for a compare event after 10 ms and the timer is restarted.

3. The input from the button is checked after the delay to confirm that the button is still down. This is a simple approach to debouncing. The processor returns to LPM4 if the button is now up, in which case the first "press" must have been noise.

4. If the press of the button was valid, Channel 1 is set up for another 10 ms delay. This time the output mode is changed to Reset so that the LED (active low) is lit when the Compare event occurs.

5. After the 10 ms event that lights the LED, Channel 1 is set up for a further 500 ms delay and the output mode is changed to Set so that the LED is turned off.

6. Timer_A is stopped after the final delay and the processor returns to LPM4 until the button is next pressed.

The action takes place in the main loop for clarity and the processor enters LPM3 to save power while waiting for the timer.

Listing 8.11: Program `doorbell.c` to light an LED for precisely 0.5s after a button is pressed, using Timer_A in the Continuous mode.

```
// doorbell.c - accurate 0.5s doorbell after debounced press
// Channel 1 of Timer_A, Continuous mode, ACLK/8 = 4KHz
// Olimex 1121STK, LED1 on P2.3 = TA1, B1 on P2.1, both active low
// J H Davies, 2007-06-19; IAR Kickstart version 3.42A
//--------------------------------------------------------------------
#include <io430x11x1.h>          // Specific device, newer format
#include <intrinsics.h>          // Intrinsic functions
//--------------------------------------------------------------------
// Duration of delays in cycles of timer clock = ACLK/8 = 4KHz
#define DELAY10ms    40          // 10ms delay for debounce and pause
#define DELAY500ms   0x0800      // Duration of buzz and dead time
//--------------------------------------------------------------------
void main (void)
{
    WDTCTL = WDTPW | WDTHOLD;     // Stop watchdog timer
// Channel 1 of timer, drives LED1 (P2.3 = TA1), active low
    TACCTL1 = OUTMOD_0 | OUT;     // Force output initially high (LED off)
// Timer from ACLK/8, no need to clear, stopped
    TACTL = TASSEL_1 | ID_3 | MC_0;
    P1OUT = BIT0 | BIT1;          // Output, high for Freq pin and TXD
    P1DIR = BIT0 | BIT1;
    P2SEL = BIT3;                 // Reroute P2.3 to Timer TA1
    P2OUT = BIT3 | BIT4;          // LEDs off (active low)
    P2DIR = BIT0 | BIT3 | BIT4 | BIT5;  // Piezo and LED outputs
    P2IES = BIT1;                 // Detect falling edge on P2.1 = B1

    for (;;) {                    // Loop forever
        P2IFG = 0;                // Clear any pending interrupts on P2
        P2IE = BIT1;              // Enable interrupts on P2.1
        __low_power_mode_4();     // LPM4 until button pressed
        P2IE = 0;                 // Disable further interrupts on P2
// Request interrupt after 10ms delay using TACCR1
        TACCR1 = TAR + DELAY10ms;    // "Compare" value for 10ms delay
        TACCTL1_bit.CCIE = 1;     // Enable interrupts on compare
        TACTL |= MC_2;            // Start timer, Continuous mode
        __low_power_mode_3();     // LPM3 until timer interrupts
        if (P2IN_bit.P2IN_1 == 0) { // Button still down? Continue if so
// Drive output low (reset) after 10ms delay using TACCR1
            TACCR1 += DELAY10ms;     // Further 10ms delay
            TACCTL1 = OUTMOD_5 | CCIE;  // "Reset" output to turn on LED
            __low_power_mode_3();    // LPM3 until timer interrupts
// Drive output high (set) after 500ms delay using TACCR1
            TACCR1 += DELAY500ms;    // Further 500ms delay
            TACCTL1 = OUTMOD_1 | CCIE;  // "Set" output to turn off LED
```

```
        __low_power_mode_3();    // LPM3 until timer interrupts
    }
    TACCTL1_bit.CCIE = 0;     // Disable further interrupts on CCIFG1
    TACTL &= ~MC_3;           // Stop timer
    }
}
// -----------------------------------------------------------------------
// Interrupt service routine for port 2; need to clear flag
// -----------------------------------------------------------------------
#pragma vector = PORT2_VECTOR
__interrupt void PORT2_ISR (void)    // Flag NOT cleared automatically
{
    P2IFG = 0;                          // Acknowledge interrupt
    __low_power_mode_off_on_exit(); // Return to main routine
}
// -----------------------------------------------------------------------
// Interrupt service routine for CCIFG1 (only source active for vector)
// -----------------------------------------------------------------------
#pragma vector = TIMERA1_VECTOR
__interrupt void TIMERA1_ISR (void) // Flag NOT cleared automatically
{
    TACCTL1_bit.CCIFG = 0;         // Acknowledge interrupt
    __low_power_mode_off_on_exit();
}
```

An obvious question about this program is why there is a 10 ms delay after the state of the button has been confirmed, before the LED is illuminated. The reason is the requirement for a *precise* duration of 0.5 s for the low pulse that drives the LED. This means that the LED must be turned on and off *directly* by the hardware of the timer, which requires a Compare event. It cannot be done from software in response to another event because the precision would be lost. Thus a dummy delay is needed before the first Compare event, for which the output mode is Reset to drive the pin low. It could be shorter than the 10 ms delay used for debouncing but I just use the same value for convenience. After the LED is illuminated, TACCR1 is increased by the value needed for a further 0.5 s and the output mode is changed to Set so that the LED is turned off. Thus a precise flash is controlled directly by Timer_A.

There are a couple of other points about the program that I should mention. The LED is connected active low and the pin should therefore be driven high as soon as it is configured as an output. This is ensured by setting the OUT bit of TACCTL1 in output mode 0 (which is the default) before the pin is rerouted to Timer_A and becomes the output TA1. The output mode is changed later to turn the LED on.

The other detail is that we need to read TAR so that TACCR1 can be set up for the first delay. This can be hazardous if the timer is running because of a possible race between ACLK and MCLK, pointed out in the section "Timer Block" on page 289. Here we do not need Timer_A while the program waits for the button to be pressed so I stop it. We can then read TAR at leisure before restarting it.

Another way of taking a reliable reading of TAR is to read it repeatedly until two successive values agree. This needs ACLK slower than MCLK, of course. A simple do-while loop for this is shown in Listing 8.12. It is essential that TAR is declared volatile in the header file (which it is) or the loop would be meaningless.

Listing 8.12: Synchronization of LastTAR with TAR in doorbel2.c.

```
do {                          // Loop until two successive reads
    LastTAR = TAR;            //   of TAR agree with each other
} while (LastTAR != TAR);     // Take care debugging this
```

Example 8.20

The program already has the convenient feature that it is triggered by edges on the input, so it responds only once if the button is pressed continuously. Add a further 0.5 s delay after the LED has gone out, during which the doorbell does not respond to the button. This helps to reduce the nuisance from irritating callers who press the button repeatedly.

8.5.3   Generation of a Precise Frequency

One of the outputs of the program in the section "Generation of Independent, Periodic Signals" on page 318, which produced several independent frequencies, drove a piezo sounder at a frequency close to 440 Hz by toggling it at twice this frequency. I mentioned that this is the note used for tuning orchestras for western music, in which case we might like to do better than "close to" 440 Hz. In fact the frequency was 443 Hz, assuming that the crystal is accurate (I ignore this issue). This is an error of about +11 cents, where 100 cents is a semitone. (Without going into musical theory, a difference in frequency of 1% is about 17 cents and an octave is 1200 cents. Musical intervals depend on *ratios* of frequencies, which means that cents are defined in terms of logarithms.) My daughter plays the oboe, which sounds the note used to tune an orchestra. She tells me that 443 Hz is not good enough, and even a school orchestra is tuned to an accuracy of better than ±5 cents. How can we generate a more precise frequency? The problem is that each interval generated by the timer must be an *integer* multiple of the period of its clock—there is no way of timing fractional cycles.

An obvious method is to run the timer from a faster clock. Suppose that we used a frequency-locked loop to set $f_{\text{SMCLK}} = 32 f_{\text{ACLK}} = 2^{20}$ Hz, which is a binary megahertz. The number of counts for events at 880 Hz would then be $2^{20}/880 = 1191.56$. That is not a good choice. Double the frequency again, which gives 2383.13 counts. Now the rounding to an integer, 2383, would give an error of only 0.1 cent, which is more than good enough. Unfortunately there are two problems. First, f_{SMCLK} must be kept accurate by a frequency-locked loop, either in hardware or software. Second, it must be kept running, which raises the power consumption. We prefer to keep only ACLK running to generate a frequency as low as 880 Hz.

One way around these problems is to adopt the idea of modulating the length of each interval. This is used by the digitally controlled oscillator (DCO), described in the section "Digitally Controlled Oscillator, DCO" on page 167. The idea is to accept that the time between any two successive changes of the output is not correct but to ensure that the *average* over a longer time is accurate. Here is how this could work for toggling the output at 880 Hz from ACLK at 32 KHz. The ideal number of cycles per interval is 32 K/880 = 37.24 to two decimal places (I describe an approach with only integers shortly):

1. The first interval is chosen to be 37 cycles, which is the integral part of 37.24. This means that the interval is too short by 0.24 cycle, which I call the *shortfall*.

2. We aim to make the next interval longer to compensate for the shortfall in the first. The target for the second interval is therefore $37.24 + 0.24 = 37.47$ cycles. Again the interval is 37 cycles and the shortfall rises to 0.47 cycles.

3. Same again for the third interval: It is 37 cycles and the shortfall is now 0.71 cycle.

4. Same again for the fourth interval: It is 37 cycles and the shortfall rises further to 0.95 cycle.

5. The target for the fifth interval is $37.24 + 0.95 = 38.18$ cycles. This time the interval is increased by 1 to 38 cycles, leaving a reduced shortfall of 0.18 to carry forward.

Thus the first five intervals are four of 37 cycles followed by one of 38 cycles of ACLK. The average number of cycles per interval is 37.2 over this range, which corresponds to a frequency of 881 Hz. This average will approach closer to 880 Hz as time goes on. The shortfall always lies between 0 and 1.

Here is another way of doing the arithmetic. The ideal number of cycles per interval is given by the division $f_{\text{timer clock}}/f_{\text{output}}$. All the numbers are integers and we can write the result as a quotient and remainder—just what we learned at school, many years ago. This can be written formally as the equation

$$f_{\text{timer clock}} = \text{quotient} \times f_{\text{output}} + \text{remainder}. \qquad (8.3)$$

For example, $32\,\text{K} = 32{,}768 = 37 \times 880 + 208$. The quotient of 37 gives the number of cycles of the timer clock in the shorter intervals and 208 is the remainder. The shortfall per interval in the previous calculation is just $(\text{remainder}/f_{\text{output}}) = 208/880 = 0.24$. It is both exact and easier to do the calculations in cycles of the timer clock. We just add the remainder to the accumulated shortfall in each interval and introduce an extra count when the shortfall exceeds the desired frequency, f_{output}. Then the accumulated shortfalls in each interval are 208, 416, 624, 832, and 1040. The last exceeds the frequency of 880 so an extra cycle is added to the interval and the shortfall is reduced by 880 to 160. It builds up again in successive intervals until it next exceeds 880 and an extra cycle is included. This approach gives exactly the same outcome as the previous method using floating-point numbers but is far more convenient and efficient.

This explanation probably sounds a lot more complicated than it is to implement. Listing 8.13 shows an interrupt service routine to handle this for the Olimex 1121STK. It drives LED1 directly, which gives a precisely timed signal that can be displayed on an oscilloscope, and toggles the piezo sounder to give a "musical" note. The quotient and remainder could be entered as expressions to be calculated by the compiler but I write the values explicitly.

Listing 8.13: Interrupt service routine for Timer_A from `freqgen1.c`, which toggles the output at a precise *average* frequency of 880 Hz.

```
#define FREQUENCY     880           // Desired frequency of events
// Generally: f_timer = QUOTIENT * FREQUENCY + REMAINDER
//   or QUOTIENT = f_timer / FREQUENCY; REMAINDER = f_timer % FREQUENCY
#define QUOTIENT    37              // values for 32KHz timer clock
#define REMAINDER   208
//-------------------------------------------------------------------
#pragma vector = TIMERA1_VECTOR
__interrupt void TIMERA1_ISR (void) // Flag NOT cleared automatically
{
    static uint16_t shortfall = 0;  // Accumulated (summed) shortfall

    TACCTL1_bit.CCIFG = 0;          // Acknowledge interrupt
    P2OUT ^= (BIT0|BIT5);           // Toggle piezo sounder
    shortfall += REMAINDER;         // Update accumulated shortfall
    if (shortfall >= FREQUENCY) {   // Over threshold for extra count?
        TACCR1 += (QUOTIENT + 1);   // Yes: extra count in output period
        shortfall -= FREQUENCY;     // Subtrct correction from shortfall
```

```
    } else {
        TACCR1 += QUOTIENT;          // No: usual count
    }
}
```

How well does this work? Unfortunately the musical quality of the note from the piezo sounder is atrocious (it is not intended for such demanding applications) and contains so many harmonics that my daughter's electronic tuner was unable to lock to the frequency. Despite this, I could discern the difference between a perfect square wave with equal intervals and the modulated wave. This was a little surprising because the intervals vary only between 37 and 38.

It was interesting to push the method further and Figure 8.8a shows the observed signal for four frequencies of the timer clock. This can be divided down to 4 KHz, in which case the interval is either four or five cycles. Now it sounds *much* worse. The frequency of the timer clock can be reduced further by dividing ACLK itself, configured with the BCSCTL1 register in the MSP430Fx1xx family. The most extreme value is a clock of 1 KHz, in which case the output is *roughly* a pattern of five intervals of one cycle followed

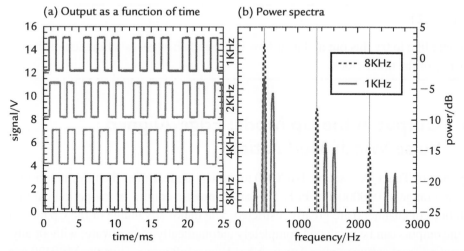

Figure 8.8: **Generation of a precise average frequency of 440 Hz using a range of frequencies for the timer clock. The data from the oscilloscope show (a) output as a function of time and (b) measured power spectra for the two extreme cases. Odd multiples of 440 Hz, expected for a square wave, are shown as vertical lines on the plot of the power spectrum.**

by one interval of two cycles. (The frequency would be 877.7 Hz if this pattern were the exact output.) The two durations are clear on the oscilloscope trace. The signal sounds terrible and is quite unusable, even if its average frequency is mathematically correct. The signal contains so many harmonics that the ear does not focus on the desired tone of 440 Hz. In fact it sounds closer to the note C than A. For those who are interested, I plotted the power spectra of the signals from timer clocks of 1 KHz and 8 KHz in Figure 8.8(b). The plot for 1 KHz is better as an example of a spread-spectrum signal than an aid to tuning.

I had to modify the program that used a 1 KHz clock because of a flaw in the design of Timer_A. This is documented as Bug TA12 in the errata to the data sheet for the F1121A. It affects Timer_A in all devices, as far as I know. The problem is that the next interrupt is lost if a capture/compare register TACCRn is incremented by only 1. There is nothing wrong in principle with doing this provided that the timer clock is slow compared with MCLK and there is a factor of 1000 between the frequencies in my program. However, the bug causes the next interrupt to be lost. A workaround is suggested in the errata. The moral of this is to check the errata if something strange seems to be happening: It may not be your program.

Example 8.21

You might like to try this out and hear how the sound changes as a function of the frequency of the timer clock.

8.6   Output in the Up Mode: Edge-Aligned Pulse-Width Modulation

The period of Timer_A is set by TACCR0 in the Up mode, rather than cycling through its natural range of 0x0000–0xFFFF. This offers precise control of the period but the capabilities of channel 0 are greatly restricted. The major advantage of the Up mode is that periodic outputs can be produced completely automatically in hardware, without any intervention from software after Timer_A has been configured. Thus the MSP430 can be left undisturbed in LPM3 if ACLK is used for Timer_A. This contrasts with the Continuous mode, where a new Compare value must be calculated in an ISR after each match has occurred. Because of the special nature of TACCR0, the notation TACCRn implies $n \geq 1$ throughout this section.

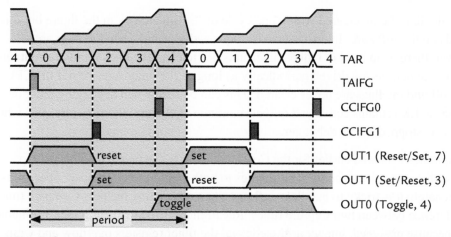

Figure 8.9: Sketch of output from channels 0 and 1 of Timer_A in an Up mode. Channel 0 is in the Toggle mode with TACCR0 = 4. The output from channel 1 is shown for both Reset/Set and Set/Reset modes with TACCR1 = 2.

Figure 8.9 shows the behavior of TAR in the Up mode with possible outputs from a normal channel (1) and the special channel 0. I have chosen TACCR0 = 4 and TACCR1 = 2. These are the main features:

- TAR counts from 0 up to the value in TACCR0, which is 4 here, and returns to 0 to start a new cycle on the next clock transition. The period is therefore TACCR0 + 1 = 5 counts.

- The flag CCIFG0 is set when TAR counts to TACCR0 and the TAIFG flag is set when TAR returns to 0, one cycle later.

- The flag CCIFG1 is set when TAR counts to TACCR1, which is 2 here.

- I show the output from channel 1, OUT1, for the two output modes that are usually used in the Up mode. It is best to think that Reset/Set, mode 7, sets the output when TAR counts to 0 and resets (clears) it when TAR counts to TACCR1. Note that the output is set when TAR counts to 0, not when it counts to TACCR0. This is not entirely clear in the family user's guides. The Set/Reset mode (3) does the opposite.

- The only output mode that changes OUT0 periodically is Toggle, mode 4, which is also shown. Note that it changes when TAR counts to TACCR0 and is therefore one count earlier than the changes in other channels. The output has twice the period of the timer because of the toggling action.

I show the flags being cleared within one cycle of the timer clock in the figure but this depends on the software. Interrupts can be requested as usual when the flags are set. However, there is no need to request interrupts nor to clear the flags if only the outputs are needed: The outputs are driven periodically as long as the timer runs even if the CPU is turned off and the flags are never cleared. This can be seen when debugging older devices because ACLK remains active and outputs continue to change automatically even when the CPU is stopped by the debugger.

The examples I showed earlier for the Continuous mode can be adapted for the Up mode with more or less trouble. This applies to both Capture and Compare operations. Unfortunately it is tricky to handle overflows when calculating the next Compare time or the difference between two Capture times. Recall that this was trivial in the Continuous mode because unsigned, integer arithmetic and the timer registers overflow and wrap around in the same way. Suppose that we are calculating the next Compare time in the Up mode. There are two ways in which values can overflow: The sum can exceed the 16-bit range of an unsigned integer or it can exceed TACCR0. Both cases must be treated correctly. Be careful! A simple case is if TACCR0 = 0x00FF, in which case the arithmetic can be performed with unsigned bytes and overflows are handled automatically.

The program in the application note *Pong Video Game Using the MSP430* (slaa177) relies heavily on Timer_A to produce video signals. It is far too intricate to explain here, sadly.

8.6.1 Uses of Channel 0 in the Up Mode

Channel 0 is special in the Up mode because its output remains constant for each period between settings of CCIFG0, as shown for Toggle mode (4) in Figure 8.9. This is the simplest way of generating a symmetric square wave of arbitrary frequency. The timer clock and TACCR0 are chosen to give *half* the desired period, the output mode for channel 0 is set to Toggle, and the signal is generated automatically. This is the ultimate simplification of the program to flash a LED at 1 Hz, started in the section "Automatic Control Flashing Light by Software Delay" on page 91. Unfortunately the TA0 output is not connected to a LED on the Olimex 1121STK (it is used for sending serial data instead) so it does not work. This code would do the job if TA0 were available:

```
    TACCR0 = 0x3FFF;          // (1/2)s period with 32KHz clock
    TACCTL0 = OUTMOD_4;       // Toggle mode, no interrupts
// Start timer from ACLK, no division, Up mode, clear, no interrupts
```

```
TACTL = TASSEL_1 | ID_0 | MC_1 | TACLR;
for (;;) {                   // Loop forever
    __low_power_mode_3();    // Remain in LPM3, CPU not needed
}                            //   (nor are interrupts)
```

It is a little pedantic to use the value 0x3FFF for TACCR0 rather than 0x4000 because the frequency is not that accurate.

Channel 0 can also be used to produce an evenly spaced but otherwise arbitrary sequence of high and low values on the output TA0 by configuring the next Compare event either to Set or Reset the output. This is done in an ISR in much the same way as in the Continuous mode but is simpler because there is no need to update the time for the next Compare event: The constant interval is instead set by TACCR0. This can be the most straightforward way of transmitting serial data, for instance.

8.6.2 Edge-Aligned PWM

Microcontrollers are often required to vary the power supplied to a load through a continuous range, not just on or off. This might seem to call for a digital-to-analog converter, or DAC, but very few microcontrollers contain true DACs. A few MSP430s provide them but most do not. The reason is that *pulse-width modulation*, or PWM, provides an adequate substitute for a DAC in most applications. It requires only a timer, which is purely digital and therefore much simpler and cheaper to fabricate than the analog circuits required for a DAC.

The idea behind PWM is very simple: The load is switched on and off periodically so that the *average* voltage has the desired value. The fraction of the time while the load is active is called the *duty cycle D*. This is illustrated in Figure 8.10. Assume that the output is driven either to ground or to V_{CC}. Then the average voltage across the output is given by

$$V_{ave} = D\,V_{CC} = \frac{t_{on}}{t_{on} + t_{off}}\,V_{CC} = \frac{t_{on}}{t_{period}}\,V_{CC} = \frac{\text{pulse width}}{\text{period}}\,V_{CC}. \tag{8.4}$$

The duty cycle is almost always varied by keeping the period constant and changing the width of the pulses, hence the name of PWM. Occasionally a different approach is used, such as keeping t_{on} constant and varying t_{off} instead. This is pulse-frequency modulation.

A second reason for preferring PWM is that the load is always either fully on or fully off, never in a partly powered state. This is more efficient and reduces the power dissipated in the switching transistors, as explained in the section "Driving Heavier Loads" on page 247.

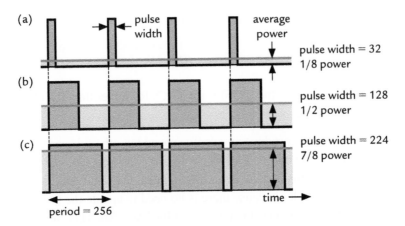

Figure 8.10: Edge-aligned pulse-width modulation showing three average powers.

Pulse-width modulation seems a crude substitute for a real analog output but is entirely satisfactory for many applications. For instance, suppose that you wish to vary the brightness of an LED. It would be obvious if PWM were used at 1 Hz because you would see the LED flash on and off. However, the variation becomes too rapid for the eye to follow if the frequency is raised above 100 Hz and only the average brightness is resolved. At least, this is true provided that you look steadily at the LED. The flashing becomes visible if you flick your eyes across the LED or wave the demonstration board about. If you try this on almost any piece of equipment you see that the LEDs flash rather than remain on continuously. Some loads, notable motors, are inductive and smooth the current waveform intrinsically. This is helpful because PWM is used widely to control motors.

In other cases a square wave is not acceptable and the PWM signal must be smoothed to bring it closer to a steady voltage. This requires an analog, low-pass filter outside the MSP430, which is easier to implement if the frequency of the PWM signal is as high as possible. With luck a simple, first-order, *RC* circuit may be sufficient. A higher-order, active filter is necessary if the ripple remains too large. In this case a "real" DAC, either internal or external, may be a better solution.

8.6.3   Simple PWM

The usual arrangement for PWM is that each output is turned on when TAR returns to 0 and turned off after a variable time that gives the desired duty cycle. This means that increasing the value in TACCRn increases the duty cycle, which makes the operation simpler to understand. Figure 8.9 shows that this needs the Reset/Set output mode (7) for loads driven active high. This is sometimes called *positive PWM*. Conversely, the Set/Reset mode (3) should be used for active low loads or negative PWM. Typically there are several outputs, each driven by a separate channel of Timer_A. Channel 0 cannot be used, of course, so Timer_A3 can support only two channels of PWM. They all are switched at the same frequency, controlled by the timer clock and the value in TACCR0. All outputs are turned on at the same time, which is why this is called *edge-aligned PWM*. They turn off at different times, according to the particular duty cycle D_n of each channel n (for $n > 0$). A timer with three channels available for PWM, such as Timer_A5, could produce the three signals shown in Figure 8.10 simultaneously. The duty cycle is given by

$$D_n = \frac{\text{pulse width n}}{\text{period}} = \frac{\text{TACCRn}}{\text{TACCR0} + 1}. \tag{8.5}$$

This equation holds for $0 \leq \text{TACCRn} \leq (\text{TACCR0} + 1)$. Larger values of TACCRn give $D_n = 1$ as are explained below—it is impossible to have $D_n > 1$.

Listing 8.14 shows very slow PWM on an Olimex 1121STK. The period is 1 Hz, determined by TACCR0 and ACLK, so that it is easy to see what is happening without an oscilloscope. The duty fractions for LED1 and LED2 are one half and one quarter, set by the values in TACCR1 and TACCR2. This board has the LEDs connected active low, which requires negative PWM and the Set/Reset output mode (3). The final step is to start the timer. Everything is automatic after that: The outputs are driven directly by Timer_A and the CPU can be stopped. No interrupts or anything else is needed.

Listing 8.14: Very slow pulse-width modulation at 1 Hz on an Olimex 1121STK in simppwm1.c. LED1 has duty cycle $D_1 = \frac{1}{2}$ and LED2 has $D_2 = \frac{1}{4}$.

```
// Set up Timer_A for PWM on active low LEDs, channels 1 and 2
    TACCR0 = 0x7FFF;              // 1s period from 32KHz timer clock
    TACCR1 = 0x4000;              // Duty cycle D = 1/2
    TACCR2 = 0x2000;              // Duty cycle D = 1/4
    TACCTL1 = OUTMOD_3;           // Set-reset mode for negative PWM
    TACCTL2 = OUTMOD_3;           // Set-reset mode for negative PWM
// Start timer from ACLK, no division, Up mode, clear, no interrupts
    TACTL = TASSEL_1 | ID_0 | MC_1 | TACLR;
    for (;;) {                    // Loop forever
        __low_power_mode_3();     // Remain in LPM3, CPU not needed
    }                             //   (nor are interrupts)
```

The LEDs turn on simultaneously when TAR returns to 0, as usual for edge-aligned PWM. LED2 turns off after 0.25 s, LED1 remains alight for a further 0.25 s, and both stay dark for the remaining 0.5 s before the next cycle starts.

If it is important that the very first cycle of PWM is correct, the OUT bit must be used to put the output into the correct initial state before starting PWM. For the usual form of edge-aligned PWM this means that the output should be turned on. The reason is that the first count of TAR is from 0 to 1 (assuming that it is cleared), so the system misses the actions that usually start a new cycle when TAR counts to 0. The outputs are active low on the Olimex 1121STK and the OUT bit should therefore be cleared. This happens to be the default so no explicit code is needed. (Often it does not matter that the first cycle is erroneous and this issue is ignored. The second and subsequent cycles are correct.)

Example 8.22

Single-step through this program. At what point does the PWM start? (The behavior may be different if you use a more modern device than the F1121A and the debugger has more control over the clocks.)

Example 8.23

How would the program need to be changed if the LEDs were driven active high rather than active low? Hint: *Two* changes should be made for each channel.

Example 8.24

Speed up the PWM so that the flashing becomes too fast for the eye to resolve. A convenient frequency is 256 Hz because it is 0x0100 in hexadecimal, which makes the changes easy to calculate. Remember to adjust the values in *all* the TACCRn registers. Does the brightness of the two LEDs look very different, given the factor of 2 in power between them?

Reduce the duty cycle of LED2 to its minimum nonzero value with the same value of TACCR0. Does the LED still appear to be illuminated?

Another important advantage of performing PWM in hardware rather than software is that it is straightforward to get the extreme duty cycles of 0 and 1. These are the two cases:

- **Zero** duty cycle is obtained by setting TACCRn = 0, consistent with equation (8.5). The set and reset events now occur simultaneously when TAR returns to 0 and the logic is designed to give zero output. The channel flag CCIFGn is set in every cycle at the same time as TAIFG and interrupts are requested if they have been enabled in TACCTLn.

- **Unity** duty cycle is obtained by making TACCRn greater than TACCR0; the obvious value from equation (8.5) is (TACCR0 + 1) but any value larger than TACCR0 has the same effect. Now the compare event never happens because TAR never counts as high as the value in TACCRn. Suppose that the output mode is Reset/Set (7). In this case the output is set every time TAR returns to 0 but the reset events never occur. Thus the output remains set continuously, giving a duty cycle of 1.

There are a couple of pitfalls with the limit of $D_n = 1$. First, the channel flag CCIFGn is never raised because the compare event does not happen. This also means that no interrupts are requested—beware if you are using the channel to generate interrupts as well as driving the output directly. The second problem arises if you are using the full range of the timer with TACCR0 = 0xFFFF. In this case it is impossible to store a larger value in the other registers TACCRn so the hardware cannot give $D_n = 1$ automatically. It must be treated as a special case. Alternatively, set TACCR0 = 0xFFFE and avoid the problem.

8.6.4 Design of PWM

There are two main parameters that must be chosen before suitable values can be selected for PWM:

- The number of desired values of the duty cycle (the resolution).
- The frequency of the output waveform.

These are linked because it is not possible to have both high resolution and a high frequency.

Suppose that the duty cycle of the output is specified in percent to the nearest integer. This means that there are 101 possible values of 0, 1, 2, . . . , 99, 100%. The simplest way of handling this is to choose TACCR0 = 99, which gives a period of 100 counts. The desired duty cycle in percent can then be written directly to TACCRn for channel n (where $n > 0$). Remember that the denominator in equation (8.5) for the duty cycle is (TACCR0 + 1).

The appropriate frequency f_{PWM} of the output waveform depends strongly on the type of load. Anything higher than about 100 Hz is sufficient for an LED. The period of the PWM is the same as that of the timer and the frequency is therefore given by

$$f_{PWM} = \frac{f_{\text{timer clock}}}{\text{period of TAR in counts}} = \frac{f_{\text{timer clock}}}{\text{TACCR0} + 1}. \tag{8.6}$$

This means that $f_{\text{timer clock}}$ must be above 100 Hz × 100 counts = 10 KHz for the LED. There would be no problem in running the timer from ACLK at 32 KHz except that 10 KHz is not readily available. The simple options are $f_{ACLK}/2 = 16$ KHz or $f_{ACLK}/4 = 8$ KHz. It would be better to choose the higher frequency to avoid visible flashing of the LED. ACLK may instead be derived from the VLO. This is slower at 12 KHz but just fast enough for this application.

This would be too slow for driving a motor because inevitably some vibration occurs at the frequency of the PWM. You might therefore choose $f_{PWM} = 20$ KHz, above the audible range. Now the frequency of the clock should be above 20 KHz × 100 counts = 2 MHz. This needs SMCLK rather than ACLK but is not otherwise a problem.

Difficulties arise if we also want to improve the resolution at the same frequency. Suppose that the duty cycle is specified to 0.1% instead of 1%, meaning 1000 values above 0 instead of 100. We have to raise TACCR0 to 999, giving a period of 1000 counts, and the timer must be clocked at 20 MHz. Too fast! The frequency of the clock limits the product of resolution and PWM frequency.

The simple approach also fails if the desired frequency is too low. In this case a larger value must be used for TACCR0 to raise the period in real time. This was done in Listing 8.14. The only "penalty" is that there is more resolution than needed. It simply means that TACCRn should be changed by a larger number than ±1 to adjust the duty cycle.

I assume that we can choose a convenient value for the range of the counter, set by TACCR0, and tolerate an error in the frequency of the PWM wave. Inevitably this is not always true. Suppose that the LED should be driven at 110 Hz to avoid being too close to harmonics of the AC mains at 50 or 60 Hz. If the timer is supplied from ACLK without division, the period should be 32 KHz/110 Hz = 297.9 counts, which could be rounded to 300 with a negligible error. The problem is that the duty cycle can no longer be specified directly in percent although the numbers are still convenient in this case.

As an example, let us drive an LED so that its intensity varies linearly from 0 to full brightness and back again to 0 with an overall period of 2 s. This is often used for the standby indicator on electronic equipment. Take $f_{PWM} = 128$ Hz, which is chosen because it is a power of 2 (easy) and is fast enough that the modulation should not be visible. Each ramp in intensity up or down takes 1 s and therefore includes 128 cycles of PWM. It is therefore convenient to choose a period of 128 counts for each cycle with TACCR0 = 127. The value in TACCR1 should be incremented by 1 after each cycle while the intensity increases and decremented by 1 after each cycle while the intensity decreases. This gives the desired ramp. The frequency of the timer clock should be $f_{\text{timer clock}} = 128 \, \text{Hz} \times 128 \, \text{counts} = 16 \, \text{KHz} = f_{\text{ACLK}}/2$, which is easy.

Listing 8.15 shows a program for an active low LED on the Olimex 1121STK. The duty cycle starts from 0 and the LED is initially turned off to avoid an erroneous flash when the program starts. The output mode is Set/Reset (3) because the load is active low. The main difference from the previous examples is that the duty cycle varies. It must be updated during every cycle of PWM and I therefore enable interrupts on CCIFG0; I explain later why I chose this rather than TAIFG or CCIFG1. The interrupt service routine contains a static variable to keep track of the direction of the ramp, up or down. There is no need to keep a copy of the duty cycle because we can safely read TACCR1 at any time in the Compare mode. Its value is incremented or decremented according to the direction of the ramp. The result is compared with the appropriate limit and the direction is reversed for the next update if the duty cycle has hit the limit.

Listing 8.15: Program `ledpwm1.c` to drive an LED on an Olimex 1121STK with PWM so that its intensity ramps linearly from 0 to full brightness and back again with an overall period of 2 s.

```
// ledpwm1.c - Ramped PWM using Timer_A in Up mode from ACLK
// PWM at 128Hz, 128 counts per cycle from ACLK/2; ramp 1s up, 1s down
// Pulse width is incremented or decremented by 1 each cycle in ramp
// Duty cycle updated by TACCR0 CCIFG interrupt
// Olimex 1121STK, LED1 on P2.3 = TA1, active low; piezo off
// J H Davies, 2007-07-18; IAR Kickstart version 3.42A
//-------------------------------------------------------------------
#include <io430x11x1.h>            // Specific device
#include <intrinsics.h>           // Intrinsic functions
#include <stdint.h>               // Integers of defined sizes
//-------------------------------------------------------------------
void main (void)
{
    WDTCTL = WDTPW | WDTHOLD;      // Stop watchdog timer
// Set up Timer_A for PWM on active low LED, channel 1
    TACCTL1 = OUTMOD_0 | OUT;      // Force output initially high (LED off)
    TACCR0 = 0x007F;              // 128Hz from 16KHz timer clock
    TACCR1 = 0x0000;              // Duty cycle D = 0 initially
    TACCTL0 = CCIE;               // Enable interrupts on compare
    TACCTL1 = OUTMOD_3;           // Set/reset mode for negative PWM
// Configure ports 1 and 2; redirect P2.3 to Timer_A
    P1OUT = BIT0 | BIT1;          // Output high for Freq pin and TXD
```

```
    P1DIR = BIT0 | BIT1;
    P2SEL = BIT3;                      // Route P2.3 to TA1
    P2OUT = BIT3 | BIT4;               // LEDs off (active low); piezo off
    P2DIR = BIT0 | BIT3 | BIT4 | BIT5;   // Piezo and LED outputs
// Start timer from ACLK/2, Up mode, clear, no interrupts
    TACTL = TASSEL_1 | ID_1 | MC_1 | TACLR;
    for (;;) {                         // Loop forever
        __low_power_mode_3();          // Enter LPM3 between interrupts
    }
}
//--------------------------------------------------------------------------
// Interrupt service routine for TACCR0.CCIFG
// Compiler warning for "if (TACCR1 > TACCR0)" can be ignored
//--------------------------------------------------------------------------
#pragma vector = TIMERA0_VECTOR
__interrupt void TIMERA0_ISR (void) // Flag cleared automatically
{
    static enum {up, down} direction = up;

    if (direction == up) {        // Increasing duty cycle?
        ++TACCR1;                 // Yes: increase TACCR1
        if (TACCR1 > TACCR0) {    // Hit 100% limit?
            direction = down;     // Yes: decrease duty cycle next
        }
    } else {
        --TACCR1;                 // Decrease duty cycle
        if (TACCR1 == 0) {        // Hit 0 limit?
            direction = up;       // Yes: increase duty cycle next
        }
    }
}
```

The compiler warns about the comparison if (TACCR1 > TACCR0) in this program. The reason is that both variables TACCR0 and TACCR1 are declared volatile and the result might therefore depend on the order in which they are read. Fortunately there is no problem here because neither variable is really volatile in the Compare mode: They change only when the program writes a new value to them. It would be different in the Capture mode, where the contents would change whenever a capture occurred. The time of this event would be unpredictable, the registers would definitely be volatile, and the warning should be taken seriously.

Example 8.25

Try this program. Does the brightness appear to rise and fall linearly?

I will tell you the answer to this one: The brightness does *not* appear to vary linearly. Instead the LED appears to be fairly bright for most of the time with only brief dark

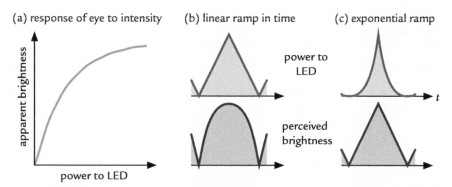

Figure 8.11: **(a) Brightness perceived by the eye as a function of power supplied to LED. Absolute power and perceived brightness as a function of time for an LED driven by a (b) linear ramp and (c) exponential ramp.**

intervals. The reason is that the response of the eye is far from linear. In fact it is close to logarithmic, which means that the eye detects changes in the *ratio* of intensities. For example, a change in absolute intensity from 1 to 2 (arbitrary units) is perceived as the same change as going from 10 to 20. The response is plotted in Figure 8.11(a), with the absolute power and perceived brightness of the LED driven by a linear ramp in (b). Another way of looking at this, which should appeal to electronic engineers, is to imagine that the eye works in decibels.

We need to correct for the logarithmic response if we want the variation of intensity to *appear* linear. This means that each step should give the same ratio of intensities. In other words, the ramp should be an exponential function, as in Figure 8.11(c). The obvious choice of ratio is 2, in which case the values for TACCR1 should be 0, 1, 2, 4, 8, 16, 32, 64, 128. This contrasts with the original linear ramp of 0, 1, 2, 3, 4, ..., 127, 128. A problem is that there are only 8 values instead of 128, so each duty cycle must be repeated for 16 periods of PWM to give the same period for the ramp. Listing 8.16 shows the interrupt service routine that updates the duty cycle. It is slightly clumsy because a new upward ramp must be started by storing 1 in TACCR1, after which it can be doubled and halved by shifting for the remainder of the pattern.

Listing 8.16: An LED driven with an exponential ramp using PWM by `ledpwm3.c`. The duty cycle changes by a factor of 2 at each step.

```
// Number of times each duty cycle is performed between updates
#define REPEATS_PER_CYCLE    16
#pragma vector = TIMERA0_VECTOR
__interrupt void TIMERA0_ISR (void)        // Flag cleared automatically
```

```
{
    static enum {up, down} direction = up;
    static uint8_t repeats = REPEATS_PER_CYCLE;

    --repeats;                              // Count repeats of each duty
    if (repeats == 0) {                     // Time to update duty cycle
        repeats = REPEATS_PER_CYCLE;        // Restart duty cycle counter
        if (direction == up) {              // Increasing duty cycle?
            if (TACCR1 == 0) {              // Yes: Restarting ramp?
                TACCR1 = 1;                 // Yes: initialize
            } else {
                TACCR1 <<= 1;               // No: double TACCR1
                if (TACCR1 > TACCR0) {      // Hit 100% limit?
                    direction = down;       // Yes: decrease duty cycle next
                }
            }
        } else {
            TACCR1 >>= 1;                   // Halve duty cycle
            if (TACCR1 == 0) {              // Hit 0 limit?
                direction = up;             // Yes: increase duty cycle next
            }
        }
    }
}
```

Example 8.26

Try this program. Does the brightness now appear to rise and fall linearly?

Unfortunately the steps of a factor of 2 in this program are rather coarse and there are too few of them. A smaller ratio is needed to get a smoother variation. The number of steps can be doubled by using a ratio of $\sqrt{2} = 2^{1/2}$ instead. It is easier to work out the values by dividing the maximum value repeatedly by the ratio, which gives 128, 91, 64, 45, 32, ..., rounded to the nearest integer. New values are interleaved between the powers of 2 used previously to double the number of values. The list can be stored in a constant array and the index stepped up and down. Finer variations can be obtained by using higher roots. For example, a ratio of $2^{1/4}$ gives 128, 108, 91, 76, 64, ... There is no need to use a root of 2 but it is reassuring to see familiar values such as 64 in the list. The ratio can be chosen to give any desired number of values. A problem is that rounding affects the smaller values badly. The solution is to use a faster clock for the timer to increase the range of counting and the resolution of the duty cycle.

There is a straightforward example of PWM to control the speed of a fan in the application note *Digital Fan Control with Tachometer Using MSP430* (slaa259). Another example is in the application note *Ultra-Low Power TV IR Remote Control Transmitter* (slaa175), which uses PWM to drive an infrared LED in the remote control. Timer_A also determines the overall length of each bit.

8.6.5 Software-Assisted PWM

None of the examples in the previous section works on the eZ430–F2013 or on the TI MSP430FG4618/F2013 Experimenter's Board because they do not have LEDs or other output devices connected to pins that can be routed to Timer_A. The timer needs assistance from software to drive a PWM signal onto other pins. Inevitably this abandons the precise timing offered by pure hardware and therefore is satisfactory only for low frequencies.

Let us adapt the simple, linear ramp in Listing 8.15 for an eZ430–F2013. This has an LED connected active high to P1.0. The timer in the F2013 is Timer_A2, which offers only the single channel 1 for PWM. Its output TA1 can be routed to three pins but these do not include P1.0. We need to emulate positive PWM, which means that the LED should be turned on when TAR returns to 0 and turned off when TAR matches TACCR1. The flags TAIFG and CCIFG1 are set at these two events and the straightforward solution is to turn the LED on and off in the corresponding interrupt service routines. I set up channel 1 for positive PWM in Reset/Set mode to make the purpose clear but the mode is irrelevant because we use only the interrupts.

A few more changes are made in Listing 8.17 because there is no watch crystal for ACLK, which must instead be provided by the VLO. I remove the division of the timer clock and reduce the number of counts per cycle to 110, which should give a ramp of 1s each way if the VLO runs at its nominal frequency of 12 KHz. I also raise the frequency of the DCO to its maximum of 16 MHz. This gives over 1000 cycles of MCLK for each cycle of ACLK, which should allow plenty of time for the interrupts to be serviced before the next count of Timer_A.

Listing 8.17: Program `ezpwm1.c` to drive the LED on an eZ430–F2013 with a linear ramp up and down using PWM. The LED is not connected directly to Timer_A and must therefore be switched on and off in interrupt service routines.

```
// ezpwm1.c - Ramped PWM using Timer_A, Up mode from ACLK = VLO = 12kHz
// 110 counts per cycle from ACLK; PWM roughly 110Hz; ramp 1s each way
// LED turned on by TAIFG and off by CCIFG1 interrupts
// Duty cycle updated by CCIFG0 interrupt
// Pulse width is incremented or decremented by 1 each cycle in ramp
// TI eZ430, LED1 on P1.0, active high, NO direct connection to Timer_A
// J H Davies, 2007-07-18; IAR Kickstart version 3.42A
//-------------------------------------------------------------------
#include <io430x20x3.h>          // Specific device
#include <intrinsics.h>          // Intrinsic functions
```

```
#include <stdint.h>                  // Integers of defined sizes
//-----------------------------------------------------------------
void main (void)
{
    WDTCTL = WDTPW | WDTHOLD;         // Stop watchdog timer
    BCSCTL3 = LFXT1S_2;              // ACLK from VLO
// Select fastest MCLK (16MHz) to ensure time for processing interrupts
    BCSCTL1 = CALBC1_16MHZ;          // Set range for calibrated 16MHz
    DCOCTL = CALDCO_16MHZ;           // Set DCO step and modulation
// Set up Timer_A for PWM on active high LED, channel 1
    TACCR0 = 110;                    // Roughly 110Hz from 12kHz timer clock
    TACCR1 = 0x0000;                 // Duty cycle D = 0 initially
    TACCTL0 = CCIE;                  // Interrupts on compare to update PWM
    TACCTL1 = OUTMOD_7 | CCIE;       // Reset-set (positive PWM), interrupts
// Configure ports 1 and 2 with pull resistors on unused pins
    P1OUT = 0;                       // Preclear output buffer
    P1DIR = BIT0;                    // Set P1.0 to output, others input
    P1REN = 0xFF & ~BIT0;            // Pull resistors on P1[7:1]
    P2SEL = 0;                       // Digital input/output rather than crystal
    P2REN = BIT6 | BIT7;             // Pull resistors on pins that exist
// Start timer from ACLK, no division, Up mode, clear, with interrupts
    TACTL = TASSEL_1 | ID_0 | MC_1 | TACLR | TAIE;
    for (;;) {                       // Loop forever
        __low_power_mode_3();        // Enter LPM3 between interrupts
    }
}
//-----------------------------------------------------------------
// Interrupt service routine for CCIFG1 and TAIFG; share vector
//-----------------------------------------------------------------
#pragma vector = TIMERA1_VECTOR
__interrupt void TIMERA1_ISR (void) // ISR for CCIFG1 and TAIFG
{
    switch (__even_in_range(TAIV, 10)) {
    case 0:                          // No interrupt pending
        break;                       // No action
    case 2:                          // Vector 2: CCIFG1
        P1OUT_bit.P1OUT_0 = 0;       // End of duty cycle: Turn off LED
        break;
    case 10:                         // Vector A: TAIFG, last value possible
        P1OUT_bit.P1OUT_0 = 1;       // Start of PWM cycle: Turn on LED
        break;
    default:                         // Should not be possible
        for (;;) {                   // Disaster. Loop here forever
        }
    }
}
//-----------------------------------------------------------------
// Interrupt service routine for TACCR0.CCIFG
//-----------------------------------------------------------------
#pragma vector = TIMERA0_VECTOR
__interrupt void TIMERA0_ISR (void) // Flag cleared automatically
{
    static enum {up, down} direction = up;

    if (direction == up) {           // Increasing duty cycle?
        ++TACCR1;                    // Yes: increase TACCR1
```

```
        if (TACCR1 > TACCR0) {    // Hit 100% limit?
            direction = down;     // Yes: decrease duty cycle next
        }
    } else {
        --TACCR1;                 // Decrease duty cycle
        if (TACCR1 == 0) {        // Hit 0% limit?
            direction = up;       // Yes: increase duty cycle next
        }
    }
}
```

If you try this, you will discover that it works—almost. The brightness ramps up and down as in the earlier program but there is a bright flash when the LED should be dark. Clearly there is a problem when the duty cycle is 0. Why is this?

A normal cycle starts by lighting the LED when TAIFG is raised and ends by extinguishing it when CCIFG1 is raised. Both events occur at the same time when the duty cycle is 0. The two interrupts share the same vector and the order of execution is set by their priorities, which are listed in Table 8.3. This shows that CCIFG1 has the higher priority and its interrupt is therefore serviced first, which turns the LED off. The interrupt for TAIFG is serviced immediately afterward and turns the LED on. This is the wrong way around. Thus the relative priorities of TAIFG and CCIFG1 cause an error for zero duty cycle. Even if the interrupts were serviced in the opposite sequence, the LED would be switched on for a very brief interval, which is undesirable.

Example 8.27

Repair this problem. Are there any similar issues for a duty cycle of unity?

The disadvantage of driving outputs from interrupt service routines is that there is a delay between the timer raising a flag and the consequent interrupt service routine. This is shown in Figure 6.6, where there is more than 1 μs between the directly driven signal OUT0 and P1.0, which is controlled by software. The delay would be shorter if the DCO were always active but would be extended if an interrupt were already being serviced or if an interrupt with higher priority were requested. The delays are longer for Listing 8.17 because its interrupt service routines are more complicated—only a single instruction is needed for the ISR in Figure 6.6.

PWM can also be produced from the Continuous mode and an example is shown for LED2 in Listing 8.10. In that case the load is connected directly to Timer_A so we only had to calculate the time for the next compare event in the ISR. It could also switch the load on and off if a direct connection were not available. Special treatment is needed for values of the duty cycle close to 0 and 1.

8.6.6   When Should the Duty Cycle Be Updated?

This seems a daft question: Surely we can just write a new value to TACCRn to update the duty cycle whenever we wish? Well, no, not if you want a perfectly clean waveform, as the example shows.

Example 8.28

Modify the program in Listing 8.15 so that the duty cycle is updated in the interrupt service routine for TAIFG rather than TACCR0. What goes wrong?

Here is the answer: You should find that the LED now flashes briefly when it should be dark. The cause of the problem is illustrated in Figure 8.12(a). The period of the

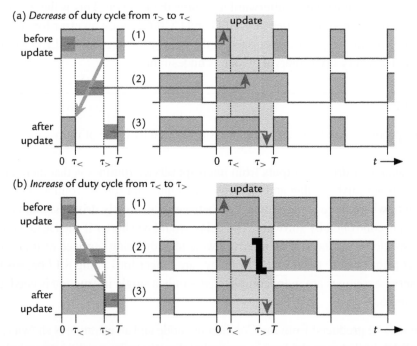

Figure 8.12: Effect of (a) decreasing and (b) increasing the duty cycle of PWM at three points in the cycle.

PWM waveform is T and I show the length of the high pulse being reduced from $\tau_>$ to $\tau_<$. Look at the detailed effect of updating the duty cycle at different points in the cycle:

1. Update between 0 and $\tau_<$, before the end of the new, shorter pulse. In this case the update happens in time for the output to be switched off at the new time $\tau_<$ and no problems arise.

2. Update between $\tau_<$ and $\tau_>$, between the ends of the new and old pulses. This is the nasty one. The output is high because the timer has not yet reached the old duration $\tau_>$. The update changes the end point to $\tau_<$, but this has already passed. The output therefore is not turned off at either $\tau_<$ or $\tau_>$ during the current cycle, which consequently has a duty cycle of unity. The output remains high until it is turned off at $\tau_<$ in the *following* cycle.

3. Update between $\tau_>$, after the end of the old, longer pulse. The output has already been turned off for the current cycle and the new time takes effect in the next cycle. No problems.

This shows that the tricky case is where the length of pulse is decreased at a time between the ends of the old and new pulses. The update is performed while TAR = 0 in the program set in Example 8.28, assuming that the interrupt service routine can be completed within one cycle of the timer clock. The problem therefore arises when TACCR1 is reduced from 1 to 0. Start with TACCR1 = 1. The LED is turned on when TAR returns to 0. At this point the TAIFG interrupt is requested and serviced, which updates TACCR1 from 1 to 0. The next compare event does not happen until TAR counts to the value of 0 in TACCR1, which does not occur until the start of the next cycle. This also means than an interrupt is missed, if these have been enabled. Oops!

The problem is less dramatic when the duty cycle is increased, as shown in Figure 8.12(b). Again the tricky case is when the duty cycle is updated between the ends of the old, shorter pulse and the new, longer pulse. The old pulse has already ended and the output was turned off at $\tau_<$. The updated value causes a second compare event to occur at $\tau_>$. This has no impact provided that interrupts are not used and the output mode is either Reset/Set or Set/Reset, and it does not request an interrupt. On the other hand, you get a nasty surprise if you use one of the toggling modes 2 or 6 because these cause the output to switch back on again at the second compare event. It will recover in the next cycle. The second interrupt may also be a problem if it is assumed that there is exactly one per cycle.

These difficulties can be avoided by updating the duty cycle only during the safe parts of the cycle. In principle this means that you should

- *Decrease* the duty cycle at the *end* of a pulse, following the CCIFGn flag.

- *Increase* the duty cycle at the *start* of a pulse, following the TAIFG flag.

Inevitably there is a catch: No CCIFGn interrupts are requested if the duty cycle is 100%. The best approach for Timer_A is to update the TACCRn registers following the CCIFG0 interrupt. This is requested when TAR counts to its maximum value of TACCR0, so there are no more compare events before the TAR returns to 0 to start the next cycle. This prevents the problems we just saw. However, it is reliable only if the updates can be completed before TAR changes. In other words, the interrupt service routine must run completely within a cycle of the timer clock. The ISR must therefore be brief (always a good idea) and MCLK should be fast.

At this point you are probably thinking, Why is this so troublesome? Surely there must be a better solution? There is, but it requires the more powerful hardware of Timer_B.

Example 8.29

Write a program to control the brightness of LED1 on the Olimex 1121STK using the buttons. The brightness should decrease gradually while B1 is down and increase while B2 is down. (What should you do if both are down?) Start with 50% power and use a linear ramp. This sort of program is often used to control the speed of a motor using PWM and is explained in the application note *PWM DC Motor Control Using Timer_A of the MSP430* (slaa120).

Care is also needed when updating PWM in the Continuous mode to ensure that the length of each cycle is constant. The following interrupt service routine for channel 0 is adapted from the course *Tips and Tricks for Timer_A* by Andreas Dannenberg and Peter Forstner:

```
#pragma vector = TIMERA0_VECTOR
__interrupt void TIMERA0_ISR (void)    // Flag cleared automatically
{
    static uint16_t PWM0Low;           // Local copy of low duration

    if (TACCTL2 & OUTMOD_4) {           // Was last action reset?
        TACCR0 += PWM0Low;             // Yes: calculate end of low output
    } else {                            // No: start new cycle (high output)
        TACCR0 += DutyCycle0;          // Calculate end of high output
        PWM0Low = TACCR0 - DutyCycle0; // Calc & store low duration now
    }
    TACCTL2 ^= OUTMOD_4;               // Toggle output mode Set/Reset
}
```

The duty cycle is controlled by the external variable DutyCycle0, which is updated in the main routine. The ISR tests whether the output has just been set or reset in the same way as the ISR for channel 2 in Listing 8.10. A new cycle has begun if the output is set and the time to the next compare event is calculated by adding DutyCycle0 to the last value of TACCR0. At the same time, the duration of the low part of the cycle is calculated and stored in a static variable PWM0Low. This value is used after the next reset event, which ensures that the cycle has the correct length even if DutyCycle0 has been changed. (I assume that interrupts are not nested so there is no danger of DutyCycle0 changing inside the ISR.)

8.7 Output in the Up/Down Mode: Centered Pulse-Width Modulation

Figure 8.13 shows the operation of the counter in the Up/Down mode:

- TAR counts from 0 up to the value in TACCR0, which is 3 here, changes direction and counts back down to 0, which is the start of the next cycle. The period is therefore $2 \times$ TACCR0 $= 6$ counts.

- The CCIFG0 flag is set when TAR counts to TACCR0 and the TAIFG flag is set when TAR returns to 0. They are equally spaced, so that TACCR0 is set halfway between settings of TAIFG.

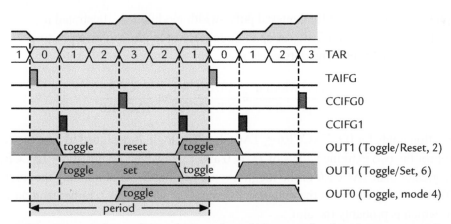

Figure 8.13: Output from channels 0 and 1 of Timer_A in the Up/Down mode. Channel 0 is in the Toggle mode with TACCR0 = 3. The output from channel 1 is shown for both the Toggle/Reset and Toggle/Set modes with TACCR1 = 1.

- The CCIFG1 flag is set when TAR counts to TACCR1, which is 1 here. This happens *twice* per cycle, once on the way up and again on the way down.

- The output from channel 1, OUT1, is shown for the two output modes that are most useful with the Up/Down mode. Both toggle the output when TAR counts to TACCR1, which means that they switch the output on at one of the matches in each Up/Down cycle and switch it off at the other. Thus all the changes in the output during normal operation occur when TAR counts to TACCR1. The second action, which occurs when TAR counts to TACCR0, sets the phase or polarity of the output as follows:

  — The output is cleared at TACCR0 in Toggle/Reset mode 2. This causes the high pulses to be centered on the points when the TAIFG flag is set. A larger value of TACCR1 causes longer pulses, so this is positive PWM.

  — Toggle/Set mode 6 does the opposite. The output can be viewed in two ways. It can be negative PWM, meaning active low pulses centered on TAIFG. Alternatively, it can be regarded as positive pulses centered on CCIFG0, midway between the pulses produced by mode 2.

- As in the Up mode, the only output mode that changes OUT0 periodically is Toggle, mode 4, which gives twice the period of the timer.

As in the Up mode, the outputs are driven periodically as long as the timer runs even if the CPU is turned off, interrupts are disabled, and the flags are never cleared.

The Up/Down mode is used for centered pulse-width modulation, illustrated in Figure 8.14. Channels 1 and 2 are set up in Toggle/Reset mode 2 to give positive pulses centered on TAIFG. The duty cycle is

$$D_n = \frac{\text{pulse width } n}{\text{period}} = \frac{2\,\text{TACCRn}}{2\,\text{TACCR0}} = \frac{\text{TACCRn}}{\text{TACCR0}}. \tag{8.7}$$

This equation holds for $0 \leq \text{TACCRn} \leq \text{TACCR0}$. The logic is designed to give the correct behavior for the two extreme values of $D_n = 0$ and 1. However, do not allow TACCRn to become greater than TACCR0 or the toggle events will not occur at all, only the reset when TAR = TACCR0. Thus the output will remain *off* with $D_n = 0$ rather than on with $D_n = 1$, which is probably the aim.

Centered PWM has two features that distinguish it from edge-aligned PWM. The first is that all the outputs do not switch on at the same time as in Figure 8.10. This avoids the large transients that can arise from the simultaneous switching of heavy loads. The second feature is that a *dead time* can be maintained between outputs. Channels 2 and 3 in Figure 8.14 show this. The pulses for channel 2 are centered on TAIFG as usual but those for channel 3 are centered on CCIFG0 instead. There is an interval between the pulses when neither is high, which is the dead time. This cannot be done for both ends of the pulses in edge-aligned PWM and calls for centered PWM instead.

The dead time is needed when the two channels drive an H-bridge or a similar, complementary pair of outputs. Look back at Figure 7.14(e). Transistors A and D might be driven by channel 2 of Timer_A with B and C connected to channel 3. It is vital that the outputs of the two channels should not be high at the same time or the transistors create a short circuit between V_{supply} and ground. Bang! The dead time between the outputs gives a brief interval for safety when both transistors are turned off.

There is no hardware in Timer_A to preserve the dead time when the duty cycle is changed. This means that the two capture/compare registers, TACCR2 and TACCR3 in Figure 8.14, must be updated in the correct sequence. Suppose that the length of the pulse on channel 2 is to be increased and that on channel 3 is to be decreased. Channel 3 should be updated first, which lengthens the dead time. Channel 2 is updated next to

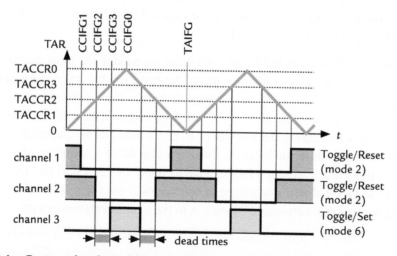

Figure 8.14: Centered pulse-width modulation from channels 1–3 of Timer_A in the Up/Down mode. Channels 1 and 2 drive active high outputs centered about TAIFG with duty cycles of one quarter and one half. The pulses from channel 3 are centered on CCIFG0 to give dead times between channels 2 and 3.

complete the change. The dead time is eliminated if a large change is made in the opposite order, with possible damage to the load under control. It is easier to use Timer_B if available.

Example 8.30

Rewrite the program in Listing 8.14 to flash the LEDs with centered PWM rather than edge alignment. First arrange that both LEDs are on at the same time as for channels 1 and 2 in Figure 8.14. Next, modify the program so that the short flash takes place while the other LED is off, like channels 2 and 3 in Figure 8.14.

8.8 Operation of Timer_A in the Sampling Mode

This is really a particular application of the Compare mode. One of the actions that takes place when a compare event occurs is to store a copy of the current capture input CCI in the latch bit SCCI. In other words, the input is sampled at the time of the compare event. This is something like a sample-and-hold circuit but it is purely digital rather than analog. It is used when a digital input needs to be sampled at precise times and the main application is to asynchronous serial communications. The details are covered in the section "A Software UART Using Timer_A" on page 590, but here is a brief description to show how Timer_A is used to receive a byte.

A byte of data comprises 8 bits but frames of 10 bits are used in the most common format for asynchronous serial communication. The two extra bits are needed to define the beginning and end of each byte to form a complete frame. The line idles high at the value for a logic 1 between transmissions. A new byte begins with a falling edge on the line, which remains low during the *start bit* of 0. The 8 data bits are transmitted next, starting with the least significant bit (lsb). Finally, a *stop bit* of 1 completes the frame and leaves the line high, ready for a falling edge to signal the next start bit. This is shown in Figure 8.15 and the following sequence of operations is used to read the signal using one capture/compare channel of Timer_A in the Continuous mode with interrupts:

- Between transmissions, the channel waits in the Capture mode for a falling edge on its input. An interrupt is requested when an edge is detected. The channel is switched to the Compare mode and the next event is set for half the length of a bit in time.

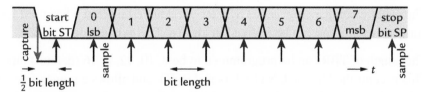

Figure 8.15: Asynchronous reception of a byte using the Capture mode to detect the initial falling edge and Sampling mode to read the bits that follow.

- The next interrupt occurs after the Compare event. It should be in the middle of the start bit (ST) and the SCCI bit is checked to confirm this. If SCCI = 0 as expected, the next Compare event is set up to occur after a further bit length. If SCCI = 1, the initial edge must have been noise and the channel is switched back to the Capture mode to await the next valid transmission.

- After the next interrupt, SCCI contains the value of the least significant bit, lsb. This is stored and the next Compare event is set up to occur after a further bit length.

- This is repeated until all 8 bits of data have been received. A final Compare event for the stop bit (SP) is set up to occur after a further bit length.

- The final sample is checked to verify that SCCI = 1 as expected for the stop bit. If this is correct, the main routine is alerted that a new byte has been received. If not, a framing error occurred and the byte is discarded. In both cases the channel returns to the Capture mode, ready for the next byte.

A different type of communication is described in the application note *Decode TV IR Remote Control Signals Using Timer_A3* (slaa134), which shows how to decode both RC5 and SIRC TV IR remote control signals.

8.9 Timer_B

Timer_B is provided on larger devices in all MSP430 families. It is closely similar to Timer_A, with a central timer block and three or seven capture/compare channels. The names of the signals and registers are the same but with a B instead of an A; for example, the counter is TBR instead of TAR. Unused bits in TACTL and TACCTLn are exploited in the corresponding registers of Timer_B to provide its additional functions. There are four differences between Timer_A and Timer_B, of which the first is probably the most important:

- The capture/compare registers TBCCRn are double-buffered when used for compare events and can be updated in groups. This is explained later.

- The length of TBR can be programmed to be 8, 10, 12, or 16 (default) bits. This is controlled by the CNTLx bits in TBCTL and allows a range of periods to be selected for the Continuous mode. Do not use this feature for the Up and Up/Down modes.

- The SCCI bit is not provided, which means that the Sampling mode is not possible. This means that Timer_B is less suitable for receiving asynchronous signals but Timer_A is always available if needed. In any case, devices that include Timer_B are likely to have a dedicated module for communications.

- All of Timer_B's outputs can be put into a high-impedance state by a high external signal applied to the TBOUTH input pin. This can be done in Timer_A only by using software to reconfigure the pins to be digital inputs.

These improvements make it much easier to produce clean, reliable PWM signals with Timer_B, avoiding the problems described in the section "When Should the Duty Cycle Be Updated?" on page 346.

Figure 8.16 shows a simplified block diagram of Timer_B for comparison with Timer_A in Figure 8.5. The comparator in the capture/compare channel of Timer_A lies directly between TAR and TACCRn. This means that any changes to TACCRn take effect *immediately*, which caused the difficulties described earlier in this chapter. Timer_B has an additional compare latch TBCLn in each channel and the comparator detects a match between TBR and TBCLn, not with TBCCRn. These latches are private to Timer_B and cannot be accessed from software, nor are they visible in the debugger. This arrangement is called *double-buffering*.

The latches are updated from TBCCRn at points in the cycle of TBR controlled by the CLLDx bits in TBCCTLn. There are four choices, which are made independently for each channel:

Immediate (0): Values are copied to TBCLn as soon as they are written to TBCCRn. Effectively there is no double-buffering. This is the default.

TBR counts to 0 (1): TBCLn is updated from TBCCRn when TBR counts to 0, which starts a new cycle. All channels are updated simultaneously. This is an obvious mode to use for edge-aligned PWM.

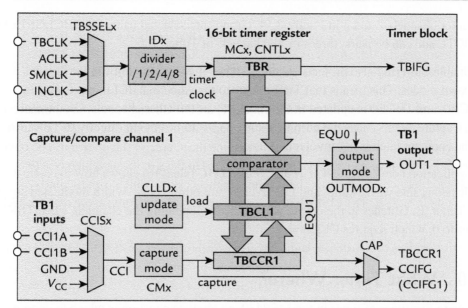

Figure 8.16: Simplified block diagram of Timer_B showing the timer block and capture/compare channel 1. Note the latch TBCL1 between the capture/compare register TBCCR1 and the comparator for the Compare mode.

TBR counts to 0, or to old TBCL0 in Up/Down mode (2): This is the same as mode 1 in the Up and Continuous modes. In the Up/Down mode, TBCLn is updated when TBR counts either to 0 or to the old value of TBCL0. This means that the update occurs at either extreme of TBR, when it is about to change direction. It gives a more rapid update of centered PWM than mode 1.

TBR counts to old TBCLn (3): TBCLn is updated from TBCCRn when TBR counts to the current (old) value of TBCLn. This means that the channel is updated at the end of the high pulse in basic, edge-aligned PWM or at both ends of the pulse in centered PWM.

Double-buffering is beneficial for PWM but is not generally appropriate for single pulses or delays. In these cases it is important that the channel is updated immediately for the next event, which requires CLLD = 0.

A further refinement is that channels can be grouped so that their latches are updated simultaneously. This is particularly useful to maintain the dead time between outputs, shown for channels 2 and 3 in Figure 8.14. Groups are selected using the TBCLGRPx bits in TBCTL and can be pairs, threes or all channels of Timer_B7.

The double-buffering also protects channel 0 when it acts as the limit in the Up and Up/Down modes. This means that TBR counts to the value in TBCL0, not to that in TBCCR0, and TBCL0 is updated in the same way as the other channels. On the other hand, Capture events cause the current value of TBR to be copied directly to TBCCRn and the compare latch is not involved.

The application note *Using PWM Timer_B as a DAC* (slaa116) shows how to create a simultaneous sine wave, ramp, and DC level by smoothing pulse-width modulated signals from Timer_B. Glitches in the output are avoided by updating the channels when TBR returns to 0, which requires CLLD = 1.

8.10 What Timer Where?

With up to five types of timer in the MSP430, an obvious question is which to choose for a particular application. Of course there are so many, widely varying applications that a simple list will not always be helpful, but here is a guide to start your selection:

Pulse-width modulation: Use Timer_B if available on your device, otherwise Timer_A. Connect the load directly to an output of the timer so that it can be driven directly by hardware.

Less regular outputs: Connect directly to an output of Timer_A or B. Use the Up mode if the intervals between changes are always the same, as in many forms of communication. The Continuous mode is easier if the intervals vary.

Inputs to be sampled at regular intervals: Connect directly to an input of Timer_A and use the Sampling mode (the Compare mode with the SCCI bit). This applies mainly to communications.

Inputs to be timed: Connect slow inputs directly to a Capture input of Timer_A or B. Fast signals should be connected to one of the timer clock inputs, such as TACLK or INCLK.

Interaction with other peripherals: Use the internal connections to other peripherals wherever possible, both for capture and compare events. This gives precise timing and saves power if the CPU need not be restarted.

Periodic software interrupts: A wide range of options are available and the selection is less clear:

- Try the **watchdog timer** if it is not needed as a watchdog and if the interval is appropriate; there is a choice of only four intervals for a given clock frequency. These are roughly 2, 16, 250, and 1000 ms from ACLK at 32 KHz, slower if VLO is used instead. Shorter intervals can be obtained by using SMCLK instead of ACLK.

- The obvious choice in a MSP430x4xx device is **Basic Timer1**, again provided that the interval is convenient. The typical range is from about 16 ms to 2 s. The **real-time clock** gives further options if available.

- If neither of these is suitable you use **Timer_A** or **B**, which can produce almost any interval desired. The snag is that this may interfere with the use of their more advanced features.

Less regular software interrupts: Use Timer_A or B, preferably in the Continuous mode.

The last resort: Use software loops. Avoid these whenever possible except for trivial cases, such as delays while a clock stabilizes.

8.11 Setting the Real-Time Clock: State Machines

Many embedded systems need to perform a number of tasks, whose sequence depends on the state of one or more signals. *State machines* provide a natural framework for constructing this sort of software and are therefore very widely used. I admit that this might not seem a natural place at which to introduce them but I have two reasons. The first is that I gave several examples of real-time clocks whose time can be set only in the debugger, which is hardly realistic. A state machine is appropriate for handling the inputs to set the time. Second, the periodic state machines that I describe here need to be driven by a periodic tick and therefore rely on interrupts from a timer.

There is a formal foundation for state machines, which is often based on the unified modeling language (UML) but I am not going to describe this. You might like to look through reference [20] and items in the section *Embedded System* in *Appendix B: Further Reading*. Properly I should use the more precise term *finite* state machines; they are also known as *automata* in computer science. More complicated systems may be designed using *state charts*, which are hierarchical extensions of state machines and share some of the properties of object-oriented languages.

8.11.1   Introduction to State Machines

Electronic engineers probably have encountered state machines in a course on digital design because they grow naturally out of synchronous counters [38]. A counter consists of flip-flops to hold the current value (the state of the counter) and combinational logic to guide the counter to the next state after a clock transition. The distinguishing feature of a synchronous counter is that all flip-flops receive the same clock and therefore change simultaneously. A standard way of designing such counters by hand is to construct a *state transition table*. I show a simple example in Figure 8.17 for a 2-bit up counter with an active high "enable" input. This is the significance of the columns in the table:

- The *present state* is the value of the counter before a clock transition.

- When a clock transition arrives, the counter changes from its present state to the *next state* according to the value of the enable input at the clock transition.

In this example the counter follows the usual sequence 0–1–2–3–0 . . . when the enable input is active (1) but remains in the same state if the enable input is inactive (0). For example, suppose that the present state of the counter is 1. It remains in this state if enable $= 0$ but changes to 2 if enable $= 1$.

An alternative way of showing the operation of the system, which should really come before the transition table, is a *state transition diagram*. This is also shown in Figure 8.17:

- States are represented by circles and labeled with the value held in the counter.

present state	enable input	next state
0	0	0
	1	1
1	0	1
	1	2
2	0	2
	1	3
3	0	3
	1	0

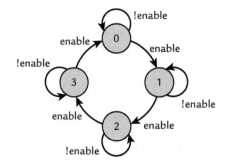

Figure 8.17: State transition table and corresponding state transition diagram for a 2-bit up counter with an enable input.

- Arrows show the transitions between states that occur at each clock cycle. An unlabeled arrow would show a transition that always occurred—if the counter had no enable input, for instance. Here the transitions are labeled to show the influence of the enable input; "!enable" means the inactive value.

Suppose again that the counter holds 1 before a clock transition. If the enable input is active, the system takes the transition to state 2. On the other hand, if the enable input is inactive, the transition labeled !enable loops back to the same state 1 so there is no change.

It is straightforward to extend this concept from hardware to software. The regular clock signal is replaced by a periodic trigger from a timer, probably as an interrupt. The state of the system is recorded in a *state variable* and software determines the next state, depending on current values of various inputs. A new and vital feature is that the software can perform a particular action associated with each transition.

For a simple example go back to the section "Read Input from a Switch" on page 80, where we illuminated an LED while a button was pressed. Figure 8.18 shows a state transition diagram for this simple system. There are only two states, LedOff and LedOn, and a single input called Button, which is active when the button is pressed. A new feature is that the transitions are labeled with two fields separated by a slash. The first field shows the value of the input required for the transition to be selected and the second shows the action that should accompany the transition. I also introduce the black blob with an arrow pointing to the LedOff state, which shows the initial state. This is how the state machine operates:

- It enters the LedOff state after initialization. The initial transition should really have an Extinguish LED action to ensure that the LED is initially switched off.

- The state machine is called periodically. While the button remains up, the state machine takes the transition labeled !Button and therefore remains in the LedOff

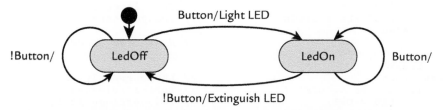

Figure 8.18: State transition diagram for an LED controlled by a button. The LED is illuminated when the button is pressed and extinguished when it is released.

state. No action is associated with this transition so nothing follows the slash in the annotation.

- If the button is pressed when the state machine is called, it performs the `Light LED` action and makes a transition to the `LedOn` state.

- The state machine remains in the `LedOn` state until the button is released, when it carries out the `Extinguish LED` action and returns to the `LedOff` state.

That is the picture: How is it turned into code? One approach is to set up a state transition table like that in Figure 8.17 with an extra column for the action (a function) associated with each transition. The table can be implemented as an array with the present state and inputs as indexes. This is concise and efficient but not easy for a human to interpret. Simple state machines are usually written in a more elementary style using `switch-case` and `if-else` statements. Listing 8.18 shows the state machine for Figure 8.18 in the interrupt service routine for channel 0 of Timer_A. I use an Olimex 1121STK but the hardware is defined using symbolic constants to keep it general. A Monitor LED flashes while the ISR is active to show when the state machine operates but it is hard to see because the operation is so brief.

Listing 8.18: Interrupt service routine for channel 0 of Timer_A from `butstmc1.c`. The state machine lights the LED while the button is pressed.

```
#pragma vector = TIMERA0_VECTOR
__interrupt void TA0_ISR (void)          // Acknowledged automatically
{
    static enum {
        LedOff,
        LedOn
    } LedState = LedOff;                  // State variable, initialized

    Monitor = ON;
    switch (LedState) {
    case LedOff:                          // LED currently off
        if (Button == ON) {              // Button pressed?
            LED = ON;                    // Yes: light LED
            LedState = LedOn;            // Change state
        } else {
                                         // No action needed
        }
        break;
    case LedOn:                           // LED currently on
        if (Button == OFF) {             // Button released?
            LED = OFF;                   // Yes: extinguish LED
            LedState = LedOff;           // Change state
        }
        break;
```

```
      default:                              // Should never happen
         LedState = LedOff;                 // Reset to initial state
         break;
      }
      Monitor = OFF;
}
```

I started by defining the state variable `LedState`. This must be `static` so that it retains its value between invocations of the function. I chose an enumeration to give helpful names to the states and initialized the state variable.

The top level of the state machine is implemented as a `switch-case` statement, which is very common usage. It is clearer than the alternative sequence of `if-else if-else`... statements but is potentially dangerous: It is painfully easy to omit a `break` statement, in which case execution falls through into the next case. Unfortunately this is perfectly legal C, so the compiler cannot stop you, but the missing break will probably wreck your intended operation. (It is almost the only aspect of C that I would like to change.) Recheck this if your state machine is misbehaving.

Within each case of the switch, the code checks the state of the `BUTTON` input and does the appropriate operations. For example, if the button is pressed in the `LedOff` state, the LED is turned on and the state is changed to `LedOn`. The "transitions" that stay in the same state require nothing to be done at all in this system. I include one empty `else` statement as a reminder of this but it is hardly necessary.

Another feature that looks unnecessary is the final default case of the switch. This is a safety feature. The state variable should never take any value other than `LedOff` and `LedOn` in normal operation but the default case is a simple way of restoring normal operation if anything should go disastrously wrong and the value of `LedState` becomes corrupted. It is always wise to include a default case in any switch.

The simple program to light an LED when the button was pressed, back in Listing 4.8, needed no state variable. Effectively it kept track of the current state by the position in the program. That worked correctly when the code was in the main function but could not be used in an interrupt service routine, because different actions must be performed each time it is called. The state machine has memory provided by the state variable, which ensures that the correct action is performed whenever the ISR is called.

The state machine in Listing 8.18 is effectively polling the input. You might reasonably point out that it would be more efficient to use interrupts for the slow input from a human operator. This was done using interrupts on digital inputs in Listing 7.1. The ISR has to perform different actions depending on whether the button is pressed or released and this could also be implemented as a simple state machine. It would be an *event-driven state machine* rather than the periodic state machine that I describe in this section because the machine is called only when an event (interrupt on change) occurs rather than regularly. We see an example of an event-driven state machine in the section "State Machines for I²C Communication" on page 559.

I describe one more simple example with the push button and LED before moving on to the clock. In this case the LED is toggled on or off when the button is pressed. (This is set as an example in the section "Read Input from a Switch" on page 80.) We need to use similar structures to control the state machine that sets the time of the clock.

The LED should be turned on when the button is next pressed, then turned off after the button has been released and pressed again. Now there are four states of the system instead of two because the LED can be either on or off when the button is either up or down. The transition diagram is shown in Figure 8.19. This seems unexpectedly complicated but the structure is needed to enforce the desired sequence of actions. It is similar in principle to the ratchet mechanism of a ballpoint pen with a button to push the writing tip in or out of the body. After the button has been depressed once to turn the LED on, it must not have any further effect until it has been released and pressed again. Thus four states are needed although there are actions associated with only two of the transitions. The other two states are needed purely to control the state machine.

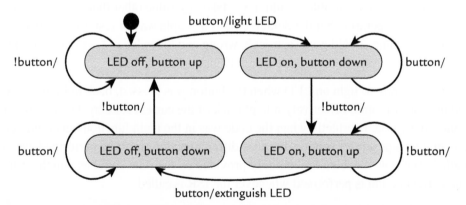

Figure 8.19: State transition diagram for an LED controlled by a button. The LED is toggled on or off each time the button is pressed.

Example 8.31

Extend Listing 8.18 to execute this periodic state machine.

8.11.2   A State Machine to Set the Time of the Clock

Now we have the tools necessary to construct a state machine to set the time of a clock on the TI MSP430FG4618/F2013 Experimenter's Board. To be specific, I use a clock with the same display as in Listing 8.3. This showed the hours and minutes in 24-hour format and toggled the colon at 1 Hz, which needs interrupts at 2 Hz. These interrupts are provided by Basic Timer1 and I use them to trigger the periodic state machine in the new design. However, I keep track of the time in software rather than using the real-time clock (RTC) module as in Listing 8.3 because the RTC is available in only a few devices at present.

It is often a good idea to start with the user's manual for a task like this. This helps to guide the design of the software so that the product is easy to use. I assume that buttons S2 and S1 are labeled *Mode* and *Advance*.

Instructions for Setting the Time

1. Hold down the Mode button until the minutes vanish and the hours flash on the display.

2. Hold down the Advance button until the hours display shows the correct time.

3. Hold down the Mode button until the hours vanish and the minutes flash on the display.

4. Hold down the Advance button until the minutes display shows the correct time.

5. Hold down the Mode button until the full display reappears. The clock restarts at the time shown with 0 seconds when the button is released.

This description might be enough for the user but we need to make it more precise before designing the software. For example, the Mode button needs to be released between the stages where it is needed so that the successive presses are distinct. It would be easiest if we require the button to be released before the following action is carried out. Thus the user must release the Mode button after it has first been pressed before the hours can be set. Another complication is that there are four possible combinations of inputs from the two buttons. In many cases we can ignore one of the buttons but the awkward conditions are when the hours or minutes are flashing after the mode button has been released. Three actions are possible here:

- The time should be advanced if only the Advance button is down.

- The state machine should move on to the next state if only the Mode button is down.

- The display should flash if neither button is down.

What if both buttons are down? I chose to give the Advance button the higher priority so that the time is advanced in this case. Another issue is that we should presumably allow the user to hold down the advance button more than once to set either the hours or minutes, in case the time had not advanced enough when he or she first pressed it.

With these thoughts in mind, I constructed the state transition diagram shown in Figure 8.20. The nine states make it more complicated than the earlier examples but the principles are the same. Let us review its operation:

Running: This is the normal state in which the clock runs, flashes the colon, and updates the display every minute. I also made it the initial state, which is questionable—should the user be alerted that the time has not yet been set? If the mode button is pressed, the state machine makes a transition to the Confirm state. The time is updated as normal even if this happens. The Advance button is ignored.

Confirm: I include this as a "guard" state in case the user presses the Mode button briefly by accident. Without it, the clock would move into the time-setting state as soon as the Mode button is found to be down when it was polled in the Running state. The button is checked 0.5 s later in the Confirm state and the machine returns to Running if the button is now up. The time is advanced so that there is no loss of accuracy. On the other hand, if the button is still down, the process of setting the time starts by erasing the minutes from the display and the machine moves on to WaitHours. You may worry that the Mode button should be checked for a longer time to avoid accidental presses. This could be done by inserting more states into the machine. Alternatively, use a counter to control the flow.

WaitHours: The clock waits for the user to release the Mode button before the hours can be set. The display is toggled on and off to show that the hours are ready to be set.

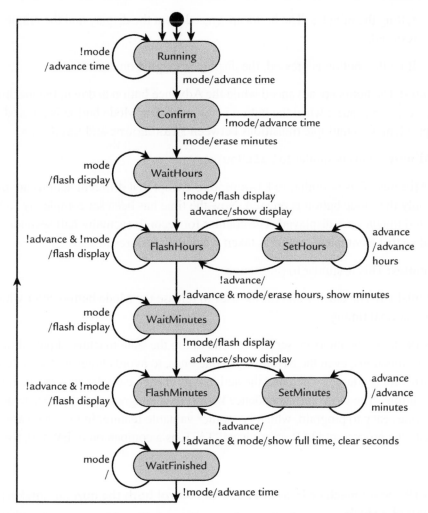

Figure 8.20: State transition diagram for setting a clock, controlled by the Mode and Advance buttons.

FlashHours: This is one of the complicated cases that I mentioned because the actions depend on both the Advance and Mode buttons:

- If the Advance button is pressed, the display is turned on and the machine moves to the SetHours state. The Mode button is ignored.

- If the Mode button is down and the Advance button is up, this shows that the hours have been set to the desired value and the machine moves on for setting the minutes. The hours are erased from the display and the minutes are restored.

- If neither button is pressed, the display is toggled on and off.

SetHours: The hours are advanced while the Advance button is down; the machine returns to FlashHours when the button is released. The Mode button is ignored. There is no problem with multiple transitions between FlashHours and SetHours.

WaitMinutes: This is similar to WaitHours.

FlashMinutes: This is similar to FlashHours except that different actions are taken when only the Mode button is pressed. Now the time has been set completely so both hours and minutes are displayed. The counter for seconds (actually half seconds) is cleared so that a complete minute is taken when the clock resumes.

SetMinutes: This is similar to SetHours.

WaitFinished: The clock waits for the user to release the Mode button, after which it resumes normal timing.

Listing 8.19 shows the interrupt service routine to run the state machine. I put all the actions into functions, even the trivial single-line ones, to avoid cluttering the picture. The overall structure is a large switch and the action within each case depends on the inputs from the buttons. The close correspondence between the diagram and the code makes state machines easy to program, which is another valuable feature. In fact there are tools to automate the process and IAR offer visualSTATE as a companion to EW430, for example.

Listing 8.19: State machine in setclk1.c to control both the normal timekeeping and setting of a clock.

```
#pragma vector = BASICTIMER_VECTOR
__interrupt void BASICTIMER_ISR (void)    // Acknowledged automatically
{
    static enum {
        Running,                          // Normal operation of clock
        Confirm,                          // Pause: mode butn still down?
        WaitHours,                        // Wait for release of mode butn
        FlashHours,                       // Flash settable hours display
        SetHours,                         // Advance hours to set time
        WaitMinutes,                      // Wait for release of mode butn
        FlashMinutes,                     // Flash settable mins display
        SetMinutes,                       // Advance minutes to set time
        WaitFinished                      // Wait for release of mode butn
    } ClockState = Running;               // State variable
```

```
switch (ClockState) {
case Running:
    AdvanceTime();                  // Always update time
    if (Mode == DOWN) {             // Mode button pressed?
        ClockState = Confirm;       // Yes: Prepare to set time
    }                               // No: Continue running clock
    break;
case Confirm:
    if (Mode == DOWN) {             // Mode button still pressed?
        EraseMinutes();             // Yes: Erase display for mins
        ClockState = WaitHours;     // Prepare to set hours
    } else {
        AdvanceTime();              // No: Update time as usual
        ClockState = Running;       // Return to timekeeping
    }
    break;
case WaitHours:
    ToggleDisplay();                // Flash display (hours only)
    if (Mode == UP) {               // Mode button released?
        ClockState = FlashHours;    // Yes: Proceed to setting hours
    }                               // No: Continue flashing display
    break;
case FlashHours:
    if (Advance == DOWN) {          // Advance button pressed?
        ActivateDisplay();          // Yes: Ensure display is visible
        ClockState = SetHours;      // Proceed to advance hours
    } else if (Mode == DOWN) {      // No: Mode button pressed?
        ShowMinutes();              // Yes: Show minutes
        EraseHours();               // Erase display for hours
        ActivateDisplay();          // Ensure display is visible
        ClockState = WaitMinutes;   // Prepare to set minutes
    } else {                        // No: Neither button pressed
        ToggleDisplay();            // Flash hours display
    }
    break;
case SetHours:
    if (Advance == DOWN) {          // Advance button pressed?
        AdvanceHours();             // Yes: Advance hours
    } else {
        ClockState = FlashHours;    // No: Return to flash hours
    }
    break;
case WaitMinutes:
    ToggleDisplay();                // Flash display (mins only)
    if (Mode == UP) {               // Mode button released?
        ClockState = FlashMinutes;  // Yes: Proceed to setting mins
    }                               // No: Continue flashing display
    break;
case FlashMinutes:
    if (Advance == DOWN) {          // Advance button pressed?
        ActivateDisplay();          // Yes: Ensure display is visible
        ClockState = SetMinutes;    // Go to advance mins
    } else if (Mode == DOWN) {      // No: Mode button pressed?
        ShowHours();                // Yes: Restore hours to display
        ActivateDisplay();          // Ensure display is visible
        ClearSeconds();             // To give clean restart from 0s
        ClockState = WaitFinished;  // Prepare to finish setting
```

```
        } else {                            // No: Neither button pressed
            ToggleDisplay();                // Flash minutes display
        }
        break;
    case SetMinutes:
        if (Advance == DOWN) {              // Advance button pressed?
            AdvanceMinutes();               // Yes: Advance minutes
        } else {
            ClockState = FlashMinutes;      // No: Return to flash minutes
        }
        break;
    case WaitFinished:
        if (Mode == UP) {                   // Mode button released?
            AdvanceTime();                  // Yes: Restart update of time
            ClockState = Running;           // Return to running state
        }                                   // No: Do nothing
        break;
    default:                                // Should never happen
        ClockState = Running;               // Emergency recovery
        break;
    }
}
```

This program works satisfactorily although it is rather tedious to update the minutes at only 2 Hz. Many products speed up the rate of change after a few seconds and that could be done here. Another annoyance is that the state machine misses the input from a button if it is pressed for less than 0.5 s between calls of the state machine. I am sure that you can think of other features that could be improved.

Example 8.32

Suppose that only one button is available to set the clock. Modify the state transition diagram in Figure 8.20 to provide a possible solution.

Checklist—Have You...

❑ Put a break statement after each case of a switch in a state machine?

CHAPTER 9

Mixed-Signal Systems: Analog Input and Output

MSP stands for "mixed signal processor" and most MSP430s are used in applications that handle both analog and digital data. We already saw how pulse-width modulation is used to provide a substitute for a "real" analog output. A few devices contain a 12-bit digital-to-analog converter (DAC12), which is used when PWM is unacceptable.

Most of this chapter is concerned with analog inputs and how these are converted to digital values that can be stored, processed, and transmitted to other systems. The MSP430 offers three methods of conversion with quite different characteristics. The first is a do-it-yourself approach that requires only a simple peripheral on the chip while the others employ a full analog-to-digital converter (ADC):

Comparator: Simple and cheap module that cannot perform a conversion by itself but is usually used with Timer_A to measure the time-constant of an external *RC* circuit. There are two versions, Comparator_A and Comparator_A+.

Successive-approximation ADC: The general-purpose type of ADC for many years. It is fast and relatively straightforward to understand. There are two versions, ADC10 and ADC12, which give 10 and 12 bits of output.

Sigma–delta ADC: A more complicated ADC that works in a quite different way to give higher resolution (more bits) but at a slower speed. There are two versions, SD16 and SD16_A, both of which give a 16-bit output.

Each module has several analog inputs. In most cases these are multiplexed to a single converter, which means that conversions are performed sequentially. A few devices have multiple ADCs so that all inputs can be converted simultaneously, which is important for metering electrical power. It is also possible to use the Schmitt trigger on the standard port inputs as a substitute for a comparator to perform measurements without a dedicated peripheral at all.

The bad news is that the ADC is only part of a complete mixed-signal system. I showed the straightforward example of a weighing machine in Figure 1.4. This has a sensor to detect the weight of the object and an amplifier to increase the magnitude of the signal before it is applied to the ADC. A real circuit also includes filtering to suppress the noise that is inevitably picked up from the environment. This should be done externally on the analog signal before conversion. Further digital filtering of the converted values may also be necessary.

The designer must consider the entire system to ensure that the components work together correctly to give the desired output and meet other specifications, such as accuracy and power demand. The range of options can seem bewildering. Here, for example, we might choose to use a sigma–delta ADC because it has differential inputs that can be connected directly to the sensor and an internal amplifier to boost the signal. Alternatively, we could use an operational amplifier and a successive-approximation ADC. The converter is cheaper and faster but this is offset by the cost of an amplifier. Yet again, it might be possible to forego a full ADC and use only a comparator, although this would probably not be the best route for this application. These issues are considered further in the section "Signal Conditioning and Operational Amplifiers" on page 475.

Table 9.1 sheds some amusing insight on the relative costs of digital and analog electronics. It shows the indicative prices of the three variants of the MSP430F200x, all of which have the same memory but offer different types of analog input. The comparison is slightly muddied because the F2001 lacks a Universal Serial Interface (USI) but I suspect

Table 9.1: Price of different variants of the MSP430F200x to show how this depends on their analog peripherals. The USI is the Universal Serial Interface, a simple module for communication over SPI and I²C.

Device	Communication	Analog input	Price
F2001	—	Comparator	$0.55
F2002	USI	10-bit successive-approximation ADC	$0.99
F2003	USI	16-bit sigma–delta ADC	$1.50

that this is a small perturbation. The price of the F2002 is almost double that of the F2001, most of which must be due to the successive-approximation ADC. Roughly two thirds of the cost of the F2003 can be ascribed to its 16-bit sigma–delta ADC, which seems astonishing for a single peripheral. On the other hand, you may be amazed that you can buy a sigma–delta ADC for so low a price as $1.50 after you have finished this chapter—let alone one with a free computer included. Not many years ago a standalone 8-bit ADC cost over $10.

9.1 Comparator_A

I concentrate on the newer Comparator_A+ module in this section but the principles apply equally well to Comparator_A.

An analog comparator compares the voltages on its two input terminals, V_+ and V_-. Its output is high if $V_+ > V_-$ and low if $V_+ < V_-$. Thus it provides a basic bridge between the analog and digital domains and acts like a 1-bit ADC. From an analog perspective it looks something like an operational amplifier (op-amp) without the usual negative feedback. In this case the high open-loop gain ensures that the output is almost always saturated at either the positive or negative supply rail. (Real op-amps are never used like this but you may have seen the same behavior due to a bad joint in the feedback circuit.) The internal design of a comparator is rather different from that of an op-amp and is optimized to switch the output between high and low as rapidly as possible, taking about 100–200 ns in Comparator_A+. This is assisted by a little positive feedback, which also gives a small hysteresis of around 1 mV.

9.1.1 Architecture of Comparator_A+

Although a comparator is fairly simple, the block diagram of Comparator_A+ in the family user's guide looks complicated because of the many options provided. These are controlled with the two peripheral registers CACTL1 and CACTL2. I show a simplified version of the module in Figure 9.1. There are a lot of multiplexers and it is worth remembering that these switch *analog* signals, not digital. The performance of all analog peripherals depends heavily on the capabilities of analog CMOS switches.

These are the main features of Comparator_A+. The entire module is switched on and off with the CAON bit. It is off by default to save current.

- The noninverting input V_+ can be connected to external signals CA0–CA2 or left without an external connection. This is selected using bits P2CA4 and P2CA0. The

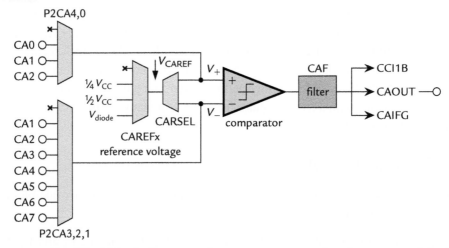

Figure 9.1: Simplified block diagram of Comparator_A+, showing some of the bits that control its operation.

strange allocation of bits maintains compatibility with Comparator_A, which had only one external input to each terminal of the comparator. An internal voltage reference can also be connected to V_+.

- Similarly, the inverting input V_- can be connected to external signals CA1–CA7 (but not CA0) or left unconnected, according to bits P2CA[3:1]. It can also be connected to an internal reference.

- The internal reference voltage V_{CAREF} can be chosen from $\frac{1}{4}V_{CC}$, $\frac{1}{2}V_{CC}$ or a nominally fixed voltage from a transistor, V_{diode}. This is selected with the CAREFx bits. It can be applied to either input of the comparator according to the CARSEL bit.

- The raw output of the comparator can optionally be filtered through an *RC* circuit to reduce oscillations in the signal, which may arise if the inputs vary slowly. This is selected with the CAF bit. The filter is off by default but should usually be enabled unless the delay that it introduces is unacceptable or any oscillation is handled in software. It extends the response time from roughly $0.2\,\mu s$ to $2\,\mu s$.

- The output is brought to an external pin CAOUT. It is also connected internally to capture input CCI1B of Timer_A, which allows precise timing without delays that would be introduced if software were needed for communication between the modules.

- The flag CAIFG is raised on either a rising or falling edge of the comparator output, selected with the CAIES bit. This can in turn request an interrupt if CAIE is set. The interrupts are maskable and GIE must therefore also be set for an interrupt to occur. Comparator_A+ has its own interrupt vector and the flag is cleared automatically when the interrupt is serviced. In summary, this is an entirely typical interrupt.

External pins must be routed to most peripherals by making the appropriate settings of the corresponding PnSEL and PnDIR bits. This holds as usual for the output CAOUT but the inputs to Comparator_A+ work in a different way. Setting a bit in the Port Disable register CAPD causes the circuits for the usual digital input and output buffers to be disconnected from the appropriate pin. This prevents possible degradation of the analog input. It also eliminates the current drawn by the digital input buffer, which rises when the input voltage is near the transition between logical 0 and 1 (see the section "Analog Aspects of 'Digital' Inputs" on page 221).

A slightly tricky aspect is that the bits of CAPD correspond to the inputs of Comparator_A+ rather than the port pins. Take the F23x0 as an example. The input CA0 can be connected to pin P2.3 in this device. Setting bit 0 of CAPD accordingly disconnects the digital input/output buffer from pin P2.3. Thus the CAPD register for Comparator_A+ overrides the registers for the port. The inputs to other analog modules may be configured in different ways.

Selecting an input to the comparator with the P2CAx bits has the same effect as using CAPD and automatically disconnects the digital input/output buffer. It might therefore seem that CAPD is superfluous. However, the input to the comparator is switched between different input pins in many applications. It is wise to use CAPD to disconnect the digital input/output buffer permanently from the pins that carry analog signals, even while they are not being measured.

Comparator_A+ offers a choice of three reference voltages, mentioned earlier. Two are derived from a potential divider across V_{CC} and ground and are accurate to within about 1% of V_{CC}. These are the most commonly used because the input voltage is also proportional to V_{CC} in many applications. The third option is the gate–source voltage of a field-effect transistor. Its nominal value is $V_{diode} = 0.55\,V$ but it varies with temperature by about $-1.2\,mV/°C$. This can also be expressed as $-2000\,ppm/°C$, where *ppm* stands for "parts per million." The voltage is also sensitive to the value of V_{CC}. This means that it is not a particularly high-quality reference compared with those provided for the full ADCs.

Nothing stops you connecting both a pin and the reference voltage to the same input of the comparator. This would clearly be a bad idea if the external signal is an input, which will fight the reference voltage. On the other hand, the external connection may be used as an output so that the reference voltage is available to other parts of the system. The data sheet does not provide a limit for the current that can be drawn (in fact it does not mention it at all) so this must be used only with high-impedance loads to avoid degrading the performance of the reference.

These are the features that I omitted from Figure 9.1:

- Setting the CAEX bit exchanges the two inputs to the comparator and inverts its output to compensate. This would have no overall effect if the comparator were perfect but real comparators have a small offset voltage. This means that the output does not switch exactly at $V_+ = V_-$ but when they differ by a small amount, between $\pm 30\,\text{mV}$. The error can be canceled by exchanging the inputs on successive measurements.

- The outputs of the two input multiplexers can be shorted together by setting the CASHORT bit. The idea is that this can be used to connect a capacitor to the input to form a sample-and-hold circuit.

The older Comparator_A has only the CA0 and CA1 inputs, normally connected to the noninverting and inverting inputs, respectively. It also lacks the internal short circuit of Comparator_A+.

9.1.2   Operation of Comparator_A+

The comparator can be used directly to compare a variable input voltage with a reference, which may be either one of the internal references or a second input. For example, the variable input might be from a temperature sensor and the comparator should detect when the equipment may freeze. Of course a microcontroller would be wasted on this task alone.

It can be useful to connect a noisy input signal to the comparator so that the MSP430 detects when the input goes through a well-defined level, such as $0.5V_{CC}$. This is more reliable than the thresholds of the Schmitt trigger on the digital inputs. There is an example in the application note *Ultrasonic Distance Measurement with the MSP430* (slaa136).

A *thermistor* is a type of temperature sensor that is often used with the comparator. It is essentially a resistor made of a material whose resistance varies strongly with temperature—the opposite of the property desired for a normal resistor. A wide variety of devices is available. Some have a resistance that increases with temperature (positive temperature coefficient, or PTC), some the opposite (NTC). There is a broad selection of operating temperature, resistance, packaging, and accuracy. They are fairly cheap and easy to use because the change in resistance is large. Unfortunately it is also strongly nonlinear. A moderately accurate equation for a NTC thermistor is

$$R(T) = R_0 \exp(x) \left(\frac{B}{T} - \frac{B}{T_0} \right), \tag{9.1}$$

where T is the *absolute* temperature in kelvins. Thermistors are usually specified by their resistance R_0 at a temperature T_0, typically 25°C. For example, a thermistor might be listed as having $R_0 = 10 \mathrm{k}\Omega$ at 25°C with $B = 3600 \mathrm{K}$. The equation shows that its resistance rises to about $30 \mathrm{k}\Omega$ at 0°C and falls to 880Ω at 100°C. Thus the resistance changes by a factor of 30 between 0°C and 100°C, a conveniently large effect for a sensor.

A simple detector for a single, fixed temperature is shown in Figure 9.2(a). It is just a potential divider between V_{CC} and ground with the midpoint connected to an input of the comparator (I show V_+ but which one does not matter for this application). The other input of the comparator is connected to an internal reference V_{CAREF}.

The voltage from the potential divider is proportional to V_{CC} in this circuit, so any changes in V_{CC} affect V_+. We do not want such variations to affect the temperature at which the output of the comparator changes. This can be achieved if the reference voltage on V_- changes in the same way. The internal references of $\frac{1}{4}V_{CC}$ and $\frac{1}{2}V_{CC}$ are derived from a potential divider and are therefore proportional to V_{CC}, just like the potential divider with

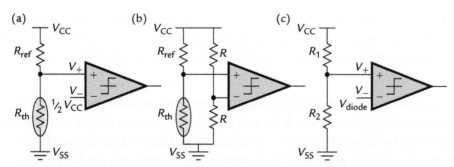

Figure 9.2: Simple circuits using a comparator. (a) Thermistor connected in a potential divider with the internal reference $0.5V_{CC}$. (b) Wheatstone bridge with two resistors R to represent the internal reference. (c) Battery voltage monitor against the fixed internal reference voltage V_{diode}.

the thermistor. We can therefore eliminate the dependence on V_{CC} by selecting one of these references. Suppose that we choose $V_{CAREF} = \frac{1}{2}V_{CC}$. Then the comparator changes output when

$$\frac{R_{th}V_{CC}}{R_{ref} + R_{th}} = \frac{V_{CC}}{2}. \tag{9.2}$$

This shows that V_{CC} cancels from the comparison. The overall circuit including the voltage reference behaves like Figure 9.2(b), which is the classic *Wheatstone bridge*. The output of the comparator changes when the resistance of the thermistor R_{th} passes through the value of the reference resistor R_{ref}. In practice you would feed the potential divider from an output pin rather than V_{CC} so that it could be switched off to conserve current.

Figure 9.2(c) shows an application where the fixed internal voltage reference is appropriate. This is a battery voltage monitor, which raises an alarm when the voltage falls too low for reliable operation of the system. For example, we might wish to check that $V_{CC} > 3.3\,\text{V}$ so that the MSP430 can operate at full speed. The diode reference gives $V_{diode} \approx 0.55\,\text{V}$ and the potential divider should give this output voltage when $V_{CC} = 3.3\,\text{V}$. This needs

$$\frac{R_2}{R_1 + R_2} = \frac{V_{diode}}{V_{CC}} = \frac{0.55}{3.3} = 0.17. \tag{9.3}$$

Thus $R_1 = 5R_2$ and we might choose values near $R_1 = 100\,\text{k}\Omega$ and $R_2 = 20\,\text{k}\Omega$. Again the potential divider should be fed from an output pin rather than V_{CC} to conserve current.

A defect of this system is that V_{diode} changes significantly with temperature. It falls to $0.48\,\text{V}$ at 85°C, in which case the warning would not occur until V_{CC} fell below $2.9\,\text{V}$. It would be safer to design the warning for this worst case, $V_{diode} \approx 0.48\,\text{V}$, rather than $0.55\,\text{V}$.

Another poor feature of these circuits is that their behavior is fixed. For example, we can not change the temperature at which the output of the temperature sensor switches because it is determined by the resistance of the reference, R_{ref}. This could be replaced by a potentiometer to make it adjustable but it would be far better to use the MSP430 to determine the value of R_{th} rather than whether it is larger or smaller than a single value. In other words, we want an analog-to-digital converter rather than a plain comparator. This goal can be achieved by adding a capacitor to form an *RC* circuit and using a timer to measure its dynamic behavior. The operation can be performed in many ways:

- A single transient of an *RC* circuit can be timed to determine *R* or *C*.

- Alternatively, the circuit can be made to oscillate and the frequency is measured rather than the duration of a single transient.

- An external voltage can be measured by comparing it with an *RC* transient.

- Alternatively, the capacitor is repeatedly charged and discharged through the resistor using rapid pulse-width modulation. The duty cycle is adjusted so that the voltage on the capacitor matches the input. This acts as a sort of integrating or sigma–delta ADC.

I assume that the aim is to measure voltage or resistance but the same methods can be used for current or capacitance. The latter is growing in importance because of the burgeoning application of touch sensors. An *RL* circuit could be used to measure inductance but this has fewer practical applications. I pick out a couple of examples; others are described in the application notes *An MSP430F11x1 Sigma–Delta Type Millivoltmeter* (slaa104) and *Economic Measurement Techniques with the Comparator_A Module* (slaa071).

The code examples for the F20x1 contain a pretty example of a battery monitor, which compares 0.25 V_{CC} with V_{diode}. This sounds straightforward until you realize that both are reference voltages, which means that only one can be selected at a time. The trick is to use an external capacitor as a sample-and-hold circuit. It is first charged to 0.25 V_{CC} by connecting it to the internal reference, then compared with V_{diode}. A limitation is that only two comparisons are possible, V_{diode} with either 0.25 V_{CC} or 0.5 V_{CC}.

9.1.3   Slope Conversion of a Resistance

Figure 9.3(a) shows a resistor and capacitor connected to the comparator so that the transient rise or fall of the voltage $V_C(t)$ across the capacitor can be measured. The technique is called *slope conversion*. The resistor is connected to a digital output pin, ideally one that can be driven directly from Timer_A. The junction of the capacitor and resistor is connected to an input of the comparator. I show V_+ but this choice is not important. The internal reference 0.5 V_{CC} is connected to the other input of the comparator. It could instead be 0.25 V_{CC} but should *not* be the fixed voltage: It is important that it is proportional to V_{CC}. Finally, the output of the comparator is connected directly to the CCI1B input to Timer_A, which allows an accurate capture of the time when the output of the comparator switches.

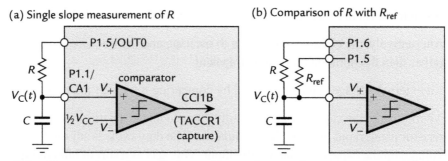

Figure 9.3: Connection of a resistor and capacitor to the comparator to form a slope ADC. (a) Single resistor and (b) reference resistor for comparison with the unknown resistor.

This is the sequence of operations to take a single measurement by discharging the capacitor through the resistor:

1. Drive the output pin high for a long time to charge the capacitor fully so that $V_C = V_{CC}$.

2. Drive the output pin low to start discharging the capacitor. Ideally this should be done by a compare event from Timer_A for the most accurate timing. If this is not possible, TAR must be read using software.

3. Wait until $V_C(t)$ falls below the reference voltage applied to the other input of the comparator. The output of the comparator switches and stimulates a capture in TACCR1.

4. The duration of the transient is given by the difference of the two times.

Figure 7.7 shows a plot of the exponential decay of the voltage across a discharging capacitor in a RC circuit. I repeat equation (7.2) that governs the voltage because we will use it extensively. The capacitor starts to discharge from V_{CC} at $t = 0$ in this equation:

$$V_C(t) = V_{CC} \exp(-t/RC). \tag{9.4}$$

Recall that the time-constant $\tau = RC$. Now suppose that the voltage is measured at two times, t_1 and t_2. The corresponding voltages are

$$V_1 = V_{CC} \exp(-t_1/RC) \quad \text{and} \quad V_2 = V_{CC} \exp(-t_2/RC). \tag{9.5}$$

Divide these two and use the properties of exponential functions to simplify the quotient:

$$\frac{V_1}{V_2} = \frac{V_{CC}\exp(-t_1/RC)}{V_{CC}\exp(-t_2/RC)} = \exp\left(\frac{t_2 - t_1}{RC}\right). \tag{9.6}$$

The initial voltage V_{CC} has vanished, which is important because it would otherwise limit the accuracy of the measurement. Next, take natural logarithms of both sides to remove the exponential function from the right-hand side. A little rearrangement gives

$$t_2 - t_1 = RC\log_e\left(\frac{V_1}{V_2}\right). \tag{9.7}$$

The left-hand side has only the difference of the times, which is important because it means that it does not matter when we started the discharge. Call the difference $T_{discharge}$. Let t_1 be the time at which the discharge started, so that $V_1 = V_{CC}$, and let t_2 be the time at which the comparator switched, so that $V_2 = 0.5\,V_{CC}$. The quotient of voltages simplifies to $V_1/V_2 = 2$ and

$$T_{discharge} = t_2 - t_1 = RC(\log_e 2). \tag{9.8}$$

Again the supply voltage has vanished because all the voltages were fractions of V_{CC}. There are no exponentials or other unpleasant functions left either, just the constant $\log_e 2 \approx 0.693$. The interval $T_{discharge}$ is calculated as $N_{discharge}/f_{timer\,clock}$, where $N_{discharge}$ is the number of counts of the timer and $f_{timer\,clock}$ is the frequency of its clock. The measured resistance is finally given by

$$R = \frac{T_{discharge}}{(\log_e 2)C} = \frac{N_{discharge}}{(\log_e 2)f_{timer\,clock}C}. \tag{9.9}$$

Thus an absolute measurement of R can be made, provided that all the parameters in this equation are known accurately. This includes C, $f_{timer\,clock}$ and the ratio of V_{CAREF} to V_{CC}.

I have used the discharge transient for illustration but the charging transient could equally well be used. In fact the capacitor has to be recharged between two discharges so we might as well make a further measurement as it charges. Wait until the capacitor has discharged fully to $V_C = 0$, then drive the output pin high to start charging the capacitor and note the time. In this direction the output of the comparator switches when $V_C(t)$ rises above the reference voltage of $0.5V_{CC}$. This should take the same time as the discharge from V_{CC} to $0.5V_{CC}$ if everything is perfect. Small errors in the reference voltage, the offset voltage of the comparator, and due to the output resistance of the port mean that the times will be slightly different in practice and can be averaged to improve the accuracy.

I chose to use $V_{CAREF} = 0.5V_{CC}$ to make the measurements of charging and discharging symmetric. It would be better to use $0.25V_{CC}$ to give a larger range if only the discharge is measured.

Slope Measurement of a Single Resistance

As an example, suppose that $R = 10\text{k}\Omega$ and $C = 10\text{nF}$. The time required to charge or discharge to $0.5V_{CC}$ is $RC(\log_e 2) \approx 69\,\mu\text{s}$. The timer clock must be much faster than this to give sufficient resolution in the measurement, which requires SMCLK rather than ACLK. I used a F2001 in a circuit without a crystal and got an accurate clock by using one of the frequencies for which calibration constants are programmed into the device. The lowest frequency is $1\,\text{MHz}$ but this would give only 69 counts and the resolution would be too poor to provide a good test of the measurement. I chose $8\,\text{MHz}$ instead to give about 555 counts.

A key assumption in the procedure just outlined is that the capacitor is fully charged before the discharge is measured. How long do we need to wait for this? The voltage across the capacitor approaches V_{CC} exponentially but never quite gets there. A reasonable condition is

$$\exp\left(-\frac{t_{\text{precharge}}}{RC}\right) < \text{resolution} \quad \text{or} \quad t_{\text{precharge}} > RC \log_e\left(\frac{1}{\text{resolution}}\right). \qquad (9.10)$$

Here we expect about 500 counts so the resolution is 1 part in 500 or 0.0002. This means that $t_{\text{precharge}} > RC \log_e(500) \approx 6.2\,RC \approx 620\,\mu\text{s}$.

There is a slightly easier way of finding this time after one measurement has been taken. The voltage across the capacitor halves from its initial value of V_{CC} to to $0.5V_{CC}$ in the discharge time $T_{\text{discharge}}$. If we wait for a further interval of $T_{\text{discharge}}$, the voltage halves again to $0.25V_{CC}$ and so on. In general, it falls to $2^{-n}V_{CC}$ after a total time of $nT_{\text{discharge}}$. For the preceding example, 1 part in 500 is close enough to 1 part in 512 or 2^{-9}, so we should wait for $9T_{\text{discharge}}$ after the measurement began or $8T_{\text{discharge}}$ after the voltage passed through $0.5V_{CC}$ and triggered the comparator.

Note that it takes far longer to charge or discharge the capacitor before the measurement than it does for the measurement itself. This is a weakness of starting from the fully charged or discharged state. It would be faster to measure the transient between two intermediate values. This could be done between $0.25\,V_{CC}$ and $0.5\,V_{CC}$ (see Example 9.2) but a larger range of voltage would be desirable.

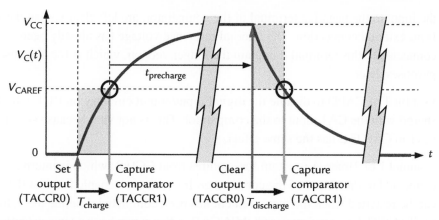

Figure 9.4: Measurement of the charge and discharge times of an *RC* circuit using Comparator_A+ and Timer_A.

Listing 9.1 shows a program to carry out measurements of the charging time followed by the discharging time. Usually the loop is paced, perhaps by using the watchdog as an interval timer, but here the measurements are repeated continuously for simplicity. Its operation is illustrated in Figure 9.4. This is how the system is set up.

- The resistor is driven from P1.5, which is connected to OUT0 from Timer_A by setting its bit of P1SEL.

- The junction of the resistor and capacitor is connected to P1.1, which can be connected to CA1. There is no need to do anything with the registers for port 1 because Comparator_A+ overrides the settings.

- Unused pins of port 1 are configured as outputs and driven low. The pins of port 2 are left as inputs and appear to be floating but there are external connections to define their voltages.

- The lines to configure Comparator_A+ are not particularly clear. I have selected the $0.5V_{CC}$ reference in CACTL1 with the CAREFx bits and have set CARSEL to apply the reference voltage to the negative terminal of the comparator. There is no need to configure interrupts because we use the direct connection to Timer_A. The comparator remains off at this point because the CAON bit is not set.

- The shorter version of the assignment to CACTL2 connects CA1 to the positive terminal of the comparator and turns on the filter. The P2CA[3:1] bits, which select the input to the negative terminal of the comparator, are left clear so that there is no external connection. The internal reference voltage has already been connected to this terminal. I explain the longer version, which is for diagnostic purposes, later.

- I set bit 1 of CAPD to disable the digital input/output circuitry on P1.1, which is shared with the CA1 input to the comparator. This is not vital because selecting CA1 in CACTL2 has the same effect.

- Channel 1 of Timer_A is set up for captures from CCI1B, which is internally connected to the output of the comparator. Interrupts are enabled so that the CPU can be restarted when the measurement has been completed. I define the value for TACCTL1 as a parameter COMP_CAP so that it can be used consistently throughout the program.

- Timer_A is started in the Continuous mode, taking its clock from SMCLK without division to give maximum resolution.

- The last step is to store a dummy value in the variable DischargeTime, which stores the duration of the discharging transient. This is needed to work out how long to wait for the capacitor to become fully discharged before the charging transient is measured.

Listing 9.1: Measurement of *RC* charge and discharge transients using Comparator_A+ in trans1.c.

```
// trans1.c - measure RC ch and discharge transient with Comparator_A+
// Timer_A channel 0 for output, channel 1 for comparator
// Homemade analog demo board for MSP430F20xx:
//   10k resistor driven from P1.5/OUT0 to capacitor
//   10n capacitor input to P1.1/CA1, routed to CA+
//   Internal reference 0.5*VCC routed to CA-
//   Internal connection of CAOUT to CCI1B on Timer_A
// Calibrated 8MHz DCO, no xtl, no ACLK, power from JTAG
// J H Davies, 2007-08-20; IAR Kickstart version 3.42A
//---------------------------------------------------------------------
#include <io430x20x1.h>            // Header file for this device
#include <intrinsics.h>            // Intrinsic functions
#include <stdint.h>                // Standard integer types
// TA1 captures, comparator (CCI1B), synchronized, interrupt, disabled
#define COMP_CAP        (CCIS_1 | SCS | CAP | CCIE)

void main (void)
{
    uint16_t ChargeTime, DischargeTime; // Measured transient times

    WDTCTL = WDTPW | WDTHOLD;       // Stop watchdog
```

```
      BCSCTL1 = CALBC1_8MHZ;          // Calibrated range for DCO
      DCOCTL = CALDCO_8MHZ;           // Calibrated tap and modulation
      P2SEL = 0;                      // Digital i/o rather than crystal
      P2OUT = 0;                      // Unused outputs driven low
      P2DIR = BIT6|BIT7;
      P1OUT = 0;                      // Unused pins are output driven low
      P1DIR = BIT3|BIT5|BIT7;         //   (Comparator_A+ overrides this)
      P1SEL = BIT5;                   // Timer_A ch 0 output to resistor
// Set up input, output and ref of Comparator_A+ (leave it off for now)
      CACTL1 = CARSEL | CAREF_2;      // 0.5*VCC reference to CA- input
      CACTL2 = P2CA4 | CAF;           // CA1 to CA+ input, output filtered
// Diagnostic: CA3 to CA- terminal to check reference externally
//    CACTL2 = P2CA4 | P2CA2 | P2CA1 | CAF;
      CAPD = BIT1;                    // Disable P1.1 digital i/o (CA1)
// Timer_A: SMCLK, no division, continuous, no need to clear, no ints
      TACTL = TASSEL_2 | ID_0 | MC_2;
      DischargeTime = 1000;           // Dummy value for first delay
      for (;;) {                      // Loop forever taking measurements
         CACTL1_bit.CAON = 1;         // Comparator on
// Charging transient: discharge C fully, Set output, wait for Comp_A+
         TACCR0 = TAR + 10*DischargeTime;    // Delay for full discharge
         TACCTL0 = OUTMOD_1;          // Set on compare to start charge
         TACCTL1 = COMP_CAP | CM_1;   // Capture comparator rising
         __low_power_mode_0();        // Wait for timer and comparator
         ChargeTime = TACCR1 - TACCR0;    // Duration of charge
// Discharging transient: charge C fully, Reset output, wait for Comp_A+
         TACCR0 = TAR + 10*ChargeTime;    // Delay to allow full charge
         TACCTL0 = OUTMOD_5;          // Reset output to start discharge
         TACCTL1 = COMP_CAP | CM_2;   // Capture comparator falling
         __low_power_mode_0();        // Wait for timer and comparator
         DischargeTime = TACCR1 - TACCR0;    // Duration of discharge
// Diagnostic: look for effect of comparator offset voltage
//       CACTL1_bit.CAEX ^= 1;        // Exchange inputs to comparator
         CACTL1_bit.CAON = 0;         // Comparator off
// Do something useful with the measurements here
      }
}
//-------------------------------------------------------------------
// ISR for CCIFG1: disable further captures and return to active mode
//-------------------------------------------------------------------
#pragma vector = TIMERA1_VECTOR
__interrupt void TIMERA1_ISR (void)  // Shared ISR for CCIFG1 and TAIFG
{
    TACCTL1 = COMP_CAP | CM_0;        // Disable further caps, clear flags
    __low_power_mode_off_on_exit();   // Return to active mode on exit
}
```

Here is the operation in detail. Each cycle discharges the capacitor, measures the charging transient, waits for the capacitor to charge "fully," and measures the discharging transient:

1. The comparator is turned on. It is turned off to save current between measurements (admittedly pointless in this program but important if the loop is paced).

2. A compare event on channel 0 of Timer_A is set up to allow time for the capacitor to discharge before the new measurement. The delay is based on the previous measurement. We calculated that we should wait about eight times the length of the measurement but I am a little more cautious and increase the multiplier to ten. There is no problem with reading TAR in this application because it uses the same clock as the CPU. The output mode is Set (mode 1), which drives P1.5 to V_{CC} after the delay to start the measurement.

3. The comparator is connected so that its input goes high when $V_C(t)$ rises through $0.5V_{CC}$. Channel 1 of Timer_A is therefore configured to capture a rising edge (CMx = 01). This also requests an interrupt. I write to the whole register, not just the CMx bits, to clear any flags that might have become set. This includes the capture overflow bit COV, which is set if the output of the comparator oscillates and causes multiple captures.

4. The CPU is turned off by entering low-power mode 0 until awakened by a capture interrupt from channel 1 of Timer_A. We cannot use LPM3 because Timer_A is running from SMCLK. There is no need to wake the CPU when OUT0 is driven high to start the measurement so interrupts were not enabled on this channel.

5. The interrupt service routine for channel 1 of Timer_A disables further, unwanted captures and the processor returns to active mode.

6. Following the interrupt, the charge time is calculated from the difference of the values in the two capture/compare registers. The compiler issues a warning because both variables are volatile but it should be safe here.

7. These steps are repeated with minor changes to measure the discharge time. Again it starts at a compare event set by TACCR0 and ends with the capture into TACCR1.

8. Further captures are disabled and the comparator is turned off when the pair of measurements is complete. I explain about CAEX later.

This measurement gives accurate timing because the signals for both the start and finish are connected directly to Timer_A. It is possible to use channel 1 for both the compares and captures but it is a little more awkward.

The program in Listing 9.1 is useless in the sense that nothing is done with the measured values so I simply set a breakpoint at the end of the main loop. This shows 540 counts for charging (ChargeTime) and 558 for discharging (DischargeTime). Their average is 549, which is remarkably close to the expected value of 555. The difference is only about 1%, which is *very* lucky when you consider the possible errors. In particular, the resistor has a tolerance of $\pm 1\%$ and I have no idea about the capacitor. The tolerance of the calibrated DCO frequency is specified as $\pm 0.2\%$, which is smaller—an interesting reflection on the accuracy of modern components. Other issues should be taken into account, such as the output resistance of P1.5, leakage current of the input P1.1, and so on.

I was puzzled by the difference between the charging and discharging times, which ought to be the same in an ideal system. My first suspect was the offset voltage of the comparator, although this should be too small. It can be tested by using the CAEX bit in CACTL1 to exchange the two inputs to the comparator and invert its output. This would have no effect if the comparator is ideal; here it changed $540 \rightarrow 535$ and $558 \rightarrow 562$. The individual readings changed by about 1% but the error in the sum is negligible. Measuring both the charging and discharging behavior largely compensates for the offset error.

The next possibility is an error in the reference voltage. I reconfigured CACTL2 so that pin CA3 is connected to the inverting terminal of the comparator, which is already connected to the reference voltage. This bought the reference voltage out to the external pin P1.3, where I could measure it. My meter showed 1.50 V compared with 3.08 V for V_{CC} (which is meant to be 3.0 V). The ratio is a little low at $1.50/3.08 = 0.487$. Then I read the data sheet (embarrassingly late) and found that the typical value is 0.48 with a range from 0.47 to 0.50. I am not sure why a one-sided tolerance is built into this reference but it explains my observations. Similarly, the $0.25\,V_{CC}$ reference is typically at $0.24\,V_{CC}$. Again, the average of the charging and discharging times largely corrects for the error in V_{CAREF}.

It is not a good idea to use the filter on the output of the comparator in this measurement because it introduces a delay of around $1\,\mu s$, which is long compared with the period of a 8 MHz clock. I disabled the filter by clearing the CAF bit and found that the counts reduced by about ten, as expected. The COV bit in TACCTL1 is set after each measurement, showing that there are multiple captures due to oscillation at the output of the comparator.

Example 9.1

Try reducing the time allowed for the capacitor to discharge or charge before the measurements. How short can you make these delays before errors appear? Is it consistent with equation (9.10)?

Example 9.2

Rewrite the program in Listing 9.1 so that it measures the time taken to charge and discharge between $0.25\,V_{CC}$ and $0.5\,V_{CC}$. Both of these points should be detected by capturing the output of Comparator_A, changing its reference voltage between the two events. The voltage at which charging starts is not important, provided that it is below $0.25\,V_{CC}$, nor does the precise time matter, so the resistor can be driven from any digital output. Similarly, the discharge can start at any voltage above $0.5\,V_{CC}$. This method has the advantage that it relies entirely on captures and therefore needs only one channel of Timer_A. It is also less sensitive to delay introduced by the filter because it affects both captures of a transient and should cancel from their difference.

Slope Measurement Using a Reference Resistance

It was a happy accident that I got such an accurate measurement using the method described in the previous section. We would like a more reliable technique without going to the expense of an accurate clock, capacitor, and reference voltage. A better solution is to compare the unknown resistance R with a known reference R_{ref} using the circuit shown in Figure 9.3(b). Accurate resistors are relatively cheap. The same sort of measurement of the discharge curve gives

$$R_{ref} = \frac{T_{ref}}{(\log_e 2)C} = \frac{N_{ref}}{(\log_e 2)\,f_{timer\ clock}\,C}, \tag{9.11}$$

which is equivalent to equation (9.9). Dividing these two equations gives

$$\frac{R}{R_{ref}} = \frac{N}{N_{ref}} \quad \text{or} \quad R = \frac{N}{N_{ref}}R_{ref}. \tag{9.12}$$

Both $f_{timer\ clock}$ and C cancel out. So has the annoying constant $\log_e 2$, which depends on the reference voltage. The final result is simple because it needs only a multiplication and division. The cancellations mean that it is not sensitive to the values of $f_{timer\ clock}$, C, nor the reference voltage of the comparator, provided that these do not change significantly between the measurements of the unknown and reference resistors. In other words, they must be stable but need not be accurate. This *ratiometric* technique is widely used in instrumentation.

Equation (9.12) can be explained simply without any of the mathematics. Suppose that the resistance in an *RC* circuit is doubled. This halves the current that flows for a particular voltage across the resistor, which in turn means that the capacitor takes twice as long to charge.

I do not explain this method further because it is described in the family user's guides and explained thoroughly in the application note *Implementing an Ultralow-Power Thermostat with Slope A/D Conversion* (slaa129). This uses the discharge time between V_{CC} and $0.25\,V_{CC}$. There is less point in measuring the charging time as well because the ratiometric method eliminates many of the potential errors. The program also uses a lookup table to convert the highly nonlinear change in resistance of a thermistor to temperature.

Figure 9.5 shows oscilloscope traces of the method in action. The capacitor is initially charged by driving both resistors high. The sample resistor is then isolated by reconfiguring its pin as an input before the pin for the 10kΩ reference resistor is pulled low to start the discharge. The time is captured when the voltage $V_C(t)$ across the capacitor falls below the reference voltage of $0.25\,V_{CC}$. Both resistors are then driven high again as outputs to recharge the capacitor before the sample is measured in the same way. I reduced the charging time to make the display clear and you can see that it is a bit short. The sample resistor was 20kΩ in this circuit and the discharge times gave $R/R_{ref} = 2$ within the resolution of the measurement. Amazing.

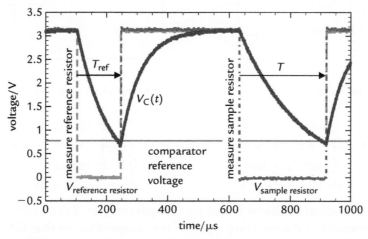

Figure 9.5: Oscilloscope traces of the discharge curve of a capacitor through a reference resistor and a sample resistor. The ratio of the discharge times gives the ratio of the resistances.

9.1.4   Relaxation Oscillator with Comparator_A

The methods that I just described use Timer_A to measure the duration of a *single RC* transient. This requires a slow charge or discharge compared with the frequency of the timer clock. The general principle of measuring slow events is illustrated in Figure 8.7(a). An alternative way of using the timer, shown in Figure 8.7(b), is to count the number of cycles of an external signal in a known time by using the signal as the timer clock and capturing a slow signal of known frequency. This can be done by making the *RC* circuit into an oscillator so that it charges and discharges repeatedly. Ideally it should do this automatically with no assistance from software so the rest of the microcontroller can enter a low-power mode for the duration of the measurement.

The circuit is shown in Figure 9.6(a). It looks rather complicated but breaks into a few simple blocks. First, the resistor R and capacitor C form the heart of the oscillator. Each cycle of the oscillation has two parts. Suppose that the capacitor is initially discharged so that $V_C = 0$.

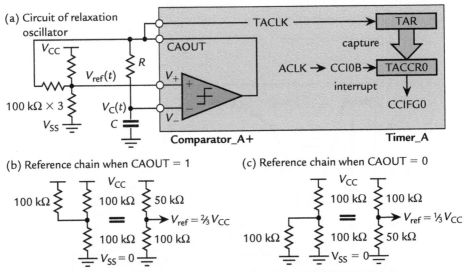

Figure 9.6: Relaxation oscillator based on Comparator_A+. (a) Circuit with the capacitor connected to the inverting input of the comparator so that CAOUT provides the correct drive for the resistor. The circuit for the threshold voltage has three equal resistors so that its output voltage is (b) $\frac{1}{3} V_{CC}$ when CAOUT $= 1$ and (c) $\frac{2}{3} V_{CC}$ when CAOUT $= 0$.

- The capacitor is charged by driving the resistor to V_{CC}. This continues until the voltage $V_C(t)$ across the capacitor reaches an upper threshold voltage, V_{ref}^{upper}.

- The capacitor is then discharged by driving the resistor to ground, V_{SS}. This continues until the voltage $V_C(t)$ across the capacitor crosses a lower threshold voltage, V_{ref}^{lower}. The cycle then repeats.

This circuit is called a *relaxation oscillator* because the *RC* circuit is continually trying to relax toward its steady state. We ensure that it never relaxes fully by switching the voltage on the resistor.

The voltage applied to the resistor must be switched between V_{CC} and V_{SS} at the end of each part of the cycle. This is done automatically by driving the resistor directly from CAOUT. In this circuit it is essential that $V_C(t)$ is connected to the inverting input of the comparator. This is because we want CAOUT to go *low* after $V_C(t)$ has risen *above* the reference voltage and vice versa.

A new feature of this oscillator is that the reference voltage must alternate between two values, unlike most of the previous circuits where it was constant. One possibility would be to use the two internal values of $0.25\,V_{CC}$ and $0.5\,V_{CC}$. The problem with this is that the MSP430 would have to become active every half cycle to reconfigure Comparator_A+. A better approach is to use an external network of resistors driven from CAOUT so that the reference voltage changes automatically every half cycle. Figures 9.6(b) and (c) help to explain its operation, assuming that all resistances are equal. The behavior does not depend on the particular value, but $100\,k\Omega$ is reasonable. When CAOUT $= 1$, two of the three resistors are connected to V_{CC} and the third goes to $V_{SS} = 0$. Effectively two of the resistors are in parallel to give $50\,k\Omega$ to V_{CC} and $100\,k\Omega$ to V_{SS}, as shown in Figure 9.6(b). Therefore $V_{ref} = \frac{2}{3}V_{CC}$ in this state. Similarly, $V_{ref} = \frac{1}{3}V_{CC}$ when CAOUT $= 0$.

Finally, the output of the oscillator is used as the clock for Timer_A by connecting CAOUT to TACLK. This must be done externally: There is no internal connection. A slow, accurate clock is needed to measure the number of cycles of the oscillator in a known interval. The simplest approach is to use ACLK because it can be captured directly by one of the channels. This was described in the section "Measurement of Frequency: Comparison of SMCLK and ACLK" on page 312. If the usual ACLK is too fast, it is possible to divide it down in the MSP430x1xx and MSP430x2xx families, provided that this does not interfere with the rest of the program. Other possibilities are to use the basic timer or watchdog timer. In these cases the capture must be done from software using the CCIS0 bit (see the section "Measurement in the Capture Mode" on page 300).

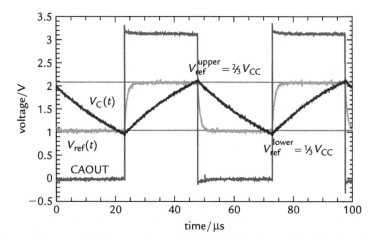

Figure 9.7: Oscilloscope traces of a relaxation oscillator based on Comparator_A+. The voltages on the two inputs are from the capacitor and reference circuit. The output voltage is CAOUT, which drives both the resistor in the oscillator and the reference circuit.

Figure 9.7 shows oscilloscope traces of the oscillator in action. The voltage $V_C(t)$ on the capacitor is made of segments of the now familiar exponential curve for charging and discharging an RC circuit. The CAOUT signal switches between V_{CC} and ground when $V_C(t)$ hits its two extremes. This in turn causes the reference voltage to alternate between $\frac{1}{3} V_{CC}$ and $\frac{2}{3} V_{CC}$, although there is enough stray capacitance in the circuit that the transitions are not abrupt.

The frequency of oscillation follows from the same equations as the transient behavior in the section "Slope Conversion of a Resistance" on page 377. The two threshold voltages in the oscillator are related by the same factor of 2 as in the discharge curve so we can use equation (9.8) for each half cycle without any changes. The charging takes the same time, so the overall period and frequency are

$$T = (2 \log_e 2) RC \approx 1.39 \, RC, \qquad f = \frac{1}{(2 \log_e 2) RC} \approx \frac{0.721}{RC}. \qquad (9.13)$$

The values in my circuit were $C = 3.3 \, \text{nF}$ and $R = 10 \text{k}\Omega$, giving $T = 46 \, \mu\text{s}$. Figure 9.7 shows a slightly longer period of $50 \, \mu\text{s}$ and the square wave on CAOUT is not quite symmetric. I am not sure why the difference is so large: It cannot be explained by the tolerance of the components, for instance, nor by the input and output resistances of the port. My $\times 10$ oscilloscope probes may be partly to blame—never forget that measurements are invasive and that low-power circuits are particularly susceptible to artifacts because of their high resistances and low capacitances.

An amusing point is that the operation of this circuit is close to that of an oscillator based on the classic 555 timer, which I mentioned at the beginning of the book in the section "Approaches to Embedded Systems" on page 2. A 555 is typically used in a relaxation circuit where the voltage on the capacitor oscillates between reference values of $\frac{1}{3} V_{CC}$ and $\frac{2}{3} V_{CC}$. The principles are the same but the details of the circuit are different; for example there are two resistors in its *RC* circuit.

9.1.5 Capacitative Touch Sensing with Comparator_A

I assume in the previous examples for Comparator_A that the capacitance is fixed and that the aim is to measure the resistance. There is no reason why the capacitance should not be measured instead. A few sensors act like capacitors but the most rapidly growing application at present is to capacitive touch sensors. These are used to replace mechanical switches in many contemporary products, of which the Apple iPod is perhaps the most famous. One advantage is higher reliability by eliminating moving parts, but the product designer probably gains most.

The sensor works by detecting the change in capacitance when a finger is brought over it. Figure 9.8 shows a simplified example. Here the capacitor is formed by two conducting pads on the bottom of an insulating sheet. The pads are often no more than enlarged tracks on a printed circuit board (PCB), with the board itself acting as the insulator (dielectric)

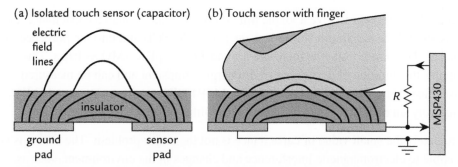

Figure 9.8: Operation of a simple capacitive touch sensor. (a) Two conducting pads on the bottom of an insulating sheet form a capacitor, whose electric field extends outside the top of the sheet. (b) A finger on top of the insulator distorts the electric field and increases the capacitance between the pads.

between the plates of the capacitor. I assume that one pad is connected to ground and the other acts as the sensor but more complicated configurations are often used in practice. An electric field is created when a voltage is applied between the plates and the field lines go from one plate to the other. Some of the lines remain in the insulator but others emerge into the air. (If you are wondering why some of the field lines end at the boundary between the insulator and air in Figure 9.8(a), it is because of the difference in dielectric constant. There should also be field lines in the air below the PCB but I omit these for clarity.) When a finger is brought close to the gap between the pads as in Figure 9.8(b), it distorts the electric field and therefore changes the capacitance. This can be detected in the same way as the change in resistance in the earlier examples. You can measure the duration of a single transient by charging or discharging the capacitance or make it into a relaxation oscillator and measure the frequency. The choice is not straightforward.

So much for the theory: How does it work in practice? A few calculations show that it is not easy. Let us estimate the capacitance of a sensor, treating it as parallel plates for simplicity (a *very* rough approximation). It might have an area of $A = (10\,\text{mm})^2$ with $d = 2\,\text{mm}$ between the plates. The dielectric constant of a PCB is about $\varepsilon_r = 4$ so the capacitance is roughly

$$C = \frac{\varepsilon_0 \varepsilon_r A}{d} \approx 2\,\text{pF}. \qquad (9.14)$$

I estimate that the capacitance of a sensor made from stripboard would change by about 5% due to a finger, making the most optimistic assumptions. Stray capacitance is clearly going to be significant and the overall change in capacitance to be detected may be only 1%.

We need a large resistor to give a viable time-constant with so small a capacitance and 5 MΩ is around the maximum before leakage across the PCB and into the ports becomes significant. The time constant is therefore around $RC = 2\,\text{pF} \times 5\,\text{M}\Omega = 10\,\mu\text{s}$. This determines the duration of a charging or discharging transient and can be measured against a fast clock. It would be good to use a MSP430F2xx with its DCO at 16 MHz. Alternatively, an oscillator could be constructed with a smaller value of resistor.

Unfortunately the small value of capacitance is not the major problem. The sensor is very susceptible to electromagnetic interference and changes in the environment, such as temperature or humidity. Careful design of the software is essential to provide a reliable response. This becomes more complicated with elaborate sensors rather than single pads. There is no space here to go into detail and I instead refer you to the application note *PCB-Based Capacitive Touch Sensing with MSP430* (slaa363). This also shows how the

Schmitt trigger and interrupts on standard digital inputs can be used for the measurement instead of a comparator.

9.2 Analog-to-Digital Conversion: General Issues

Although Comparator_A produces digital values from some types of sensor, many applications call for a "real" analog-to-digital converter (ADC). I explain some general features of ADCs here and go on to describe features of the specific types of ADC provided in the MSP430 in subsequent sections.

The purpose of an ADC is to convert an analog input, which I take to be a voltage V, into a binary value that the digital processor can handle. The input $V(t)$ is a continuous function, meaning that V can take any value within a permitted range and can change in any way as a function of time t. In contrast, the output $V[n]$ is a sequence of binary values. Each has a fixed number of bits and can therefore represent only a finite number of values. Typically the input is sampled regularly at intervals of T_s, so the continuous nature of time has also been lost. Thus the process of conversion damages both V and t and we examine the effects next.

Mixed-signal systems present serious challenges and it is no surprise that whole books are devoted to them. I find *A Baker's Dozen: Real Analog Solutions for Digital Designers* [39] to be the most accessible and the author regularly covers mixed-signal issues in her column in *Electronic Design News*. There are also good books on data conversion from Analog Devices [43, 44] and books on op-amps [45, 50] devote much space to the analog–digital interface. There is excellent material on the Web sites of all the major semiconductor manufacturers, including TI.

9.2.1 Resolution, Precision, and Accuracy

My dictionary lists *accurate* and *precise* as synonyms, which is unfortunate because their meanings are quite distinct in engineering. They must be among the technical terms that cause most misunderstanding but the difference is vital for data converters:

Accuracy: How close a measurement is to its "true" value, which would be produced by an ideal system. This is easy to define but hard to measure.

Resolution or **precision:** The number of distinct output values that a measurement can provide. Alternatively, it can be specified as the change in input that corresponds to the minimum change in output, 1 bit.

The resolution or precision can be quoted in different ways for an ADC, depending on whether you are looking at the ADC alone or its behavior in the system. Consider a 10-bit ADC, for instance. This means that its output is a binary value of 10 bits, which can represent $2^{10} = 1024$ distinct values. We also need to know the range of input voltage to determine the resolution on the input. Suppose that the range is from 0 to a full-scale value of $V_{FS} = 3\,V$. Then a change of 1 bit in the output corresponds to a change of $(3\,V)/1024 \approx 3\,mV$ on the input. This is called the *LSB* for least significant bit and is another way of defining the resolution. We could also say that the ADC converts its input to a precision of $3\,mV$.

Here is a trivial example of the contrast between precision and accuracy. A four-digit voltmeter that gives a reading of $1.234\,V$ is more precise than one with only two digits that reads $1.3\,V$. The first meter can resolve changes of $\pm1\,mV$ on its input, while the second can resolve only changes of $\pm100\,mV$. On the other hand, the first meter is less accurate if the true voltage is $1.301\,V$; it is more precise but less accurate.

Specifications of accuracy for an ADC are highly technical. I need to say a little about them but you should be aware that I am skipping over many tricky aspects; please read one of the reference books if the details are important for your application. Figure 9.9(a) shows

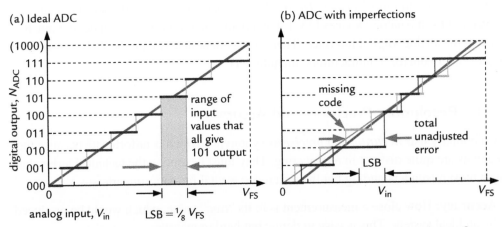

Figure 9.9: Relation between analog input V_{in} and digital output N_{ADC} (transfer function) for an (a) ideal and (b) imperfect 3-bit ADC. The straight line in (a) shows the ideal transfer function without quantization. This is repeated as the thin line in (b), while the thick line is the best fit to the staircase.

the digital output as a function of analog input to an ideal 3-bit ADC. The function is a staircase because there are only $2^3 = 8$ possible output values but any input between 0 and V_{FS} is acceptable. The width of each step is $1\,LSB = V_{FS}/8$. The reduction of a continuous input to a discrete output is called *quantization*.

We want the digital output to be as close as possible to the unquantized value of the input, which is shown as the straight line. Start with 0 V input, which naturally gives 000 output, and raise the input steadily. The output should jump from 000 to 001 when the input passes through ½ LSB = $^1\!/_{16}$ V_{FS}, because the input is then closer to 1 LSB than to 0. The output remains at 001 until the input rises above $^3\!/_2$ LSB, when the output jumps to the next value of 010. Thus a range of input values from ½ LSB to $^3\!/_2$ LSB gives the same output of 001. This is a spread of 1 LSB as expected.

This staircase continues until we reach the maximum value 111 of the output, which appears when the input passes through $^{13}\!/_2$ LSB. It ideally should continue only until the input rises to $^{15}\!/_2$ LSB, at which point the output ought to jump to the next value. However, it cannot—no more values are available with only three bits. Thus the output must remain at 111 until the input reaches its full-scale value of $V_{FS} = 8\,LSB$.

The problem is that a range of inputs of only ½ LSB gives zero output, so an interval of $^3\!/_2$ LSB is needed for maximum output so that the full range of 8 LSB is filled. It would be more convenient if a 3-bit converter could give 9 output values, 0–8, rather than only 0–7. DACs have similar behavior. The transfer function of an N-bit ADC can be summarized as

$$N_{ADC} = \text{nint}\left(2^N \frac{V_{in}}{V_{FS}}\right), \tag{9.15}$$

provided that the output does not go outside its limits of 0 and $2^N - 1$. The nint() function gives the nearest integer to its argument.

The transfer characteristic of a real ADC deviates slightly from the ideal function and some examples of errors are depicted in Figure 9.9(b):

- All the steps should have a width of 1 LSB (except at the ends) but many of them are too wide or narrow. This is called *differential nonlinearity*.

- No inputs at all give an output of 011. This is called a *missing code*. It is a serious fault and most modern converters are designed so that this cannot happen—you will see "no missing codes" on the data sheet.

- I drew the best straight-line fit to the imperfect staircase. Its slope is too high compared with the ideal line, which shows that the *gain* of the ADC is wrong. It also misses the origin, which is an *offset error*.

- This converter is at least *monotonic*, which means that the output never goes down when the input goes up.

The simplest number that characterizes the error in the transfer characteristic is the *total unadjusted error*. This is the maximum deviation between the ideal and actual staircases and is about 1½ LSB in Figure 9.9(b). For example, the F20x2 has a 10-bit successive-approximation ADC called ADC10. The data sheet quotes a typical value of ±2LSB for the total unadjusted error with a maximum of ±5LSB.

A large part of this error arises from the gain and offset voltage of the ADC. It is relatively easy to correct for these errors by calibrating the ADC at two known input voltages. Effectively this fits the best straight line to the transfer characteristic and we can deduce the corresponding staircase. The maximum deviation between this corrected staircase and the actual transfer characteristic is called the *integral nonlinearity* (INL) or integral linearity error (ILE), which is another standard parameter. It is less than ±1LSB for the F20x2.

9.2.2 Signal-to-Noise Ratios

The total unadjusted error is called a *static* parameter because in principle it is measured by sweeping the input slowly through its range. The other type of parameter is *dynamic*, where we look at the performance as a function of time.

The quantization of the input is a type of error. We can show this by plotting the continuous signal, its quantized version, and the difference, called the *quantization error*. This is shown in Figure 9.10 for a sine wave. The signal has been quantized but not sampled, so time remains continuous. I assume a 3-bit ADC again but ignore the asymmetry of the two end values to keep it simple. The sine wave lies between ±4LSB to use the full range of inputs. The quantization error lies between ±½LSB. It has a distinct pattern for this simple sine wave but looks random for a less trivial signal and is often called *quantization noise*.

Typically the quantization noise is equally likely to take any value between ±½ LSB, in which case its rms value can be shown to be $LSB/\sqrt{12}$. The sine wave has a peak

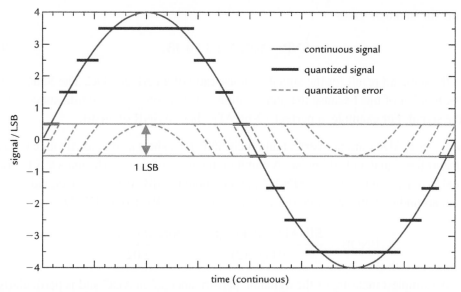

Figure 9.10: Quantization error in continuous time for a sine wave of peak amplitude 4 LSB and a 3-bit converter.

amplitude of 4 LSB so its rms value is $4\text{LSB}/\sqrt{2}$. The ratio of these two is called the *signal-to-noise ratio*, or SNR. It is always quoted in decibels (dB) and is defined by

$$\text{SNR} = 20 \log_{10} \left(\frac{\text{rms amplitude of signal}}{\text{rms amplitude of noise}} \right) \quad \text{dB} \tag{9.16}$$

$$= 20 \log_{10} \left[\frac{(4/\sqrt{2})\,\text{LSB}}{(1/\sqrt{12})\,\text{LSB}} \right] = 20 \log_{10}(4\sqrt{6}) = 20\,\text{dB}. \tag{9.17}$$

This is for a 3-bit converter, with eight possible output values, so the sine wave lies between $\pm 4\,\text{LSB}$. In general there are N bits, 2^N possible output values, and the sine wave lies between $\pm\frac{1}{2} \times 2^N\,\text{LSB}$. Then the signal-to-noise ratio is

$$\text{SNR} = 20 \log_{10} \left[\frac{(\frac{1}{2} \times 2^N/\sqrt{2})\,\text{LSB}}{(1/\sqrt{12})\,\text{LSB}} \right] = 20 \log_{10} \left(\sqrt{\tfrac{3}{2}}\, 2^N \right)$$

$$= 20 \log_{10} \left(\sqrt{\tfrac{3}{2}} \right) + 20 \log_{10} \left(2^N \right)$$

$$= 10 \log_{10}(\tfrac{3}{2}) + 20 N \log_{10}(2) \quad \text{dB}. \tag{9.18}$$

Inserting the numbers gives

$$\text{SNR} = 6.02\,N + 1.76 \text{ dB.} \tag{9.19}$$

This is the standard result for the signal-to-noise ratio of a perfect ADC. The SNR goes up with the number of bits because the quantization noise becomes less significant compared with the signal. For example, a perfect 12-bit ADC has an SNR of 74 dB.

As always, real ADCs are not perfect and their SNR falls short of this ideal value because they distort the signal and add noise. A common measure of their performance is the ratio of signal to noise and distortion (SINAD), again quoted in dB. We can turn equation (9.18) or (9.19) around to define an *effective number of bits* (ENOB) from SINAD:

$$\text{ENOB} = \frac{\text{SINAD} - 10\log_{10}(\frac{3}{2})}{20\log_{10}(2)} = \frac{\text{SINAD} - 1.76}{6.02}. \tag{9.20}$$

This is the simplest measure of the dynamic performance of an ADC and is particularly important for sigma–delta converters. One view is that it tells us the repeatability of a measurement, because noise causes the output to fluctuate even if the input is held absolutely constant. Take the 16-bit sigma–delta converter SD16_A in the F20x3 as an example. The data sheet quotes a maximum value of 87 dB for SINAD and equation (9.20) gives ENOB = 14. (You might wonder what has happened to the remaining two bits but we leave that until the section "Analog-to-Digital Conversion: Sigma–Delta" on page 438.)

9.2.3 Jitter in Timing

Accurate samples depend on accurate timing as well as the transfer characteristic. This is shown in Figure 9.11. Ideally the input $V(t)$ is sampled at time t_n when the voltage is V_n. However, errors in the timing with a spread of Δt lead to errors in the voltage of ΔV. They are related by

$$\Delta V = \left|\frac{dV}{dt}\right| \Delta t. \tag{9.21}$$

Clearly this is more important for rapidly varying signals. Accurate timing requires samples to be triggered by a timer through hardware rather than by software. The successive-approximation ADC10 and ADC12 modules in the MSP430 therefore have internal connections to Timer_A and Timer_B. This is not necessary for the sigma–delta ADCs because they are much slower.

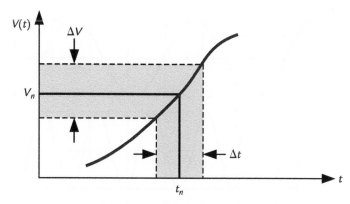

Figure 9.11: Effect of jitter in timing of sample on voltage recorded.

Audio signals provide a striking example of the importance of avoiding jitter. An audio CD contains 16-bit samples so consider a sine wave of maximum peak amplitude,

$$V(t) = (2^{15} \text{LSB}) \sin(2\pi f t). \tag{9.22}$$

Take $f = 20\,\text{KHz}$, near the upper range of audio frequencies, and suppose that the error due to jitter must be less than 1 LSB. The maximum slope of the sine wave is at $t = 0$ so we need

$$\Delta V = \left| 2^{15} \text{LSB} \times 2\pi f \cos(2\pi f t) \right|_{t=0} \Delta t < 1\,\text{LSB}. \tag{9.23}$$

Do not forget the factor of $2\pi f$ from the derivative. This needs $\Delta t < \frac{1}{4}\,\text{ns}$. It is astonishing that such precision is required to sample an audio frequency with a period of $50\,\mu\text{s}$.

9.2.4 Sampling in Time and Aliasing

We saw how finite resolution in voltage—quantization—affects the signal. Now we look at the equivalent problem in time. Let

- f be the frequency of the signal, assumed to be a simple sine wave.
- f_s be the rate at which it is sampled, with $T_s = 1/f_s$ the interval between samples.

The sampling frequency f_s is often quoted with units of "samples per second" (sps) rather than hertz (Hz) but the meaning is the same. Suppose that $f_s = 1\,\text{ksps}$ to keep the numbers simple and consider a signal with frequency $f = 310\,\text{Hz}$. Figure 9.12(a) shows a plot of the continuous signal and the discrete samples every 1 ms. There are no obvious problems.

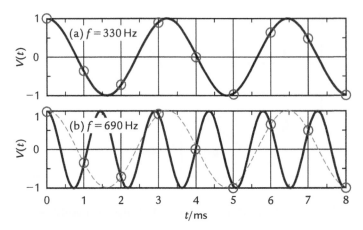

Figure 9.12: (a) A sine wave at 310 Hz with samples taken every 1 ms shown by the circles. (b) Sine waves at 310 Hz and 690 Hz sampled every 1 ms. The higher frequency is *aliased*—its samples are identical to those of the lower frequency.

Now suppose that the frequency is raised to 690 Hz, chosen because $690 = 1000 - 310$. The outcome is shown in Figure 9.12(b). The continuous input clearly has a higher frequency but the discrete values of the samples are *exactly the same* as those for the 310 Hz input. In other words, you cannot tell the difference between the signals after sampling. This, called *aliasing*, is one of the fundamental problems of sampling data in time. The same happens for frequencies of $1000 + 310 = 1310$ Hz, 2000 ± 310 Hz, and so on.

Usually we want the data to be sampled faithfully, meaning that there is no ambiguity in the sampled signal. We must therefore avoid aliasing. Suppose that the maximum frequency in the signal is f_{max}. Then the Shannon sampling theorem states that the signal can be reconstructed perfectly from discrete samples provided that the sampling rate f_s obeys

$$f_s \geq f_N = 2f_{max}. \tag{9.24}$$

(Strictly, the condition is that the sampling rate should be at least twice the *bandwidth* of the input but I assume that we are dealing with baseband signals, which go down to zero frequency.) The minimum acceptable sampling rate is called the Nyquist rate f_N and is twice the maximum frequency in the signal. Turning this around, the highest frequency that can be sampled without aliasing is half of the sampling frequency: $f \leq f_{max} = \frac{1}{2} f_s$. In the preceding example we sample the data at 1 ksps so the frequency of the input must be kept below 500 Hz to avoid aliasing.

As another example, we might want to record audio frequencies in the range 0–20 KHz. The sound must therefore be sampled at a frequency of at least 40 KHz. In practice, compact discs (CDs) use 44.1 KHz and digital audio tape (DAT) uses 48 KHz. Both obey this criterion. On the other hand, some systems such as NICAM (near instantaneous companded audio multiplex) and DAB (digital audio broadcasting) take samples at 32 KHz and therefore cannot reproduce the full range, but only up to 16 KHz. These high rates of sampling would create problems with the limited memory of small embedded systems so the bandwidth is usually reduced. A range from 300–3000 Hz is considered sufficient for "telephone quality" sound and requires sampling at only 6 ksps.

Systems that sample data may therefore need an *antialiasing filter* on the input to suppress frequencies above half the sampling rate, $f > \frac{1}{2} f_s$. It can be omitted if the signal varies very slowly compared with the sampling rate, as is often the case for embedded systems. The temperature of a building is unlikely to suffer from aliasing, for instance. A filter may still be required to suppress noise.

9.2.5 Practical Issues with ADCs

There is far more to using an ADC than simply connecting a signal to its input. Here are some of the main general issues:

Input range: You do not get good results if you connect a sensor whose output lies in the range ±10 mV directly to an ADC that expects an input to lie between 0 and 3 V. In this case the signal needs to be both amplified and shifted to match the ADC. I say more about this in the section "Signal Conditioning and Operational Amplifiers" on page 475.

Voltage reference: An ADC has no notion of absolute voltage: It compares its input with a reference voltage and the output is the ratio of the two. Thus an accurate output relies on the quality of the reference and this frequently limits the overall accuracy. Often V_{CC} can be used as the reference, which is convenient if the input is also proportional to the supply voltage, as in Figure 9.2(a).

References can be internal or external. Internal reference voltages can usually be brought out for use in the circuitry around a sensor. Do not exceed the rated current, which is usually low. Sometimes an internal reference needs an external capacitor even if the voltage is not used outside the IC.

Noise and filtering: Signals in the real world contain unwanted interference. Often this is from the AC mains at 50 or 60 Hz but another common culprit is the digital part of the system, which creates switching noise at its clock frequency. A filter is often needed to remove such noise before it reaches the ADC. (Digital filtering is also possible but it is generally a good idea to eliminate noise as early as possible.) A low-pass filter is also necessary to avoid aliasing.

Decoupling and layout: The analog supply and ground, AV_{CC} and AV_{SS}, are brought out to separate pins from the digital supply on larger devices. They should be connected to their own pair of decoupling capacitors, typically a $10\,\mu$F tantalum electrolytic in parallel with a 100nF ceramic capacitor. This helps keep noise out of the analog systems and improve the accuracy of the ADC. The effect of this can be seen in the data sheet for the F20x3 because it comes in packages with different numbers of pins: SINAD is a few dB higher in the packages where AV_{CC} and AV_{SS} are brought out separately. The layout of the PCB is also important: The general rule is to keep the analog and digital supply and ground rails separate as far as possible, joining them only at the battery or power input.

We now look in more detail at the types of ADC offered in the MSP430 and how they are used.

9.3   Analog-to-Digital Conversion: Successive Approximation

Successive-approximation converters have been the general-purpose ADCs for many years. Their resolution is typically 10–12 bits and the speed can reach megasamples per second (Msps) although those in the MSP430 are not that fast. For some reason the name is expanded into "successive-approximation register" so that it can be contracted to SAR.

SARs work by homing in on the result using binary chopping, which is a standard way of finding solutions to equations of the form $f(x) = 0$. The sequence of operations for an input voltage of $V_{in} = 0.4\,V_{FS}$ to a 4-bit SAR ADC is illustrated in Figure 9.13. Here are the steps:

1.   The input voltage V_{in} is compared with the midpoint $\frac{1}{2}V_{FS}$ of the full range.

2.   In this case $V_{in} < \frac{1}{2}V_{FS}$ so the most significant bit (msb) $= 0$.

3.   We now know that the input lies between 0 and $\frac{1}{2}V_{FS}$. The input is next compared with the midpoint of this range, $\frac{1}{4}V_{FS}$.

4.   We find $V_{in} > \frac{1}{4}V_{FS}$ so the next bit is 1.

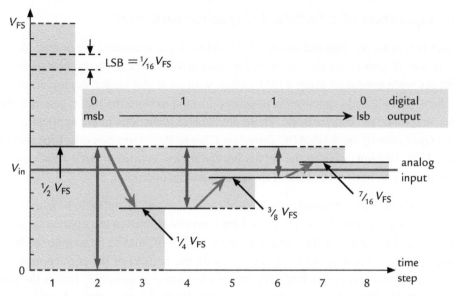

Figure 9.13: Operation of a 4-bit successive-operation ADC with an input of $V_{in} = 0.4V_{FS}$.

5. Now we know that the input lies between $\frac{1}{4}V_{FS}$ and $\frac{1}{2}V_{FS}$. The input is next compared with the midpoint of this range, $\frac{3}{8}V_{FS}$.

6. We find $V_{in} > \frac{3}{8}V_{FS}$ so this bit is 1 again.

7. Now we know that the input lies between $\frac{3}{8}V_{FS}$ and $\frac{1}{2}V_{FS}$. The input is therefore compared with the midpoint of this range, $\frac{7}{16}V_{FS}$.

8. This time $V_{in} < \frac{7}{16}V_{FS}$ so this bit is 0. It is the least significant bit (lsb) for a 4-bit converter.

This process is repeated for each bit, halving the range each time. Each bit typically requires one clock cycle (sometimes two) to make a comparison and set up the new voltage. There is also an overhead to start the conversion, particularly to sample the input voltage. The bits of output are generated in sequence, starting with the most significant bit. This is convenient for serial output, which is therefore chosen for many standalone ADCs.

Example 9.3

There is a subtle flaw in the procedure just described. Can you spot it?

9.3.1 Operation of a Switched-Capacitor SAR ADC

Two main functions are required inside a SAR ADC: logic to control the operation and some way of generating the voltages, for comparison. Back in the bipolar days the voltages were generated using a DAC, often with an R–$2R$ ladder. An ingenious arrangement of switched capacitors is now used instead to store the input, generate the voltages, and make the comparisons. There are clear explanations in the application note *The Operation of the SAR–ADC Based on Charge Redistribution* (slyt176) and in reference [52]. I do not describe the details but the tricky features of SAR ADCs are easier to appreciate if you have some idea of the circuit and its operation.

A typical, modern SAR converter uses a set of capacitors and switches to redistribute their charge, as shown in Figure 9.14. This is a 4-bit converter to match the operations shown in Figure 9.13. The values of the capacitors are C, $\frac{1}{2}C$, $\frac{1}{4}C$, and so on to give the binary chopping sequence. Their switches are labeled with the number of the corresponding bit in the output. An extra capacitor with the smallest value at switch S_0' brings the total capacitance to $2C$. The reference voltage is equal to the full-scale voltage for this circuit, $V_{\mathrm{ref}} = V_{\mathrm{FS}}$. The switches are constructed from MOSFETs like those in the input multiplexers of Comparator_A and the capacitors are also fabricated using CMOS technology. The only other active, analog component is a comparator. I do not show the clock, nor the successive-approximation register that stores the result and gives the converter its name, nor the logic needed to operate the switches.

The switches in Figure 9.14 are configured to sample the input. All the capacitors are coupled in parallel with their "top" plates tied to ground through switch S_A and their "bottom" plates connected to V_{in} through the individual switches. The important point is that the input looks like a capacitance.

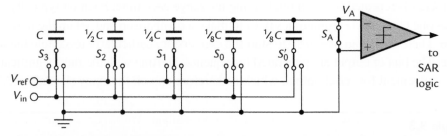

Figure 9.14: Circuit of a 4-bit, charge-redistribution SAR ADC. The switches are in the positions to sample the input.

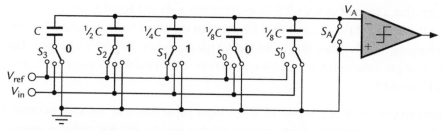

Figure 9.15: The SAR ADC with the switches in the final positions for an input of $V_{in} = 0.4\,V_{FS}$ and a binary output of 0110.

Switch S_A is opened and the individual switches S_n moved to ground to begin the conversion. The stored charge on the capacitor gives $V_A = -V_{in}$ at the inverting input of the comparator. Successive comparisons are made by moving S_n from ground to V_{ref} to give the sequence shown in Figure 9.13. The aim is to bring the voltage V_A as near as possible to 0 and the final configuration of switches gives the binary output value from the conversion. This is shown for an input of $V_{in} = 0.4\,V_{FS}$ in Figure 9.15. After the result is complete, the switches are returned to their positions in Figure 9.14 so that the converter is ready for the next sample.

9.3.2 Practical Issues with SARs

SAR ADCs are straightforward to use compared with sigma–delta ADCs but a few points need attention. They follow from the mode of operation that I just described.

Allow Time for the Input to Charge the Capacitance

The input to a charge-redistribution SAR is a capacitor. This must be charged "fully" before the conversion starts and the time required often sets the maximum speed of the converter. We addressed the same issue with Comparator_A in the section "Slope Conversion of a Resistance" on page 377, where it was necessary to charge the capacitor fully before measuring its discharge time.

How long should be allowed for this? The voltage as a function of time for charging a capacitor through a resistor was given in equation (7.1). Suppose that we use a 10-bit ADC, which means that it can resolve differences of $2^{-10}V_{FS}$ in its input. We want the error to be less than ½ LSB, which is $2^{-11}V_{FS}$. Then the time required follows from

$$\exp(-t/\tau) < 2^{-11} \approx 0.0005 \approx \exp(-7.6). \tag{9.25}$$

This means that a time of at least 7.6 time-constants τ should be allowed for the capacitor to charge. This is a great deal longer than the usual "rule of thumb" of 3τ for the charging time because that would leave the capacitor only 95% charged. That is nowhere near good enough for a 10-bit conversion. The condition is even more stringent at 9τ for a 12-bit ADC.

The time-constant $\tau = R_{in}C_I$, where C_I is the input capacitance of the SAR network ($2C$ in Figure 9.14 plus contributions from analog switches and so on). The value is given in the data sheet. The resistance R_{in} comes from the switches inside the converter and multiplexer, which is also in the data sheet, plus the output resistance of the source that drives the ADC.

Put a Capacitor across the Analog Input

The data sheet may advise you to connect a capacitor between the analog input and ground. There are two main reasons:

- The capacitor acts as a reservoir and makes it faster to charge the internal capacitance, especially if the source has a high internal resistance.

- It suppresses noise.

The value is typically 0.01–0.1 µF. There is always a disadvantage: It slows the response to changes in the input.

There is another potential problem. Often the analog input to the ADC comes from the output of an amplifier and most op-amps do not like driving capacitive loads. A small resistance, typically 10–50 Ω, should therefore be installed between the output of the amplifier and the capacitor across the input.

The Perils of Multiplexing SARs

Many SARs have multiplexed inputs, which means that the input of a single ADC can be selected from one of several analog inputs. This works well with SARs because they do not suffer from latency, which is a problem with sigma–delta converters. There is one drawback, though, which is again due to the use of capacitors. The network is returned to the sample mode after a conversion has been completed. The capacitors are still charged so the input to the network is restored to its previous value of V_{in}. This is a helpful feature if the same external input is being converted, because the input need provide only the small charge needed for the change in voltage of the capacitor since the last reading. It can be a nuisance for a different input, however. Suppose that the previous input was at 3 V but the new input is at only 0.1 V. The capacitor must discharge its 3 V into the new input to reduce its voltage. Some circuits do not take kindly to this, which is another reason for putting a small resistance in series with the inputs.

Frequency of a Conversion Clock

The operation of a SAR relies on the storage of the charge on switched capacitors. This leads to limits on the frequency of its clock:

- None of the components are ideal and the conversion must be completed before significant charge has leaked away from the capacitors. The input of the comparator draws a small current, open switches do not have infinite resistance nor do the capacitors themselves. The clock must therefore not be too slow.

- On the other hand, closed switches do not have zero resistance. Time must therefore be allowed for the charge to redistribute through them so that the voltages have stabilized before each comparison and the output of the comparator is valid. Thus the clock cannot be too fast either.

There are therefore both upper and lower limits on the clock frequency, of which the lower limit is easily overlooked. The MSP430's SAR ADCs include built-in clocks of a suitable frequency but you need to be careful if you select one of the general clocks, notably ACLK because it is usually too slow.

Reference Voltage Generator

Another issue is that the voltage reference V_{ref} is not connected to something simple, such as an input to the comparator. Instead it is connected to the capacitors through switches and sees a fluctuating load that tends to inject noise into the reference. It is not trivial to provide a good voltage reference for a switched-capacitor SAR ADC. The MSP430 provides internal references so the designers have solved this problem for you.

9.4 The ADC10 Successive-Approximation ADC

There are two SAR ADC modules available on the MSP430, the ADC10 and ADC12. The obvious distinction is the number of bits, which is given in their names, and there are several other differences. The ADC12 has a more elaborate set of registers for storing batches of results. It needs external capacitors on its voltage reference but the ADC10 does not (in fact it *must* not). Interrupts are more complicated in the ADC12 as well. I concentrate on the simpler ADC10 and describe the differences briefly in the section "The ADC12 Successive-Approximation ADC" on page 432.

9.4.1   Architecture of the ADC10

Figure 9.16 shows a simplified block diagram of the ADC10 in the F20x2; there are more inputs in larger devices. As in the case of Comparator_A+, the module looks more complicated than you might expect because of the wide range of options. I omit the connections for external references, automatic sequences of conversions, and the data transfer controller to avoid obscuring the central features, which I now describe. These are controlled in the usual way by the registers ADC10CTL0 and ADC10CTL1. Most features can be configured only while the enable conversion bit ENC is clear to ensure that the ADC is inactive.

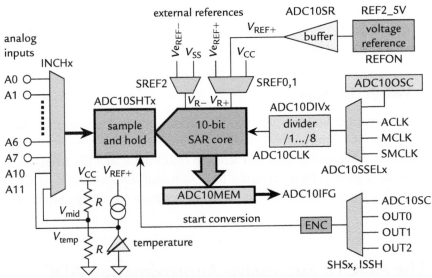

Figure 9.16: Simplified block diagram of the ADC10. The connections for external references, automatic sequences of conversions, and the data transfer controller are omitted for clarity.

Core

At the heart of the ADC10 is a 10-bit, switched-capacitor, SAR core. It is guaranteed monotonic with no missing codes. The ADC10ON bit enables the core and a flag ADC10BUSY is set while sampling and conversion is in progress. The result is written to ADC10MEM in a choice of two formats, selected with the ADC10DF bit:

- The default is straightforward, right-justified binary in the range 0x0000–0x03FF. Zero corresponds to the bottom of the input range.

- Alternatively, the result can be left-justified in twos complement format. The lowest 6 bits are always clear and bit 15 gives the sign. Zero corresponds to the middle of the input range and lower inputs give negative values.

Clock

This can be taken from MCLK, SMCLK, ACLK, or the module's internal oscillator ADC10OSC, selected with the ADC10SSELx bits. It must lie within the range 0.45–6.3 MHz for the F20x2 (but see the description of the reference buffer later). This is an unusually generous range for a SAR ADC but means that ACLK is generally too slow. The ADC10 is one of the few modules that can use MCLK; presumably this would be useful if SMCLK has been divided to too low a frequency.

The internal oscillator runs nominally at 5 MHz but the specification gives a spread of 3.7 to 6.3 MHz. It is automatically enabled when needed and disabled when conversions have finished. This makes it the most convenient source for most applications.

The frequency of the clock can be divided by 2, 3, ..., 7, 8 by configuring the ADC10DIVx bits. The output of the divider is labeled ADC10CLK and feeds both the SAR core and sample-and-hold blocks.

Sample-and-Hold Unit

This is shown separately in the block diagram but is presumably integrated into the switched-capacitor network as in Figure 9.14. However, the separate block is a useful reminder of how important it is to configure the sampling time and I illustrate this shortly. The time is chosen with the ADC10SHTx bits, which allow 4, 8, 16, or 64 cycles of ADC10CLK.

Input Selection

A multiplexer selects the input from eight external pins A0–A7 (more in larger MSP430s) and four internal connections. Two of the internal connections are for optional, external reference voltages, which share the pins for A3 and A4 in many devices. These are not shown in Figure 9.16. The other two internal connections are A10 to a temperature sensor and A11 to $V_{mid} = \frac{1}{2}(V_{CC} + V_{SS})$, which is provided to monitor the supply voltage.

Conversion Trigger

A conversion can be triggered in two ways provided that the ENC bit is set. The first is by setting the ADC10SC bit from software (it clears again automatically). This has the usual problem that the timing is not precise, which leads to errors in the converted value of a signal that varies rapidly (see the section "Jitter in Timing" on page 398). There are therefore direct connections on-chip to the output units of channels 0–2 of Timer_A. These are the same signals OUTn that can be brought out to the pins for PWM and so on; they are *not* the flags CCIFGn nor the internal signals EQUn that are set when TAR matches TACCRn. I emphasize this because you must ensure that OUTn has a rising edge when you want to start a new conversion. A PWM output may be unsuitable because it has no edges if the duty cycle can extend to either 0 or 100%. Conversion is normally triggered by a rising edge on the input but this can be inverted with the ISSH bit.

Voltage References and Buffer

The ADC10 is slightly unusual because it uses references for both ends of the range; the lower limit need not be ground (V_{SS}). This means that equation (9.15) for the digital output as a function of the analog input is replaced by the slightly more complicated relation

$$N_{ADC} = \text{nint}\left(1024\,\frac{V_{in} - V_{R-}}{V_{R+} - V_{R-}}\right), \qquad (9.26)$$

where V_{R-} and V_{R+} are the lower and upper references. There are two choices for the lower reference, selected with the SREF2 bit:

- The default is analog ground, AV_{SS}.

- An external reference V_{eREF-} can be applied directly to the SAR core without buffering.

There is more choice for the upper reference, selected with SREF0 and SREF1:

- The default is the analog supply voltage, AV_{CC}.

- An internal reference voltage V_{REF+} is enabled with REFON. It is buffered and can be chosen from either 1.5 V or 2.5 V with the REF2_5V bit. The accuracy is considerably better than the reference in Comparator_A+. For example, the higher voltage is specified as 2.50 ± 0.15 V. The stability with temperature is *much* better at ± 100 ppm/°C.

- An external reference V_{eREF+} can be applied directly to the SAR core.

- The external reference can also be buffered.

A buffer is needed between the internal reference and the core of the ADC10. This is enabled only when needed because it draws more current than any other part of the module. This shows the aggressive nature of the load presented by the switched-capacitor network. The buffer runs at lower current if the ADC10SR bit is set but the maximum value of $f_{ADC10CLK}$ is reduced from 6.3 MHz to 1.5 MHz in this case.

The supply voltage V_{CC} must be greater than 2.8 V if you want to use the 2.5 V reference. This is "obvious" but easy to forget.

The default references of V_{SS} and V_{CC} are appropriate when the input is proportional to V_{CC}, as we saw in the section "Comparator_A" on page 371. These do not need the buffer.

Finally, the buffered internal reference V_{REF+} can be routed to a pin for use outside the MSP430. No external capacitors are needed for the internal voltage reference, whether it is used internally or externally. In fact there is a limit of only 100 pF on the capacitance that may be attached to V_{REF+} when it is brought out externally. The pins for the reference inputs and output are shared with analog inputs in many devices.

Interrupts

The interrupt flag ADC10IFG is raised when the result is written to ADC10MEM except when the DTC is used, when it is set after a block has been completed (see the section "Data Transfer Controller in the ADC10" on page 428). Maskable interrupts can be requested in the usual way.

9.4.2 Timing and Current Consumption of the ADC10

Conversions take 13 cycles of ADC10CLK. The final cycle is used to copy the result into ADC10MEM and adjust its format if necessary. It is a little puzzling that 12 cycles are needed for the conversion itself because there are only 10 bits and these numbers usually match.

It is interesting to look at the currents drawn by the different blocks within the ADC10. Here are some typical figures taken from the data sheet for the F20x2.

- 0.6 mA for the SAR core.

- 0.25 mA for the internal reference alone, without the buffer.

- 0.5 mA for the buffer at reduced speed (ADC10SR $= 1$ and $f_{ADC10CLK}$ below 1.5 MHz), in which case 4.5 μs should be allowed for it to settle.

- 1.1 mA for the buffer at full speed (ADC10SR $= 0$), in which case 2 μs is sufficient for settling.

It seems strange that the internal voltage reference and its buffer draw more current than the SAR core itself. The reference is always on when REFON $= 1$ but the buffer is enabled only during conversions. It is turned on at the start of the sampling interval, which must therefore be long enough for both the input and the buffer to settle. The reference itself requires 30 μs to stabilize, which is usually far longer than a complete conversion. It is often enabled and followed by a delay of 30 μs before the core of the ADC10 is switched on. The REFON and ADC10ON bits work independently to permit this.

Both the core and internal oscillator of the ADC10 are disabled to save current when they are not needed. This makes low-power applications easy. The same is true of the reference buffer but not of the internal reference itself, because of its long settling time. You must disable this separately when it is not needed.

Each analog input to the ADC10 can be modeled as a 2 kΩ resistor in series with a 27 pF capacitor. It is vital to allow enough time for this capacitance to charge, as I emphasize in the section "Practical Issues with SARs" on page 405. Let us look at an example to see this in practice.

9.5   Basic Operation of the ADC10

We now look at some simple examples of measurements using the basic hardware of the ADC10. More complicated aspects, such as triggering from hardware and the data transfer controller, are covered in the following section.

Three steps are required to make a single conversion with the ADC10. I assume that the default references of V_{SS} and V_{CC} are suitable so that we need not worry about the settling time of the reference or buffer.

1. Configure the ADC10, including the ADC10ON bit to enable the module. The ENC bit must be clear during this operation because most bits in ADC10CTL0 and ADC10CTL1 can be changed only when ENC = 0.

2. Set the ENC bit to enable a conversion. This cannot be done while the module is being configured in the previous step.

3. Trigger the conversion, either by setting the ADC10SC bit or by an edge from Timer_A.

The last two steps must be repeated for each conversion, which requires clearing and setting the ENC bit again. This might seem a bit cumbersome if you have used an ADC in another microcontroller, where these steps are typically combined. The reason for enforcing two steps is that the ADC10 can be triggered directly from Timer_A, not just from software. A new conversion should not start until the old one has been processed and conversions reenabled by toggling ENC. (It is the same as using the "single sequence" button on a digital storage oscilloscope.)

This two-step sequence is relaxed for conversions triggered by software. In this case the first conversion can be triggered by setting the ENC and ADC10SC bits together in a single instruction. Subsequent conversions can be triggered by setting the ADC10SC bit alone, without toggling ENC.

The input to the ADC10 is selected with the INCHx bits in ADC10CTL1. You must also enable the corresponding pin for analog input by setting its bit in the ADC10AEn registers (just one for the F20x2, which has only eight external inputs). These settings override those in the port registers.

9.5.1 Single Conversion with the ADC10 Triggered by Software

I use the circuit shown in Figure 9.17(a) as a simple demonstration of the ADC10. Two resistors, one fixed and one variable, form a potential divider if pins P1.5 and P1.6 are driven to V_{CC} and ground (V_{SS}). Their junction is taken to P1.1, which can be selected as input A1 of the ADC10. There is also a 10nF capacitor from this junction to ground (this is the same circuit that I use in Figure 9.3 for Comparator_A+). An LED is connected from P2.6 to ground as a simple output. The idea is that the variable resistor mimics a thermistor and the LED should be illuminated if the temperature falls too low, when $R_2 > 10\text{k}\Omega$.

The only tricky part of this problem is to calculate the time for which the input should be sampled. I replace the real circuit with an equivalent circuit in Figure 9.17(b). This will be

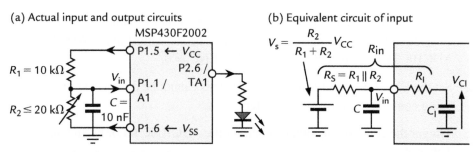

Figure 9.17: Circuit used for simple measurements with the ADC10. (a) Actual circuit, with a potential divider and capacitor on the input and LED on output. (b) Equivalent circuit for input, used to calculate sampling time.

familiar to electrical engineers as an application of Thévenin's theorem. Here is a simple justification for everybody else. After a long time, when all the capacitors have charged fully, the input voltage on P1.1 should be set by the potential divider formed by R_1 and R_2. This gives the voltage of the battery on the left of the equivalent circuit, V_s. That is fairly straightforward but it is a little harder to work out the source resistance R_s, which is in series with the battery. Very roughly, you can imagine that current flowing in or out of P1.1 and C has a choice of passing through R_1 to V_{CC} or through R_2 to V_{SS}. These resistors therefore appear to be in parallel and $R_s = R_1 \parallel R_2$.

For the rest of this calculation I pretend that *the capacitor C is not present*. (I come back to it later.) This leaves only the battery and resistance due to R_1 and R_2 in parallel outside the MSP430. Inside, the ADC10 is represented by the resistance R_I, due to the multiplexing switches, and the capacitance C_I of the SAR network. Overall, the two resistances are in series to give $R_{in} = R_s + R_I$. We now have the classic RC circuit and found in the section "Practical Issues with SARs" on page 405 that we need to allow 7.6 time-constants for charging in a 10-bit ADC. The worst case is when the resistance and capacitance take their maximum values. We should therefore assume that the variable resistor $R_2 = 20\text{k}\Omega$, in which case $R_s = R_1 \parallel R_2 \approx 7\text{k}\Omega$. The data sheet gives $R_I \leq 2\text{k}\Omega$ so we should take $R_{in} = 9\text{k}\Omega$. The data sheet also gives $C_I \leq 27\text{pF}$ and again we take this maximum value as the most pessimistic estimate. Then $\tau = R_{in}C_I = 0.24\,\mu\text{s}$ and $7.6\tau = 1.85\,\mu\text{s}$.

The sampling time is configured in terms of cycles of ADC10CLK. I use the internal oscillator ADC10OSC and we should take its maximum frequency of 6.3 MHz for safety. The number of cycles is therefore $1.85\,\mu\text{s} \times 6.3\,\text{MHz} = 12$. This is rounded up to the next available value of 16. It therefore takes longer to sample the input than to convert it.

I am sorry if that seems hard going but it is really important. The staff at the European Product Information Centre tell me that one of the most common problems they see with the ADC10 and ADC12 is that programmers do not allow enough time for sampling the input. This must be considered for *all* SAR ADCs, not just those in the MSP430. Some other microcontrollers have a fixed time for sampling, which is intended to be long enough for most applications. You should always do a calculation like the preceding one to verify that it works for you. If the required charging time is too long, the effective value of R_s can be reduced by inserting a unity-gain buffer between the source and the input to the ADC. The same issue arises with sigma—delta ADCs but these often include a buffer.

Listing 9.2 shows a program with a loop that makes single readings of the voltage in this circuit. The LED is turned on if the result is in the upper half of the range and off otherwise. The other issues in the program are much more straightforward than the calculation of the sampling time:

- The potential divider on the input is between V_{SS} and V_{CC} so it is appropriate to use these as the references as well, which is the default. The internal voltage reference and buffer are not needed.

- The simplest choice for the clock is the ADC10's internal oscillator, ADC10OSC. The other possibilities would be MCLK or SMCLK but these would remove the option of entering a low-power mode during a conversion. There is no point in dividing down the frequency of the clock: We might as well do the conversion as quickly as possible to save energy.

- The other bits in ADC10CTL1 are configured for input on channel A1 (shared with pin P1.1), triggering from ADC10SC in software, straightforward binary output, and single conversions on a single channel.

Listing 9.2: Single measurements of an external voltage using a polling loop in ad10led1.c. An LED is turned on if the voltage exceeds ½V_{CC}.

```
// ad10led1.c - measure input voltage from pot divider using ADC10
// ADC10 default references (VSS, VCC), internal clock, polling loop
// Home-made analog demo board for MSP430F20xx:
//   10k on P1.5, 20k variable on P1.6, junction to P1.1 with 10nF cap
//   LED active high on P2.6
// Default DCO, no crystal, no ACLK, power from JTAG
// J H Davies, 2007-08-30; IAR Kickstart version 3.42A
//-----------------------------------------------------------------
#include <io430x20x2.h>            // Header file for this device
#include <intrinsics.h>            // Intrinsic functions

#define LED      P2OUT_bit.P2OUT_6   // Output bit for LED
```

```
void main (void)
{
    WDTCTL = WDTPW | WDTHOLD;         // Stop watchdog
    P2SEL = 0;                        // Digital i/o rather than crystal
    P2OUT = 0;
    P2DIR = BIT6|BIT7;                // Unused outputs driven low
    P1OUT = BIT5;                     // VCC to 10k fixed resistor
    P1DIR = BIT3|BIT5|BIT6|BIT7;      // Unused outputs driven low
// VCC and VSS refs, sample for 16 cycs, int ref off, ADC on, no ints
    ADC10CTL0 = SREF_0 | ADC10SHT_2 | ADC10ON;
// Input channel 1, trigger using ADC10SC bit, no clock division,
//   internal ADC clock, single channel single conversions
    ADC10CTL1 = INCH_1 | SHS_0 | ADC10DIV_0 | ADC10SSEL_0 | CONSEQ_0;
    ADC10AE0 = BIT1;                  // Enable analog input on channel 1
    ADC10CTL0_bit.ENC = 1;           // Enable conversions
    for (;;) {                        // Loop forever taking measurements
        ADC10CTL0_bit.ADC10SC = 1;   // Trigger new conversion
        while (ADC10CTL1_bit.ADC10BUSY == 1) {
        }                            // Loop while conversion takes place
        if (ADC10MEM >= BIT9) {      // Is output in top half of range?
            LED = 1;                 // Yes: Turn LED on
        } else {
            LED = 0;                 // No: Turn LED off
        }
    }
}
```

I use a polling loop to test whether the ADC10 has finished the conversion. This seems a bit primitive and the obvious alternative is to enable the ADC10's interrupt and enter a low-power mode until the result was ready. However, the conversion takes only about 30 cycles of ADC10OSC at 5 MHz, which is 6 µs. That is only a couple of instructions with MCLK at its default speed of around 1 MHz. An interrupt would be worthwhile if MCLK were faster or the conversion slower.

Example 9.4

An even simpler way of testing whether the result is in the upper or lower half of the range is to use the twos complement output and test its sign. Try this. You need to use a cast on ADC10MEM in C because it is defined in the header file as an unsigned variable, although it can be configured in hardware to be signed.

Example 9.5

It is usually not necessary to take a new reading of the input voltage immediately after the old one as in Listing 9.2—most signals do not change that rapidly. Use the watchdog timer as an interval timer on its maximum setting to set the interval between conversions. The action could take place in the main routine but it might be better in the interrupt service routine for the watchdog so that it runs entirely in the background. Against this, does it make the ISR too long? Leave the potential divider switched on continuously.

9.5.2 A Low-Power Example with the ADC10

The last exercise was to read the voltage on the potential divider at intervals of 1 or 2 s. Both the CPU and ADC10 were shut down between readings but the potential divider remained powered. This draws at least $3\,\text{V}/30\,\text{k}\Omega = 100\,\mu\text{A}$, which is a huge current by the standards of the MSP430. Clearly we should shut down the potential divider between readings. The bad news is that this needs another look at the circuits in Figure 9.17 and this time we include the capacitor C. Now there are two charging times to consider: capacitor $C = 10\,\text{nF}$ must be charged through the potential divider when it is powered up, and $C_\text{I} = 27\,\text{pF}$ must be charged during the sampling time for the ADC10:

Charging of C: Suppose that the potential divider has been disabled for so long that C is fully discharged. We then drive P1.5 to V_{CC} in preparation for a new measurement. The equivalent circuit in Figure 9.17(b) shows that C effectively charges through $R_1 \parallel R_2 \approx 7\,\text{k}\Omega$. The time-constant is therefore about $70\,\mu\text{s}$ and we should wait the usual $7.6\tau \approx 530\,\mu\text{s}$ for "full" charging.

Charging of C_I: Suppose that the potential divider has been active for a long time so that the capacitor C is fully charged and the voltage on P1.1 is $V_{\text{CC}}R_2/(R_1 + R_2)$. When sampling starts, C_I must be charged through R_1. Now, $C \gg C_\text{I}$ so C acts like a battery, whose voltage will hardly change as it loses a tiny fraction of its charge to C_I. The time-constant is therefore set only by C_I and R_1, which gives $\tau = 54\,\text{ns}$ and $7.6\tau = 0.4\,\mu\text{s}$, which is 2.6 cycles of ADC10OSC. This should be rounded up to the next value of 4. It is smaller than the value that we calculated previously because R_1 and R_2 no longer play any part.

It takes a long time to recharge the potential divider. We need another timer for this delay and I assume that channel 0 of Timer_A is available. I left Timer_A running in the Continuous mode from the VLO for other tasks in the program; it would be easier to stop and start it if nothing else needs it. The frequency of the VLO is 12 KHz so we should allow $530\,\mu\text{s} \times 12\,\text{KHz} = 6.4$ cycles for charging, rounded up to 7.

The program is shown in Listing 9.3 and the sequence of events for each reading is as follows. All actions take place in interrupt service routines.

1. Remain in a low-power mode until awakened by an interrupt from the watchdog timer.

2. Restore power to the potential divider and set up TACCR0 for a delay of 530 μs. There is a loop to check that TAR has been read successfully because Timer_A runs from ACLK, which is not synchronous with MCLK (see the section "Timer Block" on page 289). I toggle the LED as a diagnostic to show that something is happening; a brief flash can be seen if the LED was previously off but the "antiflash" is invisible.

3. Return to low power until awakened by an interrupt from Timer_A.

4. Enable a single conversion on the ADC10. This time I use an interrupt to flag completion. The ADC10 must first be turned on while ENC remains clear, before both ENC and ADC10SC are set to start the conversion.

5. Return to low power until awakened by an interrupt from the ADC10.

6. Update the LED, disable the ADC10, remove power from the potential divider, and return to low power until the next interrupt from the watchdog. The ENC bit must be cleared before the ADC10 can be turned off.

Listing 9.3: Program `ad10led4.c` to measure the voltage from a potential divider, which is shut down between readings to save current. Each measurement is run by a sequence of interrupts from the watchdog timer, channel 0 of Timer_A and the ADC10.

```
// ad10led4.c - measure input voltage from pot divider using ADC10
// Potential divider disabled between readings set by watchdog timer
// Watchdog -> activate pot div, delay from CCIFG0 -> conversion
// ADC10 default refs (VSS, VCC), internal clock, interrupt on finish
// Homemade analog demo board for MSP430F20xx:
//   10k on P1.5, 20k variable on P1.6, junction to P1.1 with 10nF cap
//   LED active high on P2.6
// Default DCO, no crystal, ACLK from VLO, power from JTAG
// J H Davies, 2007-08-30; IAR Kickstart version 3.42A
//-------------------------------------------------------------------
#include <io430x20x2.h>         // Header file for this device
#include <intrinsics.h>         // Intrinsic functions
#include <stdint.h>             // Standard integer types

#define LED      P2OUT_bit.P2OUT_6   // Output bit for LED
#define POTDIV   P1OUT_bit.P1OUT_5   // Output bit for potential divider
// Cycles for delay = 530us * 12KHz = 6.4 -> 7
#define CHARGING    7                // Cycles of ACLK = VLO for charging
```

```
void main (void)
{
    BCSCTL3 = LFXT1S_2;                  // Select ACLK from VLO (no crystal)
// Watchdog as interval timer, clear, ACLK / 32768
    WDTCTL = WDTPW | WDTTMSEL | WDTCNTCL | WDTSSEL;
    IE1_bit.WDTIE = 1;                   // WDT interval timer interrupts
    P2SEL = 0;                           // Digital i/o rather than crystal
    P2OUT = 0;
    P2DIR = BIT6|BIT7;                   // Unused outputs driven low
    P1OUT = 0;                           // All outputs low
    P1DIR = BIT3|BIT5|BIT6|BIT7;         // Unused outputs driven low
// VCC and VSS refs, sample for 4 cycs, int ref off, ADC off, interrupts
    ADC10CTL0 = SREF_0 | ADC10SHT_0 | ADC10IE;
// Input channel 1, trigger using ADC10SC bit, 2s complement,
//   no clock division, internal osc, single channel single conversions
    ADC10CTL1 = INCH_1|SHS_0|ADC10DF|ADC10DIV_0|ADC10SSEL_0|CONSEQ_0;
    ADC10AE0 = BIT1;                     // Enable analog input on channel 1
// Timer_A from ACLK, no division, continous mode, clear, no interrupts
    TACTL = TASSEL_1 | ID_0 | MC_2 | TACLR;
    for (;;) {                           // Loop forever taking measurements
        __low_power_mode_3 ();           // LPM3 between interrupts
    }
}
//--------------------------------------------------------------------
// ISR for watchdog: enable potential divider, set up delay for charging
//--------------------------------------------------------------------
#pragma vector = WDT_VECTOR
__interrupt void WDT_ISR (void)          // Acknowledged automatically
{
    LED ^= 1;                            // Flash LED to show activity
    POTDIV = 1;                          // Apply VCC to potential divider
    do {                                 // Loop until two successive reads
        TACCR0 = TAR;                    //   of TAR agree with each other
    } while (TACCR0 != TAR);
    TACCR0 += CHARGING;                  // Delay for charging pot div
    TACCTL0 = CCIE;                      // Enable interrupts, clear any flag
}
//--------------------------------------------------------------------
// Delay here to allow potential divider to recharge
//--------------------------------------------------------------------
// Interrupt service routine for CCIFG0: start new reading
//--------------------------------------------------------------------
#pragma vector = TIMERA0_VECTOR
__interrupt void TIMERA0_ISR (void)      // Acknowledged automatically
{
    TACCTL0 = 0;                         // Disable further CCIFG0 interrupts
    ADC10CTL0_bit.ADC10ON = 1;           // Turn on ADC
    ADC10CTL0 |= (ENC|ADC10SC);          // Enable ADC10, start conversion
}
//--------------------------------------------------------------------
// Wait here for conversion to complete
//--------------------------------------------------------------------
// Interrupt service routine for ADC10: disable ADC, update LED
//--------------------------------------------------------------------
#pragma vector = ADC10_VECTOR
__interrupt void ADC10_ISR (void)        // Acknowledged automatically
```

```
{
    POTDIV = 0;                        // Remove VCC from potential divider
    ADC10CTL0_bit.ENC = 0;             // Disable conversions
    ADC10CTL0_bit.ADC10ON = 0;         // Turn off ADC
    if (((int16_t) ADC10MEM) > 0) {    // Is output positive?
        LED = 1;                       // Yes: Turn LED on
    } else {
        LED = 0;                       // No: Turn LED off
    }
}
```

I have to admit that this program is a poor design because I have not allowed for the tolerance of the frequency of the VLO, which is specified in the range 4–20 KHz. There may be insufficient time to charge the potential divider if f_{VLO} is higher than its nominal value of 12 KHz. A simple solution is to raise the CHARGING parameter to allow for the maximum frequency.

9.5.3 Temperature Sensor on the ADC10

The final introductory example uses the ADC10's temperature sensor and internal voltage reference. It would be best if the temperature could be shown on a numerical display but that is not available. Instead the MSP430 lights an LED if the temperature rises above 25°C. You might need to modify that if you live in a warmer part of the world. The readings are paced by the watchdog timer in interval mode, as in the previous example.

Look first at the performance of the sensor itself. It gives a voltage that depends on temperature and the data sheet gives a typical transfer function of

$$V_{temp} = 3.55\theta + 986 = 3.55(\theta + 278) \text{ mV}, \tag{9.27}$$

where the temperature θ is measured in celsius. The bad news is that there may be an offset of $\pm 100\,\text{mV}$, which translates into an error of nearly $\pm 30°\text{C}$. The coefficient also has a significant tolerance, quoted as $3.55 \pm 0.1\,\text{mV}/°\text{C}$. It is no surprise that the data sheet warns you that you may need to calibrate the sensor. This requires one reading at a known temperature in the range of interest, which fixes the offset. Better still, calibration at two temperatures determine both the offset and slope.

The voltage from the sensor does not exceed 1.4 V over the commercial range of temperature, so the 1.5 V reference is appropriate. Resolution would be lost if the 2.5 V reference were used instead. Combining equation (9.27) for the voltage from the sensor with the transfer function of the ADC10, equation (9.26), shows that the output N_{temp} of the ADC10 is

$$N_{temp} = \text{nint}\left[1024\,\frac{3.55(\theta+278)}{1500}\right],\tag{9.28}$$

where the voltages are all in millivolts. We want to find θ from N_{temp} and therefore need to make θ the subject:

$$\theta = \frac{1500}{3.55}\frac{N_{temp}}{1024} - 278 \approx \frac{420\,N_{temp}}{1024} - 278.\tag{9.29}$$

This can also be written as $\theta \approx 0.41\,N_{temp} - 278$, which shows that a change of one bit corresponds to about 0.4°C. In other words, this is the LSB or resolution of the sensor. It does not seem very impressive for a 10-bit ADC. The problem is that the voltage from the sensor is roughly proportional to *absolute* temperature (often abbreviated to PTAT), hence the offset of 278°C. Most of the range of the ADC is wasted because it corresponds to temperatures below the specified range of operation (-40°C to $+85$°C for standard components). The extra two bits in the ADC12 reduce LSB to roughly 0.1°C.

We also need to do a new calculation of the sampling time, now that we are using a different source. The data sheet specifies that this must be greater than 30 μs for the temperature sensor. It draws a current from the internal voltage reference, part of which is therefore enabled when the temperature sensor is selected as the input to the ADC10. We also use the internal reference for the conversion because the output of the temperature sensor is specified as an absolute voltage, not a fraction of V_{CC}. We must therefore allow time for the reference to stabilize, assuming that it is disabled between measurements to save current. This also requires 30 μs, which can presumably take place at the same time as the stabilization of the sensor. This is much slower than the earlier examples. A further allowance is needed for the buffer on the reference voltage. Its settling time depends on whether it is run at high or low current (the ADC10SR bit), which in turn depends on the frequency of ADC10CLK. We therefore need to choose this now.

The maximum frequency of the internal clock ADC10OSC without division is 6.3 MHz and the longest sampling time is 64 cycles, which gives $64/6.3 = 10$ μs. This is far too short so a slower clock is needed. One possibility would be to use SMCLK but we then have to wait in LPM0 during the conversion. The alternative is to divide ADC10OSC by 4, which extends the sampling time to 40 μs. Conveniently this also brings $f_{ADC10CLK}$ low enough that the buffer can be run with reduced drive by setting the ADC10SR bit. It saves about 0.6 mA during the conversion, which is a large fraction of the total current drawn by the ADC10 (look back at the section "The ADC10 Successive-Approximation ADC" on page 407). The buffer takes 4.5 μs to settle, which fits within the chosen sampling time.

The program is shown in Listing 9.4. The ADC10 is configured for its internal reference of 1.5 V, 64 cycles of sampling, low sampling rate, and interrupts but both the ADC and reference are left off. The internal reference is not used automatically for the conversion although it is needed to power the temperature sensor. Input channel 10 is selected for the temperature reference; the ADC10AEn registers are not needed because the input is internal. Within the main loop, the program waits in LPM3 for an interrupt from the watchdog timer. It then turns on the ADC10 and its reference, enables conversions and triggers one by software. The MSP430 returns to LPM3 while the conversion takes place, wakes on completion, and disables further conversions. The calculation of temperature from ADC10MEM is done in simple steps for clarity. It is wise to put lengthy calculations in the main function rather than an interrupt service routine so that they can be interrupted by more urgent tasks.

The variable `temperature` is defined as a 32-bit integer because the multiplication by 420 might overflow in 16-bit arithmetic. I add an offset of 512 to reduce the bias from the integer division by 1024. The point is that integer division always rounds down, so that $1023/1024 = 0$ for instance. The offset changes this so that $511/1024$ becomes $1023/1024$, which rounds to 0, but $513/1024 \rightarrow 1025/1024 \rightarrow 1$. A small bias remains after this correction because 512 is always rounded up. There are better approaches, such as "round half even" [37], but it did not seem worth the trouble here. EW430 uses a shift instead of division because 1024 is a power of 2. There is no need for the programmer to worry about this.

Listing 9.4: Program `ad10tmp1.c` to demonstrate the internal temperature sensor and voltage reference of the ADC10. An LED is turned on if the measured temperature exceeds 25°C.

```
// ad10tmp1.c - measure temperature in celsius using sensor in ADC10
// ADC10 and ref disabled between readings set by watchdog timer
// ADC10 internal 1.5V ref, internal clock/4, interrupt on finish
// Homemade analog demo board for MSP430F20xx, LED active high on P2.6
// Default DCO, no crystal, ACLK from VLO, power from JTAG
// J H Davies, 2007-08-30; IAR Kickstart version 3.42A
//------------------------------------------------------------------
#include <io430x20x2.h>          // Header file for this device
#include <intrinsics.h>          // Intrinsic functions
#include <stdint.h>              // Standard integer types

#define LED      P2OUT_bit.P2OUT_6    // Output bit for LED
```

```
void main (void)
{
    uint32_t temperature;                 // Converted value of temperature

    BCSCTL3 = LFXT1S_2;                   // Select ACLK from VLO (no crystal)
// Watchdog as interval timer, clear, ACLK / 32768
    WDTCTL = WDTPW | WDTTMSEL | WDTCNTCL | WDTSSEL;
    IE1_bit.WDTIE = 1;                    // WDT interval timer interrupts
    P2SEL = 0;                            // Digital i/o rather than crystal
    P2OUT = 0;
    P2DIR = BIT6|BIT7;                    // Unused outputs driven low
    P1OUT = 0;                            // All outputs low
    P1DIR = BIT3|BIT5|BIT6|BIT7;          // Unused outputs driven low
// Vref+ = 1.5V and VSS refs, sample 64 cycs, slow, ref + ADC off, ints
    ADC10CTL0 = SREF_1 | ADC10SHT_3 | ADC10SR | ADC10IE;
// Input channel 10 (temperature), trigger using ADC10SC bit, binary,
//    internal osc/4, single channel single conversions
    ADC10CTL1 = INCH_10 | SHS_0 | ADC10DIV_3 | ADC10SSEL_0 | CONSEQ_0;
    for (;;) {                            // Loop forever taking measurements
        __low_power_mode_3();            // Wait for WDT between measurements
        ADC10CTL0 |= (REFON|ADC10ON);    // Enable ADC10 and reference
        ADC10CTL0 |= (ENC|ADC10SC);      // Enable ADC10, start conversion
        __low_power_mode_3();            // Sleep during measurement
        ADC10CTL0_bit.ENC = 0;           // Disable further ADC10 conversions
        ADC10CTL0 &= ~(REFON|ADC10ON);   // Disable ADC10 and reference
        temperature = ADC10MEM;          // Raw converted value
        temperature *= 420;              // Scaling factor (Vref/temp coeff)
        temperature += 512;              // Allow for rounding in division
        temperature /= 1024;             // Divide by range of ADC10
        temperature -= 278;              // Subtract offset to give celsius
        if (temperature > 25) {          // Is temperature above 25 celsius?
            LED = 1;                     // Yes: Turn LED on
        } else {
            LED = 0;                     // No: Turn LED off
        }
    }
}
//-------------------------------------------------------------------
// Interrupt service routine for watchdog after sleep: return to active
//-------------------------------------------------------------------
#pragma vector = WDT_VECTOR
__interrupt void WDT_ISR (void)          // Acknowledged automatically
{
    __low_power_mode_off_on_exit();      // Return to active mode on exit
}
//-------------------------------------------------------------------
// Interrupt service routine for ADC10 after converson: return to active
//-------------------------------------------------------------------
#pragma vector = ADC10_VECTOR
__interrupt void ADC10_ISR (void)        // Acknowledged automatically
{
    __low_power_mode_off_on_exit();      // Return to active mode on exit
}
```

I found that the sensor gave a temperature of 19°C when a thermometer outside showed 22°C. It is unlikely that there is significant self-heating of the package so the error is around 3°C. This is not be good enough for a thermostat but may well be adequate for a warning that something was about to overheat. On the other hand, I was lucky and the error might have been *much* larger than this. My F2002 is in a plastic dual-in-line package (PDIP), which is rather massive and not easy to warm up with my finger. The LED lit correctly when I heated the package with an iron.

Example 9.6

If you live in the United States, you might wish to adapt the program in Listing 9.4 to use degrees Fahrenheit rather than celsius.

The battery voltage can be checked in a similar way. In this case you compare $V_{mid} = 0.5\,V_{CC}$ with the internal voltage reference. The V_{mid} signal stabilizes in $1.4\,\mu s$ but you must still wait for the reference.

Example 9.7

Adapt the program in Listing 9.4 to light the LED if the supply voltage falls below 2.7 V. This can easily be tested if your demonstration board is powered by a FET, whose output voltage can be varied from 1.8 to 3.6 V.

9.6   More Advanced Operation of the ADC10

It is time to look at some of the more advanced features of the ADC10. These include direct triggering from Timer_A and automatic storage of results. I continue to use the hardware in Figure 9.17(a) as an example.

Let us vary the intensity of the LED continuously by using PWM instead of simply switching it on and off. It is connected to P2.6 and can be driven from OUT1, which is the only channel available for PWM in the hardware with Timer_A2. To make things simple I use the Up mode with a 10-bit range, fixed with TACCR0 = 0x03FF. Then the output of the ADC10 can be copied into TACCR1 and used as the duty cycle with no calculation. The frequency of PWM should be about 100 Hz so the clock for Timer_A needs to be at around $100 \times 0x0400 = 100\,KHz$. This is too fast for ACLK (particularly with the VLO) so I divide the default SMCLK by 8 to give roughly 125 KHz.

9.6.1   Triggering the ADC10 from Timer_A

Suppose that the input varies rapidly and we wish to update every cycle of PWM. This means that the ADC10 must be triggered from Timer_A either through interrupts or directly by an OUTn signal.

The first question is which channel of Timer_A to use. Channel 1 drives the PWM and therefore produces a periodic OUT1 signal, which can be routed to the ADC10. It can also request interrupts. Unfortunately there are problems if the duty cycle is either 0 or 100%. The output becomes constant in both these cases so there are no edges on OUT1 to trigger the ADC10. Interrupts also cease with a duty cycle of 100% so that does not work either. We must use channel 0 although that also causes other difficulties, as we shall see later.

The more straightforward approach shown in Listing 9.5 uses the CCIFG0 interrupt service routine, which is called when TAR counts to TACCR0. This is the ideal point at which to update PWM, as explained in the section "When Should the Duty Cycle Be Updated?" on page 346. Unfortunately there is too little time to both perform the conversion and copy the result during a single cycle of TACLK = SMCLK/8. Instead, the ISR copies the previous converted value into TACCR1 for the next cycle of PWM. It then initiates a new conversion, whose output is used in the next ISR.

Listing 9.5: Repeated measurements of external potential to control PWM output in ad10pwm1.c. Updates are carried out in the ISR for CCIFG0 with the ADC10 triggered by software.

```
// ad10pwm1.c - PWM controlled by input voltage using ADC10
// ADC10 def refs (VSS, VCC), int clock, auto disabled between readings
// PWM updated on CCIFG0 interrupt and new conversion initiated
// Homemade analog demo board for MSP430F20xx:
//   10k on P1.5, 20k variable on P1.6, junction to P1.1 with 10nF cap
//   LED active high on P2.6 driven from OUT1 in PWM
// Default DCO, no crystal, no ACLK, power from JTAG
// J H Davies, 2007-08-30; IAR Kickstart version 3.42A
//-------------------------------------------------------------------
#include <io430x20x2.h>            // Header file for this device
#include <intrinsics.h>            // Intrinsic functions

void main (void)
{
    WDTCTL = WDTPW | WDTHOLD;      // Stop watchdog
    P2SEL = BIT6;                  // OUT1 on P2.6; no crystal
    P2OUT = 0;                     // Unused outputs driven low
    P2DIR = BIT6|BIT7;
    P1OUT = BIT5;                  // VCC to 10k fixed resistor
    P1DIR = BIT3|BIT5|BIT6|BIT7;   // Unused outputs driven low
// Timer_A for PWM at 125Hz on OUT1, SMCLK/8, Up mode
    TACCR0 = BITA - 1;             // Upper limit to match ADC10MEM
```

```
    TACCTL0 = CCIE;                    // Enable interrupt at end of cycle
    TACCR1 = BIT9;                     // About 50% to start PWM
    TACCTL1 = OUTMOD_7;                // Reset/set for positive PWM
    TACTL = TASSEL_2 | ID_3 | MC_1 | TACLR; // SMCLK/8, Up mode, clear
// ADC on, VCC and VSS refs, sample for 4 cycs, int ref off, no ints
    ADC10CTL0 = SREF_0 | ADC10SHT_0 | ADC10ON;
// Input channel 1, start using ADC10SC bit, no clock division,
//    internal ADC clock, single channel single conversions
    ADC10CTL1 = INCH_1 | SHS_0 | ADC10DIV_0 | ADC10SSEL_0 | CONSEQ_0;
    ADC10AE0 = BIT1;                   // Enable analog input on channel 1
    ADC10CTL0_bit.ENC = 1;             // Enable conversions
    for (;;) {                         // Loop forever taking measurements
        __low_power_mode_0();          // All action in ISR for CCIFG0
    }
}
// -------------------------------------------------------------------
// ISR for CCIFG0, once per PWM cycle: update PWM, start new ADC reading
// -------------------------------------------------------------------
#pragma vector = TIMERA0_VECTOR
__interrupt void TIMERA0_ISR (void) // Flag cleared automatically
{
    TACCR1 = ADC10MEM;                 // Update PWM from last ADC reading
    ADC10CTL0_bit.ADC10SC = 1;         // Trigger next conversion
}
```

9.6.2   Repeated Conversions with the ADC10

Alternatively we could trigger conversions from channel 0 of Timer_A directly. Remember that this uses the OUT0 signal, not the flag CCIFG0 nor the internal "match" signal EQU0. This is a problem because the only way of generating a periodic output from channel 0 in the Up mode is to use the Toggle output mode, which halves the frequency. Thus we can update the PWM only on alternate cycles without going to a lot more trouble.

It would be best to use one of the repeated conversion modes of the ADC10 to avoid having to toggle ENC before every conversion. Three bits control the mode. Here is how they work for hardware triggers; it is a little different if ADC10SC is used:

- **CONSEQ0** selects between a single input channel (0) or a sequence of input channels (1).

- **CONSEQ1** selects between single conversions (0), each of which must be enabled by toggling ENC, or repeated conversions (1), where only the trigger is needed each time.

- **MSC** selects whether each conversion should be individually triggered (0) or they should follow each other as rapidly as possible, without waiting for further triggers after the start of the sequence (1).

Table 9.2: Options for triggering single and repeated conversions with the ADC10 from hardware (timer) and software (ADC10SC).

Trigger source	CONSEQ1 = 0	CONSEQ1 =1	
	MSC = x	MSC = 0	MSC = 1
Hardware	Single conversions, individually triggered (toggle ENC every time)	Repeated conversions, individually triggered (toggle ENC only before start)	Multiple conversions, only first triggered
Software	Repeated conversions, individually triggered (toggle ENC only before start or with ADC10SC in all cases)		Multiple conversions, only first triggered

When a sequence of input channels is selected with the CONSEQ0 bit, the ADC10 first converts the input channel selected by INCHx, then decrements the channel until it reaches A0. I do not consider this mode further.

I summarize the behavior in Table 9.2 for a single input channel. Here we want each conversion to be triggered by OUT0 so MSC = 0. We do not want to toggle ENC before each conversion and therefore set CONSEQ1 = 1 for repeated conversions.

Each result is written to ADC10MEM as soon as it is completed, even with repeated conversions. There is nothing to stop a value being overwritten before it has been processed, nor is there any warning in the ADC10 (the ADC12 has an overrun flag). The MSC bit therefore needs to be used with particular care. It is most useful with the data transfer controller, which I describe shortly.

Listing 9.6 shows this in action. The ADC10 is configured for repeated conversions triggered by OUT0. It requests interrupts and the result is copied to TACCR1 in the interrupt service routine.

Listing 9.6: Repeated measurements of external potential to control PWM output in ad10pwm2.c. The ADC10 is triggered directly by the OUT0 signal from Timer_A and the PWM is updated in the ISR for the ADC10.

```
// ad10pwm2.c - PWM controlled by input voltage using ADC10
// ADC10 def refs (VSS, VCC), int clock, auto disabled between readings
// Repeated conversions initiated by OUT0 in toggle mode so only on
//   alternate PWM cycles; PWM updated on ADC10 interrupt
// Homemade analog demo board for MSP430F20xx:
//   10k on P1.5, 20k variable R on P1.6, junction to P1.1 with 10nF cap
//   LED active high on P2.6 driven from OUT1 in PWM
// Default DCO, no crystal, no ACLK, power from JTAG
// J H Davies, 2007-08-30; IAR Kickstart version 3.42A
```

```
//-------------------------------------------------------------
#include <io430x20x2.h>              // Header file for this device
#include <intrinsics.h>             // Intrinsic functions

void main (void)
{
    WDTCTL = WDTPW | WDTHOLD;        // Stop watchdog
    P2SEL = BIT6;                    // OUT1 on P2.6; no crystal
    P2OUT = 0;                       // Unused outputs driven low
    P2DIR = BIT6|BIT7;
    P1OUT = BIT5;                    // VCC to 10k fixed resistor
    P1DIR = BIT3|BIT5|BIT6|BIT7;     // Unused outputs driven low
// Timer_A for PWM at 125Hz on OUT1, SMCLK/8, Up mode
    TACCR0 = BITA - 1;               // Upper limit to match ADC10MEM
    TACCTL0 = OUTMOD_4;              // Toggle OUT0 to stimulate ADC10
    TACCR1 = BIT9;                   // About 50% to start PWM
    TACCTL1 = OUTMOD_7;              // Reset/set for positive PWM
    TACTL = TASSEL_2 | ID_3 | MC_1 | TACLR; // SMCLK/8, Up mode, clear
// ADC on, refs VCC, VSS, sample 4 cycles, int ref off, interrupts
    ADC10CTL0 = SREF_0 | ADC10SHT_0 | ADC10ON | ADC10IE;
// Input channel 1, start on rising edge of OUT0, no clock division,
//   internal ADC clock, single channel, repeated conversions
    ADC10CTL1 = INCH_1 | SHS_2 | ADC10DIV_0 | ADC10SSEL_0 | CONSEQ_2;
    ADC10AE0 = BIT1;                 // Enable analog input on channel 1
    ADC10CTL0_bit.ENC = 1;           // Enable ADC10 conversions
    for (;;) {                       // Loop forever taking measurements
        __low_power_mode_0();
    }
}
//-------------------------------------------------------------
// Interrupt service routine for ADC10: update PWM with new reading
//-------------------------------------------------------------
#pragma vector = ADC10_VECTOR
__interrupt void ADC10_ISR (void)    // Acknowledged automatically
{
    TACCR1 = ADC10MEM;               // Update PWM from new ADC reading
}
```

9.6.3   Data Transfer Controller in the ADC10

The program in Listing 9.6 works well but it seems pointless to restart the CPU simply to copy a word from ADC10MEM to TACCR1. Surely it would be more efficient and consistent with the low-power philosophy of the MSP430, if the ADC10 could write its result directly to TACCR1. This can indeed be done with the ADC10's *data transfer controller* (DTC). It provides a simple form of direct memory access (DMA) and the CPU is halted if necessary to avoid contention for the memory buses. The DTC uses MCLK, which may need to be restarted from the DCO if the device is in a low-power mode; it is turned off again afterward. (The storage of results is handled quite differently in the ADC12.)

The DTC allows results to be written to a single address or block of memory, anywhere in the MSP430's address space. The destination is often in RAM but can also be a peripheral register or even flash memory. Three registers control the operation of the DTC:

- **ADC10DTC0** selects options for one or two blocks in the destination and for continuous transfers. The idea behind the two-block mode is that you can process one block of data while the other is being filled.

- **ADC10DTC1** gives the number of transfers in each block. A value of 0 disables the DTC.

- **ADC10SA** is the start address for the destination. You must write to ADC10SA to enable transfers, much like setting ENC for conversions.

It is simple to configure the DTC for the PWM example. We set up ADC10DTC0 for continuous transfers, so that every conversion output is transferred, and for a single block because the destination will be only a single address. This also needs ADC10DTC1 = 1. Finally, the DTC is enabled by writing the address of TACCR1 to ADC10SA; this needs a cast to unsigned short to match the definition of ADC10SA. Casts from addresses to integers should always be done with caution. Luckily it is straightforward in the original MSP430 because everything is 16 bits wide. This is not true of the extended MSP430X, which can use 20-bit addresses. The few changes to the previous program are shown in Listing 9.7. The MSP430 now runs without the CPU at all after everything has been configured.

Listing 9.7: Changes in `ad10pwm3.c` to listing 9.6 to copy the converted result from the ADC10 to TACCR1 with the DTC. No interrupts are needed nor is the CPU after the peripherals have been configured and started.

```
// Data transfer controller: copy one result continuously to TACCR1
    ADC10DTC0 = ADC10CT;            // Continuous transfers, one block
    ADC10DTC1 = 1;                  // Single target address in memory
    ADC10SA = (unsigned short) &TACCR1; // (Starting) target address
    ADC10CTL0_bit.ENC = 1;          // Enable ADC10 conversions
    for (;;) {                      // Loop forever taking measurements
        __low_power_mode_0();       // CPU no longer needed,
    }                               //   not even for interrupts
}
```

9.6.4 Storing a Block of Readings from the ADC10 with the DTC

Another application of the DTC is to store a block of readings for subsequent processing as a batch. For example, you might wish to average a set of readings to reduce the effect of noise or to increase the effective resolution of the ADC. Listing 9.8 shows how to do this for the potential divider and LED. It replaces the single reading in Listing 9.2 with an average over 16 readings (the parameter NSAMPLES). The program takes blocks of readings continuously for simplicity but there would usually be a delay between blocks as in Listing 9.3. The 16 conversions should be made one immediately after the other, without waiting for individual triggers, so I set the MSC bit in ADC10CTL0. I also enable interrupts, which behave slightly differently when the DTC is active: An interrupt is requested at the end of the block transferred by the DTC, rather than after each conversion.

The DTC is configured for transferring a single block once. This is its default but I write the value for clarity. The number of samples goes in ADC10DTC1 and need not be repeated. However, the starting address must be written to ADC10SA before each block is converted because this is the trigger that reenables the DTC for a new set of transfers. The name of the array is the address of its first element in C, which is just what we want.

Within the main loop, each set of conversions is triggered by setting the ENC and ADC10SC bits together—recall that this is permitted for software triggers. The CPU then enters a low-power mode while the conversions take place. I could have used LPM3 but the DCO would need to be restarted every time the DTC transferred a word, so LPM0 seems more appropriate. The ADC10 requests an interrupt when the block has been filled and the MSP430 returns to the main routine in active mode. The ADC10 does not stop automatically when the DTC has filled its destination so I cleared the ENC bit. This allows the current conversion to complete normally but disables further conversions.

The average is computed in the usual way by summing the values in the array samples[] and dividing by NSAMPLES. I initialize average to NSAMPLES/2 to make a small correction for the bias introduced by integer division, as described in the section "Temperature Sensor on the ADC10" on page 420. Incidentally, the quotient NSAMPLES/2 is evaluated by the compiler, so the MSP430 need not do it.

Listing 9.8: Use of the DTC to store a block of readings from the ADC10 in ad10led5.c. An LED is turned on if the average reading is greater than $0.5\,V_{CC}$.

```
// ad10led5.c - measure input voltage from pot divider using ADC10
// Average a block of readings accumulated using DTC
// ADC10 default refs (VSS, VCC), internal clock, interrupt after block
// Homemade analog demo board for MSP430F20xx:
//    10k on P1.5, 20k variable on P1.6, junction to P1.1 with 10nF cap
//    LED active high on P2.6
// Default DCO, no crystal, no ACLK, power from JTAG
```

```
// J H Davies, 2007-09-10; IAR Kickstart version 3.42A
// ------------------------------------------------------------
#include <io430x20x2.h>          // Header file for this device
#include <intrinsics.h>          // Intrinsic functions
#include <stdint.h>              // Standard integer types

#define LED      P2OUT_bit.P2OUT_6   // Output bit for LED
#define NSAMPLES   16                // Number of samples in each block

void main (void)
{
    uint16_t samples[NSAMPLES];  // Block of samples from ADC10
    uint32_t average;            // Average value of block
    uint16_t i;                  // Loop counter

    WDTCTL = WDTPW | WDTHOLD;     // Stop watchdog
    P2SEL = 0;                    // Digital i/o rather than crystal
    P2OUT = 0;
    P2DIR = BIT6|BIT7;            // Unused outputs driven low
    P1OUT = BIT5;                 // VCC to 10k fixed resistor
    P1DIR = BIT3|BIT5|BIT6|BIT7;  // Unused outputs driven low
// VCC and VSS refs, sample 4 cycs, multiple conversions, interrupts
    ADC10CTL0 = SREF_0 | ADC10SHT_0 | MSC | ADC10ON | ADC10IE;
// Input channel 1, trigger using ADC10SC bit, simple binary,
//   internal osc undivided, single channel repeat conversions
    ADC10CTL1 = INCH_1 | SHS_0 | ADC10DIV_0 | ADC10SSEL_0 | CONSEQ_2;
    ADC10AE0 = BIT1;              // Enable analog input on channel 1
    ADC10DTC0 = 0;                // Transfer single block (default)
    ADC10DTC1 = NSAMPLES;         // Number of readings in block
    for (;;) {                    // Loop forever taking measurements
        ADC10SA = (unsigned short) samples; // Start address, enable DTC
        ADC10CTL0 |= (ENC|ADC10SC); // Enable ADC10, start conversions
        __low_power_mode_0();     // Wait for batch to complete
        ADC10CTL0_bit.ENC = 0;    // Disable further ADC10 conversions
        average = NSAMPLES/2;     // To reduce bias when dividing
        for (i = 0; i < NSAMPLES; ++i) {
            average += samples[i]; // Accumulate sum
        }
        average /= NSAMPLES;      // Arithmetic mean of samples
        if (average > BIT9) {     // Is output in top half of range?
            LED = 1;              // Yes: Turn LED on
        } else {
            LED = 0;              // No: Turn LED off
        }
    }
}
// ------------------------------------------------------------
// ISR for ADC10 at end of block: return to active mode
// ------------------------------------------------------------
#pragma vector = ADC10_VECTOR
__interrupt void ADC10_ISR (void)   // Acknowledged automatically
{
    __low_power_mode_off_on_exit(); // Return to active mode on exit
}
```

This program gives a chance to look at the performance of the ADC10 by examining samples[] in the debugger. Typically 14 of the 16 readings are identical and the others differ by only 1 or 2. I am impressed by these results given that my circuit is constructed on stripboard, which is not a wise choice for high-quality analog electronics. Of course the 10nF capacitor on the analog input helps a lot by suppressing noise.

I spotted an interesting systematic error while debugging, which I mention because it is a good reminder that a 10-bit ADC is sensitive to changes of about 1/1000, which are small. At first I set a breakpoint in the main loop, so that the program stopped after each block of readings. I then found that the first few values in samples[] were always 1 count lower than the later values. This disappeared if I removed the breakpoint and stopped the program at random. I suspect that this is caused by a small change in V_{CC} when the debugger releases the MSP430. My demonstration board is powered by the FET for convenience but a better power supply should be used for mixed-signal systems.

9.7   The ADC12 Successive-Approximation ADC

I do not describe the ADC12 in detail because it works on the same principles as the ADC10. However, some aspects of its operation are different and I pick these out. Finally, I show a simple example with the TI MSP430FG4618/F2013 Experimenter's Board.

The successive-approximation core and the choices of inputs, clocks, and triggering sources are much the same as the ADC10. The same selection of reference voltages is available but are implemented differently. Here are the principal distinctions between the ADC12 and ADC10:

- The output has 12 bits rather than 10. That is obvious but this is a good place to remind you about the importance of good analog design to avoid noise and errors if you want the lower bits to be meaningful.

- The higher precision places a stricter requirement on the sampling time as well, which should be extended to 9τ. Also, the input capacitance is higher at 40pF.

- The internal voltage reference requires an external storage capacitor. This takes a long time to charge after the reference has been turned on and 17ms should be allowed for the voltage to stabilize. That's *milliseconds*, not microseconds as for the reference and buffer in the ADC10. The capacitor is not required if you use V_{SS} and V_{CC} as references.

- The sampling time can be controlled in two ways. The first, called *pulse mode*, is the same as in the ADC10 with a selected number of cycles of ADC12CLK but a wider range of times is available. Alternatively, the sampling time can be controlled directly by the SAMPCON input from the signal that triggers the conversion. This is called the *extended sample mode*.

- Analog inputs are enabled with PnSEL rather than a separate analog enable register.

- Conversions are specified and the results stored in an entirely different way. The ADC12 has 16 memories, ADC12MEMn. Each of these has an associated control register ADC12MCTLn, which sets the references and input channel for that conversion. The two CONSEQx bits select single or multiple conversions and single or multiple channels as for the ADC10. The first memory to be used in a sequence, or the only memory for a single channel, is specified by the CSTARTADDx bits. For a sequence, the ADC12 works its way upward through the conversion memories until it encounters an end-of-sequence (EOS) bit in a control register.

- No data transfer controller is built into the ADC12 but results can be moved from the ADC12MEMn registers by the separate direct memory access controller.

- A single interrupt vector is shared by 18 flags. There is one for each conversion memory plus two overrun flags. The first is set when a new result is written to a memory before the previous one is read and the other is set when a new conversion is requested before the previous one has finished. The sources are decoded by reading the vector register ADC12IV in the same way as TAIV for Timer_A.

9.7.1 Temperature Sensor on the ADC12

I use the temperature sensor as an example again because there is no simple analog input on the MSP430FG4618/F2013 Experimenter's Board. There is a microphone but that requires the op-amps, which I have not yet covered. Sound also generates a large number of converted values, which are best written to flash memory using the DMA. The program takes a reading from the temperature sensor every 1 s, triggered by Basic Timer1, and displays the result on the LCD to a precision of 0.1°C. This matches the resolution of the ADC12 almost exactly as we saw in the section "Temperature Sensor on the ADC10" on page 420.

The sensor is identical to that in the ADC10 and again requires a sampling time of at least 30 μs. We have to reduce the frequency of the clock for the ADC10 to get so long a sampling time but there is a greater choice with the ADC12. I allow 256 cycles, which exceeds 40 µs even at the maximum value of $f_{ADC12OSC}$. This is 6.3 MHz, the same as for ADC10CLK.

The biggest practical difference is the settling time of the voltage reference, for which 17 ms should be allowed. This is the same problem that we addressed in the section "A Low-Power Example with the ADC10" on page 417, where the potential divider took over 500 µs to charge. The approach is the same: Turn on the voltage reference, set up a timer for the settling time and return to a low-power mode until interrupted by the timer. I chose the watchdog timer on the assumption that Timer_A and Timer_B might already be allocated to more demanding tasks. Unfortunately the delay is a fraction short at 16 ms, so I am living dangerously.

Most of the main routine of Listing 9.9 will be familiar from previous programs because it concerns the ports, LCD and Basic Timer1, which is set up for interrupts every 1 s. The ADC12 is configured like this:

- The SHT0x bits in ADC12CTL0 are set up for a sampling interval of 256 cycles of ADC12CLK. Memories 0–7 share this value, with another in the SHT1x field for memories 8–15; the sampling time cannot be configured for individual memories.

- The internal reference is left at 1.5 V, the default, but neither the reference nor the core of the ADC12 is turned on. Overrun interrupts can also be enabled in ADC12CTL0 but I have not done this.

- ADC12CTL1 needs more work. The first field specifies the first conversion memory to be used, or the only one in our case as we are not using a sequence. I chose memory 0.

- The sample-and-hold signal is taken from the ADC12SC bit, just as with the ADC10. I set the SHP bit, which enables the pulse mode for the sampling timer as in the ADC10.

- The clock is taken from the internal oscillator ADC12OSC without division.

- Finally, I select a single conversion of a single channel with the CONSEQx bits, which are similar to those in ADC10CTL1.

- Having specified that memory 0 is used for the conversion, we must set it up with the ADC12MCTL0 register. This contains the reference voltages and input channel used for the conversion. I chose V_{SS} and the internal reference V_{REF}, with channel 10 for the temperature sensor.

- Last of all, ADC12IE contains bits to enable interrupts from the 16 memories. I enable the interrupt from memory 0 to signal the end of the conversion.

The program works by a sequence of interrupts after the peripherals have been configured and the MSP430 enters LPM3. It is close to Listing 9.3 but here is a brief description.

1. The process for each reading starts with an interrupt from Basic Timer1. The voltage reference is turned on in the ISR and the watchdog timer is configured to give a further interrupt after 16ms.

2. The reference voltage should be stable by the time of the interrupt from the watchdog timer so the ADC12 is turned on and a new reading is enabled and triggered.

3. The final interrupt occurs when the conversion is complete. The ADC12 is disabled and turned off, as is the voltage reference. This interrupt vector is shared by all flags within ADC12 so its origin is checked by using the vector register ADC12IV. This step is necessary to clear the interrupt request, even though only one source is active. Alternatively I could have cleared the interrupt flag for the particular memory in the ADC12IFG register.

The temperature is calculated from the converted value in much the same way as for the ADC10. Its scaling factor is 4200 rather than 420 so that the result is in tenths of a degree rather than integral degrees and the division is by 4096 to reflect the greater resolution of the ADC12. A 32-bit variable is needed to store the intermediate values but the final result should fit in 16 bits and is therefore truncated before being passed to `DisplayDeciCels()`. This is adapted from the functions in the section "Decimal Display" on page 269 to show a decimal point and the °C symbol.

Listing 9.9: Program `exbdtmp1.c` to provide a thermometer on the TI MSP430FG4618/F2013 Experimenter's Board. It displays the temperature measured by the internal sensor on the LCD to a resolution of 0.1°C, updated every second. I omit the functions that initialize the ports and drive the LCD.

```
// exbdtmp1.c - Display temperature to 0.1 deg C using ADC12
// Measurements at 1s intervals set by basic timer, run by interrupts
// Delay for stabilizing reference set by watchdog interval timer
// TI MSP430FG4618/F2013 Experimenter's Board with F4619, default clocks
// J H Davies, 2007-09-12; IAR Kickstart version 3.42A
// -------------------------------------------------------------------
```

```
#include <io430xG46x.h>            // Specific device
#include <intrinsics.h>            // Intrinsic functions
#include <stdint.h>                // Integers of defined sizes
#include "LCDutils2.h"             // TI exp board utility functions
void main (void)
{
    WDTCTL = WDTPW | WDTHOLD;      // Stop watchdog timer
    IE1_bit.WDTIE = 1;            // Enable WDT interval interrupts
    FLL_CTL0 = XCAP14PF;          // Configure load caps (14pF)
    PortsInit();                   // Initialize ports
    LCDInit();                     // Initialize SBLCDA4
    DisplayHello();                // Display HELLO on LCD
// Set up ADC12 but leave both ADC and reference off
    ADC12CTL0 = SHT0_8;            // Sample 256 cycles, Vref+ = 1.5V
// Start with conversion 0, trigger from ADC12SC, use sample timer,
//   ADC12OSC undivided, single channel single conversion
    ADC12CTL1 = CSTARTADD_0|SHS_0|SHP|ADC12DIV_0|ADC12SSEL_0|CONSEQ_0;
// Conversion memory 0: end of seq, Vref and VSS refs, input channel 10
    ADC12MCTL0 = EOS | SREF_1 | INCH_10;
    ADC12IE = BIT0;                // Enable interrupts from memory 0
// Initialize basic timer: ACLK, cascaded, 1s interrupts, stopped
    BTCTL = BTHOLD | BTDIV | BT_fCLK2_DIV128;
    BTCNT1 = 0;                    // Clear counters
    BTCNT2 = 0;
    IE2_bit.BTIE = 1;             // Enable basic timer interrupts
    BTCTL &= ~BTHOLD;             // Start basic timer running
    for (;;) {                     // Measurements triggered by ints
        __low_power_mode_3();     // LPM3 between interrupts
    }
}
//-----------------------------------------------------------------------
// Interrupt service routine for basic timer, requested every 1s
// Enable ADC12 reference, allow time to stabilize using watchdog timer
//-----------------------------------------------------------------------
#pragma vector = BASICTIMER_VECTOR
__interrupt void BASICTIMER_ISR (void)  // Acknowledged automatically
{
    ADC12CTL0_bit.REFON = 1;      // Turn ADC12 reference on
// Start watchdog as interval timer, 512 cycles of ACLK, clear
    WDTCTL = WDTPW | WDTTMSEL | WDTCNTCL | WDTSSEL | WDTIS1;
}
//-----------------------------------------------------------------------
// Wait for reference voltage to stabilize
//-----------------------------------------------------------------------
// ISR for watchdog interval timer: start ADC12 conversion
//-----------------------------------------------------------------------
#pragma vector = WDT_VECTOR
__interrupt void WDT_ISR (void)       // Acknowledged automatically
{
    WDTCTL = WDTPW | WDTHOLD;      // Stop watchdog timer
    ADC12CTL0_bit.ADC12ON = 1;    // Turn ADC12 on
    ADC12CTL0 |= (ENC|ADC12SC);   // Enable, start single conversion
}
//-----------------------------------------------------------------------
// Wait for ADC12 to complete conversion
//-----------------------------------------------------------------------
```

```
// Interrupt service routine for ADC12, called when conversion complete
// Calculate temperature in tenths of celsius
//--------------------------------------------------------------------
#pragma vector = ADC12_VECTOR
__interrupt void ADC12_ISR (void)      // Flag NOT cleared automatically
{
    uint32_t temperature;              // Converted value of temperature

    ADC12CTL0_bit.ENC = 0;             // Disable ADC12 before turning off
    ADC12CTL0_bit.ADC12ON = 0;         // Turn ADC12 off
    ADC12CTL0_bit.REFON = 0;           // Turn ADC12 reference off
    if (ADC12IV != 0x0006) {           // Should never happen
//      for (;;) {                     // Loop here forever as diagnostic
//      }
    } else {                           // New temperature reading available
        temperature = ADC12MEM0;       // Raw converted value
        temperature *= 4200;           // Scaling factor (Vref/temp coeff)
        temperature += 2048;           // Allow for rounding in division
        temperature /= 4096;           // Divide by range of ADC10
        temperature -= 2780;           // Subtract offset to give celsius
        DisplayDeciCels ((int16_t) temperature);    // Display on LCD
    }
}
```

I found that the display indicated 13.7°C when the actual temperature was 22.2°C according to a very old mercury-in-glass thermometer, which was originally calibrated to this accuracy. This is an example where the precision of the measurement, 0.1°C, is much better than the accuracy of about ±9°C. The reading flickered by ±0.1°C, which showed that all bits from the ADC12 were stable except the least significant bit. The FG4618 is in a thinner package than the PDIP F2002 and responds readily to a warm finger. (I tried to warm the device further by breathing on it but this produced so much condensation that it stopped the watch crystal oscillator and ACLK.)

Several application notes show how to use and extend the ADC10 and ADC12:

- *A Simple Glass Breakage Detector Using the MSP430* (slaa351). This uses the ADC10 and op-amps integrated into the F2274 to digitize the sound picked up by a microphone. The major feature is the sophisticated digital filtering of the results in software.

- *Implementing a Smoke Detector with the MSP430F2012* (slaa335). An infrared LED and photodiode are used to detect the presence of smoke. The signal is too small to be converted directly so an external op-amp is required. Averages over four readings are taken to reduce noise, with each set stored directly to memory using the DTC.

- *Oversampling the ADC12 for Higher Resolution* (slaa323). This shows how an average can be taken over a block of measurements to improve the resolution of an ADC. You can gain roughly an extra *m* bits of resolution by averaging over 4^m samples. This technique is called *oversampling* and is exploited in sigma–delta ADCs.

- *Low-Power Tilt Sensor Using the MSP430F2012* (slaa309). A two-axis accelerometer is interfaced directly to a F2012, which illuminates six LEDs to show the direction in which the PCB is tilted. The calculation of battery lifetime is instructive. Over 80% of the current goes into the LEDs, with the MSP430 taking less than 0.5%. The expected lifetime is about 1300 hours of activity.

- *Li-Ion Battery Charger Solution Using the MSP430* (slaa287). An ADC10 is used to monitor the voltage across the battery, the charging current, and the temperature of the battery, sensed by a thermistor. Charging goes through two stages, one at constant current followed by one at constant voltage, before the battery is fully charged. The temperature sensor is needed to detect overheating, in which case the battery might catch fire.

- *Solid State Voice Recorder Using Flash MSP430* (slaa123). The sound is digitized by an ADC12 triggered by Timer_B to avoid jitter in the sampling clock. The sound is recorded in flash memory. Several external components are needed to complete the system but could be eliminated by selecting a newer device such as the FG4618.

9.8   Analog-to-Digital Conversion: Sigma–Delta

The operation of sigma–delta ADCs is entirely different from successive-approximation ADCs. Their distinctive characteristics are high precision and low speed, which makes them ideal for many applications in metering. The sigma–delta ADCs in the MSP430 work at around 1 ksps rather than 100 ksps for the SAR ADCs. Unfortunately their operation is far more sophisticated than SAR ADCs. I avoid most of the mathematics but a picture of their operation is vital to understand how to use a sigma–delta ADC. Bonnie Baker recently wrote a brief series on *Delta–Sigma ADCs in a Nutshell*, which you might like to read [62]. You can see that even the name is not standardized: It is often delta–sigma instead of sigma–delta. Greek letters are frequently used as abbreviations to give $\Sigma\Delta$ or $\Delta\Sigma$.

The basic idea behind a sigma–delta converter is to reduce the ADC itself to the simplest circuit possible. This is a 1-bit ADC—no more than a comparator—which must take

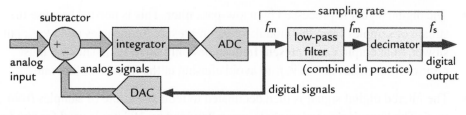

Figure 9.18: Block diagram of a sigma–delta ADC. The loop forms the sigma–delta modulator, which is followed by a digital filter.

samples at a much faster rate than the desired output to compensate for its poor resolution. It therefore produces a very rapid stream of single-bit samples, whose mean value represents the magnitude of the analog input. These bits are processed digitally to extract their average and reduce the frequency of samples to the desired rate.

9.8.1   Architecture of a Sigma–Delta ADC

Figure 9.18 shows the main blocks of a sigma–delta ADC. It falls into two main parts. The first is a feedback loop that forms the sigma–delta modulator. It handles the analog signals and does the basic analog-to-digital conversion. Its output is a stream of single bits (zeros and ones) at a high rate f_m called the *modulator* or *oversampling frequency*. This is much faster than the final frequency f_s at which samples are produced and the ratio f_m/f_s is called the *oversampling ratio*, OSR. This is a critical parameter for a sigma–delta ADC. Typical values for the MSP430 are $f_m = 1\,\text{Msps}$ and OSR = 256, giving $f_s = 4\,\text{ksps}$. Here is a little more detail of the modulator:

- The analog input goes into a difference amplifier, which subtracts (hence "delta") the current value of the output to leave the error.

- This error is integrated (much the same as summation, hence "sigma").

- The output of the integrator is converted from analog to digital in an ADC at f_m. This is performed by a 1-bit ADC, which is just a comparator.

- This digital signal is fed back and converted back to analog in a DAC so that it can be subtracted from the input, thus forming a feedback loop. The 1-bit DAC is no more than a switch.

The second part of the ADC handles purely digital signals. Its job is to take the fast stream of single bits from the modulator and convert them to a slower stream of multibit values. In principle this is done in two stages:

- The digital signal is processed by a low-pass filter. This is needed because the stream of samples from the modulator can represent frequencies up to $\frac{1}{2}f_m$ but the slower, final output can represent frequencies only up to $\frac{1}{2}f_s$. Thus we must remove frequencies above $\frac{1}{2}f_s$ to avoid aliasing at the final sampling rate.

- The filtered digital signal is then decimated to reduce the rate of samples from f_m to f_s. The term *decimated* ought to mean "divided by 10" but is used for any factor in practice.

The effective number of bits per sample rises in both these steps, which is why the lines get thicker in Figure 9.18. In practice the two operations are combined.

The simplest way of filtering (smoothing) the data and reducing the rate of samples by a factor of OSR is to calculate the average of each set of OSR samples. Suppose that OSR = 16, for instance. The sum of 16 single bits can range from 0–16, which needs four bits to represent it (ignoring one value). More generally the result needs m bits if OSR = 2^m. Thus we have turned a rapid stream of single bits into a slower stream of multibit values.

Unfortunately not all of the m bits are useful. If you have done any statistics, you will know that the standard deviation of the mean of a sample with N measurements goes down as $1/\sqrt{N}$. It is the same here and only $\frac{1}{2}m$ bits of the average are meaningful; the rest are too noisy. Sigma–delta ADCs have two ingenious features to get around this limitation. First, the stream of bits from the modulator is not random, which defeats the factor of $1/\sqrt{N}$. Second, better algorithms than a simple average are more effective at extracting the mean value of the data.

This process is illustrated in Figure 9.19 for OSR = 8. The first plot (a) shows a random sequence of zeros and ones with an average value of 0.69. The runs of zeros and ones have random lengths, some long and some short. The line with fewer steps shows the effect of averaging successive blocks of eight bits to perform the filtering and decimation. The large fluctuations are obvious.

In contrast, there are no long runs of zeros or ones in Figure 9.19(b), the output of the sigma–delta modulator: They are now all short. In other words, the fluctuations are as rapid as possible while keeping the same average. This is called *noise shaping* and it is not hard to believe that a more precise average can be obtained from this signal. The output has also been filtered more effectively than a simple average (I used a sinc^2 filter, which I explain shortly). The combination of the two improvements gives a spectacularly better output, whose variations from sample to sample are barely visible on the plot.

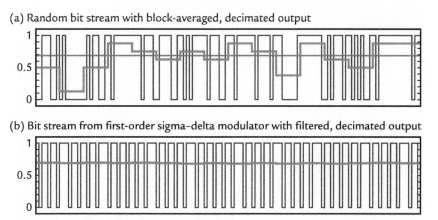

(a) Random bit stream with block-averaged, decimated output

(b) Bit stream from first-order sigma–delta modulator with filtered, decimated output

Figure 9.19: Comparison of (a) simple averaging of a random stream of bits with an average value of 0.69 and (b) a first-order sigma–delta modulator with an input of 0.69 and its filtered, decimated output. The oversampling ratio OSR = 8.

The circuit in Figure 9.18 is called a *first-order sigma–delta modulator*. More complicated feedback loops can be constructed and general-purpose converters, such as those in the MSP430, typically have second-order sigma–delta modulators. These give more useful bits for the same oversampling ratio. However, their output needs better processing than a simple average, so we look at the filtering process next.

9.8.2 Digital Filters in Sigma–Delta ADCs

The two blocks on the right-hand side of the sigma–delta ADC shown in Figure 9.18, the low-pass filter and decimator, represent operations carried out on digital signals. These signals are just sequences of numbers and the operations are calculations, which are designed to modify the input as a function of its frequency. They are therefore called *digital filters* because their function is similar to classic analog filters. They clearly work in a quite different manner from circuits with resistors and capacitors. Nonetheless, we can analyze their operation in similar ways. For example, we can put a sine wave on the input and measure the amplitude and phase of the output as a function of frequency.

I described a simple digital filter earlier, which averaged each block of OSR bits. It is also called a *sinc filter* because of its frequency response, where the sinc() function is defined by $\mathrm{sinc}(x) = \sin(x)/x$. It has the limit $\mathrm{sinc}(x) \to 1$ as $x \to 0$ and goes to 0 because of the

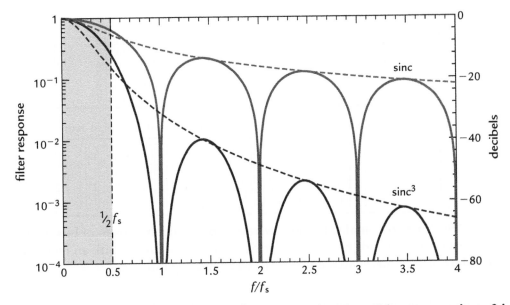

Figure 9.20: Response of sinc and sinc³ filters as a function of frequency, where f_s is the final sampling rate and ½f_s is the highest frequency before aliasing occurs. The dashed curves are the responses of simple *RC* low-pass filters that fall off in the same way at high frequency.

sine function when $x = n\pi$ for $n \neq 0$. The frequency response of the simple averager or sinc filter is roughly

$$h(f) \approx \text{sinc}\left(\frac{\pi f}{f_s}\right). \tag{9.30}$$

This is plotted in Figure 9.20. It depends only on the final sampling frequency f_s, not the modulator frequency. (I dropped a denominator that depends on the modulator frequency because it is usually unimportant.) The response is quite different from the gentle, monotonic fall of amplitude with frequency of a simple *RC* low-pass filter, shown as the upper dashed curve. Instead, there are deep notches at multiples of the sampling frequency. This type of response is therefore called a *comb* filter. The notches ideally go right down to 0, meaning that signals of frequency $f = nf_s$ are perfectly attenuated—not transmitted at all.

The notches arise because the averaging covers an exact number of cycles of these frequencies and the average of a sine wave over a complete period is 0. For example,

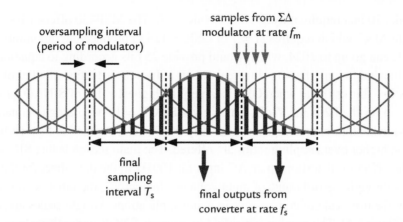

Figure 9.21: Weighted average taking with a sinc³ filter over three intervals at the final sampling rate f_s.

suppose that $f_s = 50\,\text{Hz}$, in which case the averaging is over a period $T_s = 20\,\text{ms}$. Exactly one cycle of a 50 Hz signal fits in this period, or two cycles of a 100 Hz wave, and so on. All of these therefore average to 0. This feature can be used to suppress interference from the 50 Hz mains.

Better performance can be obtained if the averaging is spread over more than a single block of OSR samples from the modulator. For a second-order sigma–delta converter, such as that in the MSP430, the averaging is usually spread over three OSR samples, which also means three periods of the output or a time $3T_s$. This is called a sinc³ filter. It is not a simple average, which weights all samples in the interval equally. Instead the samples are weighted as shown in Figure 9.21. I also plot its response in Figure 9.20. It falls off much more quickly than the simple sinc function, which means that it is better at removing unwanted signals at higher frequency. These unwanted signals include the fluctuations in the stream of bits from the modulator, so the sinc³ filter extracts a more accurate estimate of the average. (The sinc² filter that I use for Figure 9.19 lies between these two.)

9.8.3 The Final Result from a Sigma–Delta ADC

A *rough* estimate of the number of useful bits produced by an ideal second-order modulator with a sinc³ filter, the type of sigma–delta ADC in the MSP430, is

$$N_{\text{eff}} \approx \tfrac{5}{2}m \quad \text{with} \quad \text{OSR} = 2^m. \tag{9.31}$$

For example, 10 bits require only $m = 4$ or OSR $= 16$. The MSP430 offers a 16-bit sigma–delta ADC, which requires $m = 7$ or OSR $= 128$ to produce the full number of bits. In fact OSR can go up to 1024, which would provide 25 bits according to equation (9.31). However, inevitable imperfections in a real circuit reduce the effective number of bits (ENOB) below this estimate when OSR is large. The highest value of SINAD quoted in the data sheet for the F20x3 is 87 dB, which gives ENOB $= 14$ according to equation (9.20). Given this figure, you might wonder why the SD16 is claimed to be a 16-bit ADC and why the higher oversampling ratios are offered. The main reason is that SINAD includes the effect of distortion on an AC input (at 100 Hz in the data sheet for the F20x3) that sweeps through the full range of input voltages. In contrast, the input is almost constant on the timescale of conversions in many applications. Weight scales are a good example (Figure 1.4). The number of useful bits exceeds ENOB under these less demanding conditions and may even surpass 16. Perhaps this will become clearer after I have shown some practical results in Figure 9.27. I also return to this topic in the section "Weighing Machine" on page 480.

The vital point is that the number of useful bits from the sigma–delta ADCs in the MSP430 is under the control of the user through the oversampling ratio. It is *not* fixed, as for a SAR ADC. A 16-bit sigma–delta ADC produces 16 useful bits only with a sufficiently high oversampling ratio. At the same time, a large value of OSR reduces the rate at which the final values are produced because there is an upper limit on the modulator frequency f_m.

There is another important difference between SAR and sigma–delta ADCs. Suppose that you provide each with a perfectly constant input voltage and convert this to a digital value repeatedly. You would expect to get exactly the same value every time with a SAR ADC. However, the less significant bits in the output from a sigma–delta ADC *always* fluctuate (although they may not normally be visible). This is true even for an ideal converter, where the number of steady bits is given roughly by equation (9.31). This is an inevitable consequence of the way in which sigma–delta ADCs work as I demonstrate in the section "Operation of SD16_A" on page 459.

9.8.4   Other Features of Sigma–Delta Converters

Sigma–delta converters have a few peculiar features. Some of these are helpful but others may prove to be pitfalls.

Input Characteristics

The input looks like a capacitance in series with a resistance as for SAR ADCs, described in the section "Practical Issues with SARs" on page 405. Again we must check carefully that there is sufficient time for the capacitance to charge. The input is switched at a high frequency (f_m), which means that it looks like a resistance *on average* and this may be quoted on the data sheet. Do not be misled by this into neglecting the capacitance.

Antialiasing Performance

The plot in Figure 9.20 shows that the $sinc^3$ filter does not provide particularly good antialiasing. We want a filter whose amplitude is reasonably flat at low frequency before falling rapidly to a low value for $f \geq \frac{1}{2} f_s$. Unfortunately the curve is far from this ideal and falls only to -12dB at $\frac{1}{2} f_s$, which is feeble attenuation.

Differential Inputs

Most sigma–delta ADCs have *differential* inputs. This means that the ADC converts the difference in voltage between a pair of inputs, $\Delta V = V_+ - V_-$, rather than the voltage between a single input and ground. Tie V_- to ground if this feature is not wanted. However, it is often helpful and gets the best performance from the ADC. For example, the weighing machine back in Figure 1.4 uses a bridge for its sensor. This gives differential outputs naturally, which could be connected directly to a sigma–delta ADC. Of course there must be sufficient gain, which brings us to the next point.

Programmable Gain Amplifier

Sigma–delta ADCs often provide a *programmable gain amplifier* (PGA) on their inputs, which may eliminate the need for an external op-amp. Beware that these amplifiers are nothing like a classic op-amp with feedback resistors. They amplify voltage by using charges and capacitors rather than currents and voltages and their input is a capacitance, not a resistance. Do not expect a high input impedance as if there were a traditional instrumentation amplifier on the input. A separate buffer may be provided to boost the input impedance.

Latency

An unfortunate side effect of the digital filter arises because each output value depends on an average over input samples. Suppose that the input changes abruptly from one constant

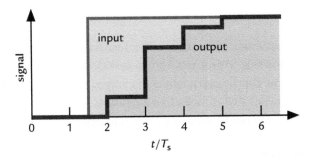

Figure 9.22: Response of a sigma–delta ADC with a sinc³ filter to a step input. It takes up to four intervals of the sampling time T_s to reach the final value.

value to another at some time. The output of a sinc³ filter does not react to this fully until at least *three* periods of T_s have passed, because it takes that long to fill the filter with the new value. A change usually occurs part way through a period so you must wait four periods for a reliable value as shown in Figure 9.22. This, called the *latency* of the converter, is one of the less pleasant features of sigma–delta ADCs. It is a particular nuisance with multiplexed inputs. The MSP430 offers an option to ignore output values before the fourth, which eliminates the problem at the cost of a delay.

Frequency Response

It follows from the slow response in time that the frequency response is also poor. Suppose that you feed a sine wave into the ADC and keep its amplitude constant while raising its frequency (below ½ f_s of course). The amplitude of the output falls roughly as

$$h(f) \approx 1 - \frac{\pi^2}{2}\left(\frac{f}{f_s}\right)^6. \tag{9.32}$$

This can be important, given that sigma–delta converters are used for high precision. You want good accuracy too. You can see this drop in Figure 9.20.

9.9 The SD16_A Sigma–Delta ADC

Fortunately it is easier to use a sigma–delta ADC than to understand its operation. The MSP430 currently offers three varieties of 16-bit sigma–delta ADC:

- The original module was the SD16, which contains three independent channels. In effect there are three complete ADCs, each with a single input. They are linked with logic so that they can sample their inputs simultaneously or with a specified delay. The F(E)42x provides this module.

- This was followed by the SD16_A, which has only a single core but a multiplexer on the input. It can therefore convert the different inputs sequentially but not simultaneously. Other enhancements include finer control over the frequency of the modulator, a high-impedance analog buffer, increased oversampling range, and an internal potential divider for measuring V_{CC}. This is now sometimes called the SD16_A1 for clarity.

- The latest module is a combination of these two. It has several SD16_A cores, each with a single input, linked by control logic. It is available only in the F47xx at the time of writing.

Most of this description applies to all versions but I pick out the SD16_A1 in detail. It is currently included in the F20x3 and F(G)42x0 although there are differences between the two implementations. In particular, the F20x3 lacks the analog input buffer.

9.9.1 Architecture of the SD16_A

Figure 9.23 shows a slightly simplified block diagram of the SD16_A. Its features are controlled by memory-mapped registers in the usual way. The names of many of the registers have a 0 at the end because they are duplicated for each channel in SD16s with multiple converters.

Input Channels

A multiplexer selects one of the eight possible input channels according to the SD16INCHx bits. The most important feature is that the inputs are *differential pairs*: The SD16_A converts the difference $\Delta V = V_+ - V_-$ rather than the potential between a single input and ground. You should ground the V_- input if your signal is single-ended rather than differential. This can be done internally, but a differential signal gives the best performance by using the full dynamic range of the converter.

Three of the channels are internal:

- A potential divider gives $\Delta V_5 = (V_{CC} - V_{SS})/11$ to monitor the supply voltage. This is particularly useful to estimate the state of a battery.

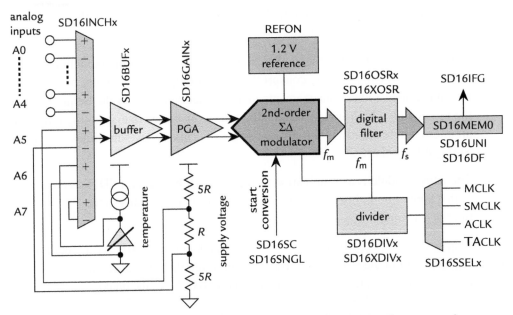

Figure 9.23: Block diagram of the SD16_A sigma–delta ADC. The external connections for the reference voltage and some aspects of the interrupts are omitted for clarity.

- Channel 6 is connected to a temperature sensor. It is very similar to that in the ADC10 and ADC12 but has a reduced coefficient because the SD16 works over a smaller range. The equation is

$$V_{\text{temp}} = 1.32\theta + 360 = 1.32(\theta + 274) \text{ mV},\qquad(9.33)$$

where the temperature θ is in celsius. The sensor is again of limited accuracy and may need calibration.

- The inputs to channel 7 are short-circuited together. This allows you to measure the offset voltage of the system (buffer, PGA, and sigma–delta converter) and subtract it from measured values.

Input Pins

The multiplexer selects the input channel but does not enable the connection between the SD16_A and the pins. This is done in several ways:

- Some larger devices have pins whose only function is as analog inputs to the SD16, so no configuration is needed.

- The negative multiplexed inputs A_{n-} are internally grounded by default. This is convenient for single-ended measurements: Activate the pin for the positive input but do nothing for the negative input to the channel.

- The bits in the analog enable register SD16AE enable input to the SD16 and override the settings in the port register. This is similar to the ADC10. Note that the bits in SD16AE correspond to pins in a port (port 1 in the F20x3 and F42x0), not to the input channels of the SD16. Thus two pins must be enabled for a differential input.

- Unfortunately this may not give enough pins. There are five possible input channels, which need ten pins. This problem is avoided in the F20x3 because two pins, P1.3 and P1.4, can be connected to one of two input channels. For example, P1.1 can be used as either A_{0-} or A_{4+}. However, the F42x0 has ten independent pins, four of which are in port 6. Their analog function is enabled with P6SEL as with the ADC12.

An antialiasing filter should be provided on the analog inputs. The family user's guide recommends a cutoff frequency of 10 KHz or below for $f_m = 1$ MHz and OSR = 256, which are typical conditions. Often the input varies so slowly that a lower frequency can be used, which also reduces noise.

High-Impedance Buffer

A high-impedance buffer can be selected if the signal does not already come from a source with a low impedance, such as an amplifier. This is like the classic voltage follower circuit with an op-amp except that it has both differential inputs and outputs. The buffer is not included in the F20x3. It can run at different currents, selected with the SD16BUFx bits, which should be chosen to match the speed of the modulator.

Programmable Gain Amplifier

The programmable gain amplifier or PGA offers a gain of 1–32 in powers of 2, chosen with the SD16GAINx bits. As I emphasized earlier, it uses switched capacitors rather than a traditional op-amp and resistors. This is important for its input characteristics, which I discuss in the next section.

The gains are not particularly accurate except for unity (no gain). For example, the 32 setting has a specified gain of 28.4 ± 1.4 in the F20x3. There may also be a significant offset voltage but this can be corrected with the aid of the short-circuit input channel 7. I show an example of this in the section "Differential Input to SD16_A" on page 465.

Reference Voltage

There is an internal 1.2 V reference, enabled with SD16REFON. It can be used by itself but you are advised to connect a 100 nF capacitor between the VREF pin and ground. The pin is not routed to the SD16 by default. You must configure it using the port registers if you wish to use the capacitor (set P1SEL.3 on the F20x3 but leave SD16AE.3 clear). It takes about 5 ms for the capacitor to charge and the voltage to stabilize.

The initial accuracy is specified as 1.20 ± 0.06 V, which is a little better than the reference in the ADC10 but is poor compared with a discrete voltage reference. On the other hand, the stability with temperature is greatly improved at 20 ppm/°C. This is better than a basic, discrete device.

The reference has an output buffer, which should be turned on with SD16VMIDON if you wish to use the reference voltage outside the MSP430. It can supply up to ± 1 mA and the capacitance should be increased to 470 nF. You might think that this would make the voltage even slower to stabilize but in fact it is much faster, only 100 μs, because the buffer provides more current than the reference itself. The buffer can be used to charge the capacitor quickly, then turned off if there is no extra load.

An external reference voltage between 1.0 and 1.5 V can also be used, in which case both the SD16REFON and SD16VMIDON bits should be clear. Note that V_{CC} is *not* available as an internal reference. You can use it by providing an external potential divider with good decoupling to remove noise.

Sigma–Delta Converter

I said plenty about this in the previous section. The SD16 has a second-order modulator with a sinc^3 digital filter. The oversampling ratio is set by the SD16OSRx bits in the SD16, which gives OSR = 32, 64, 128, and 256. The SD16_A has a further bit SD16XOSR, which can be set to give increased ratios of 512 or 1024.

The converter has a low-power mode, selected with the SD16LP bit. This halves the maximum frequency of the modulator to 0.5 MHz but has little effect on the current consumption except for high gains in the PGA. It is worthwhile only if the clock has to be slow for other considerations.

The core is shut down automatically to save current when it is not being used and does not need a bit like ADC10ON to enable it. The voltage reference is handled separately and must be enabled and disabled from software.

Clock

The clock can be taken from MCLK, SMCLK, ACLK, or unusually, the external input TACLK, which would normally go to Timer_A. The SD16 has no internal oscillator. The minimum frequency is 30 KHz, so ACLK might just be usable. The maximum frequency is 1.1 MHz, reduced to 0.5 MHz if SD16LP is set.

The frequency of the clock input can be divided by 2, 4, or 8 according to the SD16DIVx bits in the SD16. There is a second divider configured by SD16XDIVx in the SD16_A, which provides a further division by 3, 16, or 48. It is easy to miss the extra X in the name of the bit field. The final frequency of the clock is the same as that of the modulator, f_m, which is the rate at which the input is sampled. Remember that the rate of outputs is given by $f_s = f_m/OSR$.

One reason for desiring fine control over the frequency is to exploit the deep notches in the characteristic of the comb filter, shown in Figure 9.20. If interference from the 50-Hz or 60-Hz AC mains is a problem, for instance, you could arrange for f_s to be the same frequency or a submultiple. This is not always easy in practice and I describe another approach in the section "Differential Input to SD16_A" on page 465. If such interference is a serious problem, it might be better to select a standalone sigma–delta ADC whose design is optimized to reject 50 Hz or 60 Hz. Some do both, such are the wonders of digital filters.

Interrupts

There is a single interrupt vector with two sources in the SD16_A1. Interrupts are maskable and each source must be enabled in the usual way by setting the corresponding bits in the control registers:

- **SD16IFG** is set when a new converted result is available in SD16MEM0. There is a flag for each converter channel in larger SD16s.

- **SD16OVIFG** is set when an overrun occurs, which means that a new value has been written to a conversion memory before the previous value was read. It does not show which channel overflowed in SD16s with multiple converters.

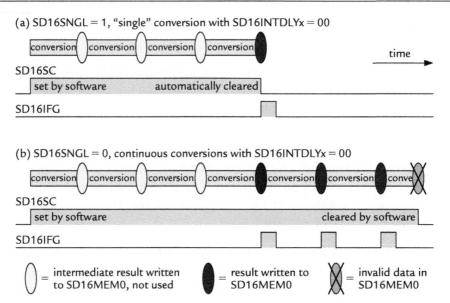

Figure 9.24: Single and continuous conversions with SD16INTDLYx = 0, so that four conversions are performed before the first interrupt is raised.

A vector register SD16IV can be used to decode the source of the interrupt in the same way as TAIV for Timer_A. However, there is an important difference: *Reading SD16IV does not clear the corresponding interrupt flag.* In practice this is not usually a problem for interrupts from SD16IFG because this flag is cleared when the conversion memory is read. (Remember that the debugger clears the flag if it reads the value to display it, which can cause the usual difficulties.) SD16IFG can also be cleared with software and this is the only option for SD16OVIFG.

SD16IFG interrupts are normally requested when each new result is produced. However, Figure 9.22 showed that sigma–delta converters exhibit latency because of the time required for a new signal to pass through the digital filter. You should therefore discard the first few values after the input channel or gain are changed. This can be done with the SD16INTDLYx bits. Interrupts are delayed until the fourth value is available if both bits are clear. This sequence is illustrated in Figure 9.24 and is the default. More generally, the first interrupt is requested after $(4 - \text{SD16INTDLYx})$ conversions.

This is particularly important if multiplexed inputs are used. You must wait until the third or fourth value appears in SD16MEM0 before the output fully reflects the new input channel. The same is true if the gain is changed. The delay specified with SD16INTDLYx therefore restarts if the input channel or gain are changed.

Warning: There is a bug (SDA3) in current versions of the SD16_A. This causes incorrect results to be produced if SD16INTDLYx is changed from its default value of 00, which gives a delay of four conversions. Leave it alone. The good news is that this bug saves you from problems introduced by latency.

Conversion Trigger

The SD16 can produce either "single" or continuous conversions, depending on the SD16SNGL bit (I explain the quotation marks shortly). Its default state is clear for continuous conversions. The two modes are illustrated in Figure 9.24:

Single conversions: Started by setting the start conversion bit, SD16SC. The SD16 performs the number of conversions specified by the SD16INTDLYx bits, raises the interrupt flag SD16IFG, clears SD16SC, and stops. Thus SD16SC is left ready to trigger another conversion. This sequence shows that the nomenclature *single conversion* is slightly misleading because the final result in SD16MEM0 is not usually the result of only a single conversion, but depends on SD16INTDLYx. In fact the SDA3 bug means that you should always allow four conversions in the SD16_A before using the result. This behavior is not entirely clear in the family user's guide at the time of writing.

Continuous conversions: Also started by setting SD16SC but in this case they continue until SD16SC is cleared in software. Conversions stop *immediately*: The SD16 does not complete the current conversion, which means that a false value may be left in SD16MEM0. Read the last value that you want before clearing SD16SC. (I found that continuous conversions can be stopped cleanly by setting the SD16SNGL bit instead of clearing SD16SC. The SD16 halts when SD16IFG is next raised, leaves a valid result in SD16MEM0 and clears SD16SC as for a single conversion. However, this is not mentioned in the family user's guide so there may be an undesirable side effect that I overlooked.) The first interrupt is delayed according to SD16INTDLYx but subsequent interrupts are requested after each conversion—the delay is applied only once unless the input channel or gain is changed.

Triggering is much less elaborate than the mechanism for the SAR ADCs and no internal connection allows conversions to be started from a timer. The SD16 is much slower than the ADC10 and ADC12 so such accurate triggering is less important. On the other hand, the interval between continuous conversions is precisely synchronized with the clock used for the SD16. SD16s with multiple converters also offer options to group the channels for synchronized conversions.

Supply Voltage

I mention this because the SD16_A needs $V_{CC} \geq 2.5\,$V in the F20x3, a more restricted range than many other peripherals. The specification for the SD16 is even more stringent at $V_{CC} \geq 2.7\,$V.

9.9.2   Range of Input Voltages and Output Formats for SD16

The relation between the analog input and digital output is a little different from the discussion in the section "Resolution, Precision, and Accuracy" on page 393 because of the differential inputs. The SD16 converts the difference $\Delta V = V_+ - V_-$ to a 16-bit digital value, which is written to SD16MEM0. This difference may be of either sign and the input accepts equal positive and negative ranges. The output can be modified if the difference is always positive but you should aim to use the full range with both signs to get the best performance from the SD16.

The range of inputs is set as usual by the voltage reference but this is modified by the gain G_{PGA} of the programmable gain amplifer. The differential input voltage must lie between $\pm V_{FSD}$, which is defined by

$$V_{FSD} = \frac{\frac{1}{2}V_{ref}}{G_{PGA}}. \tag{9.34}$$

Note the factor of ½ in front of V_{ref}: The total range of differential voltages is given by V_{ref}, but this includes equal positive and negative halves.

For the internal reference of 1.2V and $G_{PGA} = 1$, the full-scale range is $V_{FSD} = 0.6\,$V. Thus the full range of inputs to the ADC is $|\Delta V| \leq 0.6\,$V. However, the specified performance is obtained only for $|\Delta V| < 80\%\ V_{FSD}$ or about 0.5V in this case. The change in input to give a change in output of a single bit is LSB $= 1.2\text{V}/2^{16} = 18\,\mu\text{V}$.

The full-scale range is reduced in proportion as the gain is increased, so that $V_{FSD} = 18.75\,$mV for $G_{PGA} = 32$. This is illustrated in Figure 9.25 for $G_{PGA} = 1$ and 2 and the internal reference. It is worth calculating LSB at the maximum gain: LSB $= 37.5\text{mV}/2^{16} = 0.6\,\mu\text{V}$. This is sensitive indeed.

Ideally the converter responds only to the difference between its inputs, so that $V_+ = 2.0\,$V and $V_- = 1.5\,$V should give the same output as $V_+ = 0.5\,$V and $V_- = 0.0\,$V. In other words, it should reject the common-mode part of the signals. However, both voltages should be

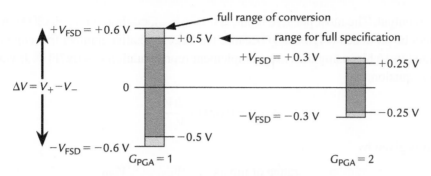

Figure 9.25: Range of inputs to the SD16 using its internal 1.2 V reference. The programmable gain amplifier is set to $G_{PGA} = 1$ and 2.

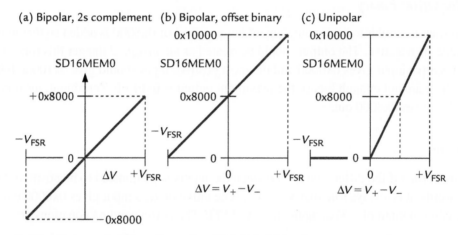

Figure 9.26: Output formats from the SD16: (a) bipolar twos complement, (b) bipolar offset binary, and (c) unipolar.

kept within the range of $AV_{SS} - 0.1\,V$ to AV_{CC} and it is a good idea to keep the common-mode voltage near the middle so that $\frac{1}{2}(V_- + V_+) \approx \frac{1}{2}(AV_{SS} + AV_{CC})$.

The digital output can be represented in three formats, sketched in Figure 9.26. These are selected with the SD16DF and SD16UNI bits.

Bipolar Twos Complement

In this format the output can take both signs, like the analog input, and must be treated as a *signed* integer. The most negative input, $\Delta V = -V_{FSD}$, gives $-0x8000$ and zero input gives zero output. The most positive input, $\Delta V = +V_{FSD}$, ought to give $+0x8000$. As usual this cannot be produced (Figure 9.9) because the largest positive number that can be represented in 16 bits using the twos-complement representation is $+0x7FFF$. If you prefer an equation,

$$N_{ADC} = 0x8000\,\frac{\Delta V}{V_{FSD}}. \tag{9.35}$$

The LSB is given by

$$LSB = \frac{\text{range of inputs}}{\text{range of outputs}} = \frac{2V_{FSD}}{2^{16}} = \frac{V_{FSD}}{0x8000}. \tag{9.36}$$

Bipolar Offset Binary

This format is similar to twos complement but an offset of $0x8000$ is added so that none of the values is negative. The output should be treated as an *unsigned* integer this time. The most negative input gives $0x0000$ and the most positive input should give $0x10000$ but again there are not enough bits so the maximum output is $0xFFFF$. Zero input, meaning $V_+ = V_-$, gives $0x8000$ out.

Unipolar

Use this format if the difference ΔV between the inputs is guaranteed to be positive. In other words, it is always true that $V_+ > V_-$. A negative or zero input gives $0x0000$ out and the maximum input of $+V_{FSD}$ again gives $0xFFFF$. The corresponding equations are

$$N_{ADC} = 0x10000\,\frac{\Delta V}{V_{FSD}} \quad \text{and} \quad LSB = \frac{V_{FSD}}{0x10000}. \tag{9.37}$$

You can see from the equations and from Figure 9.26(c) that the slope is doubled in unipolar format. Essentially this format is the same as twos complement but with the value shifted left by 1 bit (multiplied by 2) because the most significant bit is always clear for a positive number in twos-complement notation. This means that 1 bit of accuracy has been thrown away in unipolar mode by using only half the possible range of inputs.

The digital filter actually produces more than 16 bits of output. For example, there are 30 bits with OSR = 1024. By default the 16 most significant bits are written to SD16MEM0 and usually these are all that you want. If desired, the less significant bits can be read into SD16MEM0 by setting the SD16LSBACC bit. The family user's guide contains diagrams to show which bits are available as a function of OSR.

9.9.3 Input Characteristics of SD16

The input can be modeled as a capacitance in series with a resistance. You are probably used to this by now. The internal resistance is $R_I \approx 1 \, k\Omega$, which arises from the multiplexer and other analog switches as usual. The effective resistance of the source must be added to this and I take the circuit in Figure 9.17 as the example again. A difference from the SAR ADCs is that the capacitance depends on the gain of the PGA and I look at the two extreme cases. In both examples the modulator runs at $f_m = 1 \, MHz$, near its maximum frequency, because this is both the worst case and how the SD16 is usually operated.

Gain = 1

In this case the input capacitance is $C_I = 1.25 \, pF$ (it is called C_S in the family user's guide but I use the same notation as for the ADC10). The time-constant of the internal components alone is $1 \, k\Omega \times 1.25 \, pF \approx 1 \, ns$ so you might instantly assume that this is too short to worry about. Not always so—remember that this is a 16-bit converter and the criteria for charging are therefore more stringent than for the ADC10 or ADC12.

Take the usual criterion that the error should be smaller than 0.5 LSB. This means that the capacitor should charge to within $2^{-17} \approx e^{-12}$ of its full value. We must therefore allow at least 12 time-constants for charging. The capacitor charges for half of each period of the modulator clock, so the period must obey $1/f_m > 24\tau \approx 30 \, ns$. This is safely below the $1 \, \mu s$ period of the 1 MHz modulator clock.

Turning the criterion around, the time-constant must be less than $1/(24 f_m) \approx 40 \, ns$. If we include an external resistance R_s, the time-constant rises to $\tau = (R_s + R_I)C_I$. This shows that the external resistance must obey $R_s < 30 \, k\Omega$ (I round all the numbers to one significant figure). The simple test circuit with a potential divider fulfills this easily.

Gain = 32

Things get more difficult for two reasons when the gain is increased. First, we reduce the input range and thus the LSB of the converter by a factor of $32 = 2^5$. The capacitor must now be allowed to charge with an error of less than $2^{-22} \approx e^{-15}$. The period of the modulator must therefore be at least 30τ.

The second factor is that the input capacitance rises to $C_I = 10 \, pF$ for $G_{PGA} = 16$ or 32. This is one way in which the gain is increased, rather than changing the resistors as in a classic op-amp circuit. Now the resistance must obey $(R_s + R_I) \times 10 \, pF < 1/(30 f_m) \approx$

30 ns so $(R_s + R_I) < 3\,\text{k}\Omega$ and $R_s < 2\,\text{k}\Omega$. The potential divider would have too high a resistance by itself. However, the external capacitor C provides charge to the SD16_A and the time-constant reverts to $R_I C_I$ provided that $C \gg C_I$.

There are several ways of solving the problem if your source has too high a resistance to meet these conditions. The first is to use the internal, high-impedance buffer if this is available. Make sure that you configure it for the correct speed. An external buffer can be added instead. Alternatively, reduce the frequency of the modulator.

I must admit that I cut some corners in this description. The circuit of the input is slightly different from the SAR ADCs because of its differential nature. Please read the family user's guide and do the calculation there carefully if you are near the limits of accuracy. The main point is that the common-mode voltage must also be included in the calculation.

There is another way of looking at the input characteristics. Over a long interval, the repeated charging transients of the capacitor average out to a current. The input can therefore be represented by a resistance rather than a capacitance *on average*. The magnitude is $1/(f_m C_I)$ and values are tabulated in the data sheet. The worst case for the F20x3 is $75\,\text{k}\Omega$. This tells you the average loading on the external input. The resistance rises above $10\,\text{M}\Omega$ if the internal buffer is enabled.

Example 9.8

Is the average input impedance a problem for the circuit in Figure 9.17?

9.9.4   Power Consumption of SD16

The figures in the data sheets for the current drawn by the SD16 show some unexpected features. These are only rough numbers because they differ between variants of the MSP430:

- The sigma–delta core, including the PGA, draws 0.5–1.0 mA, increasing with the gain of the PGA. This is at $f_m = 1\,\text{MHz}$ and full-power mode (SD16LP = 0).

- If the clock is slowed to 0.5 MHz and the SD16LP bit is set, the current falls to 0.5–0.6 mA. There is a significant reduction for $G_{\text{PGA}} = 32$ but very little for low gain.

- The internal reference consumes about 0.2 mA and the buffer to drive the reference externally takes about the same.

- The high-impedance input buffer uses 0.2–0.5 mA. This depends on the setting chosen for its speed, which in turn depends on f_m.

These figures imply that the current drawn by the core of the SD16 is almost independent of the clock frequency f_m. This puzzles me because the analog circuit is dominated by switched capacitors and I expect that the current is roughly proportional to the frequency at which they operate. This also holds for the digital filter, which typically comprises most of the area of a sigma–delta ADC. The PGA clearly consumes a significant fraction of the current, probably in buffers that must run at "high" power to charge the internal capacitors quickly.

The conclusion is that the SD16 should be clocked as fast as possible if you want to perform a conversion with the least energy. You may be able to reduce the current by choosing a slower clock, but the longer duration of the conversion means that more charge flows in total. On the other hand, the settings that reduce current are useful if the frequency of the clock is determined by other factors.

For example, you may wish to reduce interference from the 50 Hz AC mains in Europe by averaging over a complete cycle of 20 ms. This requires $f_s = 50$ Hz. If OSR $= 256$, then $f_m = \mathrm{OSR}\, f_s = 12{,}800$ Hz. This is below the minimum specified frequency of 30 KHz and is therefore not acceptable. We could extend the duration of each measurement by increasing OSR to 1024, which requires $f_m = 51.2$ KHz. This is now within the specification but is still very slow, so you should set the SD16LP bit and run the input buffer (if used) at minimum current. One problem remains, which is how to get a clock frequency of 51.2 KHz. This is tricky, even with the extended clock dividers in the SD16_A.

9.10 Operation of SD16_A

I give a few examples of the SD16_A in action. The first input is a straightforward voltage from a potential divider. This is a single-ended signal and I therefore chose differential inputs for the next example. The final case is a cute application to make a self-dimming LED.

9.10.1 Simple, Single-Ended Input to SD16_A

Listing 9.10 shows a straightforward program to take a block of independent readings with the SD16_A in a F2003. The circuit is the same as in Figure 9.17a, with a potential divider connected to P1.1. This pin can be used as either A_{0-} or A_{4+} and is routed to the SD16_A by setting the SD16AE.1 bit. The measurement is single-ended rather than differential so I use channel 4 and ground its negative input A_{4-}. This can be taken from pin P1.2 so I ground it internally by keeping SD16AE.2 clear.

Listing 9.10: Program `sd16led3.c` to store 32 independent readings from the SD16_A in an array and light an LED if the result is in the upper half of the input range.

```
// sd16led3.c - lights LED for high input voltage using SD16A
// SD16A channel A4+ on P1.1, A4- int gnd, "single" convs (actually 4)
// Store a block of readings for statistical analysis
// Homemade analog demo board for MSP430F20xx:
//   10k on P1.5, 20k variable on P1.6, junction to P1.1 with 10nF cap
//   100n cap on P1.3 for Vref, LED active high on P2.6
// Calibrated 1MHz DCO, no crystal, no ACLK, power from JTAG
// J H Davies, 2007-10-08; IAR Kickstart version 3.42A
//-------------------------------------------------------------
#include <io430x20x3.h>            // Header file for this device
#include <intrinsics.h>           // Intrinsic functions
#include <stdint.h>               // Standard integer types

#define LED       P2OUT_bit.P2OUT_6   // Output bit for LED
#define NSAMPLES   32             // Number of samples in each block
__no_init uint16_t samples[NSAMPLES];   // Block of samples from SD16A

void main (void)
{
    uint32_t average;            // Average value of block
    volatile uint16_t i;         // Loop counter

    WDTCTL = WDTPW | WDTHOLD;     // Stop watchdog
    BCSCTL1 = CALBC1_1MHZ;        // Calibrated range for DCO
    DCOCTL = CALDCO_1MHZ;         // Calibrated tap and modulation
    P2SEL = 0;                    // Digital i/o rather than crystal
    P2OUT = 0;
    P2DIR = BIT6|BIT7;            // Unused outputs driven low
    P1OUT = BIT5;                 // VCC to 10k fixed resistor
    P1DIR = BIT3|BIT5|BIT6|BIT7;  // Unused outputs driven low
    P1SEL = BIT3;                 // Enable Vref output to capacitor
// Configure SD16A: clock from SMCLK, no division, internal reference on
    SD16CTL = SD16XDIV_0 | SD16DIV_0 | SD16SSEL_1 | SD16REFON;
// Unipolar output, "single" convs, OSR = 256, interrupts on finish
    SD16CCTL0 = SD16UNI | SD16SNGL | SD16OSR_256 | SD16IE;
// PGA gain = 1, inpt channel A4+/-, interrupt and stop after 4th result
    SD16INCTL0 = SD16GAIN_1 | SD16INCH_4 | SD16INTDLY_0;
// Enable only P1.1 for A4+; leaves A4- connected internally to VSS
    SD16AE = SD16AE1;
    for (i = 0xFFFF; i > 0; --i) {  // Delay to allow Vref to stabilize
    }
    for (;;) {                    // Loop forever
        for (i = 0; i < NSAMPLES; ++i) {   // Loop taking measurements
            SD16CCTL0_bit.SD16SC = 1;   // Trigger single conversion
            __low_power_mode_0();   // Sleep during conversion
            samples[i] = SD16MEM0;  // Store new data for analysis
        }
```

```
            average = NSAMPLES/2;        // To reduce bias when dividing
            for (i = 0; i < NSAMPLES; ++i) {
                average+= samples[i];    // Accumulate sum
            }
            average /= NSAMPLES;         // Arithmetic mean of samples
            if (average >= BITF) {       // Is average in top half of range?
                LED = 1;                 // Yes: Turn LED on
            } else {
                LED = 0;                 // No: Turn LED off
            }
        }
    }
}
//--------------------------------------------------------------------
// Minimal ISR for SD16A: clear flag and return in active mode
//--------------------------------------------------------------------
#pragma vector = SD16_VECTOR
__interrupt void SD16_ISR (void)         // NOT acknowledged automatically
{
    SD16CCTL0_bit.SD16IFG = 0;           // Clear flag, acknowledge interrupt
    __low_power_mode_off_on_exit();      // Return to active mode
}
```

You should almost always provide an antialiasing filter on the input to an ADC. This circuit is one of the exceptions because the input changes only as fast as I can twiddle the potentiometer, which is *very* slow on the timescale of a microcontroller. However, the capacitor and resistors act as a low-pass filter with a cutoff frequency of about 2 KHz. This is well within the specification, which calls for a maximum cutoff frequency of 10 KHz, but the main function of the filter here is to suppress noise. The capacitor would also be helpful to charge the input capacitance of the SD16_A rapidly at high values of G_{PGA}. (I use $G_{PGA} = 1$ so this is not an issue, as we found in the section "Input Characteristics of SD16" on page 457.)

A bad feature of the design is that the input voltage is proportional to V_{CC} but the reference is an absolute voltage from the internal generator. The converted value is therefore sensitive to changes in V_{CC}—it is not a ratiometric measurement. Unfortunately I had no choice with the hardware available.

Example 9.9

How could the design be improved to remove the dependence on V_{CC}?

I take a set of *independent* readings in Listing 9.10 to assess the performance of the SD16_A. Successive outputs from the converter do not meet this criterion because the digital filtering shown in Figure 9.21 gives each output a memory of the preceding two values. I therefore use "single" conversions with SD16INTDLYx at its default value of 0 so that each result follows four conversions, which ensures independence. The SD16_A is operated as follows:

- The SD16_A is configured for single conversions by setting the SD16SNGL bit.

- I enable interrupts and select the maximum delay of four values before the first interrupt with the confusingly named constant SD16INTDLY_0. This is a slight degree of overkill, given that the input is constant, and a delay of three values would have been sufficient to flush the digital filter. (In current devices you must not change SD16INTDLY_0 because of the SDA3 bug.)

- Within the loop that takes the samples, each conversion is started by setting SD12SC. The MSP430 enters LMP0 until an interrupt occurs.

- I clear the interrupt flag SD16IFG in the ISR to prevent another interrupt being requested immediately. Usually this would not be necessary because the flag is cleared when the memory SD16MEM0 is read, but I do not do this in the ISR. The MSP430 returns to active mode on exit.

- Back in the main routine, the latest value is stored in the array.

The values in the array are averaged in the usual way and an LED is turned on or off, depending on the result. I was really interested only in the converted values themselves but should explain a few other features of the program before analyzing the results:

- The block of samples are stored in the array samples[]. This is an external variable because it is defined outside the functions. The C language requires that external variables are initialized but it is pointless here so I have suppressed it with the key word __no_init.

  You might wonder why I define the array externally, rather than within main(). The reason is that this forces the compiler to allocate the array at the beginning of RAM, which makes it easy to find in the debugger. Otherwise it would probably be allocated on the stack, which makes debugging more tricky.

- I set bit 3 of P1SEL to enable the connection to an external 100nF capacitor, which suppresses noise in the reference voltage of the SD16_A. (I saw no change in the results when this is disabled but it is hardly a rigorous test.)

- The SD16_A is configured to use SMCLK without division. At first I was lazy and used the default output from the DCO at around 1 MHz. However, the data sheet shows that "around 1 MHz" means the range 0.8–1.5 MHz, which could take the

SD16_A outside its specification. I therefore use the calibration constants for 1 MHz stored in the device.

- I enabled the internal reference with SD16REFON. There is a simple software delay to allow plenty of time for the external capacitor to charge and the voltage to stabilize.

- The input is always positive so I chose unipolar output. I selected OSR = 256, which is the default, but later experiment by editing the SD16OSRx and SD16XOSR bits in the debugger.

- The range of voltage from the potential divider exceeds the maximum input range of the SD16_A without amplification so I left the gain of the PGA at its default value of 1. Again you can edit the SD16GAINx bits in the debugger to look at their effect.

Example 9.10

Edit the Listing 9.10 to use continuous rather than single conversions to fill the array. (This means that successive values are not independent.) Remember that clearing SD16SC may leave a meaningless value in SD16MEM0, so leave this until the last value has been stored.

I emphasized earlier that the output of a sigma–delta ADC is different from that of a SAR ADC when both are provided with a constant input voltage. We expect the same output every time from an ideal SAR ADC because the conversion process is deterministic and should go through exactly the same steps. In contrast, a sigma–delta ADC works by averaging a noisy stream of bits and there are always fluctuations if we look deeply enough into the output (assuming that the less significant bits are accessible). With this in mind, I plot a histogram of 128 conversions in Figure 9.27.

Look first at the data for OSR = 256. I superposed a Gaussian curve with the same mean and standard deviation as the measurements. It is a reasonable fit, given the small number of measurements. Clearly the result is not the same each time and there are several sources of fluctuations:

- Intrinsic noise from the sigma–delta process.

- Noise from the internal circuitry of the converter.

- Noise picked up from outside the MSP430 (this should be reduced by the capacitor across the input but it is a far from perfect filter).

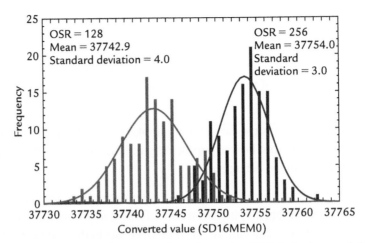

Figure 9.27: Histogram of 128 conversions from the SD16_A sigma–delta ADC with oversampling ratios of 128 and 256. The curves are Gaussian fits to the mean and standard deviation of the measurements.

The standard deviation of $3 \approx 2^{1.5}$ so we could say that there are 1½ bits of noise, leaving about $16 - 1½ = 14.5$ effective bits in the output. This is in unipolar mode, which means that half of the range has been thrown away, so we ought to add 1 bit to compensate. The effective resolution is therefore about 15½ bits for bipolar inputs. This is close to the "headline" figure of 16 bits.

The data for OSR $= 128$ show a larger standard deviation of 4, which reduces the resolution by ½ bit. This is a much smaller change than the difference of 2½ predicted by equation (9.31), which gives the effective number of bits as a function of OSR. Clearly the result is hardly limited by the sigma–delta process itself but by other sources of noise (or distortion).

The mean values of the two distributions in the plot are clearly different, which is worrying. The difference is 11 LSB, which is much larger than the standard deviations, so we cannot explain it away with statistics. Equation (9.37) shows that LSB $=$ $V_{FSD}/0x10000$ and $V_{FSD} = ½\,V_{ref} = 0.6\,V$ here, so LSB $\approx 9\,\mu V$. The means therefore differ by about $100\,\mu V$. I suspect that the difference arises from a change in input voltage caused by variation in V_{CC}, which occurs because the average current changes with OSR. To check this, I repeated the measurements using the internal temperature sensor. The difference is smaller but this test is not straightforward because the temperature of the chip may drift. This is a sharp reminder of how difficult it is to test high-resolution data converters and to get maximum accuracy from them.

I found that the results were slightly less noisy when I converted internal signals, both the temperature and V_{CC} divider. These gave 16 bits in bipolar mode for OSR = 256, which justifies the 16-bit credentials of the SD16_A. There were only 12 bits for OSR = 32, close to the prediction of equation (9.31) for $m = 5$. This shows that the resolution is limited by the sigma–delta process itself at such a low OSR rather than other sources of noise.

9.10.2   Differential Input to SD16_A

Many systems naturally provide differential inputs, often from sensors connected as a Wheatstone bridge. Two examples are the weighing machine in Figure 1.4 and the thermistor in Figure 9.2b. The differential inputs of the SD16 are ideal for handling these signals directly. Almost no extra work is needed beyond Listing 9.10. You must enable both analog inputs and in many cases the difference $\Delta V = V_+ - V_-$ can take either sign, so the format of the data in SD16MEM0 should be chosen appropriately. Not all differential inputs give both signs: The internal voltage monitor gives only positive differences, for instance. On the other hand, make sure that you choose a bipolar format to measure the input offset voltage using channel 7.

Often the voltages are small and you need to amplify them with the PGA. It may then be necessary to correct for the input offset voltage of the PGA. The problem is that the output of the PGA is not exactly 0 when it has zero (differential) input, $V_+ = V_-$. I mentioned the same issue for the comparator in the section "Slope Measurement of a Single Resistance" on page 380. The correction is made by taking a reading on channel 7, which short-circuits the inputs to the PGA, and subtracting the result from the converted value of the external voltage. In some systems this is not necessary because there is also an offset voltage from the sensor itself, which needs to be calibrated. A good example is weight scales, where the weight of the scale pan should be subtracted from the displayed value. The user presses a Tare button when there is nothing on the scales and the measured weight is subtracted from subsequent measurements. This compensates for all offset voltages.

A large oversampling ratio is often needed, both to increase resolution and to average away noise on the input. Sometimes this is not sufficient by itself and you need to take a block of readings and average them in software to improve the accuracy of the result.

I show an example of these features in Listing 9.11. It lights an LED again because there is not a lot else to do on the simple demonstration board that I use. Most of the changes from Listing 9.10 are straightforward. I use continuous conversions this time and sum the

Listing 9.11: Program `sd16dif2.c` to take the average of a block of differential readings using the SD16_A in a F20x3. The value is corrected for the input offset voltage of the PGA.

```c
// sd16dif2.c - lights LED for positive input voltage using SD16A
// SD16A channel A4+/- on P1.1/2, interrupt after continuous conversions
// Subtract PGA offset and average over block of readings
// Homemade analog demo board for MSP430F20xx:
//   10k on P1.5, 20k variable on P1.6, junction to P1.1 with 10nF cap
//   3 x 100k resistor network from VSS, VCC and P1.7, junction to P1.2
//   100n cap on P1.3 for Vref, LED active high on P2.6
// Calibrated 1MHz DCO, no crystal, no ACLK, power from JTAG
// J H Davies, 2007-10-03; IAR Kickstart version 3.42A
//------------------------------------------------------------
#include <io430x20x3.h>        // Header file for this device
#include <intrinsics.h>        // Intrinsic functions
#include <stdint.h>            // Standard integer types

#define LED     P2OUT_bit.P2OUT_6   // Output bit for LED
#define NSAMPLES   98          // Number of samples to average

void main (void)
{
    int16_t offset, difference;    // Data from SD16_A
    int32_t average;               // To accumulate average value
    uint16_t i;                    // Loop counter

    WDTCTL = WDTPW | WDTHOLD;      // Stop watchdog
    BCSCTL1 = CALBC1_1MHZ;        // Calibrated range for DCO
    DCOCTL = CALDCO_1MHZ;         // Calibrated tap and modulation
    P2SEL = 0;                    // Digital i/o rather than crystal
    P2OUT = 0;
    P2DIR = BIT6|BIT7;            // Unused outputs driven low
    P1OUT = BIT5;                 // VCC to 10k fixed resistor
    P1DIR = BIT3|BIT5|BIT6|BIT7;  // Other outputs driven low
    P1SEL = BIT3;                 // Enable Vref output to capacitor
// Configure SD16A: clock from SMCLK, no division, internal reference on
    SD16CTL = SD16XDIV_0 | SD16DIV_0 | SD16SSEL_1 | SD16REFON;
// OSR = 1024, bipolar 2s comp, continuous convs (default), interrupts
    SD16CCTL0 = SD16OSR_1024 | SD16DF | SD16IE;
    SD16AE = SD16AE1 | SD16AE2;   // Enable P1.1/2 for A4+/-
    for (;;) {                    // Loop forever taking measurements
// PGA = 32, shorted input A7 to measure PGA offset, interrupts after 4
        SD16INCTL0 = SD16GAIN_32 | SD16INCH_7 | SD16INTDLY_0;
        SD16CCTL0_bit.SD16SC = 1;  // Start new conversions
        __low_power_mode_0();      // Wait for first result
        offset = (int16_t) SD16MEM0;   // Store input offset
        SD16CCTL0_bit.SD16SC = 0;  // Halt conversions
        average = NSAMPLES/2;      // Reduce bias in division
// PGA gain = 32, external input A4+/-, interrupts after 4th conversion
        SD16INCTL0 = SD16GAIN_32 | SD16INCH_4 | SD16INTDLY_0;
        SD16CCTL0_bit.SD16SC = 1;   // Start new conversions
        for (i = 0; i < NSAMPLES; ++i) {
            __low_power_mode_0();       // Wait for next result
            average += (int16_t) SD16MEM0;  // Accumulate sum
        }
```

```
        SD16CCTL0_bit.SD16SC = 0;     // Halt conversions
        average /= NSAMPLES;          // Calculate mean
        difference = ((int16_t) average) - offset;  // Subtract offset
        if (difference > 0) {         // Is corrected output positive?
            LED = 1;                  // Yes: Turn LED on
        } else {
            LED = 0;                  // No: Turn LED off
        }
    }
}
//--------------------------------------------------------------------
// ISR for SD16A: clear flag in software (do not read SD16MEM0 here)
//--------------------------------------------------------------------
#pragma vector = SD16_VECTOR
__interrupt void SD16_ISR (void)      // NOT acknowledged automatically
{
    SD16CCTL0_bit.SD16IFG = 0;        // Acknowledge interrupt, clear flag
    __low_power_mode_off_on_exit();   // Return to active mode
}
```

results instead of storing them in an array. I was careful to read the final value from SD16MEM0 before stopping conversions by clearing the SD16SC bit.

You may have noticed one strange choice: Why did I average over 98 samples? The reason is that I was worried that the inputs might contain noise picked up from the AC mains. Differential inputs eliminate much of the interference but the SD16 is sensitive to microvolts. One way of suppressing this, which I mentioned earlier, is to choose the final sampling time T_s of the SD16 to match the period of the interference. The average of a sine wave over its period is 0, so this would eliminate the pickup. Another way of looking at this approach is that we arrange for the interfering frequency to lie in one of the notches of the comb filter in Figure 9.20.

Unfortunately the numbers are inconvenient. The frequency of the mains is 50 Hz in my part of the world, which requires $f_{SD16CLK} = 51.2$ KHz with an oversampling ratio of 1024. The divider should be 19.5 if SMCLK runs at 1 MHz. This is not available, nor could I find a combination of a calibrated frequency for the DCO and a divider that would work.

An alternative is to take samples more frequently and average them in software over one or more periods of the interference. The SD16 produces an output every 1.024 ms if $f_{SD16CLK} = 1$ MHz and OSR = 1024, so a period of the mains (20 ms) corresponds to 19.5 samples—that inconvenient number again. We could collect 39 samples for two periods instead.

A problem with this approach is that it works if the main is at 50 Hz but not where it is supplied at 60 Hz. A solution is to average over 100 ms, which includes five cycles of 50 Hz or six cycles of 60 Hz. The SD16 produces 97.7 samples in this period, which I round to 98. An obvious disadvantage is that the final output is only 10 average values per second but this is not a problem for weight scales or similar applications.

Example 9.11

Write a program for a battery voltage monitor, which lights an LED if V_{CC} falls below 3 V. Assume that there is an LED connected active high to P2.6 on a F20x3 unless you have hardware available.

Several application notes illustrate the SD16:

- *MSP430F4270 Altimeter Demo* (slaa254) was used for ATC2005. The system uses a pressure sensor with a differential output and the SD16_A in the F4270. An average over a block of readings is taken to reduce noise. A curious feature is that the readings were influenced by exposing the sensor to light. This is another reminder of the care needed to get reliable data from weak signals. An external reference voltage divided from V_{CC} is used to make the measurement ratiometric.

- *Ultra-Low Power Motion Detection Using the MSP430F2013* (slaa283) uses the differential inputs of the SD16_A in a rather different way. Both are connected through *RC* low-pass filters to the *same* output of a passive infrared (PIR) detector but the time-constants of the two filters are different. One filter is relatively fast so that it passes the signal and acts as an antialiasing filter. The other is much slower and passes only very slow variations in the offset voltage of the sensor. Thus the difference between the two inputs represents the signal with its offset voltage removed. (It is rather like a differentiator.)

- *Implementing a Single-Chip Thermocouple Interface with the MSP430x42x* (slaa216) shows how the voltage from a thermocouple can be converted directly by the SD16. Thermocouples are junctions of two dissimilar metals, which produce a voltage that varies with temperature. They are widely used in thermometers. The variation in voltage is small, around 40 μV/°C for the alloys used in the type K thermocouple described in this note. Unfortunately this coefficient varies with temperature and the voltage must be corrected for the temperature of the "cold junction" so the application is not straightforward.

9.10.3 A Self-Dimming LED on the eZ430–F2013

In many products the brightness of an LED should change in response to the level of ambient light. The relation between the brightness of the surroundings and display can vary in both directions:

- A bedside alarm clock with an LED display should be bright during daylight so that it is easy to read the time. Ideally it should become dimmer during darkness so that it does not keep you awake.

- If the display used an LCD instead, you would need to provide a backlight during darkness but could turn it off during the day.

Both applications clearly need a sensor for the ambient brightness. A dedicated sensor could of course be used but there is no need for this if you use an LED to generate the light: The LED can also act as a sensor itself. All semiconductors are sensitive to light, which is one reason why they are usually packaged in opaque material. Any diode acts like a feeble photovoltaic cell and generates a small current when light impinges on it, and this includes LEDs. Thus we can use an LED as a light sensor when it is not being used to generate light itself. The program in Listing 9.12 is based on a design ideain *Electronic Design News* [64]. A similar program was used for the MSP430 day in 2007, published as code example slac136. I chose to increase the intensity of the LED in brighter surroundings but this can easily be reversed.

How should the program be designed? The brightness of the LED is controlled by pulse-width modulation in the usual way. One possibility is to let the PWM run at a fixed duty cycle for several periods, then interrupt it to measure the intensity of the ambient light. The duty cycle is updated for the next burst of PWM. I take a different approach, which is to measure the intensity during the dark part of each cycle of PWM. This means that the duty cycle cannot be raised all the way to 100%, but the logarithmic response of the eye means that this is never noticed.

The hardware is an eZ430–F2013 stick. It has a single LED on P1.0, which cannot not be routed to an output from Timer_A so the PWM must be assisted by software as in the section "Software-Assisted PWM" on page 343. The F2013 contains a SD16_A sigma–delta ADC, which is ideal for converting the small voltages expected from the diode as a sensor.

I ran Timer_A from ACLK, which is in keeping with the low-power spirit of the MSP430 although it seems pointless when driving an LED. There is no crystal in the eZ430–F2013 so ACLK is supplied by the VLO at about 12 KHz. I like to run LEDs at a frequency of at least 100 Hz to avoid flickering, which gives about 120 clock cycles of Timer_A per period. This is rather limited resolution but could be improved only by using a faster clock.

We do not need high resolution from the ADC so I configured the SD16_A for its lowest oversampling ratio of 32. The measurement should be completed as quickly as possible and the clock is therefore taken from SMCLK at a calibrated 1 MHz. A single measurement actually requires four conversions because of the SDA3 bug in this device and therefore takes $4 \times 32 = 128\,\mu$s. This is about 1.5 cycles of ACLK, which rounds up to 2. Further time is needed for the calculation to turn the result into the next duty cycle. This requires division, which is slow. I cheated by choosing numbers that are all powers of 2 so the division can be performed rapidly by shifting. You may prefer finer control, in which case more time must be allowed for computation. I helped you a little by using the calibrated 8 MHz value for MCLK and dividing it down to 1 MHz for SMCLK. (I could have used the divider in the SD16_A instead.) The eZ430–F2013 has $V_{CC} = 3.6$ V, which permits $f_{MCLK} = 16$ MHz if you need more speed.

I have to admit that my design is a bit lax because I assume that the VLO runs at its nominal frequency of 12 KHz. In fact f_{VLO} can vary from 4 to 20 KHz over the full range of operating conditions. If the frequency is at the bottom end of the range, the PWM runs at about 30 Hz instead of 100 Hz as intended and the LED flickers annoyingly. At the other end of the range, there is insufficient time to take a reading with the SD16_A in two cycles of ACLK if the frequency is 20 KHz. The problem could be addressed by comparing the VLO with one of the built-in calibrated frequencies of the DCO and adjusting the parameters. This is demonstrated in code example slac136 but I have not done it. Sorry—bad engineering.

Returning to the SD16_A, I left its internal reference enabled continuously. It takes 5 ms to start up and each cycle of PWM takes about 10 ms so it would be possible to turn it off for about half of the time. I did not bother but it would be better to disable the reference if readings were taken less frequently. I configured the SD16_A for unipolar output because it seemed unlikely that the diode would give a negative voltage.

We need to convert the intensity to a duty cycle for PWM, which should lie in the range from 1 to PWM_MAX. I want to keep the LED on at all times, albeit faintly, which is why 0 is not permitted. The maximum is set by the number of cycles of ACLK per period of PWM and I chose 128 as the nearest power of 2. The equation is then

$$\text{dutyCycle} = \text{PWM\_MAX} \times \frac{\text{SD16MEM0} - \text{floor}}{\text{RANGE}} + 1. \qquad (9.38)$$

The diode produces a voltage even in dark surroundings, which I call `floor`, and this must be subtracted from the result of the conversion. Its value is unpredictable and the program therefore adjusts it downward whenever a new conversion has a lower value. It is initially set to a very large value, which is reduced after the first conversion. The effective value is the difference between the converted value and the floor.

The next issue is the range of inputs that correspond to the full range of duty cycles. I chose a fixed value, appropriately called RANGE, and the duty cycle takes its maximum value if the effective converted value exceeds this. It would be possible to make the upper limit self-adjusting in the same way but I decide to use a fixed range instead. The reason is that a short exposure to very bright light would render the system insensitive to weak levels of light. This is a nuisance for applications like the bedside clock.

The value of RANGE depends on the gain of the PGA and needs some experiment. I found that 16 was the maximum gain before `floor` became too large, and adjusted RANGE so that the LED reached maximum brightness at a reasonable level of indoor illumination. You may need to explore these parameters for your environment. Incidentally, a range of 4096 corresponds to a change in input of only $2\,\mathrm{mV}$ with $G_{\mathrm{PGA}} = 16$.

Timer_A runs in the Up mode as usual for PWM. Its period in TACCR0 is set by the sum of PWM_MAX, which is the maximum time for the LED to be illuminated, plus SENSE_TIME, which is the minimum dark time required for the SD16_A to take a reading. The PWM is produced by channel 1, which must request interrupts because it cannot drive the LED directly. After everything has been configured, the processor enters LPM3, with only the VLO running to provide ACLK and feed Timer_A. The action takes place in response to three interrupts per cycle of PWM, as set out in Figure 9.28. The two

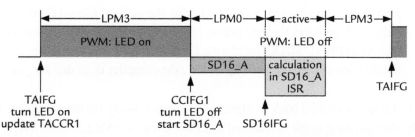

Figure 9.28: Sequence of events in each cycle of PWM in the self-dimming LED. The time required for the SD16_A and calculation are greatly exaggerated.

interrupts from Timer_A use a shared vector, which is decoded with TAIV in the usual way; CCIFG0 is not used:

TAIFG: Starts a new cycle of PWM. The LED is turned on and TACCR1 is updated from the recently calculated value of the duty cycle. It is not necessary to make the usual check for the special case of zero duty cycle because I specified that the LED should never be turned off completely.

CCIFG1: Turns the LED off to end the duty cycle, which is followed immediately by a measurement of the ambient light using the SD16_A. This could instead have been put at a fixed position in the last SENSE_TIME cycles of each period of PWM, but that would need another interrupt. The pin with the LED is switched from digital input/ output to analog input in preparation for the conversion, which is triggered by setting the SD16SC bit. A tricky feature is that the operating mode must be changed from LPM3 to LPM0 on exit from the interrupt service routine because the SD16_A requires SMCLK. I do this by using two intrinsic functions to manipulate the stacked copy of the status register: The first clears the bits for LPM3 and the second sets the bits for LPM0. This could be done in a single statement by clearing the difference between the two patterns but the two lines seem clearer.

SD16IFG: Shows that a new converted result is available. The main task is to calculate the duty cycle from this, which is done in three steps:

1. The new value is first compared with floor, which contains the lowest observed value. If the new value is lower, floor is updated and the duty cycle is set to its minimum value of 1.

2. The second test is whether the new value is so high that the duty cycle should take its maximum value.

3. Finally, the duty cycle is calculated using the general formula if the new value falls between the two limits. The parameter DIVISOR is defined as (RANGE/PWM_MAX) for clarity. I arranged for it to be a power of 2 so that the division can be replaced by a shift (the compiler does this for you).

The LED is also switched back to digital input/output, ready for the next cycle of PWM. Finally, the operating mode is returned to LPM3 because SMCLK is no longer needed. Again I use a pair of statements for clarity. The MSP430 remains in LPM3, with only Timer_A active, until the next TAIFG interrupt starts a new cycle. See Listing 9.12.

Listing 9.12: A self-dimming LED on the eZ430–F2013 provided by `eZ4301ed2.c`. The LED is driven with PWM, whose duty cycle is controlled by the ambient light. This is measured using the LED and SD16_A during dark intervals of the PWM cycle.

```c
// eZ4301ed2.c - self-dimming LED on P1.0 in eZ430-F2013 using SD16A
// PWM controlled by software, about 100Hz from ACLK = VLO,
//    SD16A measures light during off phase of PWM
// Calibrated 8MHz DCO, no crystal, ACLK from VLO, power from JTAG (SBW)
// J H Davies, 2007-09-30; IAR Kickstart version 3.42A
// -----------------------------------------------------------------
#include <io430x20x3.h>                 // Header file for this device
#include <intrinsics.h>                 // Intrinsic functions
#include <stdint.h>                     // Standard integer types

#define LED_OUT      P1OUT_bit.P1OUT_0  // Output to LED on P1.0
#define LED_ANALOG   SD16AE_bit.SD16AE0 // Enable analog input from LED
#define RANGE        4096               // Dynamic range of values from SD16
#define PWM_MAX      128                // Maximum value of duty cycle
#define DIVISOR      (RANGE/PWM_MAX)    // For converting SD16 -> PWM
#define SENSE_TIME   2                  // Cycles of TACLK needed for SD16

uint16_t dutyCycle = PWM_MAX;           // Duty cycle computed from SD16
                                        // Start at maximum (but LED off)
void main (void)
{
    WDTCTL = WDTPW | WDTHOLD;           // Stop watchdog
    BCSCTL1 = CALBC1_8MHZ;              // Calibrated range for DCO
    DCOCTL = CALDCO_8MHZ;               // Calibrated tap and modulation
    BCSCTL2 = DIVS_3;                   // SMCLK = DCO / 8 = 1MHz
    BCSCTL3 = LFXT1S_2;                 // Select ACLK from VLO (no crystal)
    P2SEL = 0;                          // Digital i/o rather than crystal
    P2REN = BIT6|BIT7;                  // Pull Rs on unused pins (6 and 7)
    P1REN = ~BIT0;                      // Pull Rs on all pins except 0
    P1DIR = BIT0;                       // To drive LED on P1.0
    LED_OUT = 0;                        // LED initially off (active high)
// Configure SD16A: clock from SMCLK, no division, internal reference on
    SD16CTL = SD16XDIV_0 | SD16DIV_0 | SD16SSEL_1 | SD16REFON;
// Unipolar, single convs, OSR = 32, interrupts on finish
    SD16CCTL0 = SD16UNI | SD16SNGL | SD16OSR_32 | SD16IE;
// PGA gain = 16, input channel A0+/-, result after 4th conversion
    SD16INCTL0 = SD16GAIN_16 | SD16INCH_0 | SD16INTDLY_0;
// Timer_A for software-assisted PWM using channel 1, up to TACCR0 mode
    TACCR0 = PWM_MAX + SENSE_TIME;      // Overall period
    TACCR1 = dutyCycle;                 // Initial duty cycle
    TACCTL1 = CCIE;                     // Interrupts on compare
// Start Timer_A from ACLK, undivided, up mode, clear, interrupts
    TACTL = TASSEL_1 | ID_0 | MC_1 | TACLR | TAIE;
    for (;;) {                          // Loop forever
        __low_power_mode_3();           // All action in interrupts
    }
}
// -----------------------------------------------------------------
// Interrupt service routine for CCIFG1 and TAIFG; share vector
// -----------------------------------------------------------------
#pragma vector = TIMERA1_VECTOR
__interrupt void TIMERA1_ISR (void) // Shared ISR for CCIFG1 and TAIFG
```

```
{
    switch (__even_in_range(TAIV, TAIV_TAIFG)) {    // Acknowledges int
    case 0:                         // No interrupt pending
        break;                      // No action
    case TAIV_TAIFG:                // TAIFG vector
// Start PWM duty cycle by turning LED on and updating duty cycle
        LED_OUT = 1;                // Turn LED on; duty cycle always > 0
        TACCR1 = dutyCycle;         // Update duty cycle from SD16 reading
        break;
    case TAIV_CCIFG1:               // CCIFG1 vector
// Finish PWM duty cycle by turning off LED, then measuring light level
        LED_OUT = 0;                // End of duty cycle: Turn off LED
        LED_ANALOG = 1;             // Switch LED to SD16A input A0+
        SD16CCTL0_bit.SD16SC = 1;   // Start SD16A conversion
// Change mode from LPM3 to LPM0 on exit to provide SMCLK for SD16
        __bic_SR_register_on_exit(LPM3_bits);
        __bis_SR_register_on_exit(LPM0_bits);
        break;
    default:                        // Should not be possible: ignore
        break;
    }
}
//-------------------------------------------------------------------
// ISR for SD16A: compute new duty cycle in range [1, PWM_MAX]
//-------------------------------------------------------------------
#pragma vector = SD16_VECTOR
__interrupt void SD16_ISR (void)    // Acknowledged when SD16MEM0 read
{
    static uint16_t floor = 0xFFFF - RANGE; // Dark reading from SD16

    LED_ANALOG = 0;                 // Switch LED back to digital output
    if (SD16MEM0 < floor) {         // Update floor if new reading is lower
        floor = SD16MEM0;
        dutyCycle = 1;              // Minimum value; never go down to 0
    } else if (SD16MEM0 >= (floor + RANGE - DIVISOR)) {
        dutyCycle = PWM_MAX;        // Maximum value (saturates)
    } else {
        dutyCycle = (SD16MEM0 - floor) / DIVISOR + 1;
    }
// Change mode from LPM0 to LPM3 on exit: SMCLK no longer needed
    __bic_SR_register_on_exit(LPM0_bits);   // (not really necessary)
    __bis_SR_register_on_exit(LPM3_bits);
}
```

It works—which I find somewhat amazing—although it is not easy to see that an LED gets brighter when you shine light on it. The most obvious problem is that the LED flickers in a dim room. This is because a change in duty cycle of ±1 is a large factor when the value is small and the eye is very sensitive to this. This could perhaps be solved by using a larger oversampling ratio to reduce noise in the sigma–delta conversion or by further filtering in software but I suspect that it really needs finer PWM.

Example 9.12

Modify Listing 9.12 so that the intensity of the LED responds to the ambient light in the opposite way: The LED gets brighter as the surroundings become darker. It is a lot easier to test this in action. You might wish to allow the LED to turn off completely in this case but need to modify the PWM to allow for a duty cycle of 0. Do not lose the interrupts that stimulate the SD16_A.

9.11 Signal Conditioning and Operational Amplifiers

All the examples that I have described in this chapter use signals connected directly to the input of the ADC. Only an antialiasing filter is needed (which I do not always show). The PGA in the SD16, coupled with the high resolution of sigma–delta converters, is often sufficient to convert small voltages directly. Of course it is not always so simple and an external amplifier or something more complicated may be needed. I look at two examples to examine whether direct connection to a MSP430 is sufficient or whether further conditioning of the signal may be needed.

9.11.1 Thermistor for the Range 5–30°C

Suppose that we wish to use a thermistor to measure temperature. I first consider the range 5–30°C with a resolution of 0.1°C. Thermistors were introduced in the section "Operation of Comparator_A+" on page 374 and equation (9.1) gives their resistance as a function of temperature. I assume that we have a 10 kΩ thermistor, which means the value of its resistance at 25°C, with $B = 3600$ K. The resistance as a function of temperature is plotted in Figure 9.29. It falls from 24 kΩ at 5°C to 8 kΩ at 30°C and the plot shows that the variation is not linear even over this narrow range.

The simplest way of getting a voltage from the thermistor is to put it in a potential divider, as in Figure 9.30(a). I chose the obvious value of 10 kΩ for the reference resistor in series with the thermistor (we come back to this) and the resulting voltage V_{in} from the divider is also plotted in Figure 9.29. A pleasant feature of this circuit is that the change in voltage is much more linear than the change in resistance. It falls from $0.705\,V_{CC}$ to $0.450\,V_{CC}$ over the range.

We want to measure a range of 25°C with a resolution of 0.1°C, which requires 250 intervals or 251 points, including both ends. This can be done in several ways. I assume

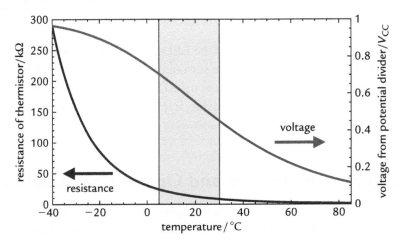

Figure 9.29: Plot of the resistance of a 10 kΩ thermistor (at 25°C) as a function of temperature and the output voltage of a potential divider with a 10 kΩ resistor.

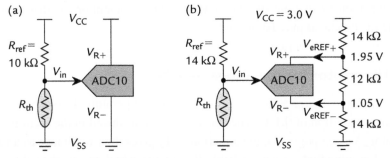

Figure 9.30: Connection of a thermistor to an ADC10. (a) Potential divider using the default references, V_R + = V_{CC} and V_R- = V_{SS}. (b) Conversion of a reduced range by using external references for both V_R+ and V_R-. I do not show the decoupling capacitors that should be connected to the inputs and external references.

first that the ADC works on inputs from V_{SS} to V_{CC} and that the voltage from the potential divider is a perfectly linear function of temperature, which is probably good enough.

- An amplifer could be used to expand and shift the spread of voltages from the potential divider so that it spans the input range of the ADC from V_{SS} to V_{CC}. (I ignore the problem that no practical amplifier can do this.) The ADC need resolve only 251 values so an 8-bit converter is sufficient. This approach was often necessary in the past, when converters with greater resolution were expensive.

- The potential divider can be connected directly to the ADC. In this case the range from $0.450\,V_{CC}$ to $0.705\,V_{CC}$ must be resolved into 250 intervals, so each interval is $\text{LSB} = (0.705 - 0.450)/250 = 0.00102\,V_{CC}$. Now we need about $1/0.00102 \approx 1000$ intervals. This implies a 10-bit ADC so the MSP430's ADC10 is a perfect fit. Of course we would waste three quarters of its full input range.

Even the simplest ADC available in the MSP430, the ADC10, has good enough performance for a direct connection. (We also have the option of the comparator for this particular application.)

I mentioned in the section "Architecture of the ADC10" on page 408 that the ADC10 has the unusual feature of reference voltages for both ends of its range, V_{R+} and V_{R-}. This can be used to restrict the range of the converter to the voltages produced by the potential divider. Before doing that, it would be a good idea to ensure that we exploit the largest possible range of voltages from the divider. This is controlled by the resistor R_{ref} in series with the thermistor, which was arbitrarily chosen to be $10\text{k}\Omega$—the specified resistance of the thermistor at 25°C. A little algebra shows that the optimum choice is $R_{ref} = (R_{min}R_{max})^{1/2} = 14\text{k}\Omega$, where R_{min} and R_{max} are the minimum and maximum values of the resistance of the thermistor over the range of operation. This makes the spread as large as possible and also places it symmetrically between V_{SS} and V_{CC}. The voltage from the potential divider now goes from $0.37\,V_{CC}$ to $0.63\,V_{CC}$; the range is only slightly increased because $10\text{k}\Omega$ was already close to optimum.

We can now use external reference voltages to restrict the ADC10 to convert values between $0.35\,V_{CC}$ and $0.65\,V_{CC}$, which allows a small margin outside the expected range of inputs. These values are generated by the potential divider shown in Figure 9.30(b). Decoupling capacitors should be provided on the external reference voltages but are not shown. A simple potential divider may not be good enough in practice because there is an internal buffer for V_{eREF+} but not for V_{eREF-}. This trick appears to improve the resolution by a factor of just over 3. Unfortunately there is a catch: The difference $V_{eREF+} - V_{eREF-} = 0.30\,V_{CC} = 0.9\,\text{V}$ but the data sheet specifies that this should be greater than $1.4\,\text{V}$ for full accuracy. Thus it is possible to improve the resolution by only a factor of about 2.

Different issues come into play if the ADC is an SD16 rather than an ADC10. Resolution is unlikely to be a problem, given 16 bits. Suppose first that we again use the thermistor in a potential divider, connect it to the positive input of the SD16, and ground the negative input. This is a single-ended measurement and the SD16 would be configured in unipolar model. The specification requires $|\Delta V| \leq 0.5\,\text{V}$ for full performance, which needs a larger

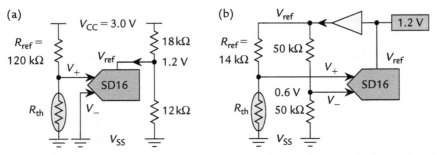

Figure 9.31: Connection of a thermistor to an SD16. (a) Single-ended measurement of a potential divider using an external reference, all supplied from V_{CC}. (b) Differential measurement to use the full range of inputs, supplied from the internal 1.2 V reference through the buffer. I do not show the decoupling capacitors that should be connected to the inputs and references.

resistance than before in series with the thermistor. A value of 120 kΩ does the job for $V_{CC} = 3.0$ V, as shown in Figure 9.31a.

The next issue is that the input is proportional to V_{CC} but the internal reference is a fixed voltage. This gives poor accuracy if V_{CC} varies: The input and references voltages should be proportional to each other to give a ratiometric measurement. One solution is to use an external reference voltage derived by a potential divider from V_{CC} as shown in Figure 9.31(a). As usual I omit the decoupling capacitors for clarity but they are essential. The reference voltage should lie in the range 1.0–1.5 V and I use 1.2 V to match the internal value. If V_{CC} varies, perhaps due to the decline in voltage of a battery during its lifetime, you should ensure that V_{ref} remains within specification at all times.

An alternative is to use the internal reference for both the ADC and the sensor. The 1.2 V reference can be driven outside the MSP430 by using its internal buffer as in Figure 9.31(b). This can provide up to 1 mA of current but do not forget the capacitor of at least 470 nF. This circuit has the advantage that the voltages automatically fall into the specified ranges but the cost is the current drawn by the reference and buffer.

I revert to $R_{ref} = 14$ kΩ in this circuit to get the maximum change in voltage from the potential divider. The measurement has been made differential by connecting V_- to a potential divider that gives the midpoint between V_{SS} and V_{CC}. This arrangement is effectively a Wheatstone bridge and makes the best use of the SD16. The differential input lies in the range $|V_+ - V_-| < 0.16$ V so the PGA could be used with a gain of 2 to improve the resolution further.

9.11.2   Thermistor for the Range −40°C to +85°C

Now suppose that the range of temperature to be measured is extended to −40°C to +85°C, the full specified range of the MSP430, with the same resolution of 0.1°C. This requires 1250 intervals so we need an ADC with at least 11 bits. You might at first be tempted to negotiate for a slightly worse resolution and get away with 10 bits but this approach is thwarted by nonlinearity. The variation in resistance of the thermistor is now huge, from 291 kΩ to 1.3 kΩ. Fortunately the potential divider still helps us greatly but the curve in Figure 9.29 is clearly not linear. It flattens near the extreme temperatures and the worst case is at low temperature. (You might not expect this because the resistance of the thermistor varies most rapidly here.) The voltage spans most of the range from V_{SS} to V_{CC} so a straightforward amplifier is not of much help, although a more complicated circuit might be used to correct for the nonlinearity. The circuit in Figure 9.30(b) to restrict the range of conversion also is useless.

Look at the worst case, which is the lowest temperature. The value of V_{in}/V_{CC} from the potential divider with $R_{ref} = 10\,k\Omega$ changes from 0.9668 at −40°C to 0.9646 at −39°C, a fractional change of 0.0022. We want to resolve 0.1°C, which is a change of 0.00022. Taking the reciprocal shows that the ADC must resolve 4570 intervals. Thus the nonlinearity increases the required resolution by a factor of 4. An ADC12 would not be quite good enough, so we might have to negotiate for a reduced resolution at low temperature. The optimum value of R_{ref} rises to 20 kΩ but does not help significantly.

We need more resolution to meet the specification fully, which means that an SD16 must be chosen instead of an ADC12. Either of the circuits in Figure 9.31 can be used. The series resistor in Figure 9.31(a) must be raised to 1.5 MΩ to keep the input voltage below 0.5 V. This is a rather high value and might bring problems of noise and leakage of current. Looking at the bridge in Figure 9.31(b), the 14 kΩ resistor should be raised to 20 kΩ to keep the range of V_+ centered on V_-. The input now lies in the range $|V_+ - V_-| < 0.52\,V$, which slightly exceeds the specification of 0.5 V. Therefore, there is a marginal loss of accuracy at extreme temperatures.

These solutions ensure that the measurement has sufficient resolution over the desired range of temperature but do not handle the nonlinear relation between input voltage and temperature. The most general method of handling this problem is to use a *lookup table*. This contains pairs of values that give the temperature for a number of selected input voltages. It is not feasible to store all possible inputs so intermediate values are treated by linear interpolation. An alternative approach is to find a mathematical function to describe the curve. This needs to be fairly straightforward in practice, certainly nothing like the exponential function in equation (9.1), but may be worth considering if a simple polynomial is satisfactory.

9.11.3   Weighing Machine

The second example is the weighing machine in Figure 1.4. Its sensor is a Wheatstone bridge of four elements, whose resistance changes when a load is placed on the scales. They are arranged so that the resistance of two opposite elements rises while the resistance of the other two falls. The differential output voltage ΔV is given by

$$\frac{\Delta V}{V_{CC}} \equiv \frac{V_+ - V_-}{V_{CC}} = \frac{R + \Delta R}{(R + \Delta R) + (R - \Delta R)} - \frac{R - \Delta R}{(R + \Delta R) + (R - \Delta R)} = \frac{\Delta R}{R}. \tag{9.39}$$

This is small compared with the individual values of V_+ and V_-, which are close to $0.5\,V_{CC}$. A typical value is $\Delta R/R = 1\%$ for a full load of 1000 g. Suppose that we wish to resolve differences of 10 g, which might be appropriate for domestic kitchen scales. This requires only 100 intervals, which sounds trivial.

This system clearly calls for a differential measurement but suppose first that we had only an ADC with a single-ended input and a range from V_{SS} to V_{CC}. In this case we would have to convert V_+ and V_- separately using two channels of the ADC and subtract the values. Now, the maximum value of the differential voltage ΔV is $V_{CC}/100$ so each input changes by half of this, or $V_{CC}/200$. We need to resolve 100 intervals, each of which is therefore $V_{CC}/20,000$. Even a 14-bit ADC is not good enough to provide a resolution of 100 values in the desired output. It is always difficult to measure a small difference between two large values accurately.

The signal needs to be amplified for this type of ADC, as shown in Figure 1.4. An *instrumentation amplifier* presents a high impedance on its two inputs and amplifies the difference in voltage between them. It is typically based on three op-amps but you need not worry about this because a broad selection of packaged instrumentation amplifiers is available. The differential gain may be fixed, set by an external resistor or programmed in some other way. Only a 7-bit ADC is needed if the gain is chosen so that output fills most of the range between V_{SS} and V_{CC}.

Better still, we might be able to avoid an external amplifier by using the differential input of an SD16 instead. The output should be bipolar because the sensors typically have a fairly large offset voltage, which may be of either sign. Suppose that we run the bridge and V_{ref} at 1.2 V using the ratiometric approach in Figure 9.31(b). The maximum value of ΔV is then 12 mV and we need to resolve 1% of this or 0.12 mV. For comparison, the LSB is 18 μV with 16-bit resolution and the PGA set for unity gain, which is smaller than the resolution required by a factor of more than 6. Thus the SD16 appears to meet the requirements easily, which shows the advantage of choosing an appropriate type of ADC for the job.

Actually the margin is not as large as it seems. Look back at the histogram of the output from the SD16_A for OSR = 256 in Figure 9.27. This shows that the effective resolution is about 15½ bits for bipolar inputs. The problem is that this is based on the standard deviation of 3, which is a rather small measure of the fluctuations. Clearly a lot of fluctuations are greater than ±3. You would not want the display on a set of kitchen scales to flicker continually and might therefore prefer a more conservative measure of the fluctuations, such as their full width (peak-to-peak value). Unfortunately this is hard to define precisely because there is always the chance of a bigger fluctuation if you wait long enough. However, a value of 16 looks reasonable for Figure 9.27. This is 2^4 so the "useful" number of bits is $16 - 4 = 12$ in unipolar mode, equivalent to 13 in bipolar mode. This is sometimes called the number of *flicker-free bits* and is about 2½ bits smaller than the effective resolution calculated from the standard deviation.

If we assume 13 flicker-free bits, the corresponding value of LSB is $(1.2 \text{V})/2^{13} = 0.15$ mV. This is slightly *larger* than the resolution of 0.12 mV required, so we are back in trouble again. Fortunately the remedy is straightforward. The input voltage from the bridge is small, $|\Delta V| \leq 12$ mV, and therefore uses only a small part of the SD16's full range of 0.5 V. We can therefore boost the signal with the PGA and make better use of the ADC. Suppose that the maximum gain is selected. This is nominally 32 but typically 28.4 according to the data sheet. The maximum input is then expanded to $12 \times 28.4 = 340$ mV and the system can resolve $(340 \text{mV})/(0.15 \text{mV}) = 2300$ values. We could now display flicker-free values of weights to the nearest 1 g instead of 10 g if desired. The gain would need to be calibrated and the offset eliminated by a tare measurement.

A more demanding specification is addressed in the application note *MSP430F42x Single Chip Weigh Scale* (slaa220). It is necessary to improve the resolution beyond 16 bits by using the extra bits produced by the digital filter and further averaging in software. The F42x contains an SD16 and drives an LCD display to integrate the complete system. It therefore integrates all the components required for the old-fashioned system shown in Figure 1.4.

9.11.4   Single-Supply Operational Amplifiers

A circuit based on an operational amplifier is needed if it is not possible to connect a signal directly to the input of an ADC. The most obvious reason is that the magnitude of the signal is too small but another common problem is that the input resistance of the ADC is too low and places an excessive load on the sensor. The input must also be capable of charging the input capacitance, particularly for SAR ADCs. Another situation is where a small differential signal must be extracted from a large common-mode voltage as in the weighing machine. Finally, the signal may need more aggressive filtering than a simple, first-order *RC* circuit can provide. I do not go into the details because they would easily fill a whole chapter. Besides, several excellent books and articles are available, listed in the section "Analog and Mixed-Signal Systems" on page 648 of "Further Reading."

To highlight just one point, systems with the MSP430 are likely to use *single-supply* op-amps. Most courses on electronics concentrate on op-amps with dual supplies and it is not trivial to change standard circuits to single supplies. For example, Figure 9.32(a) shows a classic inverting amplifier. It has dual supplies at the traditional values of $\pm 15\,$V and the signals are centered around ground at $0\,$V. The gain is given by $-R_2/R_1$ for an ideal op-amp.

Dual supplies are a nuisance in a digital system with a single, positive supply voltage and we want to use the same supply for both the analog and digital components. A wide selection of op-amps now are designed to work from a single supply of $+3\,$V, which suit the MSP430. I redrew the inverting amplifier trivially for a single supply in Figure 9.32(b). *This does not work.* The input lies between $V_{SS} = 0\,$V and V_{CC}, which is positive, so the output of an inverting amplifier should be negative. It is impossible for the op-amp to produce this if it has only a positive supply voltage. Therefore the straightforward circuit for an inverting amplifier cannot be used. This sounds obvious but it is an easy mistake to make with so familiar a circuit.

One solution is to add a bias voltage to the noninverting input of the op-amp as shown in Figure 9.32(c). This gives

$$V_{\text{out}} = 10(\tfrac{1}{2}V_{CC} - V_{\text{in}}). \tag{9.40}$$

Now the output remains safely positive for a range of inputs just below $\tfrac{1}{2}V_{CC}$.

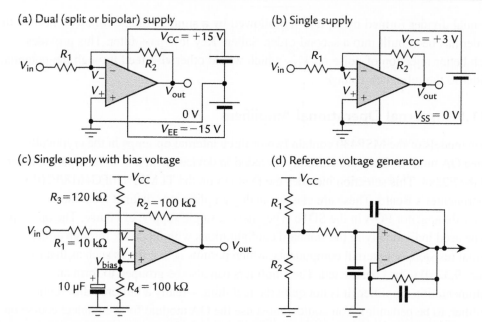

Figure 9.32: Some basic circuits using op-amps. Classic inverting amplifier using (a) dual supplies and (b) redrawn mindlessly for a single supply. (c) Inverting, single-supply amplifier with a bias voltage. (d) Reference voltage generator based on a low-pass filter.

Example 9.13

If you studied circuits with op-amps, it should be straightforward to derive equation (9.40) for the circuit in Figure 9.32(c). Make the usual assumptions for an ideal op-amp. For what range of inputs does it work correctly, taking $V_{CC} = 3\,V$?

This example shows that it may be better to concentrate on the lack of a voltage reference near the midpoint of the supplies rather than the "single supply" aspect. Another way of designing circuits is to provide such a midpoint reference and use it in the same way as ground in a circuit with dual supplies. Unfortunately a simple potential divider is rarely "stiff" enough unless it supplies only the input to an op-amp as in Figure 9.32(c). Even in this case it must be filtered with a fairly large capacitor. (Do not forget a capacitor across the power pins of the op-amp, as well.) Sometimes the reference voltage output of an ADC can be used. Either 1.2V from the SD16 or 1.5V from the ADC10 or ADC12 would be convenient. Otherwise, another op-amp must be used; Figure 9.32(d) shows a recommended circuit [46]. You might have expected to see the potential divider formed by R_1 and R_2 followed by a simple voltage follower. Instead the divider has been built into a second-order, Sallen–Key low-pass filter. This provides much better rejection of noise on V_{CC}, which would otherwise feed through the potential divider to the reference voltage.

9.11.5   Internal Operational Amplifiers

A few models of the MSP430 contain two or three internal op-amps in the cryptically named OA module. They are currently included in devices with a G in their part number and the F22x4. This selection includes the FG4618 on the TI MSP430FG4618/F2013 Experimenter's Board. These are classic, analog amplifiers as distinct from the switched-capacitor PGA in the SD16. They have a single supply, of course. The op-amps can be used individually in the standard configurations with feedback through either internal resistors or external components, which permits active filters such as that in Figure 9.32(d) to be constructed. The amplifiers can also be grouped to form an instrumentation amplifier (it is not quite the real thing—really a buffered differential amplifier, to be pedantic). You could almost use the OA module for a complete course on op-amps.

The outputs of the amplifiers in most devices have internal connections to inputs of the ADC12. They can also be brought out externally by using the pins assigned to the ADC12's inputs. All the usual, digital features of these pins remain available. This means that the output of an op-amp can generate an interrupt or capture event if the pin provides the appropriate function.

Internal connections allow the op-amps to provide the front end for an ADC, including a low-pass filter to eliminate noise and aliasing. They can also be used to filter and buffer the output of a DAC12. All these features are exploited on the MSP430FG4618/F2013 Experimenter's Board. The inputs can go from rail to rail if desired (except in the F22x4). Experts might like to know that the input range is extended by using a charge pump rather than two complementary input stages. This requires extra current and the feature can therefore be disabled if it is not needed. Similarly, there is a choice of gain–bandwidth products, which can be selected to save current when high speed is unnecessary. The output from a DAC12 can be used as the bias voltage in inverting amplifiers such as that in Figure 9.32(c).

I use an op-amp in the example for the next section but do not describe the OA module further because it is included in only a few devices and the details vary between them. The *User's Guide* for the MSP430FG4618/F2013 Experimenter's Board (slau213) illustrates the amplifiers in an audio signal chain. Another example is provided by the application note *A Simple Glass Breakage Detector Using the MSP430* (slaa351), which shows how the two op-amps in the MSP430F2274 can be used as an inverting amplifier and antialiasing filter.

9.12 Digital-to-Analog Conversion

Some MSP430s provide the DAC12 module, which is a digital-to-analog converter. Its architecture is based on a ladder of resistors and is conceptually like a potential divider with a large number of settings. Selecting a particular output requires only a particular configuration of switches; it is not like most analog-to-digital converters, where some sort of searching process is required. Thus the core of a DAC responds almost instantly to a change in digital input. However, the voltage from the potential divider must be buffered before it is delivered to the load and the speed of the amplifier limits the rate at which the output of the DAC12 can respond. As usual, greater speed comes with the cost of a higher current consumption and the DAC12 offers a selection of speeds. At maximum current the output can switch between its maximum and minimum values in about $15\,\mu s$, but this is extended to $100\,\mu s$ at the lowest current. It is much quicker to switch between adjacent output voltages, $1–5\,\mu s$.

The output voltage of an N-bit DAC is related to its digital input N_{DAC} by

$$V_{out} = \frac{N_{DAC}}{2^N} V_{FS}, \tag{9.41}$$

which is analogous to equation (9.15) for an ADC. The input must lie in the range from 0 to $(2^N - 1)$, which is 0x0000–0x0FFF for the DAC12 in 12-bit mode. This means that V_{out} cannot quite reach its full-scale value V_{FS}, just as an ADC cannot give the ideal digital output for its full-scale input voltage. The DAC12 can also be operated in an 8-bit mode with inputs of 0x00–0xFF. There is a further option for twos-complement input rather than straightforward binary. The range of inputs in the 8-bit mode is then −0x80 for zero output to +0x7F for maximum output and similarly in 12-bit mode.

A DAC needs a reference voltage, just like an ADC, and the DAC12 borrows its reference from the ADC in the same device. This is usually an ADC12 but occasionally an SD16. It also needs a trigger, which can be taken in hardware from Timer_A or B for precision. Input data can be read directly from memory using the DMA controller, again like the ADC12.

The DAC12 is used where PWM cannot provide a satisfactory analog output. Here are two contrasting examples. The first is when an absolutely constant voltage is needed, such as the bias voltage for an op-amp. The alternative would be to filter a PWM wave heavily, but this itself requires an op-amp with resistors and capacitors to act as an active filter.

The other extreme is when the output varies too rapidly for PWM. Speech is an example. The bandwidth for "telephone quality" speech—comprehensible but otherwise of minimal fidelity—is generally taken to be 300–3000 Hz and must therefore be sampled at 6 KHz to obey the Shannon theorem (equation (9.24)). Suppose that we use 12-bit samples to match the DAC12. Then each period of PWM must contain 2^{12} cycles of the timer clock, whose frequency would need to be $2^{12} \times 6\,\mathrm{KHz} \approx 25\,\mathrm{MHz}$. This is a little too fast for current versions of the MSP430 but may be possible by the time that you read this.

At present, the DAC12 is appropriate. Its output needs to be treated against aliasing in the same way as inputs to an ADC. A filter should therefore be provided to attenuate frequencies above 3 KHz. (Technically this is called a *reconstruction filter* and the topic is treated in books on digital signal processing.) Higher-quality audio systems use sigma–delta DACs to give 16 or more bits of resolution. The application note *Speech and Sound Compression/Decompression with MSP430* (slaa361) illustrates audio processing including an example for the TI MSP430FG4618/F2013 Experimenter's Board.

Occasionally you may need a DAC to produce only a small number of voltages. In this case you can build your own using resistors and ordinary digital outputs. The idea is described in Section 3.6.1 of *Application Reports* (slaa024).

9.12.1   Direct Digital Synthesis of a Sound Wave

Listing 9.13 shows direct digital synthesis of a sound wave using the DAC12 in the FG4618 on the MSP430FG4618/F2013 Experimenter's Board. This sounds fancy but is in fact quite straightforward: The DAC12 repeatedly converts a table of values stored in memory, which specify the desired waveform. I chose a pure sine wave for simplicity but the same technique can be used for any periodic signal. The system also uses an op-amp to amplify the output and provide sufficient current to drive a headphone. Successive values could be written to the DAC12 using software in an ISR but this is wasteful for so trivial a task and I instead use the direct memory access controller. This means that the sound is produced with no need for the CPU after the MSP430 has been configured. I do not describe the DMA but it works in much the same way as the ADC10's data transfer controller, which I covered in the section "Data Transfer Controller in the ADC10" on page 428. Timer_A controls the updating of the DAC12.

A single period of the signal is stored as a table in the array wave[]. It is a sine wave, which I generated simply with a spreadsheet. The const key word tells the compiler that this can be stored in ROM and need not take up precious RAM. The data are clearly redundant because of the high symmetry of the sine wave. It would be necessary to store only a quarter of a period if the update were performed in software because the rest of the wave could be computed trivially.

Here is how the modules are configured and operated. The program is an extended version of one of the code examples for the FG4618.

Operational Amplifier

The MSP430FG4618/F2013 Experimenter's Board is designed so that OA2 can drive the headphone output (curiously labeled MIC OUT). The noninverting input of the op-amp is taken from channel 0 of the DAC12 (OAP_2 bit). The signal lies between 0.0 and 1.5 V so I configure OA2 as a noninverting amplifier (OAFC_4) with a gain of 2 (OAFBR_2) to give the largest possible output. (It is still rather faint.) There is no need for rail-to-rail inputs if the maximum signal is only 1.5 V so I disable this feature to save current (OARRIP). Setting the OAADC1 bit connects the output to the A5 input of the ADC12 and brings the signal out to the corresponding pin, P6.5. I configure the op-amp for maximum speed and current (OAPM_3) to avoid distortion, although this is probably unnecessary. There is no need to choose a connection for the inverting input of the op-amp (OANx) because it is used only for feedback in this configuration.

DAC12

We use channel 0 of the DAC12, or DAC12_0 for short. I configure it in the 8-bit mode (DAC12RES), which reduces the storage required for the table. The resolution is so coarse that 12 bits seemed pointless. I use the bipolar range (DAC12DF) so that the values go positive and negative, like a familiar sine wave. The buffers run at maximum speed (DAC12AMP_7) although this may again be excessive.

Several bits are needed to configure the voltage reference. I use the 1.5 V internal reference from the ADC12 (DAC12SREF_0). The full-scale output of the DAC12 can either be equal to the reference voltage, which I chose with DAC12IR, or amplified to three times this. The reference itself must be enabled with the REFON bit in a register for the ADC12.

The output of the DAC12 can have a fairly large offset voltage and there is therefore an option for calibration. I do this by setting the DAC12CALON bit and looping until it clears to show that the process is complete. You are advised to avoid activity on the ports and CPU during calibration. This means that a low-power mode should be used, but I have not done this.

The DAC12 is triggered in hardware for precision. I chose rising edges of the OUT1 signal from Timer_A (DAC12LSEL_2). This is similar to triggers for the ADC10, described in the section "Architecture of the ADC10" on page 408. Again the trigger is the output signal, *not* the interrupt flag. The DAC12 must be enabled by setting the DAC12ENC bit before it will accept triggers. I do not do this until all the peripherals are configured.

The data register DAC12_0DAT is double-buffered when the DAC12 is triggered from hardware. This means that new values written to DAC12_0DAT are not copied into the core of the DAC until a trigger arrives. The output of the DAT is then updated and the DAC12IFG flag is raised to show that the channel is ready for new data to be written. An interrupt can also be requested if the new value is generated in software but I use DAC12IFG as a signal to the DMA controller instead, in which case interrupts must *not* be enabled. The DAC updates only after it has received both new data and a trigger, which creates a small problem for starting it.

DMA Controller

In Listing 9.8, I showed how to use the data transfer controller to store a block of readings from the ADC10 in RAM without using the CPU. Here we use channel 0 of the DMA controller to do the opposite: It transfers a block of data to the DAC12, one item after each trigger, and repeats the block after it has reached the end. This mode is called *repeated single transfers* (DMADT_4). The module is configured for a single destination address (DMADSTINCR_0) and to increment the source address (DMASRCINCR_3). Both the source and destination are bytes (DMADSTBYTE and DMASRCBYTE) and the size of the block is specified in DMA0SZ.[1] Again I do not enable the module by setting the DMAEN bit until everything has been configured.

The address of the destination and the first item in the source are written to DMA0DA and DMA0SA. As usual the address must be cast into an integer type to avoid a compilation error. Usually I use `uint16_t`, which is appropriate for the original MSP430 because both integers and addresses have 16 bits. However, the FG4618 has the newer MSP430X architecture. This supports 20-bit addresses so a 16-bit integer may not be large enough to hold a valid pointer. The DMA controller works over the full range of addresses and therefore has 32-bit registers for DMA0DA and DMA0SA, although the top 12 bits are

---

[1] The header file lacks a definition for this register at the time of writing so I provide one.

currently reserved. I am cautious and cast the addresses to uintptr_t, which is defined in stdint.h as "an integer type large enough to be able to hold a pointer."

Each transfer is triggered when the DAC12IFG flag in DAC12_0 is raised (DMA0TSEL_5). The DMA automatically clears the flag when the transfer starts. An option can be configured in DMACTL1 for transfers to wait until the CPU fetches the next instruction. This is important if the destination is in flash memory but would be disastrous here because the CPU is turned off.

Timer_A

Updates of DAC12_0 and therefore changes in the output voltage must occur at precise times to avoid distortion of the wave. This is achieved by using a connection in hardware to the OUT1 signal from Timer_A, which must therefore be set up to produce a square wave. I clocked Timer_A from ACLK, which is provided by a 32 KHz crystal on the board. This allows the MSP430 to be left in LPM3. The shortest period is two cycles, giving 16 KHz, and DAC12_0 is updated at this rate. The lookup table contains 32 values so the final frequency of the sound is 512 Hz, which is close to the musical note C at one octave above middle C. The frequency can be changed by adjusting the period of Timer_A but this is rather coarse with ACLK; it would be better to use SMCLK for finer control. You could also change the length of the lookup table.

Listing 9.13: Direct digital synthesis of a sine wave by dacdma2.c on the TI MSP430FG4618/F2013 Experimenter's Board. This uses the DAC12, OA, and DMA modules with the reference voltage provided by the ADC12. I omit the function to initialize the ports.

```
// dacdma2.c - Convert stored wave repeatedly using DAC12 and DMA
// DMA triggered by DAC12_0, which is itself triggered by Timer_A.OUT1
// DAC12_0 output buffered by OA2 (gain = 2) to drive headphone
// TI MSP430FG4618/F2013 Experimenter's Board with F4619, default clocks,
//       LCD not used
// J H Davies, 2007-10-13; IAR Kickstart version 3.42A
//-------------------------------------------------------------------------
#include <io430xG46x.h>              // Specific device
#include <intrinsics.h>             // Intrinsic functions
#include <stdint.h>                 // Integers of defined sizes
#include "LCDutils.h"               // TI exp board utility functions
// Add definition that seems to be missing from io430xG46x.h
__no_init volatile unsigned long DMA0SZ @ DMA0SZ_;

// Lookup table, sine wave, full cycle, bipolar, signed bytes (+/-127)
const int8_t wave[] = {    0,    25,    49,    71,    90,   106,   117,   125,
                         127,   125,   117,   106,    90,    71,    49,    25,
                           0,   -25,   -49,   -71,   -90,  -106,  -117,  -125,
                        -127,  -125,  -117,  -106,   -90,   -71,   -49,   -25};
```

```
void main (void)
{
    WDTCTL = WDTPW | WDTHOLD;          // Stop watchdog timer
    FLL_CTL0 = XCAP14PF;              // Configure load caps (14pF)
    PortsInit();                      // Initialize input/output ports
// Configure op-amp 2 as a buffer for the DAC12 output
    OA2CTL1 = OAFC_4 | OAFBR_2 | OARRIP;      // Noninv, gain = 2, no RRIP
// Noninverting input from DAC0, fast (high current), external output A5
    OA2CTL0 = OAP_2 | OAPM_3 | OAADC1;
    ADC12CTL0 = REFON;               // ADC12 1.5V ref on, needed for DAC
// Configure DAC12_0 for audio output to buffer (OA2); do not enable
// Internal Vref, 8-bit, TA1 trigger, gain = 1, max speed, 2s complement
    DAC12_0CTL = DAC12SREF_0 | DAC12RES | DAC12LSEL_2 | DAC12IR
                 | DAC12AMP_7 | DAC12DF; // DAC12 not yet enabled!
    DAC12_0CTL_bit.DAC12CALON = 1;   // Start calibration
    while (DAC12_0CTL_bit.DAC12CALON == 1) {    // (Should stop CPU)
    }                                // Wait for calibration to complete
// Configure DMA chan 0 to copy wave[] table to DAC12_0; do not enable
    DMACTL0 = DMA0TSEL_5;            // Trigger channel 0 from DAC12_0IFG
    DMACTL1 = 0;                     // DMA occurs immediately (default)
// Repeated single transfers, fixed dest, increment source, bytes
    DMA0CTL = DMADT_4|DMADSTINCR_0|DMASRCINCR_3|DMADSTBYTE|DMASRCBYTE;
    DMA0SA = (uintptr_t) wave;       // Source address: wave array
    DMA0DA = (uintptr_t) &DAC12_0DAT;   // Destination: DAC12 data
    DMA0SZ = sizeof(wave) / sizeof(int8_t); // Number of transfers
// Configure Timer_A to produce square wave on OUT1 to trigger DAC12_0
    TACCR0 = 1;                      // Period = 2 (highest frequency)
    TACCR1 = 1;                      // To generate OUT1 for DAC12_0
    TACCTL1 = OUTMOD_7;              // Reset/set for active high PWM
// Everything is now configured: start the action
    DMA0CTL_bit.DMAEN = 1;           // Enable DMA channel 0
    DAC12_0CTL_bit.DAC12ENC = 1;     // Enable DAC12 channel 0
    DAC12_0CTL_bit.DAC12IFG = 1;     // Stimulate DMA to write DAC12_0DAT
// Start Timer_A from ACLK, undivided, Up mode, clear
    TACTL = TASSEL_1 | ID_0 | MC_1 | TACLR; // Start triggers to DAC12_0
    for (;;) {                       // Enter LPM3, only ACLK needed
        __low_power_mode_3();        // (Interrupts enabled but not used)
    }                                // CPU no longer needed
}
```

Sequence of Operations

This is the sequence of operations in each cycle after the processor has entered LPM3:

1. The OUT1 signal from Timer_A goes high.

2. This stimulates DAC12_0 to copy the new data from DAC12_0DAT to its core, which updates the output voltage. The DAC then sets DAC12IFG to show that it is ready for the next value.

3. This triggers a transfer by the DMA controller, which clears DAC12IFG and copies the next byte in the lookup table to DAC12_0DAT. It returns to the start of the table after it has reached the end.

None of these operations needs the CPU although the DCO must be started so that MCLK is available when the DMA controller operates the buses.

I mentioned earlier that there is a small problem in starting the system because DAC12_0 does not update its output until it has received both new data and a trigger. The trigger is no problem because Timer_A produces these regularly. Unfortunately the DMA controller does not send any data until it sees that DAC12IFG has been raised, which does not happen until an update has occurred.... The sequence must be started by using software either to write the first value to DAC12_0DAT or to set DAC12IFG and stimulate the DMA controller. I do the latter.

The output at 512 Hz sounds like a pure tone in my headphone. You might find the high quality surprising if you look at the oscilloscope trace of the output in Figure 9.33: It is far from a smooth curve with only 32 samples per cycle. Reconstruction adds unwanted Fourier components at frequencies of $nf_s \pm f$ to a signal of frequency f, where $f_s = 16$ KHz is the sampling frequency. The first unwanted tones are therefore around 16 KHz, which is unlikely to be audible through poor-quality headphones (and aging ears).

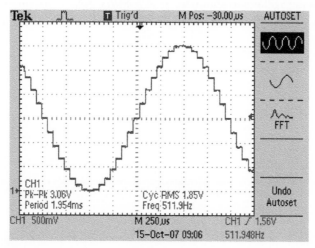

Figure 9.33: Oscilloscope trace of the output from the DAC12. The signal is a sine wave at 512 Hz with 32 samples per cycle.

I reduced the frequency by writing larger values to TACCR0 in the debugger. The musical quality obviously declined although the trace on the oscilloscope was identical after I adjusted the time base. Take TACCR0 = 7, for example, which gives a period of 8 and reduces all frequencies by a factor of 4. The desired note is now two octaves lower at 128 Hz. More significant, the first unwanted tones have come down to near 4 KHz, which is clearly audible as distortion of the sine wave. The signal could be improved by passing it through a low-pass (reconstruction) filter to reduce the amplitude of the unwanted Fourier components. There is an option for this on the MSP430FG4618/2013 Experimenter's Board.

CHAPTER 10

Communication

The term *communication* covers an enormous range of possibilities in embedded electronic systems. At one extreme we are continually promised the "Internet refrigerator," which will monitor its contents and reorder groceries to keep itself stocked. (It has not arrived in my kitchen yet.) At the other limit, a small microcontroller might need to store data in an external memory chip, which requires communication across a few millimeters of printed circuit board. This grabs fewer headlines but is more typical of small embedded systems, where the MSP430 is likely to be found. There are plenty of intermediate examples too. Information might be downloaded from a datalogger by connecting it to a personal computer. In the past this would probably have been done with a serial (RS-232); the interface is termed a universal asynchronous receiver/transmitter (UART) cable, which is easy for a MCU. Now it is more likely to use a universal serial bus (USB), which is convenient for the user but much harder work for the embedded system. Short-range wireless communications are also growing in importance with the rise of ZigBee and other low-power systems, designed with embedded systems in mind.

I devote most of this chapter to three common types of communication that current MSP430s can handle directly. All of them are *serial* communications, meaning that a single bit is transferred at a time. They are

- Serial peripheral interface (SPI).

- Inter-integrated circuit (I²C) bus.

- Asynchronous serial communication (usually, if inaccurately, called RS-232).

Most MSP430s contain modules to handle straightforward communications in hardware. Even the 14-pin MSP430F20xx devices except for the F20x1 have a module for SPI and

I²C. If no hardware is available, slow communication can be achieved by driving the pins using software. This is affectionately (or derisorily) known as *bit-banging*. Timer_A was designed with this in mind and is particularly helpful for asynchronous serial communication. It can also be used for other simple interfaces that the modules do not handle such as the Dallas one-wire bus, which is featured on the Olimex 1121STK.

Of the three types of communication that I cover in detail, SPI and I²C share many characteristics while asynchronous communication is rather different. The major practical difference is that SPI and I²C are typically used between a microcontroller and other devices on the same PCB, while asynchronous communication is used to exchange data with other equipment such as a PC. Thus asynchronous systems must conform fully to established standards so that they work reliably and must be protected against electromagnetic interference and other hazards. This requires special interface circuits. On the other hand, simple wires are usually sufficient for SPI and I²C provided that all devices work at the same voltage.

From a theoretical view, the difference lies in the way in which timing in managed. All digital communication needs a clock, which tells the transmitter when it may place the next bit on its output and informs the receiver when the data are valid and should be read. SPI and I²C are *synchronous*, which means that a clock signal is sent along with the data. The device that generates the clock is called the *master* and other devices are *slaves*. An extra wire carries the clock. In contrast, no clock signal is transmitted in asynchronous communication. Information must therefore be sent in separate frames, each of which is short enough that the separate clocks in the transmitter and receiver remain synchronized. In practice the frames are almost always single bytes. Transmission and reception are essentially separate processes and can take place together, giving full-duplex communication. Only one wire is needed for each direction of transmission (plus ground, of course, which is never counted).

SPI and I²C have similar applications. The major difference between them is that I²C is a true bus, which is designed to accommodate a large number of devices. Transactions follow a protocol that starts with an address to select a particular slave and includes acknowledgment bits to confirm successful delivery. There is only one wire for data, giving a total of two with the clock. Data can travel in either direction but only in one way at a time, which is called *half-duplex transmission*.

In contrast, SPI uses two lines for data so that information can be sent simultaneously in both directions. In fact both processes *must* occur together because of the concept that underlies SPI. The full version of SPI includes a further line that is used to select a particular slave and gives a total of four wires for the interface. There is no control of transmission in software—no addresses or acknowledgment. Thus SPI needs more wires than I²C and offers less sophistication but is simpler and faster. This makes SPI more suitable when large amounts of data have to be transferred. SPI and I²C are often used to communicate with

- Port expanders to increase the effective number of pins for digital input and output.

- ADCs and DACs.

- Sensors with digital outputs, such as thermometers.

- External memory (dataflash, EEPROM).

- Real-time clocks.

- Other processors.

10.1   Communication Peripherals in the MSP430

Currently three types of communication peripherals are offered in different variants of the MSP430. Smaller devices had none at all in the past but the USI was introduced to provide an economical interface for cheaper members of the MSP430F2xx family. Here is a quick summary of their principal features.

10.1.1   Universal Serial Interface

The universal serial interface (USI) is a lightweight module, which is included in the small F20x2 and F20x3 devices. It is much less grand than its name might imply. For a start, it handles only synchronous communication—SPI and I²C. It includes a shift register, clock generator, bit counter, and a few extra items used to assist I²C. This is sufficient to run SPI easily provided that a $\overline{SS}$ signal is not needed. I²C is inevitably more complicated and needs considerable help from software. The USI is a great advance on bit-banging, particularly for a slave, despite its evident limitations. Unfortunately the I²C mode is a little buggy at the time of writing and needs considerable care.

10.1.2   Universal Serial Communication Interface

Recent, larger devices in the MSP430F2xx and MSP430F4xx families contain one or more *universal serial communication interface* (USCI) modules. The hardware handles almost all aspects of the communication, unlike the USI, so the software needs only to provide the data to transmit and store the received data in normal operation. Typically this requires only a couple of small interrupt service routines. The USCI can use direct memory access although I do not describe this.

Each USCI contains two channels, A and B. These are largely independent but share a few registers and interrupt vectors:

Asynchronous channel, USCI_A: Acts as a universal asynchronous receiver/ transmitter (UART) to support the usual RS-232 communication. It can detect the baud rate of an incoming signal, which enables its use on a local interconnect network (LIN). The output signal can be modulated for an infrared diode to work with IrDA and the incoming signal can be decoded to match. Finally, it can also handle SPI. This looks out of place in the asynchronous channel because SPI is synchronous but it is both simple and widely used, hence its inclusion.

Synchronous channel, USCI_B: Handles both SPI and I²C as either master or slave. It contains a full state machine to run I²C communications in compliance with the specification from NXP (formerly Philips). This includes multiple masters and clock stretching.

Some devices have more than one USCI, in which case the modules are called USCI_A0, USCI_A1, and so on. There is a small difference because the interrupt flags and enable bits for the "0" modules are in the special function registers IFG2 and IE2, while those for the "1" modules are in their own registers, UC1IFG and UC1IE. Although the module is called USCI, the names of its registers and bits start with only UC.

10.1.3   Universal Synchronous/Asynchronous Receiver/Transmitter

The universal synchronous/asynchronous receiver/transmitter (USART) is an older module, which has been superseded by the USCI. In most cases it provides only asynchronous communication and SPI but it also handles I²C in a few devices.

10.1.4   Bit-Banging

You have to resort to bit-banging if no peripheral is available for communication. This means that every aspect of the signals is handled in software, assisted where appropriate by Timer_A or B to ensure precise timing.

Synchronous masters: Easy to bit-bang because the master generates the clock and therefore has full control over the timing. A timer can be used for slow communication but often the software requires so many clock cycles that the MSP430 can simply run at full speed, particularly for SPI. Actions are triggered by edges of the clock, which does not need a precise frequency, so there is no problem if there are pauses when the software has more to do.

Synchronous slaves: More difficult. The problem is that the slave must react quickly when a clock transition arrives from the master. It may be possible to use interrupts on the input ports but the latency may be too long, in which case a polling loop with the MSP430 running at full power is unavoidable. Nevertheless it can be done, at least for fairly slow rates. Some extra hardware almost always is needed to detect start and stop conditions for an I²C slave.

Asynchronous communication: Straightforward with the aid of Timer_A, which was designed with this in mind. Its sampling mode is particularly helpful for reception.

The introduction of the USI should reduce the need for bit-banging synchronous protocols.

10.2 Serial Peripheral Interface

The serial peripheral interface was introduced by Motorola and is the simplest synchronous communication protocol in general use. Its versatility led to widespread adoption in a much wider range of applications than straightforward communication. The only problem is that it is not a fixed standard like I²C. There are plenty of options within "standard" SPI and innumerable variations that go beyond this. You must always read the data sheet closely for a device that uses SPI and ensure that you understand the details of the protocol precisely. This will emerge as we explore the details. In the absence of a firm specification I have taken the behavior of the SPI modules in Freescale (formerly part of Motorola) microcontrollers as the definition of ideal SPI. The data sheet for the MC9S08GB60 is particularly clear.

The concept of SPI is shown in Figure 10.1 for the minimal system of two devices. I chose a particular configuration for the interface and explain the myriad options later. One device is the master and the other the slave. The master provides the clock for both devices and a signal to select (enable) the slave, but the path followed by the data is identical in each. In its full form SPI requires four wires (plus ground, which is essential but never counted) and transmits data simultaneously in both directions (full duplex) between two devices. Motorola's nomenclature for the two data connections is "master in, slave out" (MISO) and "master out, slave in" (MOSI). This is admirably clear and makes the functions

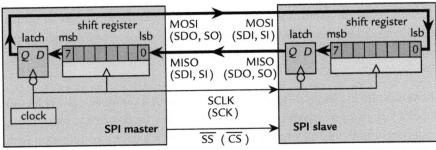

Figure 10.1: Serial peripheral interface between a master and a single slave. Conceptually the whole system is a shift register. I show a selection of the more common names used for the signals and choose a particular set of options for the protocol.

unambiguous. The two MISO pins should be connected together and likewise the two MOSI pins. Other terms are widely used, such as SDI, SI, or DIN for serial data in and SDO, SO, or DOUT for serial data out. In this case you connect an input on one device to an output on the other. There is similar variety in the names for the clock signal including SCLK (most popular), SPSCK, and SCK. The final signal selects the slave. This is usually active low and labeled $\overline{SS}$ for slave select (Motorola), $\overline{CS}$ for chip select, or $\overline{CE}$ for chip enable. A slave should drive its output only when $\overline{SS}$ is active; the output should float at other times in case another slave is selected. In some modes of SPI, the first bit should be placed on the output when $\overline{SS}$ becomes active to start a new transfer.

The concept of SPI is based on two shift registers, one in each device, which are connected to form a loop. The registers usually hold 8 bits. Each device places a new bit on its output from the most significant bit (msb) of the shift register when the clock has a negative edge and reads its input into the lsb of the shift register on a positive edge of the clock. Thus a bit is transferred in each direction during each clock cycle. After eight cycles the contents of the shift registers have been exchanged and the transfer is complete. Transmission and reception are clearly inseparable: You cannot do one without the other, at least in principle. Thus a byte must be transmitted in order to receive a byte.

One byte is the most common length of a transfer but any number of bits can be transmitted and a word of 16 bits is natural with the MSP430. Many data converters also use SPI with 16-bit transfers. The JTAG interface, which is used to program and debug the MSP430, is based on SPI and transfers thousands of bits.

Data are transferred in both directions with the full SPI but many systems do not use the complete interface. For example, an external DAC configured as an SPI slave may never need to return digital data to a microcontroller so there is no need for the MISO connection. However, the same steps always take place internally: The only difference is that the output of the DACs shift register never leaves the chip. This contrasts with asynchronous communication, where transmission and reception are independent. The $\overline{SS}$ line can sometimes be omitted if only two devices are connected, in which case the slave's $\overline{SS}$ pin should be tied to ground so that it is enabled permanently. This is not always permitted, depending on the function of $\overline{SS}$.

Reading and writing are triggered by edges of the clock and the timing is usually straightforward. The clock need not have a particular frequency and nothing goes wrong if different cycles vary in duration, provided that minimum setup and hold times are observed. These are usually so short that the MSP430 has trouble violating them. An exception to this rule is the time between $\overline{SS}$ becoming active and the first clock transition. The reason is that $\overline{SS}$ often has more significance than merely the start of transmission: It may stimulate a new conversion in an ADC, for instance.

Another reason for the $\overline{SS}$ signal is that the clock alone does not provide quite enough time points for a complete transaction. Supposed that we wanted to send only a single bit. This requires three steps:

1. Put data on output.

2. Read data from input.

3. Remove data from output.

A single cycle of the clock provides stimuli for only two of these so the third is taken from $\overline{SS}$. It can either start or end the transmission and this option is specified by the *clock phase* bit, CPHA in Motorola's notation:

CPHA = 0: A transition on $\overline{SS}$ starts the transaction and causes the first bit of data to be placed on the outputs. The inputs are read and clocked into the shift register on the leading (first) edge of the first clock pulse. The second bit is put on the output at the trailing (second) edge of the first clock pulse and this continues until the last bit is read on the leading edge of the last clock pulse. The outputs are removed after the trailing edge of the last clock pulse or when $\overline{SS}$ becomes inactive. In summary, data are

- Read on the leading edge of each clock pulse.

- Written on the trailing edge of each clock pulse.

It is essential that $\overline{SS}$ goes inactive between transfers because the first output is stimulated by $\overline{SS}$ becoming active.

CPHA = 1: The first bit of data is placed on the output following the leading edge of the first clock pulse. It is read on the trailing edge of the pulse. This continues until the last bit has been read on the trailing edge of the last clock pulse. Thus data are

- Written on the leading edge of each clock pulse.

- Read on the trailing edge of each clock pulse.

The outputs are removed when $\overline{SS}$ becomes inactive. Thus $\overline{SS}$ is needed only to control when the slave should drive its output, not to provide any timing. If there is only one slave, whose output can remain active at all times, the $\overline{SS}$ signal need not be used.

Thus CPHA controls whether writing and reading take place on the leading and trailing edges of the clock pulses or vice versa. This is the first of many options that control the configuration of SPI. A related option is the *clock polarity*, selected with the CPOL bit:

CPOL = 0: Clock idles low between transfers.

CPOL = 1: Clock idles high between transfers.

There is no fundamental difference between these two polarities: One is just the complement of the other. Combinations of the two bits CPOL and CPHA give four standard modes for the clock in SPI, which are listed in Table 10.1. Modes 0 and 3 seem to be the most widely used. The data are read on rising edges and written on falling edges in both of these, which is consistent with Figure 10.1.

Warning: Motorola's notation CPHA and CPOL is used almost universally but not in the documents for the MSP430. The user's guides use the following names instead:

- CKPL = CPOL for the polarity.

- CKPH = $\overline{CPHA}$ for the phase, *which is inverted*.

Table 10.1:   The four standard modes for the clock in SPI.

Mode	CPOL/CKPL	CPHA/$\overline{CKPH}$
0	0	0
1	0	1
2	1	0
3	1	1

The second definition is a particular nuisance because of the inversion. The user's guides also use SIMO and SOMI instead of MOSI and MISO, which is quaint but less troublesome.

Figures 10.2 and 10.3 illustrate complete waveforms for a 4-bit transfer in modes 0 and 3. These show the different relations between the clock and $\overline{SS}$. In both cases $\overline{SS}$ becomes active one half cycle before the first edge on the clock and is released to the inactive level one half cycle after the last edge on the clock. The slave may show spurious data on its

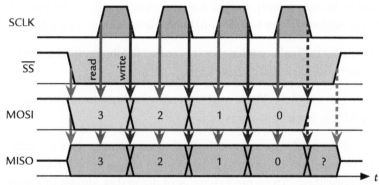

Figure 10.2: A complete transfer of 4 bits using SPI in mode 0 (CPHA = 0, CPOL = 0). The first bit is placed on the outputs when $\overline{SS}$ becomes active (low). Inputs are read on rising edges of the clock and subsequent bits are placed on the outputs on falling edges of the clock.

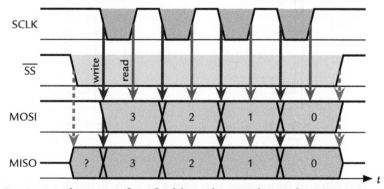

Figure 10.3: A complete transfer of 4 bits using SPI in mode 3 (CPHA = 1, CPOL = 1). Bits are placed on the outputs on falling edges of the clock and read on rising edges of the clock.

output at either the start or end of the transmission, shown by the "?", but there are no rising edges that would cause these values to be read so they have no practical impact.

I mentioned earlier that the output from the slave (MISO) goes to a high-impedance state when $\overline{SS}$ is inactive. This is shown by the "floating" level on the diagrams. It is not important if only two devices are connected but becomes vital if more than one slave shares a bus.

That is not the end of the options. Here are two more. You can see why it is important to study the data sheet to check the format used by a device.

- The data in SPI are usually shifted with the most significant bit (msb) first but it can be the other way; the bidirectional three-wire variant uses lsb first.

- The slave select signal (or whatever it is called for the particular device) is usually active low but occasionally active high.

Some devices use extra signals. There may be read/write lines for memories, "start conversion" for ADCs, interrupts, alerts, and so on.

I described how SPI is used to connect a single slave to a master. It is not really a bus because it lacks protocols to control and acknowledge transactions but can be extended to handle more than one slave. There are two ways of doing this, shown in Figure 10.4.

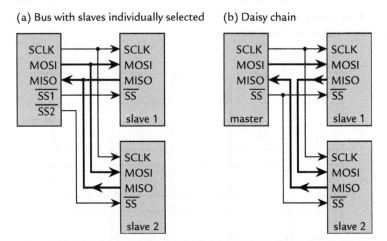

Figure 10.4: Two ways of connecting two slaves to a single master using SPI. (a) A slave can be selected individually by providing separate $\overline{SS}$ lines. (b) All slaves can be connected in a "daisy chain," in which case they must all be updated together.

In both cases the master provides the clock to all the slaves. Slaves can be addressed individually in the first configuration. All the MOSI pins are connected together, as are the MISO pins, but each slave's $\overline{SS}$ pin is connected to a separate pin on the master. Slaves must ignore data on MOSI when their $\overline{SS}$ pin is idle, despite the activity on the clock, and must leave their MISO pins in a high-impedance state. In other words, the MISO pins must have three-state outputs.

The alternative is to connect all the devices in a "daisy chain," as shown in Figure 10.4(b). In this configuration the MOSI pins are *not* all connected together, nor the MISO pins, but rather the MISO pin of a slave is connected to the MOSI pin of the next slave in the chain. The MOSI pin of of the final slave is connected to MOSI on the master. Effectively all the shift registers inside each device are connected into a single, long loop. In principle the master needs to contain a shift register whose length is the sum of the lengths of the registers in all the slaves but in practice this is mimicked in software. The slaves must allow data to be clocked through them transparently while $\overline{SS}$ is held active and react to the new data only when $\overline{SS}$ is released. It is not possible to update an individual slave; data must be written to the complete chain. JTAG works like this and discrete shift registers are often daisy-chained to form a large input/output port.

SPI is sometimes called a *four-wire* bus. You may also encounter Microwire, which is a similar interface introduced by National Semiconductor. It is also based on the concept of a shift register but only one data line is active at a time—it is not full-duplex. A disadvantage of SPI is that it uses four precious pins and a *three-wire* bus is sometimes used to save a pin. Usually this means that the two lines for data, MOSI and MISO, are multiplexed onto a single, bidirectional line for serial data (SDA). This abandons the full-duplex nature of SPI and a protocol is needed to ensure that the master and slave do not attempt to transmit simultaneously. The clock usually uses mode 0 and data are transmitted with the lsb first. Having said that, the term *three-wire* is also used to describe SPI without the $\overline{SS}$ line, which is entirely different. The MSP430 user's guides employ *three-wire* or *three-pin SPI* in this sense.

SPI is straightforward and easy to use, once you are sure of the configuration, because there are no complicated protocols to observe. A disadvantage is that there is no built-in acknowledgment unless the master checks the data returned by the slave. It is fast because the data and clock lines are always driven actively, unlike I²C, and can run at tens of megahertz (but not on a MSP430). The clock does not need a precise or stable frequency because everything is triggered by edges. It rapidly becomes clumsy for multiple slaves but is the simplest serial interface for connecting a single device to a microcontroller.

Many devices are available in versions with either an SPI or an I²C interface but SPI is preferred when large volumes of data must be transferred. Examples of this include

- Large memories, particularly removable flash memory such as MultiMedia or Secure Digital cards. See the application note *Interfacing the MSP430 with MMC/SD Flash Memory Cards* (slaa281).

- Graphical displays, such as some LCDs.

- Interfaces to fast networks such as ethernet, CAN, and USB.

As an opposite example, SPI can also be used to connect simple shift registers such as the 74HC164 or '165 to the MSP430. This is an economical way of adding extra input and output pins and is illustrated in the code examples.

The simplicity of SPI makes it straightforward to bit-bang a master and the application note *Implementing a Direct Thermocouple Interface with MSP430x4xx and ADS1240* (slaa125) illustrates this for a sigma–delta ADC that uses SPI. A slave is a little more difficult and speed may be a serious problem.

Checklist—Have You . . .

❑ Confirmed that all devices are using the same configuration for SPI, particularly CPHA and CPOL?

❑ Remembered that CKPH $= \overline{\text{CPHA}}$?

❑ Checked which signals are used if the interface has fewer than four wires?

10.3   SPI with the USI

We start with the simplest communication module, the USI in its SPI mode. This function is selected by clearing the USII2C bit in the register USICTL1. Figure 10.5 shows a block diagram of the USI, which is very close to the schematic model of SPI in Figure 10.1. The USI is controlled through memory-mapped registers in the usual way. These can be addressed either as bytes or words but the symbolic constants in the current version of the header files work only for bytes. (The current header files also lack bit fields for the USI, which is a minor annoyance.)

The USI uses pins P1.5–7 for SCLK, SDO, and SDI on the F20x3. They are enabled for the USI with the USIPE5–7 bits in USICTL0, not with P1SEL. The input functions remain available even when the USI has control of the pins. Thus interrupts can be requested when a new transfer starts, for instance. One standard connection is missing: The USI does not provide $\overline{SS}$ as either an input or output. This must be managed by using a standard input/output pin and driving it from software as I show in the examples.

The path followed by data is highlighted in Figure 10.5 and is almost identical to the simple model of SPI. There is an additional buffer in the path, which is enabled by the USIOE bit. This can be used to disable the output of a slave when its $\overline{SS}$ input is inactive. Another feature is that the latch is level-sensitive or transparent rather than edge-triggered. Its input comes from the internal shift clock, which is the same as SCLK in SPI modes 0 and 3. When the shift clock is low, the gate (G) input is active and the latch is transparent. This means that changes on D pass immediately to the output Q. We see the effect of this later. The latch holds its value when the shift clock goes high and the rising edge triggers both the bit counter and the shift register, so that a new bit is shifted in to the register from SDI. Setting the USIGE bit overrides the shift clock and makes the latch transparent all the time, which is useful for controlling the value on SDO before and after the clocked transfer.

The shift register USISR is directly accessible: It is not buffered in any way. Received values must be read from USISR before they are overwritten, either by the next value to send or more data received. There is no overrun flag to warn you that the value was not

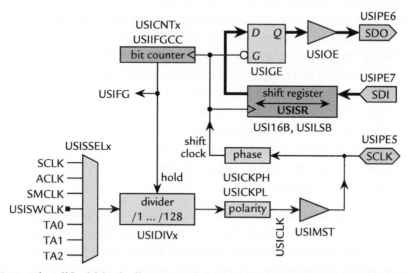

Figure 10.5: Simplified block diagram of the USI in SPI mode (USII2C = 0) with the principal bits that configure it. The path for data through the shift register is emphasized with heavy lines.

read in time. Data written to USISR take effect immediately. The register shifts left by default, which causes data to be transferred with the msb first. This is normal for SPI but can be reversed with the USILSB bit to send the lsb first. The register is 8 bits long by default, which is a common length on SPI, but can be extended to 16 bits with the USI16B bit. It is not necessary to transfer all the bits in the register, but the values need to be adjusted if the full register is not used. This is explained in the user's guide. The full shift register is accessed as the word USISR when more than 8 bits are transferred or the lower byte is addressed as USISRL for 8 bits or fewer.

There are many options for the source of the clock. These include SMCLK and ACLK as usual. The outputs from Timer_A can also be used, which offers finer control over the frequency of the clock. Yet another option is to drive the clock from software by toggling the USISWCLK bit. The frequency of the clock can be divided by powers of 2 up to 128 with the USIDIVx bits and its polarity is controlled by USICKPL, which is equivalent to CPOL in the standard notation. The clock is driven onto the SCLK pin if the module acts as an SPI master, in which case the USIMST bit should be set. The source of the clock is the only difference between a master and slave in the USI. The phase of the clock is selected with USICKPH before it is used for the shift register, output latch, and bit counter. Remember that this is the inverse of the usual CPHA notation.

Transfers are controlled through the bit counter in the USICNT register. The USICNTx field specifies the number of bits to transfer, which is the same as the number of clock cycles needed. The notation is a little unclear: USICNTx means the five least significant bits of the USICNT byte register. Writing a nonzero value to USICNTx normally clears the USI interrupt flag USIIFG and triggers a new transfer. (This behavior can be modified by setting the USIIFGCC bit, which is useful mainly for I²C as I explain later.) The clock is started in a master while a slave waits for its SCLK input from the master's clock. Once the clock is active, new bits are shifted into the register from the input on rising edges of the shift clock and the output is updated from the register when the shift clock is low and the latch is transparent. The bit counter is decremented on rising edges, at the same time as the input is sampled.

This continues until the final rising edge of the shift clock, which causes the last bit from the input to enter the shift register and USICNTx decrements to 0. At this point the USIIFG flag is raised, an interrupt can be requested, and the clock is stopped in a master after it has completed the cycle (there is still a trailing edge to come if CPHA = 0). The same actions occur if USICNTx is cleared from software, unlike Timer_A and Timer_B where flags are set only after normal counting. No further decrements occur after

USICNTx has reached 0; it does not wrap around. This concludes the transfer and the received data should be read immediately.

The USI has a single interrupt vector, which is associated with the USIIFG flag in SPI mode. The interrupt is maskable and is enabled in the usual way with the USIIE bit. USIIFG is *not* cleared automatically when its interrupt is serviced. As I mentioned earlier, it is cleared by the hardware when a nonzero value is written to USICNTx in preparation for the next transfer. This is useful in a slave or in a master where multiple transfers are sent in succession but the bit must be cleared using software in an ISR for a master when there is a delay before the next transfer.

The USI does not have an overall "enable" bit. It is entirely digital and presumably consumes negligible current when its clock is stopped. The module is reset by setting the USISWRST bit. This is its default state after a power-up clear (PUC) and USISWRST should be kept set until the USI has been fully configured. A peculiarity of this module is that USIIFG is *set* after a PUC and is held set while the USI remains in its reset state. The reason is to show that the shift register is ready for new data as soon as it is released from reset. You may therefore need to clear USIIFG after configuring the USI but before interrupts are enabled.

You may have noticed that I do not mention $\overline{SS}$. The USI contains no hardware to deal with it and $\overline{SS}$ must be handled in software if desired.

10.3.1 Loopback Test of SPI Using the USI

Systems that communicate are particularly difficult to debug because there are at least two devices, any of which might be responsible for a fault. It is therefore important to start by testing the simplest configuration possible. This helps to identify trivial errors. Perhaps the most irritating fault is when it turns out that everything is working correctly inside the chip but there is no signal on the pins because they have not been set up properly. It is easily done and the problem becomes obvious as soon as you connect an oscilloscope.

I start with a "loopback test" of SPI in the F2013 on the TI MSP430FG4618/F2013 Experimenter's Board. This means that the output is simply connected back to the input rather than to a second device. Some modules, including the USCI, provide an internal loopback where the signal never leaves the module, which is useful for a first test. The USI does not offer this but it is only a matter of putting a jumper between the pins for SDI and SDO, as shown in Figure 10.6. Always use an oscilloscope to check the signals and confirm that all is well before attempting real communication.

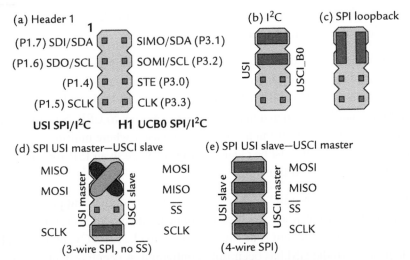

Figure 10.6: Header 1 on the TI MSP430FG4618/F2013 Experimenter's Board, showing the arrangements of jumpers for different types of communication between the F2013 and the FG4618. Wires rather than jumpers are needed for the "crossover" connection in (d) to use the F2013 as an SPI master.

Listing 10.1 shows a program to test the SPI in a F2013 with an external loopback. It sends only 8 bits to give a clearer picture on an oscilloscope. The device must be a master and the USI must therefore drive SCLK. The USI does not provide a $\overline{\text{SS}}$ signal so I chose clock mode 3 (Table 10.1), for which $\overline{\text{SS}}$ is not essential. This has CPOL = 1 and CPHA = 1 in Motorola notation, which means CKPL = 1 and CKPH = 0 for the MSP430. I want an SPI clock with a frequency near 100 KHz and obtain this by taking SMCLK from the DCO at its calibrated value of 12 MHz and dividing by 128. This gives about 94 KHz. The idea is to run MCLK as fast as possible, consistent with a 3 V supply, to minimize interrupt latency and make the results easier to interpret. I could have used a channel of Timer_A to get a frequency closer to 100 KHz but it did not seem worthwhile.

I configured P1.4 as an output and drove it low while the USI is active to act as $\overline{\text{SS}}$. The other input/output pins of the F2013 are left as inputs with pull resistors. Most of the configuration for the USI is explained in the previous section but here are a few extra details:

- The USIMST bit is set because this is an SPI master.

- The USISWRST bit is kept set during the write to USICTL0 so that the USI is held in its reset state to prevent any unwanted activity.

- The USICKPH bit, which selects the phase of the clock, is kept clear in this mode because it is equivalent to CPHA $= 1$ (sigh).

- I mentioned that USIIFG is held set while the USI remains in its reset state. There is no point in writing anything to it while USISWRST $= 1$ but I set USIIFG in the write to USICTL1 as a reminder of this feature.

- The clock is taken from SMCLK (USISSEL_2) and divided by 128 (USIDIV_7). It idles high in mode 3, selected with USICKPL.

- The USI is now configured so I release it from reset by clearing USISWRST. (There are no bit fields in the current version of the header file, hence the mask.) This leaves USIIFG raised so I clear it immediately to avoid an unwanted interrupt, which would otherwise occur as soon as the intrinsic function `__low_power_mode_0()` puts the processor into LPM0 with interrupts enabled.

The program operates as follows. It keeps SMCLK running at all times because it is needed for the watchdog timer. Thus the low-power mode is LPM0.

- The MSP430 is awakened every 0.7 ms by the watchdog timer running as an interval timer. The interrupt service routine drives $\overline{SS}$ low and copies a new value for transmission into the shift register. I use the byte USISRL for the register because the data have only 8 bits. The final step is to start transmission by writing the bit count of 8 to USICNT. The processor then returns to LPM0 during the transfer.

- The USI requests an interrupt when its bit count reaches 0. Its interrupt service routine drives $\overline{SS}$ high to signal the end of the transfer and clears the interrupt flag, which must be done by software. The hardware can clear the flag automatically when a new value is written to USICNT but that does not work in this program: It would start another transfer immediately instead of waiting for the next interrupt from the watchdog timer. Finally, the ISR stores the received data and calculates a new value to transmit by incrementing the received value. The data are trivial but the different patterns on the oscilloscope help to identify features of the transaction.

Listing 10.1: Loopback test of an SPI master by `usispiloop4.c`, using the USI in the F2013 on the TI MSP430FG4618/F2013 Experimenter's Board.

```
// usispiloop4.c - test SPI using USI with external loopback
// 8-bit transfer, SPI mode 3 (CPOL = CPHA = 1), every 0.7ms after WDT
// F2013 on TI MSP430FG4618/F2013 Experimenter's Board:
```

```c
//    SPI clock = SMCLK / 128 = 94KHz, so about 85us for transfer
//    SCLK on P1.5, MOSI (SDO) on P1.6, MISO (SDI) on P1.7
//    P1.4 used as _SS (nSS) but TIMING IS NOT PRECISE (done in software)
// Calibrated 12MHz DCO, no crystal, no ACLK
// J H Davies, 2007-11-02; IAR Kickstart version 4.09A
//-------------------------------------------------------------------
#include <io430x20x3.h>               // Header file for this device
#include <intrinsics.h>               // Intrinsic functions
#include <stdint.h>                   // Integers of defined sizes

#define nSS     P1OUT_bit.P1OUT_4     // Output pin for _SS (active low)

uint8_t RXdata, TXdata = 0x5A;        // Received and transmitted data,
                                      //   subsequent values incremented
void main (void)
{
    BCSCTL1 = CALBC1_12MHZ;           // Calibrated range for DCO
    DCOCTL = CALDCO_12MHZ;            // Calibrated tap and modulation
// Watchdog as interval timer, clear, SMCLK / 8192 = 1.5KHz (0.7ms)
    WDTCTL = WDTPW | WDTTMSEL | WDTCNTCL | WDTIS0;
    IE1_bit.WDTIE = 1;                // WDT interval timer interrupts
    P2SEL = 0;                        // Digital i/o rather than crystal
    P2DIR = 0;                        // Inputs
    P2REN = BIT6 | BIT7;              // Pull resistors on unused pins
// USI will override these settings for SCLK, SDI and SDO (but not nSS)
    P1OUT = BIT4;                     // nSS high (inactive)
    P1DIR = BIT4;                     // nSS output, others input
    P1REN = BIT0|BIT1|BIT2|BIT3;      // Pull resistors on unused pins
// Enable SDI, SDO, SCLK, msb first, master, enable output, latch data
    USICTL0 = USIPE7 | USIPE6 | USIPE5 | USIMST | USIOE | USISWRST;
// Write then read (CPHA = 1 -> CKPH = 0), SPI not I2C, enable interrupt
    USICTL1 = USIIE | USIIFG;         // Can't clear USIIFG in reset mode
//   SCLK = SMCLK / 128, clock idles high (CPOL = CKPL = 1)
    USICKCTL = USIDIV_7 | USISSEL_2 | USICKPL;
    USICTL0 &= ~USISWRST;             // Release USI from reset
    USICTL1 &= ~USIIFG;               // Avoid unwanted interrupt
    for (;;) {                        // Loop forever with interrupts
        __low_power_mode_0();         // LPM0 between interrupts
    }
}
//-------------------------------------------------------------------
// ISR for watchdog: activate nSS, start new SPI transfer
//-------------------------------------------------------------------
#pragma vector = WDT_VECTOR
__interrupt void WDT_ISR (void)       // Acknowledged automatically
{
    nSS = 0;                          // Lower nSS (make active)
    USISRL = TXdata;                  // Load shift register for transfer
    USICNT = 8;                       // Start SPI to transfer 8 bits
}
//-------------------------------------------------------------------
// ISR for USI: deactivate nSS, acknowledge, store received data
//-------------------------------------------------------------------
#pragma vector = USI_VECTOR
__interrupt void USI_ISR (void)       // NOT acknowledged automatically
{
```

```
    nSS = 1;                      // Raise nSS (make inactive)
    USICTL1 &= ~USIIFG;           // Clear flag (no bitfield?)
    RXdata = USISRL;              // Store received data (low byte)
    TXdata = RXdata + 1;          // Slightly different next time!
}
```

An oscilloscope trace of the signals is shown in Figure 10.7(a). Compare these observations with the theoretical Figure 10.3. I chose a value of 0x5A to transmit, which highlights the important features:

- The clock SCLK idles high as expected for CPOL = 1 and has eight low pulses.

- The data on SDO change at the falling edges as expected for CPHA = 1 (CKPH = 0). Presumably the input is read on the rising edges of SCLK but that cannot be seen.

- The $\overline{SS}$ signal goes low well before the clock starts. There is a surprisingly long delay between this edge and the first edge of the clock, about 1.5 cycles of SCLK.

- $\overline{SS}$ returns high immediately after the last rising edge of the clock. This is because the interrupt flag USIIFG is set at this point, as soon as the last bit has been read

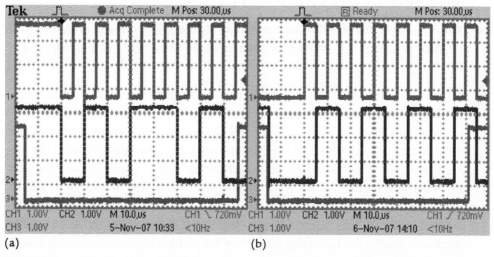

(a)   (b)

Figure 10.7: Transfer of 8 bits using SPI and the USI in (a, left) mode 3 with data of 0x5A and (b, right) mode 0 with 0x55. The traces are SCLK (top), SDO, and $\overline{SS}$ produced by software (bottom). I adjusted the frequency of the DCO slightly so that SCLK runs at 100 KHz.

into the shift register and the received value is complete. It would be better if there were a delay of about 0.5 cycle of SCLK as in Figure 10.3.

- MOSI ideally stays high before and after an SPI transmission but SDO does not. The USI simply leaves it at the value given by the final bit transmitted, which is low here, and this remains on SDO until the first bit of the next transmission. This is partly like the output from the slave, MISO, in Figure 10.3, where "?" is the last bit of the previous transmission. The USI does not contain the hardware necessary to give the ideal behavior expected for an SPI master. This should not be a problem because there is no clock and therefore nothing should read its input.

Figure 10.7(b) shows corresponding traces for mode 0, which should be compared with Figure 10.2. I simply toggled the values of the CKPH and CKPL bits using the debugger to get these results and transmit a value of 0x55 to show the behavior at the end of the transfer. This is a little harder to understand than mode 3.

- The clock SCLK idles low for CPOL = 0 and has eight high pulses.

- The value of SDO first changes just after $\overline{SS}$ goes low, when the value to transmit is written to USISRL. This causes the output to change *immediately* because the latch between the shift register and the SDO pin in Figure 10.5 is transparent between transfers when CPHA = 0 (CKPH = 1).

- This value remains on SDO for longer than the other bits because of the time required to start SCLK. Subsequent changes take place on the falling edges of SCLK.

- A final change in SDO occurs at the last falling edge of SCLK. The output has already been updated eight times, once when $\overline{SS}$ went low and seven times afterward on the clock edges. There are therefore no more data to send. An ideal SPI master ought to drive SDO high at this point but again the USI lacks the hardware to do this. Instead it simply updates the output from the next available bit in the shift register, which is the first bit that was shifted in from SDI. This corresponds to the "?" for the slave in Figure 10.2. In this example the output is driven low again. This behavior is not elegant but has no effect on the transfer because it is after the last rising edge of SCLK, which is when the last bit was read into the shift register. Again this value stays on SDO until the next transmission starts. In fact the output latch remains transparent because SCLK and the shift clock idle low in this mode.

- As in mode 3, $\overline{SS}$ goes low well before the first clock edge and just before SDO first changes. The delay is one instruction because the signal is driven by software. The behavior of $\overline{SS}$ at the end is unsatisfactory because it again goes high just after the last rising edge of the clock, when the last bit is read into the shift register. In mode 0 there is still one falling edge of the clock to come, so the rising edge of $\overline{SS}$ is about one cycle of SCLK too early.

Example 10.1

Modify Listing 10.1 to use SPI mode 0 rather than mode 3 and drive SDO high after transmission, which is the ideal behavior for SPI. Do not worry about $\overline{SS}$. It is harder to do this for mode 3 but feel free to try. You may find the USIGE bit helpful to control the output latch of the USI.

10.4 SPI with the USCI

Both modules USCI_A and USCI_B can operate SPI, although USCI_A is nominally for asynchronous operation. They work in exactly the same way and have the same set of registers, except that the interrupt flags and enabling bits share the same register for the A and B channels of a USCI. I focus on USCI_B0 in the FG4618, which is used to communicate with the F2013 on the TI MSP430FG4618/F2013 Experimenter's Board.

The USCI is far more sophisticated than the USI, which makes it impossible to draw a meaningful, comprehensive block diagram. Most of the configuration bits affect a state machine that controls the operation of the interface. I therefore drew a highly simplified diagram of USCI_B0 in Figure 10.8, which brings out the differences compared with the USI. The module is configured in master mode; it takes its clock from the CLK input in the slave mode.

The main pins are labeled SOMI, SIMO, and CLK with a prefix such as UCB0 to identify the particular module. They should be routed to the USCI by setting bits in the PnSEL registers. For example, this needs bits 1–3 of P3SEL for UCB0 in the FG4618. The direction is controlled by the USCI and may change if the module is switched between acting as a master and slave. There is a fourth pin labeled STE, which I mention later.

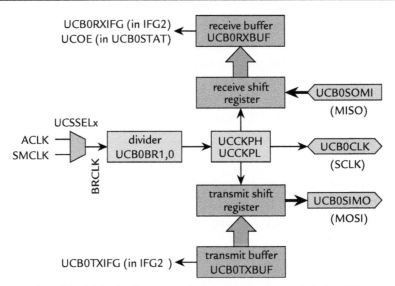

Figure 10.8: Simplified block diagram of the USCI_B0 module in SPI master mode.

There are separate shift registers for transmitting and receiving, in contrast with the single register in a loop used in the USI and the conceptual model of SPI. Moreover, these registers are double-buffered and the user has no direct access to the shift registers themselves. This means that a byte is moved from the receive shift register to RXBUF as soon as reception is complete, which leaves the shift register ready to accept the next transfer. Similarly, a byte written to TXBUF remains in its buffer until the previous byte has been transmitted, at which point it is moved to the transmit shift register. This relaxes considerably the constraints on handling interrupts in the USI, where the shift register must be read and updated rapidly between transfers. Although there are separate registers, reception and transmission are not independent because of the nature of SPI. A byte must be sent in order to receive a byte, even if there is nothing connected to the output pin.

The shift registers hold bytes in the USCI so there is no option for 16-bit transfers. Instead, the UC7BIT can be set for 7-bit transfers, although I never encountered anything that uses this (it is more applicable to UART mode). The order of sending bits is selected with UCMSB, *which is clear by default for lsb first.* You almost certainly need to set this bit for SPI. Master mode is selected with UCMST.

There are fewer choices than in the USI for the source of the clock, called BRCLK. This can be taken from either SMCLK or ACLK, chosen with UCSSELx. The frequency is then divided by the value in the UCB0BR1:UCB0BR0 registers, considered as a word UCBRx.

They are not initialized so be sure to write something to both of them. None of this applies to a slave, where the clock is taken from SCL. The polarity and phase are controlled by UCCKPH and UCCKPL as in the USI, with the same annoyance that UCCKPH = $\overline{\text{CPHA}}$.

A helpful feature for low-power applications is that the USCI automatically activates SMCLK if BRCLK is taken from this: There is no need for the user to change the low-power mode explicitly when a transaction is started. SMCLK is distributed to the whole MSP430 when it is activated, not just to the USCI, so other peripherals that use SMCLK also become active. The device reverts to its previous low-power mode when SMCLK is no longer needed. This feature does not work for ACLK.

The module should be put into synchronous mode with UCSYNC. This might look pointless but is necessary because USCI_A can operate in either asynchronous or asynchronous mode and the registers are almost identical in the A and B channels. Unlike the USI, the USCI offers a choice of SPI modes. The default is three-pin SPI, meaning that there is no $\overline{\text{SS}}$ signal. However, it can also use an input called slave transmit enable (STE), which can be either active high or low according to UCMODEx. Typically this is active low and is used to enable the output if the USCI is a slave, in which case UCMODEx = 10. It can also be used to control a master in systems where there may be more than one SPI master, although this is most unusual and I do not discuss it further.

The USCI is reset with the UCSWRST bit, which works in the same way as the USISWRST bit in the USI. It is set after a PUC and should be kept set until the module is fully configured and ready to go. An internal loopback connection is available by setting the UCLISTEN bit. This permits a convenient first test of a program to ensure that everything inside the MSP430 has been configured correctly.

A transfer begins when a value is written to the transmit buffer TXBUF. The clock is started in a master, while a slave waits for a clock signal on its CLK input, and the flag UCBUSY is raised to show that the module is busy. No further action is required for the byte to be sent and the corresponding byte received. Flags show the state of the transmit and receive buffers.

> **Transmit interrupt flag, TXIFG:** Raised when the buffer TXBUF is ready to accept another byte. A maskable interrupt is requested if enabled with TXIE. The flag is automatically cleared when a value is written to TXBUF. This flag is *set* after the module has been reset to show that TXBUF is empty and ready for data. An interrupt may therefore be requested as soon as it is enabled. It does not follow that transmission is complete if TXIFG is set: It just means that the buffer is ready to accept another byte.

To see how this works, suppose that the module has just been reset, in which case all the buffers are empty and TXIFG is set. You can then write the first byte for transmission to TXBUF, which clears the TXIFG flag. This value is immediately moved to the shift register for transmission, which raises the flag again to show that you can safely write the second byte to TXBUF. The second value waits in the buffer until the first byte has been transmitted, at which point it is moved to the shift register and TXIFG is raised to show that the buffer is ready for the next byte. Data should be written to TXBUF only when TXIFG is set, which shows that the buffer is empty. You should therefore test the flag if you do not use the interrupt.

Receive interrupt flag, RXIFG: Raised when a new value has been received and moved from the shift register to RXBUF. This indicates that a transaction is complete. A maskable interrupt is requested if enabled with RXIE and RXIFG is automatically cleared when the value is read from RXBUF. A new byte can safely be received into the shift register without corrupting the value of the previous byte, which is now in RXBUF. However, RXBUF must be read before the latest byte has been received completely or the previous value is overwritten. The UCOE flag indicates such an overrun error. It is cleared automatically when RXBUF is read and must *not* be cleared by software, unlike some other overrun flags.

The four flags for each USCI (both TXIFG and RXIFG for each of channels A and B) are located in the special function register IFG2 for USCI 0 but in a dedicated register UC1IFG for USCI 1. Similarly, the corresponding enable flags are in either IE2 or UC1IE. There is a single interrupt vector for both transmit flags in an USCI (both channels A and B) and a second vector for both receive flags. You must therefore check the source of the interrupt in the interrupt service routine if both the A and B channels of an USCI are active. There is no interrupt vector register like TAIV: It is simply a matter of testing the flags.

Warning: You cannot enable the receive and transmit interrupts while the USCI_B0 is held in its reset condition with the UCSWRST bit because UCB0RXIE and UCB0TXIE are kept clear. Follow the recommended procedure in the user's guide and enable these interrupts only after UCSWRST has been cleared.

10.4.1   Loopback Test of SPI in USCI_B0

Listing 10.2 shows a test of the SPI using USCI_B0 in the FG4618 on the TI MSP430FG4618/F2013 Experimenter's Board. Its overall structure is very similar to that of Listing 10.1 for the USI but the details make almost every line different. I again drive an

$\overline{SS}$ signal from software, mainly to show the timing of the receive interrupt. The FLL+ is configured to run with a multiplier of 184 from ACLK at 32 KHz, which gives $f_{MCLK} = 6.03$ MHz. This is close to the maximum frequency with a 3 V supply. There is the usual delay loop to allow the FLL+ to stabilize.

The configuration of USCI_B0 should be clear from the previous section. It is set up for mode 3 of SPI and I use the internal loopback. I change these bits later using the debugger to investigate mode 0 and an external loopback. The SPI clock is taken from SMCLK and divided by 60 so it should be very close to 100 KHz.

I trigger transmissions using Basic Timer1 rather than the watchdog timer because it gives finer control over the intervals. Each transmission should take around 80 μs so I arrange for interrupts at 4 KHz or an interval of 244 μs. The timer runs from ACLK so that the MSP430 can be left in LPM3 between transmissions.

The action is triggered by two interrupts, the first from Basic Timer1 to start a transmission and the second from the SPI receiver to conclude the transfer:

1. Following the interrupt from Basic Timer1, I test the UCB0TXIFG flag in the ISR for Basic Timer1 to ensure that USCI_B0 is ready to accept new data for transmission. (Something would be seriously wrong if this flag were not set.) The $\overline{SS}$ output is driven low to signal the start of transmission, which is triggered by writing to UCB0TXBUF.

   The MSP430 would normally return to its previous low-power mode, LPM3, after the end of this ISR. However, SMCLK is needed for the SPI clock and automatically is kept running by the USCI while it is needed. There is no need for software to change the mode to LPM0 as in the USI.

2. The $\overline{SS}$ output is driven high after the interrupt from UCB0RXIFG to mark the end of the transmission. Reading the received data from UCB0RXBUF clears the flag automatically and I update the value to be transmitted next time (trivially).

   In principle I should have checked the source of the interrupt, which could have come from either USCI_A0 or USCI_B0. There is no point here because USCI_A0 is inactive, but perhaps I am a little lazy.

Listing 10.2: Loopback test of an SPI master by `uscispiloop1.c`, using USCI_B0 in the FG4618 on the TI MSP430FG4618/F2013 Experimenter's Board.

```
// uscispiloop1.c - test SPI using USCI_B with loopback (int and ext)
// 8-bit transfer, SPI mode 3 (CPOL = CPHA = 1), every (1/4)ms after BT
// FG4619 on TI Experimenter's Board, 32KHz crystal, DCO at 6MHz
```

```
//    SPI clock = SMCLK / 60 = 100KHz, so about 80us for transfer
//    SCLK on P3.3, MOSI on P3.1, MISO on P3.2
//    P3.0 used as _SS (nSS) but TIMING IS NOT PRECISE (done in software)
// J H Davies, 2007-11-05; IAR Kickstart version 4.09A
//-----------------------------------------------------------------------
#include <io430xG46x.h>              // Specific device
#include <intrinsics.h>             // Intrinsic functions
#include <stdint.h>                 // Integers of defined sizes
#include "LCDutils2.h"              // TI exp board utility functions

#define nSS     P3OUT_bit.P3OUT_0   // Output pin for _SS (active low)

uint8_t RXdata, TXdata = 0x5A;      // Received and transmitted data,
                                    //   subsequent values incremented
void main (void)
{
    volatile uint16_t i;            // Loop counter to stabilize FLL+

    WDTCTL = WDTPW | WDTHOLD;       // Stop watchdog timer
    FLL_CTL0 = DCOPLUS | XCAP14PF;  // FLL+ divider, 14pF load caps
    SCFQCTL = 91;                   // f_DCO = 2(91 + 1)f_ACLK = 6.03MHz
    SCFI0 = FLLD_2 | FN_3;          // Multiply by 2, 2 - 17MHz range
    do {                            // Wait until FLL has locked
        for (i = 0xFFFF; i > 0; --i) {
        }                           // Delay for FLL+ to lock
        IFG1_bit.OFIFG = 0;         // Attempt to clear osc fault flag
    } while (IFG1_bit.OFIFG != 0);  // Repeat if not yet clear
    PortsInit();                    // Initialize ports
    LCDInit();                      // Initialize SBLCDA4
    DisplayHello();                 // Display HELLO on LCD
    P3OUT_bit.P3OUT_0 = 1;          // High for nSS inactive
    P3DIR_bit.P3DIR_0 = 1;          // Enable for nSS output
    P3SEL = BIT1 | BIT2 | BIT3;     // Route pins to USCI_B for SPI
// SPI mode 3 needs CPOL=CKPL=1, CPHA=1 -> CKPH=0; msb first, master,
//    8 bit (default), 3-wire (default, mode 0), synchronous
    UCB0CTL0 = UCCKPL | UCMSB | UCMST | UCMODE_0 | UCSYNC;
    UCB0CTL1 = UCSSEL1 | UCSWRST;   // Clock from SMCLK; hold in reset
    UCB0BR1 = 0;                    // Upper byte of divider word
    UCB0BR0 = 60;                   // Clock = SMCLK / 60 = 100KHz
    UCB0STAT = UCLISTEN;            // Internal loopback
    UCB0CTL1 &= ~UCSWRST;           // Release from reset
    IE2_bit.UCB0RXIE = 1;           // Enable interrupts on receive
// Basic Timer: hold, counter 2 from ACLK, period = 8 (cntr 1 not used)
    BTCTL = BTHOLD | BT_fCLK2_ACLK | BT_fCLK2_DIV8;
    BTCNT2 = 0;                     // Clear counter
    BTCTL &= ~BTHOLD;               // Start basic timer
    IE2_bit.BTIE = 1;               // Enable basic timer interrupts
    for (;;) {                      // Transmissions triggered by BT
        __low_power_mode_3();       // LPM3 between interrupts
    }
}
//-----------------------------------------------------------------------
// ISR for basic timer: check TXIFG, activate nSS, start new SPI transfr
//-----------------------------------------------------------------------
#pragma vector = BASICTIMER_VECTOR
__interrupt void BASICTIMER_ISR (void)  // Acknowledged automatically
```

```
{
    if (IFG2_bit.UCB0TXIFG == 1) {    // Ready for new data?
        nSS = 0;                      // Lower nSS (make active)
        UCB0TXBUF = TXdata;           // Load shift register for transfer
    }                       // SMCLK remains active for USCI automatically
}
//-----------------------------------------------------------------
// ISR for USCI_A,B0 RX: deactivate nSS, store rec'd data (acknowledges)
//-----------------------------------------------------------------
#pragma vector = USCIAB0RX_VECTOR
__interrupt void USCIAB0RX_ISR (void)    // Not acknowledged automaticaly
{
    nSS = 1;                          // Raise nSS (make inactive)
    RXdata = UCB0RXBUF;               // Store received data, clears flag
    TXdata = RXdata + 1;              // Slightly different next time
}
```

I looked at the signal on the oscilloscope again and Figure 10.9 shows traces that correspond to those in Figure 10.7 from the USI. Overall they are very similar—as they should be—but there are two significant differences. First, the clock in the USCI starts more promptly after $\overline{SS}$ goes low. This ensures that the first pulse in mode 0 from the USCI has the same duration as the others, unlike the USI where it is much longer. The second difference is at the end of the transfer in mode 0. The USI gives an extra, spurious transition on the final falling edge of the clock because the same shift register is used for both transmission and reception. The USCI is better behaved: Nothing happens at this point.

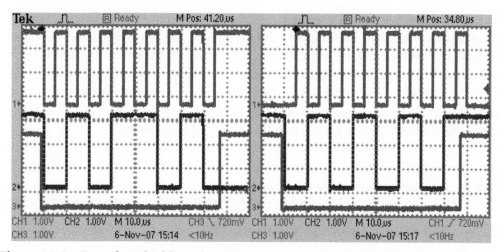

Figure 10.9: Transfer of 8 bits using SPI and the USCI_B in (left) mode 3 with data of 0x5A and (right) mode 0 with 0x55. The traces are SCLK (top), SDO, and $\overline{SS}$ produced by software (bottom).

The behavior of the $\overline{SS}$ signal is similar in the USI and USCI. Remember that this is driven by software, not hardware. In mode 0 it goes high before the final falling edge of the clock in both modules. This is because the (receive) interrupt flag is raised after the final rising edge, when reception is complete. Thus a delay is needed to bring it consistent with the behavior expected from the theoretical Figure 10.2.

10.5   A Thermometer Using SPI in Mode 3 with the F2013 as Master

A clear way of demonstrating the SPI on the MSP430FG4618/F2013 Experimenter's Board is to use the temperature sensor in the SD16A in the F2013 and display its reading on the LCD driven by the FG4618. One MSP430 must be the master and the other the slave, which gives two options. I start with the less obvious configuration with the F2013 as master. I also use mode 3, in which case there is no need for $\overline{SS}$ and only three wires are required.

Warning: The required connections on header 1 are shown in Figure 10.6(d) and need wires rather than jumpers for the two data lines. The problem is that the pins of the USI are designated as input and output, while those on the FG4618 are master and slave. Thus the pins on the FG4618 change direction if the device runs as a slave rather than a master, but those on the F2013 do not. A crossover is therefore needed if the F2013 is the SPI master.

The F2013 takes a reading of the temperature after a delay, converts it to hundredths of a degree, and transmits a 16-bit value over SPI as a master. The FG4618 receives the value and displays it on the LCD. In return, it sends a multiplier for the next delay, which depends on the state of the two buttons S1 and S2. An apparent incompatibility is that the USI in the F2013 can send 16 bits of data but the USCI in the FG4618 can handle only bytes. This is easy to solve because everything is triggered by the clock in mode 3 (it does not need $\overline{SS}$ for the first output, like mode 0), so the USCI can simply receive the 16-bit message as two 8-bit bytes and reassemble them. Its double-buffering gives plenty of time to process the first byte while the second byte arrives. Let us look at the two programs, which are very similar to those for the loopback tests.

10.5.1   SPI Master in Mode 3 with the USI

Listing 10.3 is heavily based on previous programs to measure the temperature and the code for the USI from Listing 10.1. The DCO runs at 1 MHz and is divided down by a factor of 16 for the SPI clock, which is therefore a little slower than before. The temperature is computed to a resolution of 0.01°C. This makes changes obvious but is hardly justified by the absolute accuracy of the sensor without calibration. I raise OSR to 1024 for the SD16_A, which reduces the intrinsic noise from the sigma–delta process to about ±0.01°C.

Each cycle of operation employs a familiar sequence of interrupts from the watchdog timer, SD16A and USI:

1. The delay between readings is given by (delay + 1) intervals of the watchdog timer, where delay is the value sent by the FG4618 and the basic interval is about 0.7 s. Note that delay is an external variable because it is needed in two interrupt service routines. A new "single" conversion is started on the SD16A when the time is up and I illuminate an LED to show that the system is active. The low-power mode is changed from LPM3 to LPM0 so that SMCLK remains running for the SD16A when the ISR has finished.

2. The temperature is computed from the output of the SD16A using the same steps as in Listings 9.4 and 9.9 although the constants are of course different. I noticed that the compiler performed the division by 0x10000 by taking the top word of the 32-bit value, which is the efficient thing to do. The result is truncated to 16 bits and written to the shift register USISR. Finally, the SPI is started by writing the number of bits to USICNT. I set the USI16B bit because the whole register must be used for 16 bits, not just the lower byte as in Listing 10.1.

3. The ISR for the USI first clears the flag, which does not happen automatically, and stores the received value into delay. This is only 8 bits and is therefore taken from the lower byte of the shift register. Finally, the LED is extinguished to show that the activity is over and the MSP430 is returned to LPM3 on exit. It then waits until the next interrupt from the watchdog timer.

Listing 10.3: SPI master on a F2013 using `usispimaster1.c`. It transmits a 16-bit value for the temperature after a delay that it receives from the slave.

```
// usispimaster1.c - measure temperature and transmit as master over SPI
// 16-bit transfer, SPI mode 3 (CPOL = CPHA = 1), msb first
// Delay between readings received from SPI in multiples of 0.7s
// Vref for SD16A left running continuously for simplicity
// F2013 on TI Experimenter's Board:
//   SPI clock = SMCLK / 16 = 62.5KHz, so about (1/4) ms for transfer
//   SCLK on P1.5, MOSI (SDO) on P1.6, MISO (SDI) on P1.7
//   LED on P1.0 to show activity (needs JP2)
// Calibrated 1MHz DCO, no crystal, ACLK from VLO at 12KHz
// J H Davies, 2007-11-07; IAR Kickstart version 4.09A
//-----------------------------------------------------------------
```

```c
#include <io430x20x3.h>          // Header file for this device
#include <intrinsics.h>          // Intrinsic functions
#include <stdint.h>              // Integers of defined sizes

#define LED       P1OUT_bit.P1OUT_0   // Output pin for LED (active high)

uint8_t delay = 0;               // Wait (delay+1) watchdog intervals
                                 //   between readings and transfers
void main (void)
{
    BCSCTL3 = LFXT1S_2;          // Select ACLK from VLO (no crystal)
    BCSCTL1 = CALBC1_1MHZ;       // Calibrated range for DCO
    DCOCTL = CALDCO_1MHZ;        // Calibrated tap and modulation
// Watchdog as interval timer, clear, ACLK / 8192 = 1.5Hz (0.7s)
    WDTCTL = WDTPW | WDTTMSEL | WDTCNTCL | WDTSSEL | WDTIS0;
    IE1_bit.WDTIE = 1;           // WDT interval timer interrupts
    P2SEL = 0;                   // Digital i/o rather than crystal
    P2DIR = 0;                   // Inputs
    P2REN = BIT6 | BIT7;         // Pull resistors on unused pins
// USI will override these settings for SCLK, SDI and SDO
    P1OUT = 0;                   // LED low (inactive)
    P1DIR = BIT0;                // LED output, others input
    P1REN = BIT1|BIT2|BIT3|BIT4; // Pull resistors on unused pins
// Configure SD16A: clock from SMCLK, no division, internal reference on
    SD16CTL = SD16XDIV_0 | SD16DIV_0 | SD16SSEL_1 | SD16REFON;
// Unipolar output, "single" convs, OSR = 1024, interrupts on finish
    SD16CCTL0 = SD16UNI | SD16SNGL | SD16OSR_1024 | SD16IE;
// PGA gain = 1, temperature sensor A6+/-, interrupts after 4th result
    SD16INCTL0 = SD16GAIN_1 | SD16INCH_6 | SD16INTDLY_0;
// No external inputs
    SD16AE = 0;
// Enable SDI, SDO, SCLK, msb first, master, enable output, latch data
    USICTL0 = USIPE7 | USIPE6 | USIPE5 | USIMST | USIOE | USISWRST;
// Write then read (CPHA = 1 -> CKPH = 0), SPI not I2C, enable interrupt
    USICTL1 = USIIE | USIIFG;    // Can't clear USIIFG in reset mode
//  SCLK = SMCLK / 16, clock idles high (CPOL = CKPL = 1)
    USICKCTL = USIDIV_4 | USISSEL_2 | USICKPL;
    USICTL0 &= ~USISWRST;        // Release USI from reset
    USICTL1 &= ~USIIFG;          // Avoid unwanted interrupt
    for (;;) {                   // Loop forever with interrupts
        __low_power_mode_3();    // LPM3 between interrupts
    }
}
//-------------------------------------------------------------------
// ISR for watchdog: activate SD16 to read temperature, change to LPM0
//-------------------------------------------------------------------
#pragma vector = WDT_VECTOR
__interrupt void WDT_ISR (void)  // Acknowledged automatically
{
    if (delay > 0) {             // Waited for enough intervals?
        --delay;                 // No: decrement count
    } else {                     // Yes: start conversion and TX
        LED = 1;                 // Turn LED on, show start
        SD16CCTL0_bit.SD16SC = 1;   // Trigger new conversion
// Change from LPM3 to LPM0 on exit to provide SMCLK for SD16 and USI
```

```
            __bic_SR_register_on_exit(LPM3_bits);
            __bis_SR_register_on_exit(LPM0_bits);
        }
}
//-------------------------------------------------------------------------
// Wait for SD16A to complete conversion
//-------------------------------------------------------------------------
// ISR for SD16A: compute temperature to 0.01C, transmit result over SPI
//-------------------------------------------------------------------------
#pragma vector = SD16_VECTOR
__interrupt void SD16_ISR (void)        // Acknowledged when SD16MEM0 read
{
    uint32_t temperature;               // Converted value of temperature
                                        //    in units of 0.01degC
    temperature = SD16MEM0;             // Raw converted value
    temperature *= 45454;               // Scaling factor (Vref/temp coeff)
    temperature += 0x8000;              // Allow for rounding in division
    temperature /= 0x10000;             // Divide by range of SD16A
    temperature -= 27300;               // Subtract offset to give celsius
    USISR = (uint16_t) temperature;     // Copy lower word for transmission
    USICNT = USI16B | 16;               // Start SPI to transfer 16 bits
}
//-------------------------------------------------------------------------
// Wait for USI to complete transmission
//-------------------------------------------------------------------------
// ISR for USI: clear flag, store received delay, return to LPM3
//-------------------------------------------------------------------------
#pragma vector = USI_VECTOR
__interrupt void USI_ISR (void)         // NOT acknowledged automatically
{
    USICTL1 &= ~USIIFG;                 // Clear flag, acknowledge interrupt
    delay = USISRL;                     // Lower byte gives next delay
    LED = 0;                            // Turn LED off, show finish
// Change mode from LPM0 to LPM3 on exit: SMCLK no longer needed
    __bic_SR_register_on_exit(LPM0_bits);   // (not really necessary)
    __bis_SR_register_on_exit(LPM3_bits);
}
```

10.5.2 SPI Slave in Mode 3 with USCI_B0

Listing 10.4 for the slave using the USCI is pleasantly short (admittedly because a lot of the tedious material is in LCDutils.c). The USCI_B0 is configured in much the same way as in Listing 10.2 but is shorter because no clock is needed for a slave. I enable interrupts on both reception and transmission this time. An important feature is that these are not linked in a fixed sequence, like those in most previous programs, but work independently.

Transmitter Interrupts

Recall that these are requested whenever data has been moved to the shift register for transmission and the buffer UCB0TXBUF is ready for a new value. I keep these interrupts enabled all the time so that the buffer is refilled immediately, the shift register is always loaded, and the channel is constantly ready to transmit when the slave starts the clock for a new transfer. Remember that transmission and reception occur simultaneously in SPI, which means that a value is received only if the transmitter is active. There is no need to check UCB0TXIFG, as in Listing 10.2, because this flag calls the interrupt.

The transmitted value is loaded from the push buttons S1 and S2. These are active low and the value from the port, P1IN, is therefore complemented before the two bits are masked to give a value in the range 0–3.

Each reading from the F2013 occupies 2 bytes so it might be better to construct a 16-bit value to send. I simply send single bytes and the receiver extracts one of them—we are using only 2 bits in any case. It would be essential to synchronize transmission and reception if 16 bits were used to ensure that the bytes were received in the correct order.

Receiver Interrupts

These are requested when a complete byte has been received and moved from the shift register to the buffer UCB0RXBUF. The 16-bit message from the F2013 arrives in two successive bytes, which must be reassembled into a word. The static variable bytesReceived keeps track of this: The first byte is moved into the upper byte of RXdata and the second byte is inserted into the lower byte using an OR operation. The result is displayed on the LCD with a resolution of 0.01°C by the function DisplayCentiCels().

I add a couple of lines to flash an LED while the word is being received, which should happen at the same time as the LED on the F2013 is alight.

Listing 10.4: SPI slave on a FG4618 using `uscispislave1.c`. It receives a 16-bit value for the temperature in two bytes and returns bytes to specify the delay between readings.

```
// uscispiloop1.c - receive temperature over SPI using USCI_B
// Send next delay value from buttons on P1.0,1 (active low)
// 2 x 8-bit transfers, SPI mode 3 (CPOL = CPHA = 1), slave mode
// FG4619 on TI Experimenter's Board, 32KHz crystal, 1MHz DCO (default)
//    SCLK on P3.3, MOSI on P3.1, MISO on P3.2
//    Display temperature on LCD, flash LED to show activity
// J H Davies, 2007-11-09; IAR Kickstart version 4.09A
// ------------------------------------------------------------------
#include <io430xG46x.h>              // Specific device
#include <intrinsics.h>              // Intrinsic functions
#include <stdint.h>                  // Integers of defined sizes
#include "LCDutils2.h"               // TI exp board utility functions
```

```c
#define LED       P5OUT_bit.P5OUT_1    // Output pin for LED (active high)
                                       // Pin is output low by default
void main (void)
{
    volatile uint16_t i;               // Loop counter to stabilize FLL+

    WDTCTL = WDTPW | WDTHOLD;           // Stop watchdog timer
    FLL_CTL0 = XCAP14PF;               // 14pF load caps, 1MHz default
    do {                               // Wait until FLL has locked
        for (i = 0x2700; i > 0; --i) {  // One loop should be enough
        }                              // Delay for FLL+ to lock
        IFG1_bit.OFIFG = 0;            // Attempt to clear osc fault flag
    } while (IFG1_bit.OFIFG != 0);     // Repeat if not yet clear
    PortsInit();                       // Initialize ports
    LCDInit();                         // Initialize SBLCDA4
    DisplayHello();                    // Display HELLO on LCD
    P3SEL = BIT1 | BIT2 | BIT3;        // Route pins to USCI_B for SPI
// SPI mode 3 needs CPOL=CKPL=1, CPHA=1 -> CKPH=0; msb first,
//   slave (default), 8 bit (default), 3-wire (default, mode 0), synch
    UCB0CTL0 = UCCKPL | UCMSB | UCMODE_0 | UCSYNC;
// No need to configure UCB0CTL1, UCB0BR0,1 because clock not used
    UCB0CTL1 &= ~UCSWRST;              // Release from reset
    IE2 |= (UCB0TXIE|UCB0RXIE);        // Enable interrupts on RX and TX
    for (;;) {                         // Transmissions triggered by master
        __low_power_mode_3();          // LPM3 between interrupts
    }
}
//--------------------------------------------------------------------
// ISR for USCI_A,B0 TX: load buffer with delay value from push buttons
//--------------------------------------------------------------------
#pragma vector = USCIAB0TX_VECTOR
__interrupt void USCIAB0TX_ISR (void)   // Not acknowledged automaticaly
{
    UCB0TXBUF = (~P1IN) & (BIT1|BIT0);  // Mask buttons, active low
}                                       // Writing to UCB0TXBUF clears flag
//--------------------------------------------------------------------
// ISR for USCI_A,B0 RX: store received data, send to LCD when complete
//--------------------------------------------------------------------
#pragma vector = USCIAB0RX_VECTOR
__interrupt void USCIAB0RX_ISR (void)   // Not acknowledged automaticaly
{
    static int16_t RXdata;             // Received data, both bytes
    static uint8_t bytesReceived = 0;  // Count bytes in message

    if (bytesReceived == 0) {
        bytesReceived = 1;
        RXdata = UCB0RXBUF << 8;        // Store received data, clears flag
        LED = 1;                        // Mark first byte
    } else {
        bytesReceived = 0;
        RXdata |= UCB0RXBUF;            // Merge received data, clears flag
        DisplayCentiCels (RXdata);      // Display temperature to 0.01oC
        LED = 0;                        // Mark end of transmission
    }
}
```

There is a danger with this system because the message is received as two separate bytes: It is possible for the master and slave to lose synchronization, so that the two bytes received are the second byte from one message and the first byte from the next. This is unlikely to happen in normal operation but will probably occur while debugging. A good solution would be for the master to produce an $\overline{SS}$ signal, which would reset bytesReceived in the slave at the start of each message. This could easily be done in an ISR using an interrupt on the input pin. The MSP430FG4618/F2013 Experimenter's Board has $\overline{SS}$ connected to P3.0, which unfortunately does not have interrupts, so I do not implement this feature. We see it in action for the slave in the next section.

10.6   A Thermometer Using SPI in Mode 0 with the FG4618 as Master

Usually the thermometer would be the slave and the FG4618 would be the master, so we now look at this configuration. Many sensors use mode 0 of SPI and I therefore chose this, in which case an $\overline{SS}$ signal is essential to start each transaction. Normally it is necessary to toggle $\overline{SS}$ between bytes to start each one. However, the USI in the F2013 can send 16 bits in a single operation so I keep $\overline{SS}$ low until the full word has been received. It is then driven high to mark the conclusion of the complete transaction. The slave releases its SDO pin when $\overline{SS}$ is high so that multiple devices can share the SPI, as in Figure 10.4(a). This system uses the full four-wire SPI and all four jumpers must therefore be placed on H1, as in Figure 10.6(e).

The main issue in designing this system is how to synchronize the measurements of temperature with the communication. Ideally we want the data to be as fresh as possible, so a measurement should be taken immediately before the data are transmitted. That is easy when the thermometer is the master but more difficult now that it is the slave. The same issue arises with any data converter that uses SPI. SAR ADCs with SPI output often use the $\overline{SS}$ input to change the analog input from track to hold and start a new conversion. The first few bits transmitted over SPI are usually zeros until the msb becomes available from the SAR logic. An example is described in the application note *Interfacing Low Power Serial (SPI) ADCs to the MSP430F449* (slaa234).

This approach is not applicable to sigma–delta ADCs because they are much slower and we have to wait until the end of the conversion for the result, which does not emerge in the stepwise fashion of a SAR ADC. I therefore chose to start a new conversion as soon as $\overline{SS}$ goes high after a transmission is completed. The slave then waits in a low-power mode until the master shows that it wants new data by driving $\overline{SS}$ low.

10.6.1   SPI Master in Mode 0 with USCI_B0

For a change I wrote the master program in Listing 10.5 with much of the action in the main loop rather than interrupt service routines. The main loop is paced by interrupts from Basic Timer1 at intervals of 0.5 sec. The code for the SPI handles a message with an arbitrary number N_MESSAGE of bytes rather than the 2 bytes needed for this application. In fact this is not so useful for mode 0 because $\overline{SS}$ should really be toggled between bytes but would be helpful for mode 3. There is a buffer and a counter for reception and transmission. They are defined before main() so that they are available in all functions, including the interrupt service routines.

An SPI transaction starts by filling the buffer for transmission and initializing the counters. There are no useful data to send in this program but SPI requires a transmission in order to receive data and the USCI is stimulated to transmit by writing data to TXBUF. The pin for $\overline{SS}$ is driven low to alert the slave and interrupts are enabled for the USCI transmitter, which starts the SPI clock and the transfer. At the end of the transfer, $\overline{SS}$ is raised and the received data are processed. I flash an LED during the process to provide the usual reassurance that something is happening. The detailed handling of each byte is performed in interrupt service routines in a similar manner to the USCI slave.

Transmitter Interrupts

The next byte is moved to the buffer UCB0TXBUF, ready for transmission when the shift register is emptied. Interrupts are disabled when the final byte enters UCB0TXBUF, which will stop the transaction after this byte has been sent.

This interrupt is active only briefly for a message of 2 bytes. The shift register is empty when the first interrupt occurs because the last transmission was long ago, so the first byte moves from UCB0TXBUF to the shift register straight away. A second interrupt is therefore requested immediately and the second byte is copied to UCB0TXBUF. Therefore the complete message probably enters USCI_B0 before the first bit is transmitted. An interrupt would not be requested for a third byte until the first has been completely transmitted.

Receiver Interrupts

I left the receiver interrupts enabled all the time, which should be safe with SPI because the number of bytes received must be the same as the number transmitted. For the same reason I did not check that there is space in the buffer RXdata[] for the received byte.

This is perhaps a little reckless, given the number of viruses and worms that have exploited similar practice in Web browsers and the like. When the buffer is full, the low-power mode is cleared and control returns to the main loop.

Listing 10.5: SPI master on a FG4618 using `uscispimaster1.c`. It uses SPI mode 0 and Provides an $\overline{SS}$ signal to enable the slave.

```
// uscispimaster1.c - receive temperature over SPI using USCI_B
// 2 x 8-bit transfers, SPI mode 0 (CPOL = CPHA = 0), master mode
// FG4619 on TI Experimenter's Board, 32KHz crystal, 1MHz DCO (default)
//   SCLK on P3.3, MOSI on P3.1, MISO on P3.2
//   Display temperature on LCD, flash LED to show activity
//   P3.0 used as _SS (nSS) output in software before word, not bytes
// J H Davies, 2007-11-11; IAR Kickstart version 4.09A
//-------------------------------------------------------------------
#include <io430xG46x.h>                // Specific device
#include <intrinsics.h>                // Intrinsic functions
#include <stdint.h>                    // Integers of defined sizes
#include "LCDutils2.h"                 // TI exp board utility functions

#define N_MESSAGE   2                  // Number of bytes in message
uint8_t RXcount, TXcount;             // Counters for bytes
uint8_t RXdata[N_MESSAGE], TXdata[N_MESSAGE];   // Recd and transd data
#define nSS        P3OUT_bit.P3OUT_0   // Output pin for _SS (active low)
#define LED        P5OUT_bit.P5OUT_1   // Output pin for LED (active high)
                                       // Pin is output low by default

void main (void)
{
    volatile uint16_t i;               // Loop counter to stabilize FLL+
    int16_t Temperature;               // Value received over SPI

    WDTCTL = WDTPW | WDTHOLD;           // Stop watchdog timer
    FLL_CTL0 = XCAP14PF;               // 14pF load caps; 1MHz default
    do {                               // Wait until FLL has locked
        for (i = 0x2700; i > 0; --i) { // One loop should be enough!
        }                              // Delay for FLL+ to lock
        IFG1_bit.OFIFG = 0;            // Attempt to clear osc fault flag
    } while (IFG1_bit.OFIFG != 0);     // Repeat if not yet clear
    PortsInit();                       // Initialize ports
    LCDInit();                         // Initialize SBLCDA4
    DisplayHello();                    // Display HELLO on LCD
    P3OUT_bit.P3OUT_0 = 1;             // High for nSS inactive
    P3DIR_bit.P3DIR_0 = 1;             // Enable for nSS output
    P3SEL = BIT1 | BIT2 | BIT3;        // Route pins to USCI_B for SPI
// SPI mode 0 needs CPOL=CKPL=0, CPHA=0 -> CKPH=1; msb first, master,
//   8 bit (default), 3-wire (default, mode 0), synchronous
    UCB0CTL0 = UCCKPH | UCMSB | UCMST | UCMODE_0 | UCSYNC;
    UCB0CTL1 = UCSSEL1 | UCSWRST;      // Clock from SMCLK; hold in reset
    UCB0BR1 = 0;                       // Upper byte of divider word
    UCB0BR0 = 10;                      // Clock = SMCLK / 10 = 100KHz
    UCB0CTL1 &= ~UCSWRST;              // Release from reset
    IE2_bit.UCB0RXIE = 1;              // Enable interrupts on RX all time
    BTCTL = BTHOLD | BT_ADLY_500;      // Hold, period = 500ms from ACLK
    BTCNT2 = 0;                        // Clear counter
    BTCTL &= ~BTHOLD;                  // Start basic timer
```

```
      IE2_bit.BTIE = 1;                 // Enable basic timer interrupts
      for (;;) {                        // Transmissions triggered by BT
         __low_power_mode_3();          // Wait for BT (needs only ACLK)
// Set up SPI transaction, to be performed using interrupts
         LED = 1;                       // Show start of activity
         for (i = 0; i < N_MESSAGE; ++i) {
            TXdata[i] = 0xAA;           // Make outgoing message
         }                              //    (rather boring here!)
         RXcount = TXcount = 0;         // Initialize message byte counters
         nSS = 0;                       // Lower nSS (make active)
         IE2_bit.UCB0TXIE = 1;          // Enable TX interrupts, starts SPI
         __low_power_mode_0();          // Wait for SPI (needs SMCLK)
// SPI transaction completed           (but USCI would keep SMCLK active)
         nSS = 1;                       // Raise nSS (make inactive)
         Temperature = (RXdata[0] << 8) | RXdata[1]; // Assemble word
         DisplayCentiCels (Temperature); // Display temperature to 0.01oC
         LED = 0;                       // Show end of activity
      }
}
//-----------------------------------------------------------------------
// ISR for basic timer: return to main routine
//-----------------------------------------------------------------------
#pragma vector = BASICTIMER_VECTOR
__interrupt void BASICTIMER_ISR (void)  // Acknowledged automatically
{
    __low_power_mode_off_on_exit(); // Return to start new SPI message
}
//-----------------------------------------------------------------------
// ISR for USCI_A,B0 TX: load next byte for TX, halt interrupts at end
//-----------------------------------------------------------------------
#pragma vector = USCIAB0TX_VECTOR
__interrupt void USCIAB0TX_ISR (void)   // Acknowledge by write to TXBUF
{
    UCB0TXBUF = TXdata[TXcount++];  // Send next byte, update counter
    if (TXcount >= N_MESSAGE) {     // Sent complete message?
       IE2_bit.UCB0TXIE = 0;        // Disable further interrupts
    }
}
//-----------------------------------------------------------------------
// ISR for USCI_A,B0 RX: store data, return when all bytes received
//-----------------------------------------------------------------------
#pragma vector = USCIAB0RX_VECTOR
__interrupt void USCIAB0RX_ISR (void)   // Acknowledge by read of RXBUF
{
    RXdata[RXcount++] = UCB0RXBUF;  // Store recd data, update counter
    if (RXcount >= N_MESSAGE) {     // Received complete message?
       __low_power_mode_off_on_exit(); // Return to process message
    }
}
```

10.6.2 SPI Slave in Mode 0 with the USI

Listing 10.6 shows the corresponding program for the slave. The MSP430 spends most of
its time in LPM4 with no clocks running at all. Even the transaction over SPI takes place

in LPM4 because the USI module is a slave and takes its clock from the SCLK input. The main feature of this program is the handling of $\overline{SS}$, which is performed by interrupts. The signal comes in to P1.4 and I therefore enabled interrupts on change for this pin. The action depends on the sign of the edge.

Falling Edge on $\overline{SS}$

This starts a new transfer over SPI. I designed the slave to release its output SDO after a transaction so it must first reenable the output by setting the USIOE bit. The value to be transmitted is already waiting in the shift register USISR so the only other task is to enable the USI by writing the number of bits to USICNT. This cannot be done earlier or the USI would respond to a clock during a message that was intended for another slave.

These steps must be performed before the first edge arrives on SCLK from the master so you should check whether there is sufficient time, including the interrupt latency and the time required to restart the DCO from LPM4.

Rising Edge on $\overline{SS}$

This signals the end of the transfer. I first disabled SDO so that other slaves can drive their outputs when selected. The next step, writing 0 to USICNT, is unnecessary if the transfer is completed successfully. I include it to stop the USI if the master raises $\overline{SS}$ unexpectedly. The next job is to take a new reading of the temperature for the next transmission. I disabled interrupts on the $\overline{SS}$ input so that the slave does not respond to a new request from the master until it is ready. A weakness of SPI is that the master cannot tell whether the slave is prepared to respond—I²C handles this much better. The operating mode is changed from LPM4 to LPM0 to provide SMCLK for the SD16A.

The interrupt service routine for the SD16A executes the usual calculation. It then copies the result to USISR and reenables interrupts on $\overline{SS}$ so that everything is ready for the next transmission. The reason for setting bit 15 in the "diagnostic" line becomes clear when we look at the oscilloscope traces. Finally, the ISR restores LPM4.

Listing 10.6: SPI slave on a F2013 using `usispislave1.c`. It is enabled by $\overline{SS}$ and transmits a 16-bit value for the temperature. Most time is spent in LPM4 with no clocks running, even during the SPI transfer.

```
// usispislave1.c - measure temperature and transmit as slave over SPI
// 16-bit transfer, SPI mode 0 (CPOL = CPHA = 0) with _SS, msb first
// New reading taken after previous value transmitted; Vref kept active
// F2013 on TI Experimenter's Board:
```

```
//    SCLK input on P1.5, MISO (SDO) on P1.6, MOSI (SDI) on P1.7
//    P1.4 is _SS (nSS) with interrupts; SDO released when _SS inactive
//    LED on P1.0 to show activity
// Calibrated 12MHz DCO, no crystal, no ACLK
// J H Davies, 2007-11-11; IAR Kickstart version 4.09A
//-------------------------------------------------------------------
#include <io430x20x3.h>             // Header file for this device
#include <intrinsics.h>             // Intrinsic functions
#include <stdint.h>                 // Integers of defined sizes

#define LED      P1OUT_bit.P1OUT_0  // Output pin for LED (active high)
#define nSS      BIT4               // Input pin for _SS on port 1

void main (void)
{
    WDTCTL = WDTPW | WDTHOLD;       // Stop watchdog
    BCSCTL1 = CALBC1_12MHZ;         // Calibrated range for DCO
    DCOCTL = CALDCO_12MHZ;          // Calibrated tap and modulation
    P2SEL = 0;                      // Digital i/o rather than crystal
    P2DIR = 0;                      // Inputs
    P2REN = BIT6 | BIT7;            // Pull resistors on unused pins
// USI will override these settings for SCLK, SDI and SDO
    P1OUT = 0;                      // LED low (off)
    P1DIR = BIT0;                   // LED output, others input
    P1REN = BIT1|BIT2|BIT3;         // Pull unused pins
    P1IE = nSS;                     // Interrupts for _SS
    P1IES = 0;                      // Rising edge first to initialize
    P1IFG = nSS;                    // Simulate interrupt from nSS
// Configure SD16A: clock from SMCLK/12 = 1MHz, internal reference on
    SD16CTL = SD16XDIV_1 | SD16DIV_2 | SD16SSEL_1 | SD16REFON;
// Unipolar output, "single" convs, OSR = 1024, interrupts on finish
    SD16CCTL0 = SD16UNI | SD16SNGL | SD16OSR_1024 | SD16IE;
// PGA gain = 1, temperature sensor A6+/-, interrupts after 4th result
    SD16INCTL0 = SD16GAIN_1 | SD16INCH_6 | SD16INTDLY_0;
// No external inputs
    SD16AE = 0;
// Enable SDI, SDO, SCLK, msb first, slave, output not enabled, latch on
    USICTL0 = USIPE7 | USIPE6 | USIPE5 | USISWRST;     // Hold in reset
// Read then write (CPHA = 0 -> CKPH = 1), SPI not I2C, no interrupts
    USICTL1 = USICKPH | USIIFG;     // Can't clear USIIFG while reset
// No clock generator for slave, clock idles low (CPOL = CKPL = 0)
    USICKCTL = 0;
    USICTL0 &= ~USISWRST;           // Release USI from reset
    for (;;) {                      // Loop forever with interrupts
        __low_power_mode_4();       // LPM4 between interrupts, no clock
    }
}
//-------------------------------------------------------------------
// ISR for _SS on P1.4: enable SDO and SPI on falling edge (_SS active)
// Disable SDO and _SS, take new reading of temperature on rising edge
//-------------------------------------------------------------------
#pragma vector = PORT1_VECTOR
__interrupt void PORT1_ISR (void)   // Not acknowledged automatically
{
    P1IFG &= ~nSS;                  // Clear flag, acknowledge interrupt
    if ((P1IES & nSS) != 0) {       // Falling edge, _SS now active
```

```
        USICTL0 |= USIOE;              // Enable slave output
        USICNT = USI16B | 16;          // Start SPI to transfer 16 bits
        LED = 1;                       // Turn LED on, show activity
    } else {                           // Rising edge, _SS now inactive
        USICTL0 &= ~USIOE;             // Disable slave output
        USICNT = USI16B | 0;           // Abort any transfer in progress
        P1IE &= ~nSS;                  // Disable _SS during conversion
        SD16CCTL0_bit.SD16SC = 1;      // Trigger new conversion for temp
// Change mode from LPM4 to LPM0 on exit to provide SMCLK for SD16
        __bic_SR_register_on_exit(LPM4_bits);
        __bis_SR_register_on_exit(LPM0_bits);
    }
    P1IES ^= nSS;                      // Toggle edge sensitivity
}
//---------------------------------------------------------------------
// Wait for SD16A to complete conversion of temperature
//---------------------------------------------------------------------
// ISR for SD16A: compute temperature (units of 0.01C), move to USI
//---------------------------------------------------------------------
#pragma vector = SD16_VECTOR
__interrupt void SD16_ISR (void)       // Acknowledged when SD16MEM0 read
{
    uint32_t temperature;              // Converted value of temperature
                                       //    in units of 0.01degC
    temperature = SD16MEM0;            // Raw converted value
    temperature *= 45454;              // Scaling factor (Vref/temp coeff)
    temperature += 0x8000;             // Allow for rounding in division
    temperature /= 0x10000;            // Divide by range of SD16A
    temperature -= 27300;              // Subtract offset to give celsius
    USISR = (uint16_t) temperature;    // Copy lower word for transmission
//  USISR |= BITF;                     // Diagnostic: Set first bit of TX
    P1IE |= nSS;                       // Reenable interrupts for _SS
    LED = 0;                           // Turn LED off, show finish
// Change mode from LPM0 to LPM4 on exit: clocks no longer needed
    __bic_SR_register_on_exit(LPM0_bits);
    __bis_SR_register_on_exit(LPM4_bits);
}
```

Figure 10.10(a) shows a complete transaction with $\overline{SS}$ magnified to show its relation with all the other signals. The final bit transmitted by the USI over MISO is meaningless in mode 0, because it is the first bit received from MOSI, and I chose this to be a one to highlight the behavior at the end of the transfer. The SDO (MISO) signal decays when $\overline{SS}$ becomes inactive (high), showing that the slave has correctly released the line. On the MSP430FG4618/F2013 Experimenter's Board the MISO line drifts to V_{SS} when there are no active outputs, rather than to a voltage midway between V_{SS} and V_{CC} as I show in Figure 10.2.

The low idling value makes it awkward to see when MISO becomes active again because the first bit in the temperature is always 0. I therefore added the line in Listing 10.6 to set the msb of the temperature (it might have been better to change the sign of the

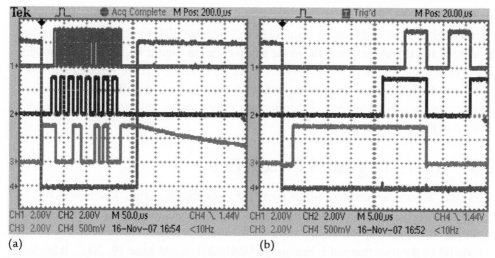

Figure 10.10: Transfer of 16 bits over SPI between the FG4618 as master and F2013 as slave in mode 0. The traces from top to bottom are SCLK, MOSI, and MISO; $\overline{SS}$ is magnified to show its relation with all the other signals. (a, left) Complete transaction, showing decay of MISO after the slave has released its output when $\overline{SS}$ becomes inactive. (b, right) Details of start to show when MISO becomes active.

temperature by taking the twos complement). Now we can see clearly that MISO becomes active again just after $\overline{SS}$ goes low. This is some time before the first bit, which is also 1, appears on MOSI. This part of the signals is expanded in Figure 10.10(b). MOSI goes high one half cycle of SCLK before the first rising edge of the clock because SCLK is available to the master. MISO goes high about 2.5 µs after the falling edge of $\overline{SS}$. This is the time required to request an interrupt from the port, restart the DCO of the F2013 from LPM4, and execute the code in the interrupt service routine.

One reason why the output of the slave is driven earlier than that of the master is the different frequencies of MCLK: 12 MHz in the slave but only 1 MHz in the master. I reduced the frequency of MCLK in the slave to match that of the master, which slowed the reaction to the falling edge of $\overline{SS}$. MISO then became active just after MOSI and only about 2 µs before the first rising edge of SCLK, when the inputs are read. It is safer to ensure that the slave has a faster MCLK than the master to ensure that MISO is valid well before it is required.

I did one more experiment and enabled the internal pull-up on P1.6, which is the pin used for MISO, with the idea of seeing this line idle high rather than low between transmission. The results were strange because of the problems that I mentioned in the section "Digital Input and Output: Parallel Ports" on page 208. Enabling the pull resistor removes the full drive from the pin, so the sharp edges in Figure 10.10 become rounded because of the time required to charge the capacitances in the circuit. Another curious effect is that USIOE ceases to have any influence. Nor is the line pulled up when idle, so the experiment was not a success. I mention this only because pull-ups work well with I²C, as we see in the next sections.

10.7   Inter-integrated Circuit Bus

The I²C bus was introduced by Philips (now NXP) Semiconductors. It was widely adopted and has become even more popular since its patents expired in 2006. It is a true bus, unlike SPI, with a specification and user manual that can be downloaded from NXP. Revision 03 of the user manual is document UM10204, dated June 19, 2007. It is clearly written and a lot easier to read than you might expect. The I²C bus uses only two, bidirectional lines:

- Serial data (SDA).

- Serial clock (SCL).

Of course there must be a connection for ground as well. It is often called the *two-wire interface*.

Thus I²C provides the full functionality of a bus while using fewer lines than SPI. Inevitably there are penalties. The first is that it is slow, only 100 kbit/sec in standard mode, because of the electrical arrangements needed to avoid damage if two nodes attempt to transmit simultaneously. Second, a protocol must be observed: You cannot merely transmit the data and nothing more, as in SPI. More hardware is needed than a simple shift register and transmissions must be controlled by logic such as a state machine. This may be implemented either in hardware, as in the USCI_B, or in software for the USI.

Transfers on the bus take place between a master and a slave. Each slave has a unique address, which is usually 7 bits long. The master starts the transfer, provides the clock, addresses a particular slave, manages the transfer, and finally terminates it. There may be more than one master on the bus although only one can be in control at a time. I do not describe multimaster buses in any detail because most systems with I²C have only a single master and a few slaves—sometimes just one. SPI would be simpler in this case but I²C saves pins.

10.7.1   Hardware for I²C

The electronic interface to the I²C bus is shown in Figure 10.11 for a master and two slaves. A full-featured slave has the same hardware as a master but most are simpler and cannot drive the clock line SCL. On the other hand, slaves must always be able to drive SDA even if they receive only data as in a DAC, for instance.

Digital outputs are normally driven actively for both their binary values, either to V_{SS} for logic 0 or to V_{CC} for logic 1. This was described in the section "Digital Outputs" on page 238. Problems clearly arise if two such outputs are connected together on a bus and attempt to drive it to different values. The I²C bus avoids this by using active devices to pull the lines of the bus down but not up. Pull-up resistors R_p keep the lines at V_{CC} when none of the drivers is active. The devices must therefore have *open-drain* (or open-collector) outputs. This means that there is only an n-channel MOSFET between the output and ground as in Figure 10.11, not the usual complementary pair as in Figure 7.1.

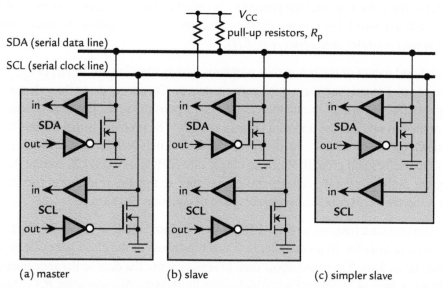

Figure 10.11: Electronic interface to the I²C bus. The two lines of the bus, SCL and SDA, are bidirectional and pulled up to V_{CC} with resistors R_p. (a) A master device can read and write both SCL and SDA independently. (b) A slave may have an identical interface but (c) most slaves cannot drive SCL.

Each line of the bus works like this. The pull-up resistor R_p holds the line at V_{CC} when there is no activity, so both the clock and data idle high. If a single n-MOSFET is turned on, its line is pulled down to V_{SS} to give a logical 0. A wasted current flows to ground through the pull-up resistor as for push-button inputs (see Figure 4.4). Nothing changes if a second n-MOSFET is turned on because it has the same effect as the first one. There are no short circuits between V_{CC} and V_{SS} or anything like that because there are no transistors connected between the outputs and V_{CC}.

Contention arises if one device tries to write a 1 to the bus and a second writes a 0. The second device "wins" because its output actively pulls the line down, while the device with an output of 1 does nothing itself, but relies on the pull-up resistor to provide the voltage. The value of 1 is therefore called recessive while 0 is dominant. The arrangement is also called *wired and* because the value on a line is 1 only if all the devices write a 1 to it: The line is pulled down to 0 if one or more devices writes a 0. A device that writes to the bus when there is any possibility of contention must monitor the value on the bus. It has lost control and must relinquish the bus if it detects any difference between the value on the bus and the value that it writes. This happens only if more than one master attempts to start a transfer at the same time and the procedure is called *arbitration*. It ensures that only one master is left in control, but I do not describe multimaster buses further. A similar approach is used in the controller area network (CAN).

The wired-and nature of the bus avoids problems with contention but brings two disadvantages. I mentioned the wasted current already. This depends on the value of the pull-up resistors R_p, which should therefore be as large as possible. The second issue is the speed of the bus and has the opposite influence on R_p. The problem is that the bus has capacitance C_b, which arises from every input and output as well as between the wires of the bus themselves and ground. When an output turns off its n-MOSFET, current flows through the pull-up resistor to charge this capacitance so that the voltage can rise toward V_{CC}. It is the classic exponential charging curve of equation (7.1) again.

The maximum clock frequency on the I²C bus in "standard" mode is $f_{SCL} = 100$ KHz. Each cycle therefore lasts for 10 μs and there are two transitions of the clock per cycle, so each half cycle takes 5 μs. This must allow time to read or write the data so the specification requires a rise time of less than 1 μs, which sets a limit on the time-constant $R_p C_b$. Suppose that $C_b = 100$ pF. Then $R_p < (1$ μs$)/100$ pF $= 10$ kΩ. Often the capacitance is higher and typical values of R_p are 1–10 kΩ. At the bottom of this range, the current through R_p when the bus is pulled down rises to $V_{CC}/R_p = 3$ mA for $V_{CC} = 3$ V. This is the maximum permitted by the specification and is more than an MSP430 draws under many conditions.

It is much faster to pull the bus down to V_{SS} because the resistance of the n-MOSFET when it is turned on is usually much smaller than R_p. If none of the slaves can drive SCL, there is no possibility of contention on this line. The master could then use a standard digital output, driving the line to both V_{SS} and V_{CC}, and the pull-up resistor could be omitted. This would give sharper transitions on SCL and save the wasted current through the pull-up resistor. SDA is always bidirectional so this approach is not applicable.

There are faster modes for I²C. The "fast" mode permits f_{SCL} to reach 400 KHz by allowing higher currents and therefore smaller values for the pull-up resistors. Fast mode plus can run at 1 MHz. The high-speed mode reaches 3.4 MHz by including an active pull-up for the clock and numerous other changes to the specification.

10.7.2 I²C Protocol

A simple example of a transfer on I²C is shown in Figure 10.12, where the master reads a single byte from the slave. The slave has presumably been configured to respond in this way by writing a command in a previous transfer.

Transfers consist of a sequence of 8-bit bytes, which are sent with the msb first and must be acknowledged to confirm successful reception. The recipient does this by writing a further acknowledgment (A) bit of 0 to SDA.

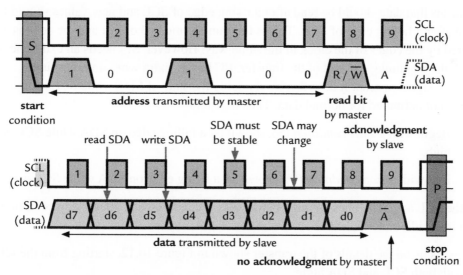

Figure 10.12: Simple transfer over I²C. The master writes an address, which is acknowledged by the slave, and reads a single byte from the slave.

The first byte in a message specifies the address of the slave that should respond. The address contains only 7 bits because the final bit of the byte (the lsb), called $R/\overline{W}$, shows whether the master wishes to write subsequent bytes to the slave or read data back from it:

- $R/\overline{W} = 0$ means that the master will *write* data to the slave. The master is the transmitter and the slave is the receiver. In this case the master transmits everything in the transfer except the acknowledgment bits, which the slave must provide for every byte.

- $R/\overline{W} = 1$ means that the master would like to *read* data from the slave. The direction of transmission therefore changes after the address byte so that the slave becomes the transmitter and the master is the receiver. The master must acknowledge each byte except the last.

In all cases the master provides the clock for the whole transfer.

Look next at the level of individual bits. The timing relationship between the data on the SDA line and the clock on SCL is

- Data on SDA must be stable while SCL is high.

- The state of SDA may change only while SCL is low.

This means that data should be read after a rising edge of SCL and new values should be written after a falling edge of SCL. This is the same as mode 3 of SPI and the actions are triggered by the edges of the clock in the same way. However, SPI needs an extra line $\overline{SS}$ to mark the beginning and end of the transfer. I²C has no extra wires and must use a distinctive signal instead. It does this by changing the state of SDA while SCL is high, which is not permitted for normal data. Thus transfers

- Begin with a *start condition* (S), shown by a falling edge on SDA while SCL is high.

- End with a *stop condition* (P), shown by a rising edge on SDA while SCL is high.

Both the SCL and SDA lines are high before a start condition and after a stop condition, which is the idle state of the bus.

We now focus on the details of the transfer shown in Figure 10.12, starting from the idle state with both SCL and SDA high:

1. The master sends a start condition (S) by pulling SDA low while SCL is high.

2. The master starts the clock and puts the first bit of the address on SDA after SCL has gone low.

3. The value on SDA is valid after SCL has gone high and is read by all slaves on the bus.

4. The last two steps are repeated until all 7 bits of the address have been sent.

5. The final bit of the first byte specifies the direction for the rest of the transfer. Here it is $R/\overline{W} = 1$, which shows that the master wishes to read data from the slave.

6. The ninth bit is the acknowledgment (A or Ack), which is low and is sent by the slave that recognizes its address.

7. The master must check that a slave acknowledges the address and abort the transfer if the low bit is missing.

8. The next 8 clock cycles are used to transmit 1 byte of data from the slave to the master. The master continues to provide the clock.

9. The ninth bit would normally be an acknowledgment but this is the exceptional case: The master does *not* acknowledge the final byte that it wishes to read in a transfer. This signals to the slave that the master has received sufficient data. Here the master expects only a single byte so it does not pull SDA low. This is a "not acknowledgment" signal ($\overline{A}$ or Nack).

10. There is a final cycle of the clock to set up the stop signal. The master pulls SDA low after the falling edge of the clock, which is the normal time for changing SDA. It releases it again after the final rising edge of the clock to give a rising edge on SDA while SCL is high, which provides the stop signal (P).

The success of I²C rests on this simple, elegant protocol.

A slave that needs more time to prepare data for a transfer can *stretch* the clock by holding SCL low after the master has driven it low, so that SCL cannot rise when the master releases it. The master must stop its internal clock and wait until the slave releases SCL. Clearly the slave needs a driver for SCL, as in Figure 10.11(b). This is uncommon and clock stretching is rarely used. However, masters must handle it if they comply with the specification for I²C. This is particularly important with the USI because it exploits clock stretching to allow time for software to prepare the next part of a transaction. The master has control of the clock and can therefore stretch it whenever it wishes.

I²C Addresses

Each device on an I²C bus needs a unique 7-bit address. There are far too few of these for each device to be given a unique address during manufacture, as in ethernet, or for each distinct type of component to have its own address. Moreover, it is common to have several devices of the same type on a single bus—a set of temperature sensors in different parts of a computer, for instance. The general practice is therefore that the more significant bits of the address are hardwired inside a component to identify its function, while the less significant bits pick out each individual device. For example, many ADCs and sensors use the set of addresses 1001xxx. In the past it was common to select the lower bits by tying pins high or low. The Burr–Brown TMP275 temperature sensor uses the voltage on pins A0–A2 to set the lower 3 bits, for instance. Three address pins are a luxury that cannot be afforded in modern devices with tiny packages. It is possible to get four addresses from the state of a single pin, as in the TMP102. Other products come in variants that are identical apart from a different I²C address, which is fixed during manufacture. Make sure that you order a selection of addresses if the devices are to share a bus.

Addresses of the form 0000xxx and 1111xxx are reserved. The most important is the "general call" with an address of 0000000 and $R/\overline{W} = 0$. This is used to configure and reset the system but I do not describe it further. There is is an extension to 10-bit addressing but few I²C networks come close to exhausting the usual 7-bit addresses so this is rarely used.

Combined Transfers on I²C

Each transaction proceeds in a single direction determined by the $R/\overline{W}$ bit after the master has placed the address of the slave on the bus. At first glance this seems to rule out more complicated transactions, where the master first transmits a command and then receives the resulting data. The solution is to use two transfers. First is a "master transmit" where the master sends the address with $R/\overline{W} = 0$ followed by the command. This is followed by a second "master receive" transfer, where the master sends the address again but with $R/\overline{W} = 1$. The slave responds by sending the data. This is called a *combined transfer*.

An obvious problem is that the master needs to retain control of the bus for the whole transaction, but another master could attempt to win it after the first transfer. As usual there is an elegant solution with I²C. The master omits the usual stop condition (P) at the end of the first transfer before it starts the second transfer with the usual start condition (S). This is called a *repeated start condition* (Sr). The bus is considered idle only after a stop condition, so its absence after the first part of the transaction ensures that the original master retains control.

Systems Built on I²C

Several systems have architectures based on I²C. Probably the most common is the system management bus (SMBus), which is widely used in PCs and other electronic systems whose status needs to be monitored, such as smart batteries. SMBus employs I²C for its low-level communication but with a tighter specification. A major difference is that the clock in I²C can run at an arbitrarily low frequency but SMBus includes a timeout feature to prevent a faulty device from paralyzing the bus. This requires a minimum frequency of 10 KHz. SMbus also has an Alert line, which acts rather like an interrupt. There are other minor differences but in practice most devices are compatible with both I²C and SMBus.

Bit-Banging I²C

There are few extra issues if you need to drive an I²C bus using software and the standard hardware of a MSP430. I am not going to explain this in detail because the support for I²C in the USI means that it should rarely be necessary to resort to bit-banging.

The first issue is the open-drain output needed for the lines of the bus. In many cases there is only a single master, in which case a normal output can be used for SCL. This is never true for SDA but fortunately there is a simple way to simulate an open-drain output. This is to keep the corresponding bit in the output register clear, $PnOUT.x = 0$, and to drive the output by switching the direction of the pin using $PnDIR.x = \overline{SDA}$:

- When the transmitted data should be $SDA = 0$, the pin is switched to output mode with $PnDIR.x = 1$ and it pulls SDA to ground.

- When the transmitted data should be $SDA = 1$, the pin is switched to input mode and the pull-up resistor on SDA gives V_{CC}.

Another issue is the detection of start and stop conditions. Remember that these are signaled by changing SDA while SCL is high, when SDA should normally be stable. The transitions can be detected using a port interrupt on SDA but unfortunately this responds to *any* edge on SDA, not just those when SCL is high. There seems to be no efficient way of filtering the desired transitions using the usual, internal port logic but an external, edge-triggered flip-flop helps.

Several application notes show how to handle I²C in software:

- *Interfacing the MSP430 and TMP100 I²C Temperature Sensor* (slaa151) covers the most common case, where the MSP430 acts as a master and reads data from a single slave.

- *I²C Interfacing of the MSP430 to a 24xx Series EEPROM* (slaa115) is similar but with a different type of slave.

- *Software I²C Slave Using the MSP430* (slaa330) shows how an infrared receiver and keypad provide the data for an I²C slave. There is a careful analysis of the timing requirements to ensure that the slave can run at $f_{\text{SCL}} = 100$ KHz with MCLK at 8 MHz.

- *MSP430 SMBus* (slaa073) describes both a master and slave using software for SMBus rather than I²C.

Checklist—Have You ...

❏ Remembered to include the pull-up resistors on SCL and SDA in your circuit?

❏ Checked that all slaves have distinct addresses, particularly with devices whose address is fixed?

10.8   A Simple I²C Master with the USCI_B0 on a FG4618

In this section and the next we explore the two programs needed to make a thermometer using I²C on the TI MSP430FG4618/F2013 Experimenter's Board. This is equivalent to the programs in the section "A Thermometer Using SPI in Mode 0 with the FG4618 as Master" on page 526, where we used SPI. The FG4618 is the master, which receives 2 bytes of data from the F2013 as slave. I cover the USCI_B0 in the FG4618 in this section, because it is more straightforward, and leave the trickier USI to the following section. Code libraries are available for the USCI_B, described in two application notes:

- *Using the USCI I2C Master* (slaa382).

- *Using the USCI I2C Slave* (slaa383).

It is more difficult to test I²C than SPI because there is no useful equivalent of a loopback, given the bidirectional nature of the bus. A very basic test of an isolated master can be made by transmitting an address for a slave. The address is not acknowledged, of course, because there is no slave, and the master should react accordingly. This is not as daft as it sounds because the test fails if the outputs are not enabled or there are no pull-up resistors on the bus, which is less embarrassing if detected early.

The connections for I²C need a little care on the TI MSP430FG4618/F2013 Experimenter's Board. The jumpers should be connected as in Figure 10.6(b). *There are no pull-up resistors on the board* so these must be provided by activating the internal weak pull-ups in the F2013. This is easy to forget and the value of the resistors is a little high for a good waveform, as we see shortly.

I am not going to describe the hardware of the USCI_B in the I²C mode because I think that it is of limited value. Most of the important functions are hidden in a box called *state machine*, which makes it more profitable to read the text of the user's guides and the explanations of the registers. Inevitably it is more complicated to handle I²C than SPI because transactions must be controlled at three levels:

- Individual bits, much as in SPI.

- Individual bytes, each of which is followed by an acknowledgment bit associated with each.

- Overall transaction, including the start and stop bits, recognition of a slave's address, direction of transmission, and so on.

There is a big difference between the USI and USCI_B for I²C. The USCI_B has a state machine in hardware, which carries out all of the operations in the lower two levels and some at the highest level. When the device is a slave, it automatically recognizes its own address, requests an interrupt, and configures itself as a transmitter or receiver according to the R/$\overline{\text{W}}$ bit. As a master, it generates start and stop conditions when required. In both cases the acknowledgment bits are sent and received automatically. In contrast you are mostly on your own with the USI because it handles only the lowest level of the process.

The USCI_B uses some extra registers in the I²C mode and many of the bits are renamed in the registers that are also used for SPI. This makes the header files confusing, because they require two sets of definitions for many registers. For example, bit 7 of UCB0CTL0 is called UCCKPH and sets the phase of the clock in SPI mode, but becomes UCA10 for 10-bit addressing in I²C mode. These are the extra registers:

Own address, UCB0I2COA: Holds the device's own I²C address for when it is addressed as a slave. There is also a bit called UCGCEN, which can be set to enable the device to respond to a general call.

Slave address, UCB0I2CSA: Specifies the slave to be addressed when this device is a master. It needs a separate register because a device might function as either a master or slave on a multimaster system.[1]

Interrupt enable, UCB0I2CIE: Contains bits to enable four extra interrupts used in I²C mode to show changes of state: not-acknowledge, stop condition, start condition, and arbitration lost. The corresponding flags UCNACKIFG, UCSTPIFG, UCSTTIFG, and UCALIFG are in UCB0STAT.

The UCB0STAT register also holds the UCSCLLOW bit, which is set when another device on the bus holds SCL low. It is handy when debugging a system where the bus has frozen because a slave refuses to release SCL.

The USCI has two interrupt vectors, both of which are shared with the USCI_A. One is used for reception and the other for transmission in the SPI mode (see the section "SPI with the USCI" on page 513) but they are allocated differently for I²C:

- The vector for transmission (USCIAB0TX_VECTOR in the header file) is used for *both* the transmit and receive flags UCB0TXIFG and UCB0RXIFG. These are in the same registers as in SPI mode. It makes sense to share a vector for these flags because I²C is only half duplex: Data are transferred in only one direction at a time, unlike SPI and UART modes.

- The vector that is normally used for reception (USCIAB0RX_VECTOR) is instead used for the four state-change flags in UCB0STAT.

I work through a program rather than explain all the details of the module first. Let me say straight away that the programs in this section and the next are written to bring out the lower-level details of the I²C transaction and are *not* structured in a way that I would recommend in practice. I describe better approaches in the section "State Machines for I²C Communication" on page 559.

Before looking at the software, it is a good idea to know what we expect the transaction to look like. Figure 10.13 shows an oscilloscope trace of the master FG4618 reading 2 bytes from the slave F2013 on the MSP430FG4618/F2013 Experimenter's Board. This is similar to the idealized sketch in Figure 10.12 except that 2 bytes are received instead of 1 and the master therefore acknowledges the first byte. I used a fictitious value of

---

[1] There is a small bug in the data sheet and header file for the FG461x at the time of writing: Both UCB0I2COA and UCB0I2CSA are listed as bytes rather than words.

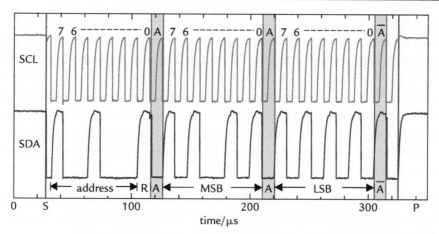

Figure 10.13: Oscilloscope trace of a transfer over I²C on an MSP430FG4618/F2013 Experimenter's Board between the USCI_B0 in the FG4618, which acts as master, and the USI in the F2013, which is the slave. The master writes an address with a read bit, which is acknowledged by the slave, and reads 2 bytes of 0xA5 and 0xAA from the slave.

0xA5AA for the data to emphasize various features of the protocol. Remember that the master provides the clock throughout the transaction. It runs close to 100 KHz here, which is the maximum for the standard mode of I²C. This is the sequence of events:

1. The master generates a start condition (S).

2. The master sends the address, 0b1001000, followed by $R/\overline{W} = 1$ to show that it wishes to read data from the slave.

3. The slave acknowledges the address by sending a low bit for A.

4. The slave sends the first byte of data, 0xA5.

5. The master acknowledges the receipt of the byte by sending a low bit for A.

6. The slave sends the second byte of data, 0xAA.

7. The master does not acknowledge this byte to show that it received sufficient data, so there is a high, not-acknowledgment bit $\overline{A}$.

8. The master sends a stop condition (P) to conclude the transaction and relinquish control of the bus.

An obvious feature of the traces is the asymmetric shape of the pulses on both SCL and SDA. The falling edges are pulled down actively by the open-drain transistors and are therefore fast. In contrast, the rising edges depend on the current that flows through the pull-up resistors to charge the capacitance of the bus. Thus the rising edges obey a classic *RC* charging curve, whose time-constant in this system is just fast enough to meet the I²C specification for the clock at 100 KHz in the standard mode. You can make the rising edges sharper by installing pull-up resistors with a smaller value but the inevitable cost is a higher current. I point out a couple of more subtle features of the traces as I explain the programs in detail.

The ports and clock are configured in much the same way as in the programs that used the SPI with the minor difference that only two pins are needed for I²C. Here is how the USCI_B0 is set up:

Control register 0, UCB0CTL0: You must choose UCMODE_3 for I²C and UCSYNC for synchronous mode. The UCMST bit configures the module as a master. Further bits can be used for 10-bit addressing and multiple masters, which I ignore.

Control register 1, UCB0CTL1: This selects the source of the clock, UCSSEL1 for SMCLK. The UCSWRST bit should remain set to hold the module in its reset condition until everything has been configured. The UCTR bit determines whether the module is the I²C transmitter or receiver and is left clear here for the receiver. We use other bits in this register later to trigger different steps in the transaction.

Baud rate control register, UCB0BR0 and UCB0BR1: Select the baud rate in the same way as for SPI. I chose a divider of 10 to give about 100 KHz. This value must be at least 4 (or 16 if there are multiple masters).

Own address register, UCBI2COA: This device is only a master so there is no need to define its own address, which is 0 by default. Nor do we use general calls, so the default state of clear for the UCGCEN bit is appropriate. The register could have been left in its default state but I include it for completeness.

Listing 10.7 does not use interrupts for I²C so I left UCB0I2CIE in its default state with all state-change interrupts disabled. I did not enable the transmit and receive interrupts in IE2 either. The slave address could have been set up in the initialization because there is only one slave in this system, but I did this later as if it might change.

Listing 10.7: I²C master by `usci2cmaster1.c`, using the USCI_B0 in the FG4618 on a TI MSP430FG4618/F2013 Experimenter's Board. It reads two Bytes from the slave.

```c
// usci2cmaster1.c - receive temperature over I2C using USCI_B0
// Master mode, receive two bytes from slave; needs pull-ups on SCL, SDA
// Simple control flow for I2C, all in main routine, no interrupts
// FG4619 on TI Experimenter's Board, 32KHz crystal, 1MHz DCO (default)
//   SCL on P3.2, SDA on P3.1
//   Display temperature on LCD, flash LED to show activity
// J H Davies, 2007-12-01; IAR Kickstart version 4.09A
//-----------------------------------------------------------------
#include <io430xG46x.h>              // Needs corrected version for USCI!
#include <intrinsics.h>             // Intrinsic functions
#include <stdint.h>                 // Integers of defined sizes
#include "LCDutils2.h"              // TI exp board utility functions

#define SLAVE_ADDRESS   0x48        // I2C address of thermometer
#define LED       P5OUT_bit.P5OUT_1 // Output pin for LED (active high)
                                    // Pin is output low by default
void main (void)
{
    volatile uint16_t i;            // Loop counter to stabilize FLL+
    int16_t Temperature;           // Value received over I2C

    WDTCTL = WDTPW | WDTHOLD;       // Stop watchdog timer
    FLL_CTL0 = XCAP14PF;           // 14pF load caps; 1MHz default
    do {                           // Wait until FLL has locked
        for (i = 0x2700; i > 0; --i) {  // One loop should be enough
        }                          // Delay for FLL+ to lock
        IFG1_bit.OFIFG = 0;        // Attempt to clear osc fault flag
    } while (IFG1_bit.OFIFG != 0); // Repeat if not yet clear
    PortsInit();                   // Initialize ports
    LCDInit();                     // Initialize SBLCDA4
    DisplayHello();                // Display HELLO on LCD
    P3SEL = BIT1 | BIT2;           // Route pins to USCI_B for I2C
// 7-bit addresses (default), single master, master mode, I2C, synch
    UCB0CTL0 = UCMST | UCMODE_3 | UCSYNC;
// Clock from SMCLK, receiver (default), hold in reset
    UCB0CTL1 = UCSSEL1 | UCSWRST;
    UCB0BR1 = 0;                   // Upper byte of divider word
    UCB0BR0 = 10;                  // Clock = SMCLK / 10 = 100KHz
    UCB0I2COA = 0;                 // Ignore genl call; own address = 0
    UCB0CTL1 &= ~UCSWRST;          // Release from reset
// Set up basic timer for interrupts at 2Hz (500ms)
    BTCTL = BTHOLD | BT_ADLY_500;  // Hold, period = 500ms from ACLK
    BTCNT2 = 0;                    // Clear counter
    BTCTL &= ~BTHOLD;              // Start basic timer
    IE2_bit.BTIE = 1;             // Enable basic timer interrupts
    for (;;) {                      // Transfers triggered by BT
        __low_power_mode_3();       // Wait for BT (needs only ACLK)
        LED = 1;                   // Show start of activity
        UCB0I2CSA = SLAVE_ADDRESS; // Slave to be addressed
        UCB0CTL1 |= UCTXSTT;       // Send Start and slave address
        while ((UCB0CTL1 & UCTXSTT) != 0) {  // Wait for address to be sent
        }
```

```
        if ((UCB0STAT & UCNACKIFG) != 0) {   // Address NOT acknowledged?
            UCB0CTL1 |= UCTXSTP;      // Send Stop condition and finish
        } else {                      // Address acknowledged: receive
            while (IFG2_bit.UCB0RXIFG == 0) {
            }                         // Wait for first byte
            Temperature = UCB0RXBUF << 8;    // MSB of temperature
            UCB0CTL1 |= UCTXSTP;      // Send Stop condition after byte
            while (IFG2_bit.UCB0RXIFG == 0) {
            }                         // Wait for second byte
            Temperature |= UCB0RXBUF;    // LSB of temperature
            DisplayCentiCels (Temperature); // Display temp to 0.01oC
        }
        LED = 0;                            // Show end of activity
    }
}
//---------------------------------------------------------------------------
// ISR for basic timer: return to main routine
//---------------------------------------------------------------------------
#pragma vector = BASICTIMER_VECTOR
__interrupt void BASICTIMER_ISR (void)  // Acknowledged automatically
{
    __low_power_mode_off_on_exit(); // Return to start new I2C message
}
```

I²C transactions in the main loop are triggered by interrupts from the Basic Timer1 every 0.5 s. In this program I include polling loops to check when each part of the transaction has been completed but you would almost certainly use interrupts in practice. Here are the main steps:

1. The address of the slave for this transaction is written to UCBI2CSA; this can be done at any time (and does not vary in this program).

2. The transaction is started by setting the UCTXSTT bit in UCB0CTL1. You would also set the UCTR bit if the master was also to transmit the data, but it is left clear here because the master receives data from the slave. The USCI_B automatically checks that the bus is available, generates a start condition (S), transmits the address, and receives an acknowledgment bit. It clears the UCTXSTT bit when this is complete.

3. If a slave does not acknowledge the address, the UCNACKIFG flag in the status register UCB0STAT is set. An interrupt is also requested if the corresponding enable bit UCNACKIE is set in UCB0I2CIE. In this program I responded by setting the UCTXSTP bit in UCB0CTL1, which causes a stop condition to be generated and conclude the transaction.

4. If the address is acknowledged, the USCI_B proceeds automatically to receive the first byte of data. The receive flag UCB0RXIFG is raised when the byte is

complete and has been copied to the receive buffer UCB0RXBUF. An interrupt can be requested if enabled. As usual the data should be read before the next byte is complete, which clears UCB0RXIFG. The timing is fairly relaxed because of the double buffering and the USCI_B helps further by holding SCL if UCB0RXBUF has not been read in time. Thus there is no need for an overrun flag.

5. The USCI_B automatically sends an acknowledgment bit (A) and continues to receive the next byte. Here the second byte is also the last and I therefore set the UCTXSTP bit while this byte is being received. This tells the USCI_B to send a not-acknowledge bit ($\overline{\text{A}}$) after reception and generate a stop condition (P) rather than attempting to receive a further byte.

6. The receive flag UCB0RXIFG is again raised when the second byte is complete, which finishes the transaction.

Easy! There are clear diagrams in the user's guide to show possible variations including combined transfers. The only tricky issue is receiving a single byte. The UCTXSTP bit must be set while this byte is being received but there is no UCB0RXIFG flag from the previous byte to trigger this. Instead you must wait until UCTXSTT clears, check that UCNACKIFG is clear to show that the address has been acknowledged, then set UCTXSTP while the first byte is being received. This is trivial with polling but more awkward in an interrupt-driven framework.

10.9   A Simple I²C Slave with the USI on a F2013

The USI is quite different from the USCI_B because the hardware handles only the bits and may need some help even with this. These are the principal extra features of the USI in I²C mode:

• The outputs for SCL and SDA become open drain. They must be enabled with USIPE6 and USIPE7 as for SPI but USIPE5 is not needed.

• There is hardware to detect start and stop conditions. A flag is raised when each condition is detected and an interrupt can be requested on start conditions.

• The interrupt vector has two flags. One is USIIFG as in SPI mode. The extra flag is USISTTIFG, which is set when a start condition is detected.

• Lost arbitration is detected and flagged.

- Several features support clock stretching. When the USI is a master, it allows slaves to stretch the clock provided that USIDIVx > 0. As a slave, SCL is automatically held low to stretch the clock when any of these conditions is true:

  — USIIFG = 1.

  — USISTTIFG = 1.

  — USICNTx = 0.

  Normally USIIFG is cleared when a nonzero value is written to USICNTx and SCL would then be released. This is undesirable if more processing is needed before the I²C transaction should be resumed, which is why the USIIFGCC bit in USICNT can be set to prevent automatic clearing of USIIFG. SCL continues to be held low, stretching the clock, until USIIFG is cleared with software. If the clock must not be stretched, the USISCLREL bit in USICNT should be set to release SCL. This bit is automatically cleared when a start condition is detected and must therefore be repeatedly set when necessary.

- There is a similar feature to insert delays when the USI is a master. In this case it leaves the clock in its idle, high state while USIIFG = 1, which shows that the USI is waiting for software to react.

Clock stretching in a slave and the delays in a master are important because the software must perform most of the tasks carried out by the state machine in the USCI_B. It is barely possible to run an I²C slave at 100 KHz with the USI without stretching the clock, as we shall see. In fact the USI is designed on the assumption that the clock *will* be stretched when it is a slave and it is more difficult to use without this.

The USI does not handle the acknowledgment bits of I²C automatically. This means that the complete transfer of each byte must be treated as two transfers with the USI. Take the byte for the address and direction as an example. This is received as an 8-bit transfer in hardware. Software must then check the address. If it matches, it must set up the USI to send a 1-bit transfer with a 0 bit for the acknowledgment. Similarly, after sending a byte of data as an 8-bit transmission, the USI must be changed to receive mode for a single acknowledgment bit from the master.

Several bits must be given specific values when the USI is used for I²C:

- USII2C = 1 to activate the extra hardware.

- USICKPL = 1 and USICKPH = 0 so that the clock behaves as in mode 3 of SPI.

- USI16B $=0$ and USILSB $=0$ so that only bytes are sent, msb first, as expected for I²C.

These are fairly obvious but it would be easy to assume that setting the USII2C bit would enforce the other settings automatically. Here is a quick reminder of how the shift register works with this configuration of clock:

- A new bit enters the lsb of the shift register on a positive edge of the clock and the contents of the register shift left.

- The output latch is transparent while the clock is low, so that changes in the msb of the shift register appear immediately on SDA if the output is enabled with USIOE.

- The output remains constant while the clock is high and the latch holds its value.

It is particularly important to remember these aspects when switching between reception and transmission. This must be done only while SCL is low to avoid violating the rule that SDA should not change while SCL is high. A code library is described in the application note *Using the USI I²C Code Library* (slaa368).

I have to admit that the USI has given me more trouble than any other module described in this book. I encountered minor problems with SPI and had several, serious difficulties with I²C. There is no doubt that it can be made to work (see the TI code examples, libraries, and application notes) and we should be grateful to avoid bit-banging in a $1 microcontroller like the F2002. However, I strongly recommend that you test programs that use the USI with special care. This is particularly important for buses with several devices, where slaves that were not addressed can freeze the bus if you are not careful to ensure that they release SCL. It is also possible to get poor-quality pulses that give unreliable communication. These are obvious on the screen of an oscilloscope but hard to detect from the received data alone.

After that caution, I describe a program to provide a slave transmitter for the thermometer. It is designed to work without stretching the clock if the processor is "fast" and the I²C bus is "slow." I clarify the limits on "fast" and "slow" later. The clock is stretched if the processor is slower. Again this program uses polling rather than interrupts so that you can see how each step in the transaction is handled.

The overall structure is that the program waits in the lowest-power mode, LPM4, until a start condition is received. This requests an interrupt, which wakes the MSP430. The USI receives an address and the R/$\overline{\text{W}}$ bit, checks these, and returns to LPM4 unless the address matches that of the slave and the master requests a read. In this case the slave sends an acknowledgment bit and 2 bytes of data. The master should acknowledge the first byte of data and the transmission is abandoned if this is missing. Finally, the SD16 takes a new reading of the temperature. The I²C transaction is handled using polling in the main routine except for the initial start condition.

The program has significant weaknesses due to its structure. For example, the interrupt on a start condition is enabled only when the program expects one. However, the master can abandon a transaction and commence a new one by issuing a start condition at any time. The slave should automatically reset itself if this happens. We see how to do better in the section "State Machines for I²C Communication" on page 559.

Listing 10.8 shows the program, which was used with the DCO running at 12 MHz for the oscilloscope traces in Figure 10.13. I already explained most of the configuration of the USI for I²C. The output is initially disabled because USIOE is clear by default so that the module is ready to receive a transfer. The address of the slave is defined as a constant and I create a further constant for the address followed by R/W̄ = 1. This program cannot respond to a write instruction (R/W̄ = 0) so it should not be acknowledged. I also define a constant for SCL because we need to check the state of the SCL line to avoid changing SDA while SCL is high. The rest of the setup is familiar from previous programs.

Listing 10.8: I²C slave thermometer in `usi2cslave7.c`, using the USI in the F2013 on a TI MSP430FG4618/F2013 Experimenter's Board to send two bytes to the master. This program does not stretch the clock if the processor is sufficiently fast.

```
// usi2cslave7.c - transmit temperature, I2C slave, optional clk stretch
// 2 bytes, MSB first, new reading after transmission; Vref kept active
// Simple control flow for I2C, mostly in main routine, Start interrupts
// F2013 on TI Experimenter's Board:
//   SCL on P1.6, SDA on P1.7, LED on P1.0 to show activity
// Calibrated 12MHz DCO, no crystal, no ACLK
// J H Davies, 2007-12-16; IAR Kickstart version 4.09A
//-----------------------------------------------------------------
#include <io430x20x3.h>          // Header file for this device
#include <intrinsics.h>          // Intrinsic functions
#include <stdint.h>              // Integers of defined sizes

#define LED       P1OUT_bit.P1OUT_0   // Output pin for LED (active high)
#define SCL       P1IN_bit.P1IN_6     // I/O pin for SCL (I2C clock)
#define SLAVE_ADDRESS   0x48          // I2C address of this thermometer
#define READ_ADDRESS    ((SLAVE_ADDRESS << 1) | BIT0)
                                      // Address plus bit R/_W = 1
void main (void)
{
    uint8_t   tempMSB, tempLSB;       // Temperature bytes to transmit
    uint32_t  temperature;            // Converted value of temperature
                                      //   in units of 0.01degC

    WDTCTL = WDTPW | WDTHOLD;         // Stop watchdog
    BCSCTL1 = CALBC1_12MHZ;           // Calibrated range for DCO
```

```
        DCOCTL = CALDCO_12MHZ;            // Calibrated tap and modulation
        P2SEL = 0;                        // Digital i/o rather than crystal
        P2DIR = 0;                        // Inputs
        P2REN = BIT6 | BIT7;              // Pull resistors on unused pins
// USI will override these settings for SCL, SDA
        P1OUT = ~BIT0;                    // LED low (off) others high
        P1DIR = BIT0;                     // LED output, others input
        P1REN = ~BIT0;                    // Pull unused pins up
// Configure SD16A: clock from SMCLK/12 = 1MHz, internal reference on
        SD16CTL = SD16XDIV_1 | SD16DIV_2 | SD16SSEL_1 | SD16REFON;
// Unipolar output, "single" convs, OSR = 1024, interrupts on finish
        SD16CCTL0 = SD16UNI | SD16SNGL | SD16OSR_1024 | SD16IE;
// PGA gain = 1, temperature sensor A6+/-, interrupts after 4th result
        SD16INCTL0 = SD16GAIN_1 | SD16INCH_6 | SD16INTDLY_0;
// No external inputs
        SD16AE = 0;
// Enable SDA and SCL, hold USI in reset state
// Defaults: msb first, slave, enable output latch, disable data output
        USICTL0 = USIPE7 | USIPE6 | USISWRST;
// I2C not SPI, can't clr USIIFG; Defaults: CKPH = 0 (CPHA = 1), no ints
        USICTL1 = USII2C | USIIFG;
//  No clock generator for slave, clock idles high (CKPL = CPOL = 1)
        USICKCTL = USICKPL;
        USICTL0 &= ~USISWRST;            // Release USI from reset
        for (;;) {                        // Loop forever
// Clean up after previous transaction
            LED = 0;                      // LED off while waiting for I2C
            USICNT = USISCLREL;           // Release SCL: we are not active
            USICTL1 &= ~USIIFG;           // Clr any pending flag (necessary?)
// Take new temperature reading with SD16
            SD16CCTL0_bit.SD16SC = 1;     // Trigger new conversion for temp
            __low_power_mode_0();         // LPM0 during conversion
            temperature = SD16MEM0;       // Raw converted value
            temperature *= 45454;         // Scaling factor (Vref/temp coeff)
            temperature += 0x8000;        // Allow for rounding in division
            temperature /= 0x10000;       // Divide by range of SD16A
            temperature -= 27300;         // Subtract offset to give celsius
//          temperature = 0xA5AA;         // Dummy value to transmit
            tempMSB = (uint8_t) (temperature >> 8); // Extract bytes
            tempLSB = (uint8_t) temperature;  //     to transmit
// Data ready: activate interrupt to wait for Start condition
            USICTL1 &= ~USISTTIFG;        // Clr any pending Start interrupt
            USICTL1 |= USISTTIE;          // Enable Start interrupts
            __low_power_mode_4();         // Wait for Start interrupt, LPM4
// Start condition detected: USI now receiving address from master
            LED = 1;                      // LED on while I2C active
            while ((USICTL1 & USIIFG) == 0) {
            }                             // Wait for 8 bits to be received
// Check address: is it ours (with read direction, not write)?
            if (USISRL != READ_ADDRESS)  // Not us?
                continue;                 // Abandon transaction: restart loop
// Send acknowledgment bit (low) after waiting for SCL to go low
            while (SCL == 1) {
            }                             // Wait for SCL to go low
            USISRL = 0;                   // Low bit to send
            USICTL0 |= USIOE;             // Enable output
```

```
            USICNT = 1;                          // Send 1 bit, clr IFG, release SCL
            while ((USICTL1 & USIIFG) == 0) {
            }                                    // Wait for bit to be sent
// Send first byte of data (MSB of temperature) after waiting for SCL
            while (SCL == 1) {                   // Wait for SCL to go low
            }
            USISRL = tempMSB;                    // Send MSB of temperature
            USICNT = 8;                          // Send 8 bits, clr IFG, release SCL
            while ((USICTL1 & USIIFG) == 0) {
            }                                    // Wait for bits to be sent
// Check for acknowledgment: receive Ack/Nack bit
            while (SCL == 1) {                   // Wait for SCL to go low
            }
            USICTL0 &= ~USIOE;                   // Disable output
            USICNT = 1;                          // Rec 1 bit, clr IFG, release SCL
            while ((USICTL1 & USIIFG) == 0) {
            }                                    // Wait for bit to be received
            if ((USISRL & BIT0) != 0)            // Received Nack instead of Ack?
                continue;                        // Abandon transaction: restart loop
// Acknowledgment received: send second (last) byte of data
            while (SCL == 1) {                   // Wait for SCL to go low
            }
            USISRL = tempLSB;                    // Send LSB of temperature
            USICTL0 |= USIOE;                    // Enable output
            USICNT = 8;                          // Send 8 bits, clr IFG, release SCL
            while ((USICTL1 & USIIFG) == 0) {
            }                                    // Wait for bits to be sent
// Receive final not-acknowledgment (and ignore it)
            while (SCL == 1) {                   // Wait for SCL to go low
            }
            USICTL0 &= ~USIOE;                   // Disable output
            USICNT = 1;                          // Rec 1 bit, clr IFG, release SCL
            while ((USICTL1 & USIIFG) == 0) {
            }                                    // Wait for bit to be received
    }                                            // Repeat loop (start with clean up)
}
//-------------------------------------------------------------------
// ISR for USI: Start condition, set up USI to receive address
//-------------------------------------------------------------------
#pragma vector = USI_VECTOR
__interrupt void USI_ISR (void)          // Not acknowledged automatically
{
    USICTL1 &= ~USISTTIE;                       // Disable further Start interrupts
    USICNT = 8;                                 // Read 8 bits, clear USIIFG
    __low_power_mode_off_on_exit();             // Return to main function
    USICTL1 &= ~USISTTIFG;                      // Acknowledge int'pt, release SCL
}
//-------------------------------------------------------------------
// ISR for SD16A: return to main function to compute temperature
//-------------------------------------------------------------------
#pragma vector = SD16_VECTOR
__interrupt void SD16_ISR (void)         // Not acknowledged automatically
{
    SD16CCTL0_bit.SD16IFG = 0;                  // Clear flag, acknowledge interrupt
    __low_power_mode_off_on_exit();             // Return to main function
}
```

After initialization, the program takes a reading of temperature and splits the result into two bytes for transmission. This saves time when handling the I²C transaction and the whole program is written with this aspect in mind. Here is the operation of the main loop:

1. The first two lines for the USI "clean up" after the previous transaction. It would be more logical if they were at the end of the loop but I used the `continue` statement to abort a transaction and this starts a new iteration. The USISCLREL bit is set to prevent this slave holding SCL low because it should take no further part in I²C activity until the next start condition.

2. Interrupts on start are enabled after the temperature has been converted. I cleared the flag before this to avoid a possible interrupt from an old start condition.

3. The processor waits in LPM4 until a start condition is detected. This raises the flag USISTTIFG, wakes the processor with an interrupt, and clears USISCLREL to enable clock stretching now that the slave is active again. The USI is prepared to receive the byte containing the address and R/$\overline{\text{W}}$ bit by writing 8 to USICNT. Clearing USISTTIFG also releases SCL so that activity on the I²C bus resumes. I disabled further interrupts on start conditions because there is no way of handling them within the structure of this program.

4. Back in the main loop, we poll the USIIFG flag to wait until all 8 bits have been received. This flag is set just after the rising edge of SCL, so SCL is now high and SDA must not be changed.

5. The received address and R/$\overline{\text{W}}$ bit are checked against the constant READ_ADDRESS. If these do not match, the `continue` statement causes this iteration of the main loop to be abandoned and a new one starts. The cleanup code at the start of the loop ensures that the bus is not held any further by the USI.

6. A low acknowledgment bit must be sent if the address and direction match. This is done by setting up the USI to transmit a single bit of 0. However, we must not enable the output until SCL has gone low so there is another polling loop to wait for this. I then cleared USISRL (easier than clearing the msb alone), wrote 1 to USICNT to send a single bit, and enabled the output by setting USIOE. The output latch is transparent when SCL is low so the msb of USISRL appears on the output as soon as USIOE is set; it does not wait for an edge of the clock.

   You might reasonably ask, Why not write to USISRL and USICNT immediately after finding that the address matches, and delay only the setting of

USIOE to enable the output until SCL has gone low? In principle this should work but unfortunately there is a bug (USI4) in the USI at the time of writing. This causes a glitch to appear on SCL if it is high when a value is written to USICNT. I therefore take a cautious approach and do nothing to the USI until SCL has gone low.

7. The first byte of data is sent after the acknowledgment bit. Again there is a polling loop to wait for SCL to go low because writing the data to USISRL would otherwise change the output immediately. This is not permitted while SCL is high.

8. The master should acknowledge the data so the USI must be set up to receive a single bit after sending the byte. Yet again this cannot be done until SCL has gone low to avoid disturbing SDA, so there is another polling loop. The output is then disabled, ready to receive the bit.

9. If the first byte is acknowledged, the second byte is sent in the same way; if not, the program continues to the next iteration of the main loop.

10. Last, the program receives the final not-acknowledgment bit from the master. I am not sure what the slave is expected to do if an acknowledgment is received instead of a not-acknowledgment after it transmitted all of its data and therefore ignore the possibility. This concludes the main loop, which restarts with the cleanup code.

Whew! This is a great deal more effort than the USCI_B. Perhaps the ugliest feature of the program is the polling loops that wait for SCL to go low. These are required only because of my desire to construct a slave that would function perfectly without stretching the clock. This enabled me to get the traces of a "textbook" I²C transaction in Figure 10.13.

How fast can SCL run before it becomes necessary to stretch the clock? The absence of any buffering means that the software must react to the end of one USI transfer and prepare for the next within one cycle of SCL. This gives about 10 μs for a 100 KHz bus, ignoring finer details such as the data setup time. The most challenging operation is probably to respond to the address and issue an acknowledgment if it matches. I list the time required for each step in Table 10.2. This corresponds to Listing 10.8 except that I make the more realistic assumption that the processor is put into a low-power mode with the DCO off (LPM3 or LPM4) while it receives the address and is reactivated by an interrupt. On the other hand, I did not allow for extra instructions that would be needed for a decision if the

Table 10.2: Time required to respond to address received over I²C and set up USI to send acknowledgment.

Operation	Time
Wake from LPM4, start DCO	2 μs
Accept interrupt	6 cycles
Compare USISRL	5 + 2 cycles
Test SCL	5 + 2 cycles
Clear USISRL	4 cycles
Write to USICNT	4 cycles
Set USIOE	4 cycles
Total	2 μs + 32 cycles

slave could respond to both values of the R/$\overline{\text{W}}$ bit or for the overhead needed to run a state machine.

The table shows that it takes about 2 μs + 32 cycles to complete the operations, which must fit into the 10-μs period of SCL. This needs $f_{\text{MCLK}} \geq 4$ MHz. In practice you should make an allowance for jitter on the clock and consider the rise and setup times carefully. A helpful aspect is that the acknowledgment bit is low and it is fast to pull SDA down—there is no need to allow for the slow rise due to the pull-up resistor. Thus the practical minimum is probably around 5 MHz but I did not test this. Figure 10.13 looks good with the DCO at 12 MHz. Nevertheless, the delays due to software can be seen on close inspection. The falling edge of the R/$\overline{\text{W}}$ pulse at the end of the address byte from the master is nearly 2 μs after the corresponding edge on SCL because of the instructions that are executed before USIOE is set and the slave drives SDA low for the acknowledgment pulse. Similarly, the msb of the first byte of data sent by the slave is a little short. (The USCI_B also has delays, although they cannot be seen on the scale of Figure 10.13. The final falling edge of SDA, needed for the stop condition, is 1 μs after the final falling edge of SCL. It presumably takes one cycle of SMCLK for the state machine in the USCI_B to begin the stop condition.)

The conclusion is that it is hard work for the processor to run an I²C transaction without stretching the clock and needs roughly $f_{\text{MCLK}} > 50 f_{\text{SCL}}$ or $f_{\text{MCLK}} > 5$ MHz for the maximum speed in the standard mode, $f_{\text{SCL}} = 100$ KHz. The clock is stretched if MCLK is too slow. This estimate is pessimistic for Listing 10.8 because it does not use a low-power

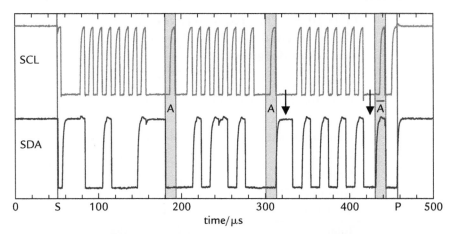

Figure 10.14: Oscilloscope trace of a transfer over I²C on an MSP430FG4618/F2013 Experimenter's Board between the USCI_B0 in the FG4618, which acts as master, and the USI in the F2013, which is the slave. The F2013 runs at 1 MHz rather than 12 MHz, as in Figure 10.13, and therefore has to stretch SCL. The arrows point to two pulses that exist only while SCL is low and are therefore meaningless.

mode during I²C transactions and there is no delay for MCLK to restart. I could reduce f_{MCLK} to about 3 MHz before stretching was visible on the oscilloscope.

An example of a complete transaction with $f_{MCLK} = 1$ MHz is shown in Figure 10.14. Each transfer by the USI runs at the full clock speed of 100 KHz but there are clear gaps between these transfers where SCL is low and the clock has been stretched.

I changed the data to 0x5A55 to show another feature, which may perplex you when you see similar traces on an oscilloscope. Arrows point to two pulses in SDA, both of which look like good data but neither of which has any effect. If you examine the traces closely, you will see that both are restricted to times when SCL is low and go away when SCL returns high.

The first of these pulses follows the acknowledgment bit of 0 for the first byte. This bit is sent by the master, which releases SDA when SCL goes low so that the slave can transmit the next byte. The voltage on SDA then rises toward V_{CC} as the bus is pulled up by its resistor. The next bit is also a 0 from the slave because it is the msb of the second byte of data, 0x55, but this does not appear on SDA until a few lines of software have written the byte to USISRL and enabled the output of the USI. This gives plenty of time for a well-developed high pulse. It becomes narrower if f_{MCLK} is raised but a narrow glitch is clearly visible even at 12 MHz. This sort of feature is almost inevitable with I²C, where control of a bus line is handed over from one device to another.

Example 10.2

Can you explain the second (low) pulse, just before the final not-acknowledgment pulse ($\overline{A}$)?

10.10   State Machines for I²C Communication

Nobody would write a real program for communication using polling loops in the main function. This prevents the processor from doing anything useful and wastes power, particularly in a slave that could spend much of its time in LPM4 with no clocks running at all. The loops that wait for flags to be raised should be replaced by interrupts but these must form part of a more complicated framework than in most previous programs because of the protocol to be followed for I²C communication. A more advanced structure is required for the program and a popular choice is a state machine.

I introduced periodic state machines in the section "Introduction to State Machines" on page 358. They were called regularly by using a timer and the actions depended on the values of inputs when the machine was called. This structure is not appropriate for handling communications, where each action should be triggered by an interrupt that arrives at an unpredictable time. Instead we use an *event-triggered* state machine. This idea lies behind both programs that I describe in this section but the approach is quite different for the USI and USCI_B because the hardware of the state machine is already built into the USCI_B. I look first at the USI while its operation is still fresh in your mind.

10.10.1   State Machine for an I²C Slave Using the USI

The loops that wait for USIIFG to become set should be replaced by interrupts. In most previous programs, the same actions were always executed in response to a particular interrupt. In this case, however, there is a sequence of different actions and we must ensure that the interrupt service routine performs the correct one. The polling programs handled this by stepping through the code in the main function. This obviously cannot be done in an ISR because it always enters at the start and therefore needs some sort of memory that tells it what to do next time it is called. This is the function of the event-triggered state machine. Its state transition diagram looks much like that of a periodic state machine but each transition is triggered by a particular event (interrupt).

I take a pragmatic approach to constructing a state machine for the USI, which follows the code examples provided by TI for the F20xx. I drew a flow diagram for a slave transmitter and divided it into blocks, which are separated where the program waits for USIIFG to become set. The intervals between blocks are polling loops in the earlier program but now correspond to interrupts. Each block becomes a state of the machine. They are given names and the usual state variable records the state to be called when the next interrupt arrives. The result is shown in Figure 10.15. I did not label the transitions because almost all are triggered by USIIFG. The exceptions are the start condition at the top and the initial transition, which puts the machine into the Idle state. The actions take place "within" each state rather than being associated with transitions as in the periodic machine.

When an interrupt from USIIFG arrives, the ISR enters the state to which the state variable points and carries out the appropriate actions. The final step before leaving the ISR is to store the next state in the state variable. The flow down the left of the diagram closely resembles the structure of Listing 10.8 except for the generalization to longer messages, which requires a loop around ReceiveAck and CheckSendByte. A minor difference is that the state machine "cleans up" at the various points where it detects the end of the transaction rather than at the start of the main loop. Another change is that it takes a new reading of the temperature only after a successful conclusion of the transaction.

The state machine should never enter the Idle state during normal operation and I emphasize this by using thinner lines for the arrows that lead to it. If something goes wrong and the machine enters the Idle state, the circular line shows that further interrupts from USIIFG cause it to stay there. This may not be useful but at least no harm is done.

Interrupts from USIIFG cause the machine to step through the states of a transaction. An interrupt is also requested when a start condition is detected and the USISTTIFG flag is raised. This is something like a reset signal and marks the beginning of a new I²C transaction, regardless of the previous state of the machine. I could have added further arrows labeled USISTTIFG that leave each state and point to ReceiveAddress. This would clutter the diagram and obscure the usual flow so I instead drew the start condition as a "higher-level" event to show that USISTTIFG takes priority over USIIFG.

This structure gives much more robust behavior than Listing 10.8 if there is a problem with a transaction. A start condition always causes a transition to ReceiveAddress, whenever it occurs, which enables the state machine to reset itself and recover from many errors. If there were also interrupts on stop conditions, these could induce transitions to a state that cleaned up after the previous transaction, which might have been incomplete.

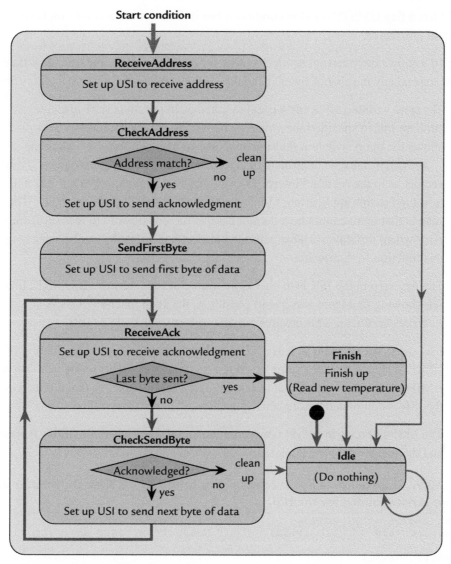

Figure 10.15: State machine for a slave I²C transmitter using the USI. It checks the address and sends any number of bytes. All transitions except those from the initial state and a start condition are triggered by USIIFG.

The USI has a flag USISTP for stop conditions but it does not request interrupts so we cannot add this feature.

Listing 10.9 shows the interrupt service routine for the USI, which implements a state machine to send any number of bytes. I should explain a couple of general features first:

- The state variable I2CState is enumerated as in previous state machines but here I have specified the values to be successive *even* numbers. This allows the main switch in the state machine to be made faster by using the __even_in_range() intrinsic function. Normally this is used to decode interrupt vectors as in the section "Interrupts from Timer_A" on page 296 but it is not restricted to this application. I took this trick from the TI code examples. The only snag is that there cannot be a default case so there is no way to recover from an error where the state variable gets corrupted out of range. It may be safer to reserve this function for hardware.

- The first step in the ISR is to check the source of the interrupt. If USISTTIFG is set, showing that there was a start condition, the state machine restarts at ReceiveAddress. This ensures a higher priority for USISTTIFG.

- I set the USIIFGCC bit in USICNT, which prevents the automatic clearing of USIIFG when a nonzero value is written to USICNTx.[2] This holds SCL low until USIIFG is cleared by the final statement in the ISR. It ensures that all the work of the ISR is completed before activity is allowed to resume on the I²C bus.

- The MSP430 rests in LPM4 between interrupts except when the SD16_A is active and the temperature is calculated.

Listing 10.9: State machine in `usi2cslvsm4.c` to run an I²C slave transmitter in the interrupt service routine for the USI.

```
#pragma vector = USI_VECTOR
__interrupt void USI_ISR (void)          // Not acknowledged automatically
{
    static uint8_t TXcount;              // Counter for bytes transmitted
    static enum {                        // States of state machine
        Idle = 0,
        ReceiveAddress = 2,
        CheckAddress = 4,
        SendFirstByte = 6,
        ReceiveAck = 8,
```

---

[2] This is called USIIFGCC in the user's guide but USIFGCC in the EW430 header file at the time of writing.

```
        CheckSendByte = 10,
        Finish = 12
    } I2CState = Idle;                  // State machine initially idle

    if (USICTL1 & USISTTIFG) {          // Start condition received
        LED = 1;                        // LED on when I2C active
        I2CState = ReceiveAddress;      // Start state machine
    }
// Select state of machine, perform actions, move on to next state
    switch (__even_in_range(I2CState, Finish)) {
    case Idle:                          // Should never happen
        break;
    case ReceiveAddress:                // Prepare to receive address
        USICNT = USIFGCC | 8;           // Stop auto clr USIIFG, read 8 bits
        USICTL1 &= ~USISTTIFG;          // Clear flag to acknowledge Start
        I2CState = CheckAddress;        // Check received address next
        break;
    case CheckAddress:                  // Check received address
        if (USISRL != READ_ADDRESS) {   // Was it our address?
            USICNT |= USISCLREL;        // No: Release hold on SCL
            I2CState = Idle;            // Return to idle state
            LED = 0;                    // LED off when I2C inactive
        } else {                        // Yes: send acknowledgment bit
            USISRL = 0;                 // Clear bit to be sent (msb)
            USICTL0 |= USIOE;           // Enable output
            USICNT |= 1;                // Send 1 bit
            TXcount = N_TRANSMIT - 1;   // Initialize byte counter
            I2CState = SendFirstByte;   // Send first byte next
        }
        break;
    case SendFirstByte:                 // Send first byte of data
        USISRL = TXdata[TXcount];       // Load most significant byte
        USICNT |= 8;                    // Send 8 bits
        I2CState = ReceiveAck;          // Receive acknowledgment next
        break;
    case ReceiveAck:                    // Receive acknowledgment
        USICTL0 &= ~USIOE;              // Disable output to receive Ack
        USICNT |= 1;                    // Receive 1 bit
        if (TXcount == 0)        {      // Has last byte been sent?
            I2CState = Finish;          // Yes: transmission finished
        } else {
            --TXcount;                  // No: Update byte counter
            I2CState = CheckSendByte;   // Check acknowledgment next
        }
        break;
    case CheckSendByte:      // Check acknowledgment and send next byte
        if ((USISRL & BIT0) != 0) { // Received Nack or Ack?
            USICNT |= USISCLREL;        // Nack: Release hold on SCL
            I2CState = Idle;            // Return to idle state
            LED = 0;                    // LED off when I2C inactive
        } else {                        // Ack: send next byte
            USISRL = TXdata[TXcount];   // Load next byte of data
            USICTL0 |= USIOE;           // Enable output
            USICNT |= 8;                // Send 8 bits
            I2CState = ReceiveAck;      // Receive acknowledgment next
```

```
        }
        break;
    case Finish:
        USICTL0 &= ~USIOE;          // Disable output to release SDA
        USICNT |= USISCLREL;        // Release hold on SCL
        I2CState = Idle;            // Return to idle state
        LED = 0;                    // LED off when I2C inactive
        SD16CCTL0_bit.SD16SC = 1;   // Trigger new conversion for temp
// Change mode from LPM4 to LPM0 on exit to provide SMCLK for SD16
        __bic_SR_register_on_exit(LPM4_bits);
        __bis_SR_register_on_exit(LPM0_bits);
        break;
    }
    USICTL1 &= ~USIIFG;             // Clear flag, release SCL if held
}
```

Oscilloscope traces produced by this program show more stretching that those for
Listing 10.8. This is expected because the DCO must be restarted after every interrupt and
the state machine introduces a further overhead. There might be problems if MCLK were
too fast because the program does not check that SCL is low before changing SDA. I used
polling loops to wait for SCL in Listing 10.8 but you should never do this in an interrupt
service routine. If the master fails and SCL remains low, the slave becomes stuck in its
polling loop and this cannot be interrupted if it is in an ISR itself.

10.10.2   State Machine for an I²C Master Using the USCI_B

It does not take much to set up a state machine for the USCI_B because it is built into the
hardware already. In essence we can let the hardware get on with the job and provide
assistance when requested by an interrupt. The only other action required is to start each
transaction by setting the UCTXSTT bit. The USCI_B0 has six interrupts in the I²C mode,
which I introduced in the section "A Simple I²C Master with the USCI_B0
on a FG4618" on page 542. Remember that both the receive and transmit flags are
associated with the USCIAB0TX vector, which handles transmission alone in other modes
of the USCI. Only two interrupts are relevant to a master receiver:

Received character, UCB0RXIFG: Set when a complete byte has been received and is
available in RXBUF. The byte should be stored, which clears the flag. Another action
may be needed, depending on the number of bytes received in the message:

- Early bytes are simply stored and no further action is needed.

- The UCTXSTP bit must be set while the final byte is being received, so that it is
  not acknowledged and is followed by a stop condition. This should be done after
  RXIFG has been raised for the next-to-last byte.

- The message is complete and should be processed after the last byte has been received.

This is the only interrupt that is active during normal reception. It is associated with the USCIAB0TX vector. Remember to enable this interrupt *after* UCSWRST has been cleared to release the USCI_B from reset.

Not-acknowledgment, UCNACKIFG: Set when an acknowledgment was expected but not received. It is associated with the USCIAB0RX vector. For a master receiver this can arise only if no slave acknowledges the address, in which case a stop condition or repeated start should be sent in response.

Listing 10.10 shows a program based on this strategy. For generality it is designed to receive an arbitrary number of bytes (2 or more) although there are only 2 for the thermometer. The ISRs wake the main function at the end of a transaction and pass back a value in I2CStatus to show the outcome. The main function can then take the appropriate action, which might be a lengthy process. It would be a bad idea to do this in an ISR because that would delay the response to the next interrupt, which must often be handled promptly for communication.

Listing 10.10: Interrupt-driven I²C master receiving using the USCI_B0 in usci2cmastsm1.c.

```
// usci2cmastsm1.c - Receive temperature over I2C using USCI_B0
// Master mode, receive two bytes from slave; needs pull-ups on SCL, SDA
// State machine to control flow for I2C interrupts; display in main
// FG4619 on TI Experimenter's Board, 32KHz crystal, 1MHz DCO (default)
//   SCL on P3.2, SDA on P3.1
//   Display temperature on LCD, flash LED to show activity
// J H Davies, 2008-01-08; IAR Kickstart version 4.09A
//-------------------------------------------------------------------
#include <io430xG46x.h>              // Needs corrected version for USCI
#include <intrinsics.h>              // Intrinsic functions
#include <stdint.h>                  // Integers of defined sizes
#include "LCDutils2.h"               // TI exp board utility functions

#define N_MESSAGE    2               // Number of bytes in message
uint8_t RXcount;                     // Counter for bytes
uint8_t RXdata[N_MESSAGE];           // Received data, [0] = LSB (last)
enum {
    I2CIdle,
    I2CSuccess,
    I2CNack
} I2CStatus = I2CIdle;               // Return status of I2C to main
#define SLAVE_ADDRESS   0x48         // I2C address of thermometer
#define LED      P5OUT_bit.P5OUT_1   // Output pin for LED (active high)
                                     // Pin is output low by default

void main (void)
{
```

```
    volatile uint16_t i;              // Loop counter to stabilize FLL+
    int16_t Temperature;             // Value received over I2C

    WDTCTL = WDTPW | WDTHOLD;         // Stop watchdog timer
    FLL_CTL0 = XCAP14PF;             // 14pF load caps; 1MHz default
    do {                            // Wait until FLL has locked
        for (i = 0x2700; i > 0; --i) {  // One loop should be enough
        }                          // Delay for FLL+ to lock
        IFG1_bit.OFIFG = 0;        // Attempt to clear osc fault flag
    } while (IFG1_bit.OFIFG != 0);  // Repeat if not yet clear
    PortsInit();                    // Initialize ports
    LCDInit();                      // Initialize SBLCDA4
    DisplayHello();                 // Display HELLO on LCD
    P3SEL = BIT1 | BIT2;            // Route pins to USCI_B for I2C
// 7-bit addresses (default), single master, master mode, I2C, synch
    UCB0CTL0 = UCMST | UCMODE_3 | UCSYNC;
// Clock from SMCLK, receiver (default), hold in reset
    UCB0CTL1 = UCSSEL1 | UCSWRST;
    UCB0BR1 = 0;                     // Upper byte of divider word
    UCB0BR0 = 10;                    // Clock = SMCLK / 10 = 100KHz
    UCB0I2CSA = SLAVE_ADDRESS;       // Slave to be addressed
    UCB0I2COA = 0;                   // Ignore genl call; own address = 0
    UCB0I2CIE = UCNACKIE;            // Enable interrupts on Nack
    UCB0CTL1 &= ~UCSWRST;           // Release from reset
    IE2_bit.UCB0RXIE = 1;           // Enable interrupts on Receive
// Set up basic timer for interrupts at 2Hz (500ms)
    BTCTL = BTHOLD | BT_ADLY_500;    // Hold, period = 500ms from ACLK
    BTCNT2 = 0;                      // Clear counter
    BTCTL &= ~BTHOLD;               // Start basic timer
    IE2_bit.BTIE = 1;               // Enable basic timer interrupts
    for (;;) {                       // Transfers triggered by BT
        __low_power_mode_3();       // Wait for BT (needs only ACLK)
        switch (I2CStatus) {
        case I2CIdle:               // Should never happen
            break;
        case I2CSuccess:            // Successful reception
            Temperature = (RXdata[1] << 8) | RXdata[0]; // Assemble word
            DisplayCentiCels (Temperature); // Display temperature
            break;
        case I2CNack:               // Address not recognised
            DisplayErr();           // Display error message
            break;
        default:                    // Should never happen
            I2CStatus = I2CIdle;    // Emergency reset
            break;
        }
    }
}
//-----------------------------------------------------------------
// ISR for basic timer: start I2C transfer
//-----------------------------------------------------------------
#pragma vector = BASICTIMER_VECTOR
__interrupt void BASICTIMER_ISR (void)  // Acknowledged automatically
{
    LED = 1;                         // Show start of activity
    RXcount = N_MESSAGE;            // Initialize byte counter for MSB
```

```
    UCB0CTL1 |= UCTXSTT;                    // Send Start and slave address
}
//--------------------------------------------------------------------
// ISR for UCB0RXIFG: store received byte, send Stop condition, display
// Shared with UCA0TXIFG and UCB0TXIFG, neither active
//--------------------------------------------------------------------
#pragma vector = USCIAB0TX_VECTOR
__interrupt void USCIAB0TX_ISR (void)       // Acknowledge by read of RXBUF
{
    if (RXcount > 0) {                      // Byte expected?
        RXdata[--RXcount] = UCB0RXBUF;      // Yes: Update ctr, store byte
        if (RXcount == 1) {                 // Last byte but one
            UCB0CTL1 |= UCTXSTP;            // Send Stop cond after byte
        } else if (RXcount == 0) {          // Last byte
            I2CStatus = I2CSuccess;            // Tell main() of success
            __low_power_mode_off_on_exit(); // Return to main to display
            LED = 0;                        // Finished
        }
    } else {                               // No: Should never happen
        IFG2_bit.UCB0RXIFG = 1;            // Clear flag to acknowledge
    }
}
//--------------------------------------------------------------------
// ISR for UCB0NACKIFG (not-acknowledge): abort transaction
// Shared with UCA0RXIFG and other UCB0 state-change flags, all inactive
//--------------------------------------------------------------------
#pragma vector = USCIAB0RX_VECTOR
__interrupt void USCIAB0RX_ISR (void)       // Not acknowledged automaticaly
{
    UCB0STAT &= ~UCNACKIFG;                 // Clear Nack flag
    UCB0CTL1 |= UCTXSTP;                    // Send Stop condition to finish
    I2CStatus = I2CNack;                    // Tell main() of problem
    __low_power_mode_off_on_exit();         // Return to main for display
    LED = 0;                               // Finished
}
```

Both vectors for the USCI have multiple sources but only one interrupt is active for each so there is no need to check the flag. This would be essential if USCI_A0 were also running.

10.11 A Thermometer Using I²C with the F2013 as Master

To complete the tour of the I²C modules, I now describe a thermometer where the USI in the F2013 is the master and the USCI_B0 in the FG4618 is the slave. This is roughly equivalent to the system in the section "A Thermometer Using SPI in Mode 3 with the F2013 as Master" on page 520 except that I do not include the transmission of the delay from the FG4618 to the F2013. The general idea is that the F2013 wakes periodically, reads the temperature, and transmits 2 bytes to the FG4618, which displays it. Few changes are needed for the USCI_B0 but some new aspects enter the program for the USI.

10.11.1 I²C Slave with the USCI_B

The overall structure of the program is the same as in the section "State Machine for an I²C Master Using the USCI_B" on page 564: The USCI_B does most of the work, helped by software when it requests an interrupt. A rather different set of interrupts is relevant to a slave. These are the four with the actions taken when each arises:

- **Start, UCSTTIFG** initializes the counter for bytes received.

- **Stop, UCSTPIFG** triggers the display of the temperature.

- **Receive, UCB0RXIFG** stores the received byte from UCB0RXBUF.

- **Transmit, UCB0TXIFG** loads a byte to transmit into UCB0TXBUF; I take this from the push buttons, as in Listing 10.4.

The last item may be unexpected: Why bother to handle transmit interrupts when we intend to only receive data? The reason is the automatic operation of the USCI_B. When it receives an address over the bus that matches its own address in UCB0I2COA it checks the R/$\overline{\text{W}}$ bit and continues according to the action requested by the master. There is no way of restricting it to handle transmission or reception alone. Defensive programming requires the program to handle both cases or the bus will freeze if an unexpected request for transmission arrives. I sent the state of the push buttons S1 and S2, which could be used to set the next delay of the F2013.

Listing 10.11 shows the interrupt service routines. All the action occurs here except for driving the display, which is in the main function so that it can be interrupted if necessary. The USCI_B can run as a slave in LPM4 but the LCD needs ACLK so this program spends most of its time in LPM3. There is no need to configure a clock for the USCI_B as it is only a slave.

Listing 10.11: Configuration of the USCI_B0 and interrupt service routines from the I²C slave in usci2cslvsm2.c.

```
    P3SEL = BIT1 | BIT2;              // Route pins to USCI_B for I2C
// 7-bit addresses, single master, slave mode (all defaults), I2C, synch
    UCB0CTL0 = UCMODE_3 | UCSYNC;
// No need to configure UCB0CTL1, UCB0BR nor UCB0I2CSA for a slave
    UCB0I2COA = OWN_ADDRESS;          // Ignore genl call; own address
    UCB0CTL1 &= ~UCSWRST;             // Release from reset
```

```
    UCB0I2CIE = UCSTTIE | UCSTPIE;      // Enable interrupts on Start, Stop
    IE2 |= (UCB0RXIE|UCB0TXIE);         // Enable interrupts on RX and TX
    for (;;) {                          // Transfers triggered by BT
        __low_power_mode_3();           // Need ACLK for LCD
        Temperature = (RXdata[1] << 8) | RXdata[0]; // Assemble word
        DisplayCentiCels (Temperature); // Display temperature
    }
}
//-----------------------------------------------------------------------
// ISR for UCSTTIFG (Start): start transaction, reset byte counter
//     UCSTPIFG (Stop): return to main() to display temperature
// Shared with UCA0RXIFG and other UCB0 state-change flags, all inactive
//-----------------------------------------------------------------------
#pragma vector = USCIAB0RX_VECTOR
__interrupt void USCIAB0RX_ISR (void)   // Not acknowledged automaticaly
{
    if (UCB0STAT & UCSTTIFG) {          // Start condition?
        UCB0STAT &= ~UCSTTIFG;          // Yes: Clear start flag
        RXcount = N_RECEIVE;            // Initialize byte ctr for MSB
        LED = 1;                        // Show start of transaction
    } else {                            // Must be Stop condition
        UCB0STAT &= ~UCSTPIFG;          // Clear stop flag
        __low_power_mode_off_on_exit(); // Return to main, display temp
        LED = 0;                        // Show end of transaction
    }
}
//-----------------------------------------------------------------------
// ISR for UCB0RXIFG: store received byte; reading UCB0RXBUF clears flag
//     TXIFG: send value from pushbuttons; writing UCB0TXBUF clears flag
//     UCA0TXIFG: not active
//-----------------------------------------------------------------------
#pragma vector = USCIAB0TX_VECTOR
__interrupt void USCIAB0TX_ISR (void)   // Not acknowledged automaticaly
{
    if (IFG2_bit.UCB0RXIFG) {           // Byte received? (MSB first)
        if (RXcount > 0) {              // Yes: Byte expected?
            RXdata[--RXcount] = UCB0RXBUF;  // Update ctr, store byte
        } else {                        // Too many bytes, "impossible"
            IFG2_bit.UCB0RXIFG = 1;     // Clear flag to acknowledge
        }
    } else {                            // No: Provide byte to transmit
        UCB0TXBUF = (~P1IN) & (BIT1|BIT0);  // Mask buttons, active low
    }
}
```

10.11.2   I²C Master with the USI

This requires more effort, as you would expect. The high-level structure of the program is the same as Listing 10.3 with activity triggered by the watchdog counter in interval mode. There is a state machine to run the I²C transaction with the USI, which is similar to that in Figure 10.15. The major new feature is that the master must generate the start and stop conditions. The USI does not do this automatically and software is therefore required.

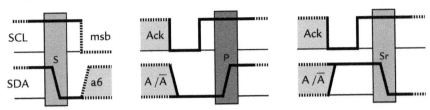

Figure 10.16: Start, stop, and repeated start conditions on the I²C bus.

I sketched the start, stop, and repeated start conditions in Figure 10.16 as a reminder. The start condition is a falling edge on SDA while SCL is high, when SDA is normally stable. This means that a start condition cannot be generated by normal operation of the shift register. Instead, we have to use the USIGE bit to override the usual behavior of the transparent latch in the output (see Figure 10.5). Setting this bit transfers the msb of the shift register to the output immediately despite the high state of the clock. Thus we can force a start condition onto SDA with these three lines of C:

```
USISRL = 0;              // Low msb for start condition
USICTL0 |= USIGE;        // Open output latch, drive 0 on SDA
USICTL0 &= ~USIGE;       // Hold 0 in output latch
```

The final line restores the normal behavior of the latch, which holds its value until SCL next goes low. This is performed in the state SendAddress of the machine.

A little more effort is needed to generate a stop condition, which is a rising edge on SDA while SCL is high. The start condition is easy because we know that SDA is already high, assuming that the bus is idle (which should really be checked). A stop condition follows an acknowledgment bit, which could be high or low. We must therefore send a dummy bit of 0, using the shift register and USICNT in the usual way, so that SDA is low in preparation for the stop condition. The dummy bit is sent in state PrepareStop of the machine and the final stop condition is generated in SendStop. A repeated start is produced in a similar way.

A further issue is how to start the state machine for a new transaction. This is done by the distinct interrupt USISTTIFG from a start condition in the slave, which is treated with a higher priority than the usual interrupts from USIIFG. I copied this approach by setting the flag USIIFG in software to request an interrupt and start the state machine, while passing the variable I2CCommand to force the machine into the state SendAddress and start a new transmission. Another possibility would have been to set USISTTIFG from software. The state machine is shown in Listing 10.12 and is based on the TI code examples for the F20xx.

Listing 10.12: State machine from `usi2cmastsm1.c` to transmit data over I²C as a master using the USI.

```
#pragma vector = USI_VECTOR
__interrupt void USI_ISR (void)        // Not acknowledged automatically
{
    static uint8_t TXcount;            // Counter for bytes transmitted
    static enum {                      // States of state machine
        Idle = 0,
        SendAddress = 2,
        ReceiveAddressAck = 4,
        CheckSendData = 6,
        ReceiveDataAck = 8,
        PrepareStop = 10,
        SendStop = 12
    } I2CState = Idle;

    if (I2CCommand == I2CSendData) {   // Start new transaction
        LED = 1;                       // LED on while I2C active
        I2CState = SendAddress;        // Start state machine
        I2CCommand = I2CNothing;       // Acknowledge command
    }
    switch (__even_in_range(I2CState, SendStop)) {
    case Idle:                         // Should never happen!
        break;
    case SendAddress:                  // Send Start, prepare address
        USISRL = 0;                    // Low msb for Start condition
        USICTL0 |= USIOE;              // Enable output, grab SDA
        USICTL0 |= USIGE;              // Open output latch, drive 0 on SDA
        USICTL0 &= ~USIGE;             // Hold 0 in output latch
        USISRL = SLAVE_ADDRESS << 1;   // Address with Write bit
        USICNT = USIFGCC | 8;          // Stop auto clr USIIFG, send 8 bits
        TXcount = N_TRANSMIT - 1;      // Initialize byte counter
        I2CState = ReceiveAddressAck;  // Check address acknowledg next
        break;
    case ReceiveAddressAck:            // Receive acknowledgment after addr
        USICTL0 &= ~USIOE;             // Disable output, release SDA
        USICNT |= 1;                   // Receive 1 bit
        I2CState = CheckSendData;      // Check Ack, send data byte next
        break;
    case CheckSendData:                // Send byte of data if last Ack'd
        if ((USISRL & BIT0) != 0) {    // Received Nack or Ack?
            USISRL = 0;                // Nack: abandon TX, prepare Stop
            USICTL0 |= USIOE;          // Enable output
            USICNT |= 1;               // Send 1 bit, pull SDA low
            I2CState = SendStop;       // Send Stop next
        } else {                       // Ack: send byte of data
            USISRL = TXdata[TXcount];  // Load byte of data
            USICTL0 |= USIOE;          // Enable output
            USICNT |= 8;               // Send 8 bits
            I2CState = ReceiveDataAck; // Receive ack next
        }
        break;
    case ReceiveDataAck:               // Receive acknowledgment after data
        USICTL0 &= ~USIOE;             // Disable output to receive Ack
        USICNT |= 1;                   // Receive 1 bit
```

```
        if (TXcount == 0) {          // Was that the last byte?
            I2CState = PrepareStop;    // Yes: finished, send Stop next
        } else {
            --TXcount;                // No: Update byte counter
            I2CState = CheckSendData;  // Check Ack, send data next
        }
        break;
    case PrepareStop:                // Prepare to send Stop condition
        USISRL = 0;                  // Prepare to pull SDA low for Stop
        USICTL0 |= USIOE;            // Enable output
        USICNT |= 1;                 // Send 1 bit, pull SDA low
        I2CState = SendStop;         // Send Stop next
        break;
    case SendStop:
        USISRL = 0xFF;              // High msb for Stop condition
        USICTL0 |= USIGE;           // Open output latch, drive 1 on SDA
        USICTL0 &= ~USIGE;          // Hold 1 in output latch
        USICTL0 &= ~USIOE;          // Disable output to release SDA
        USICNT |= USISCLREL;        // Release hold on SCL
        I2CState = Idle;            // Return to idle state
        LED = 0;                    // LED off when I2C inactive
// Change mode from LPM0 to LPM3 on exit: SMCLK no longer needed
        __bic_SR_register_on_exit(LPM0_bits);   // (not really necessary)
        __bis_SR_register_on_exit(LPM3_bits);
        break;
    }
    USICTL1 &= ~USIIFG;            // Clear flag, release SCL if held
}
```

Example 10.3

Rewrite Listing 10.12 to use USISTTIFG to start a transaction rather than I2CCommand.

The I²C specification requires certain hold and setup times around start, stop, and repeated start conditions. For example, there should be a delay of $t_{HD;STA} \geq 4\,\mu s$ between a start condition and the first falling edge on SCL. The MSP430 executes only one typical instruction in this time with $f_{MCLK} = 1$ MHz so there is no chance of violating the specification. I therefore ignore the issue but you would need to check more carefully if MCLK were faster.

Figure 10.17 shows oscilloscope traces of this program in action. The long gaps due to the state machine are obvious. They stretch the clock when it is *high* in a master, which greatly elongates the last bit of each byte and the acknowledgment bits. This contrasts with the traces in Figure 10.14 where the USI was a slave and the clock was stretched when low.

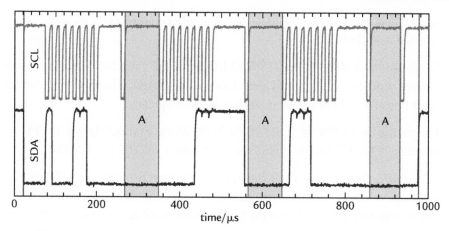

Figure 10.17: I²C master transmitting two bytes using the USI.

Example 10.4

Why are all the acknowledgment bits low in Figure 10.17, even the last one?

Example 10.5

Extend Listing 10.12 to handle a combined transaction where the master first sends 2 bytes to the slave, then sends a repeated start and receives a byte for the next delay. This requires several extra steps:

- Send a repeated start condition instead of the stop condition.

- Send the address of the slave with $R/\overline{W} = 1$.

- Check that the address is acknowledged.

- Receive a byte from the slave and store the data.

- Send a not-acknowledgment bit to show that no more data is expected.

- Send a stop condition.

The slave in Listing 10.11 handles this transaction without modification, which is convenient.

Several application notes show how to use I²C and related protocols with the communication modules:

- *Implementing SMBus Using MSP430 Hardware I²C* (slaa249) describes several SMBus protocols using the I²C interface in the USART. These include the extra Alert line and a check on the minimum clock frequency.

- *Interfacing an EEPROM to the MSP430 I²C Module* (slaa208) uses the I²C interface in the USART to communication with a 24xx65 EEPROM.

- *CEC-to-I²C Bridge with the MSP430* (slaa377) uses the USI in a MSP430F2013 to bridge between I²C and the consumer electronics control (CEC) interface. This is a single-wire bus defined within the high-definition media interface (HDMI) used for televisions and monitors.

10.12 Asynchronous Serial Communication

A decade ago, every personal computer had serial ports to connect extra equipment such as keyboards, mice, and modems. They have now been displaced by the universal serial bus in consumer applications but old-fashioned serial communication remains popular for connecting embedded systems. The main reason is simplicity. Asynchronous serial communication can be managed in hardware by a peripheral called a universal asynchronous receiver/transmitter, which is not complicated and is therefore built into many microcontrollers. Even if this is not available, it can be emulated with a timer assisted by software. USB is *much* more difficult to handle.

In practice it is not a big problem to use an asynchronous serial link to a personal computer because USB–serial converters are readily available. They provide "virtual COM ports" under Windows, which appear much like the real hardware on older machines. There is an abundance of information on this and other aspects of serial ports in Axelson's book [55]. This was recently updated to a second edition, which shows the enduring nature of the serial port.

Asynchronous serial communication usually requires only a single wire for each direction plus a common ground. Most general-purpose connections are full duplex, meaning that data can be sent simultaneously in both directions. These act independently, unlike SPI. There are usually no further control lines, nor is there anything like the protocol required to run a I²C bus; characters are simply sent when required. It really is straightforward, which explains its continuing popularity. Issues such as the detection and correction of errors are usually handled by the application that supervises the communication. For example, a block of data may be followed by a checksum to confirm that it has been received correctly. Asynchronous links usually connect only two pieces of equipment but a few buses use asynchronous communication.

10.12.1 Format of Data for Asynchronous Transmission

I described the usual format of asynchronous data in the section "Operation of Timer_A in the Sampling Mode" on page 352. Data are sent in short *frames*, each of which typically contains a single byte. Two examples are shown in Figure 10.18. The line idles high and each frame contains

- One low start bit (ST).

- Eight data bits, usually lsb first.

- One high stop bit (SP).

The bits are either high or low and have no gaps between them, a format known as *non-return to zero* (NRZ). They are usually sent with lsb first, which is the reverse order compared with I²C or the usual sequence on SPI. The format of the frame is called 8-N-1 because there are 8 bits of data, no parity bit, and 1 stop bit. You may occasionally encounter other formats. For example, the basic ASCII code has only 128 values so 7 bits of data were often sent in the past. The eighth bit was sometimes used for parity as a simple check for errors in transmission. A parity bit may also be added to 8-bit data and the MSP430 bootstrap loader uses this format.

A minimum of 1 stop bit is needed to separate each frame and provide a high level before the falling edge of the next stop bit. Slower systems may specify more stop bits, often 1.5 or 2, so that they can keep up with the flow of data but this is now rare.

The *baud rate* gives the frequency at which bits are transmitted on the line. It is the inverse of the bit period and the name is used to distinguish it from the rate at which useful data

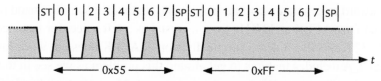

Figure 10.18: Two asynchronous bytes that carry the data 0x55 and 0xFF. Each frame is delimited by a start (ST) bit and a stop (SP) bit.

are communicated. Each 8 bits of data are accompanied by a start and stop bit so the maximum data rate is only 8/10 of the baud rate. A common speed for embedded systems is 9600 baud but both higher and lower rates are also used.

No clock is transmitted in asynchronous communication so the transmitter and receiver must run independently at nearly the same baud rates. How close do they have to be? Here is a reminder of the basic procedure for receiving a frame in a UART, which was sketched in Figure 8.15.

1. Start timing at the falling edge that begins a start bit.

2. Sample the input after half a bit period to confirm a valid start bit.

3. Sample the input after a further complete bit period to read the first bit (lsb).

4. Repeat this until all 8 bits have been received, finishing with the msb.

5. Wait a further bit period and check that the input is high as expected for the stop bit. A *framing error* occurs if this bit is low.

The final sample is taken 9½ bit periods after the initial falling edge and must lie within the stop bit. The permissible error is therefore about ±0.5 bit period in 9.5 periods or ±5%. The baud rates of the receiver and transmitter must be the same within this tolerance. There may be errors in both rates, which add together in the worst case, so each should be accurate to within about ±2%.

Figure 10.18 shows 2 bytes, one of which is easy to receive and the other is difficult. The first carries the value 0x55 (lsb first), which gives the maximum number of transitions within the frame. An advanced receiver can resynchronize to these transitions, which makes it easy to match the frequency of the transmitter. In fact a byte of 0x55 is used in a local interconnect network so that a receiver can synchronize its clock to that of the transmitter. I say a little about LIN later. In contrast, the byte of 0xFF has no transitions at all after the rising edge of the start bit. The receiver is on its own and accurate frequencies in the UARTs are essential.

It is often awkward to generate accurate frequencies that match standard baud rates. We might like to use ACLK at 32 KHz from a crystal as a reference, for instance. Unfortunately 32 K/9600 = 3.41 so simple division of ACLK does not work for 9600 baud. The USCI_A includes features to get around this problem as we see in the next section. A crystal with a special frequency may be needed for very high baud rates, where division of MCLK or ACLK is inaccurate.

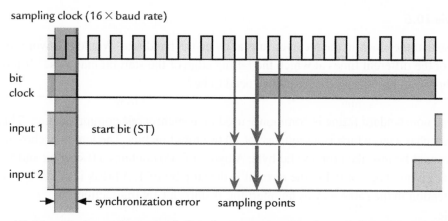

Figure 10.19: Clocks and received data in a UART with the sampling clock running at 16 times the baud rate. Reception starts when a falling edge at the beginning of a start bit is detected. The two input signals correspond to the latest and earliest edges that would trigger reception to start at the same time. The low start bit is followed by a high lsb of the data.

The clock in the receiver must run faster than the baud rate so that it can start timing promptly when a falling edge arrives to signal a new frame. This is because the start bit can arrive at any time, with no relation to the clock inside the receiver. Typically the receiver uses a clock at 16 times the baud rate, which is illustrated in Figure 10.19. The frequency of the bit clock is the same as the baud rate and this clock is started when a falling edge at the beginning of a start bit is detected. This edge may have occurred at any time during the preceding cycle of the sampling clock so there is a synchronization error of up to ½ a period of the sampling clock.

The input is sampled half-way through the bit. Only one sample is needed in principle but most UARTs take more to reduce the impact of noise, which may cause brief incorrect values. Often three samples are used, separated by periods of the sampling clock. Ideally all three values should be the same. If they differ, the value shared by two is taken and the different, single value is rejected (there are only two possible values). This is called a *majority vote*. Sometimes a noise flag is set if the majority vote is needed. The exact position of the samples relative to the start of the bit depends on the synchronization error and I have shown the signals for the two extreme cases in Figure 10.19. In both cases all the sampling points are well inside the bit and it should be received correctly.

Example 10.6

Suppose that the clock in the receiver runs at twice the baud rate. Would this work, if you allow for the different times at which the falling edge of the start bit may arise? If it does work, how accurate would the frequency need to be?

Only one nonstandard frame is commonly used in asynchronous communication. This is the *break* character, which is a sequence of 10 or more low bits. The length is chosen so that it cannot be mistaken for a valid byte: A byte of 0x00 contains a 0 start bit and 8 data bits of 0, giving 9 bits of 0, but the next bit is the stop bit of 1. A break is typically used to gain attention in the same way as an interrupt. It starts a LIN transaction, for example.

The bootstrap loader that is built into most variants of the MSP430 uses half-duplex, asynchronous serial communication. Each character contains 8 data bits and an even parity bit. Transmission starts at 9600 baud but this can be raised to increase the speed of programming. Details are in the application notes *Features of the MSP430 Bootstrap Loader* (slaa089) and *Application of Bootstrap Loader in MSP430 with Flash Hardware, Software Proposal* (slaa096).

10.12.2 Interface Standards and RS-232

Sometimes you can connect two devices directly for asynchronous communication, which may be called TTL RS-232. The signal voltages are close to V_{SS} for 0 and V_{CC} for 1 as for SPI and I²C. This works correctly provided that both devices run from the same supply voltage or are at least compatible. However, asynchronous communication is usually used between separate pieces of equipment. In this case the connection must obey established standards and be protected against common faults. Therefore an interface is needed between the UART in the microcontroller and the outside world. This is where RS-232 comes in.

RS-232 is an ancient standard, originally intended for connecting equipment such as a teletype (data terminal equipment) to a modem (data communication equipment). For many years it was officially RS-232-C, published in 1969, but the current version is ANSI/TIA/EIA-232-F, which is maintained by the Telecommunication Industry Association. I am lazy and continue to call it RS-232.

RS-232 specifies the following voltages for the signals, which may come as an unpleasant surprise:

- Logical 1 is represented by a voltage between −15 V and −3 V. This is also called *mark*.

- Logical 0 is represented by a voltage between +3 V and +15 V, also called *space*.

- The crossover region between ±3 V does not correspond to valid data.

- Connections must be resistant to short circuits and to voltages of ±25 V.

Clearly this presents problems for the MSP430, whose natural range of voltages lies wholly within the crossover region. The original recommendation was for signals to be at ±12 V but many systems now use lower voltages, often ±5 V. This is closer to the voltages used for digital electronics but the negative voltage is problematic. External circuits for a transmitter and receiver or *transceiver* are therefore needed to connect the MSP430 or any other microcontroller to an RS-232 port. There are several ways in which they can be implemented:

- Special transceiver ICs are available that generate the voltages needed for RS-232 transmission and provide interfaces between the external and internal signals, including isolation and protection. Perhaps the best known family is the MAX232 and descendants. The original MAX232 was intended for 5 V systems but newer devices work from lower voltages. Typically they include charge pumps to produce voltages of ±2 V_{CC}, which provide the RS-232 levels. Three or four capacitors are needed for the charge pumps, which is a painless way of gaining the extra voltage. The SoftBaugh demonstration board shown in Figure 3.3 includes a MAX3221 for its RS-232 interface.

- If this is too expensive, numerous simple circuits use a couple of transistors and a few passive components for the receiver and transmitter. They "steal" the voltage needed for RS-232 from the line driven by the other transmitter, which must therefore come from a full-strength port. This should be true for a computer but not necessarily for another embedded system—two thieves cannot steal the voltage from each other. This approach is used in the Olimex 1121STK.

- An opto-isolator can be used to provide greater separation between the microcontroller and the RS-232 circuits. The input to the opto-isolator drives an

LED, which shines onto a phototransistor that provides the output. Thus there is no electrical connection between the two sides. The TI MSP430FG4618/F2013 Experimenter's Board has an opto-isolated RS-232 connection. Circuits are still needed to drive the lines and this board takes its power from the other transmitter.

There is no universally agreed physical connection for RS-232 although many systems use the nine-way D-type connector introduced with the IBM PC/AT. This offers plenty of pins beyond the two signals and ground needed for the data alone. The extra wires were used to control the interface between the teletype and modem in the days when equipment had trouble keeping up with a signal at 110 baud. They have names like Request to Send and Clear to Send but are now rarely used for these purposes. Many systems ignore them and others use them for nonstandard purposes. Axelson [55] explains how to deal with equipment that needs the extra signals.

Improvements and extensions to RS-232 have been made over the years. A weakness of RS-232 is that it relies on single-ended signals, which means the voltage between a single wire and ground. RS-422 uses *balanced* or *differential signalling*, where the signal is the *difference* in voltage between a pair of wires. For example, the voltages on the two lines might be $(+3\,V, -3\,V)$ for a 0 and $(-3\,V, +3\,V)$ for a 1. This enables reliable communication over a longer distance than RS-232 or a higher baud rate at short distances. The serial ports on the Apple Macintosh used RS-422 before the days of USB. *Differential signaling*, is also employed in USB, CAN, and many other interfaces.

Up to 10 receivers are permitted on RS-422 but only a single transmitter. RS-485 is a further extension to multiple transmitters, which makes the system into a bus. It is used in industrial control and other applications that require long networks but do not demand great speed. Axelson [55] covers RS-485 as well and the analog aspects are described in the application note *422 and 485 Standards Overview and System Configurations* (slla070).

I already mentioned the *local interconnect network*. This is a low-cost network based on standard asynchronous communication, which is designed so that nodes do not need accurate clocks and crystals. Frames start with a break followed by a synchronization byte of 0x55, which each receiver uses to match its baud rate to that of the transmitter. The rest of the frame is a sequence of bytes sent in the usual asynchronous format. It resembles a read or write transaction in I²C but includes a checksum. LIN was originally intended for automotive applications, where it complements faster but more demanding buses such as the controller area network, but has been adopted elsewhere because it needs little more than a UART in each node. Full information is on the Web at www.lin-subbus.org.

10.13 Asynchronous Communication with the USCI_A

The USCI_A has several modes of operation. First, it can be used for SPI in the same way as USCI_B by setting the UCSYNC bit in UCA0CTL0. If this bit is clear, there are four asynchronous modes. These are selected with the UCMODExx bits, also in UCA0CTL0.

- Standard UART mode, UCMODExx = 00.

- Multiprocessor modes, UCMODExx = 01 or 10. These are used to detect addresses when more than two devices are used on a bus, such as RS-485.

- Automatic baud rate detection, UCMODExx = 11. This is particularly intended for LIN.

A further option is to encode the output for IrDA, which is a format for infrared transmission and is chosen with the UCIREN bit. It is described in the application note *Implementing IrDA with the MSP430* (slaa202), which covers both the USCI_A and bit-banging with Timer_A. I describe only the standard UART mode. This is configured in a similar way to SPI with one major exception, which is the baud rate. The exact frequency of the clock is not important for SPI, because it is synchronous, but is critical for asynchronous communication. The USCI_A offers sophisticated options for setting the baud rate, which merit a separate section.

10.13.1 Setting the Baud Rate with the USCI_A

The most complicated aspect of configuring the USCI_A is setting the baud rate. For a start, there are three clocks in the USCI_A:

- **BRCLK** is the input to the module (SMCLK, ACLK, or UCA0CLK).

- **BITCLK** controls the rate at which bits are received and transmitted. Ideally its frequency should be the same as the baud rate, $f_{\mathrm{BITCLK}} = f_{\mathrm{baud}}$.

- **BITCLK16** is the sampling clock in oversampling mode, with a frequency $f_{\mathrm{BITCLK16}} = 16 f_{\mathrm{BITCLK}}$.

The periods of these clocks are $T_{\mathrm{BITCLK}} = 1/f_{\mathrm{BITCLK}}$ and so on. There are two modes for setting the baud rate, illustrated in Figure 10.20. These are selected with the UCOS16 bit in the modulation control register UCA0MCTL.

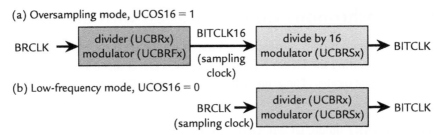

Figure 10.20: Clock generator in USCI_A showing (a) oversampling mode with two dividers and intermediate clock BITCLK16, and (b) low-frequency mode with a single divider.

Oversampling mode, UCOS16 = 1: BRCLK is first divided to give BITCLK16, which is further divided by a factor of 16 to give BITCLK. BITCLK16 is used to control the sampling of received bits as in Figure 10.19.

Low-frequency mode, UCOS16 = 0: BRCLK is used directly as the sampling clock and is divided to give BITCLK. In general $f_{\text{BRCLK}} \neq 16 f_{\text{BITCLK}}$ so the sampling differs from Figure 10.19.

Simple division of BRCLK by an integer does not give a sufficiently accurate baud rate in many cases. Each divider therefore has a modulator to permit finer control over the average frequency of BITCLK. I explain this shortly. Several registers and bit fields control the division:

UCA0BR0 and **UCA0BR1:** Together act as UCBRx, which provides the main divider of BRCLK as in SPI mode (see the section "SPI with the USCI" on page 513).

UCBRFx: Modulates the divider that gives BITCLK16 in oversampling mode. This allows the ratio $f_{\text{BRCLK}}/f_{\text{BITCLK16}}$ to be set more accurately than an exact integer would permit (it can be specified to the nearest 1/16).

UCBRSx: Modulates the divider that gives BITCLK in a similar way.

All this complication is needed because the baud rate is rarely close to being a perfect factor of f_{BRCLK}, the frequency of the clock supply to the module. At this point you have a choice of two ways forward. The user's guide contains tables of values for UCBRx, UCBRFx, and UCBRSx that can be used for both modes with common baud rates and clock frequencies. In most cases you can simply copy these and skip the rest of this section. If you wish to know more, read on. I first describe the oversampling mode, then the low-frequency mode.

Oversampling Mode of the Baud Rate Generator

This requires a fast input clock BRCLK, whose frequency must be at least 16 times the baud rate. There are two dividers, the first of which is adjustable using UCBRx to give the intermediate, sampling clock BITCLK16. The second makes a further division by 16 to give BITCLK, whose frequency should match the baud rate as closely as possible. See Figure 10.20(a).

As an example, suppose that $f_{baud} = 9600\,\text{Hz}$ and $f_{BRCLK} = 2^{20}\,\text{Hz}$ (a binary megahertz), which could be produced using the FLL to lock the frequency of the DCO to 32 times that of a 32 KHz crystal. We want $f_{BITCLK16} = 16 f_{baud} = 153, 6\,\text{KHz}$. Thus the divider should use a factor of $2^{20}/153, 600 = 6.83$ to two decimal places. This is not possible, of course. The simplest approach would be to use the closest integer of 7, which would give $f_{BITCLK} = 9362\,\text{Hz}$. This is too low by about 2.5% compared with 9600 baud, which might just be acceptable. However, we would not be so lucky if f_{BRCLK} were a decimal megahertz and this is why the divider includes a modulator.

The modulator works on the same principle as the DCO, described in the section "Digitally Controlled Oscillator, DCO" on page 167, and I showed a similar approach using software in the section "Generation of a Precise Frequency" on page 326. In USCI_A, the idea is to allow the duration of each cycle of BITCLK16, $T_{BITCLK16}$, to vary in such a way that the total duration of 16 cycles is as accurate as possible. Now $16\,T_{BITCLK16} = T_{BITCLK}$, the period of BITCLK, which should match the baud rate. The example needs $f_{BRCLK}/f_{baud} = 2^{20}/9600 = 109.2$ cycles of BRCLK to each cycle of BITCLK. This can be rounded to 109 with an error of only 0.2%.

The 109 cycles of BRCLK that make one cycle of BITCLK must be divided into 16 cycles of BITCLK16 in as uniform a way as possible. In this case,

- 13 cycles of BITCLK16 have length $7\,T_{BRCLK}$.

- The remaining three cycles of BITCLK16 have length $6\,T_{BRCLK}$.

The USCI_A is configured to do this by writing a divider of 6 to UCBRx and a modulation index of 13 to UCBRFx. In general this means that each set of 16 cycles of BITCLK16, which form one cycle of BITCLK, contains

- $(16 - \text{UCBRFx})$ cycles of BITCLK16 with a duration of $\text{UCBRx} \times T_{BRCLK}$.

- UCBRFx cycles of BITCLK16 with a duration of $(\text{UCBRx} + 1) \times T_{BRCLK}$.

Cycles of BITCLK16 with these two durations are distributed to minimize the errors in the sampling times. These move slightly from their ideal positions in Figure 10.19 but remain close to the center of the received bit. We can write an equation to find UCBRx and UCBRFx more generally:

$$\text{nint}\left(\frac{f_{\text{BRCLK}}}{f_{\text{baud}}}\right) = (16 - \text{UCBRFx})\text{UCBRx} + \text{UCBRFx}(\text{UCBRx} + 1)$$

$$= 16\,\text{UCBRx} + \text{UCBRFx}, \tag{10.1}$$

where nint() returns the nearest integer to its argument. Thus UCBRx is the quotient and UCBRFx is the remainder when nint($f_{\text{BRCLK}}/f_{\text{baud}}$) is divided by 16.

In most cases this procedure brings f_{BITCLK} close enough to f_{baud}. If it does not, it is also possible to modulate the second division from BITCLK16 to BITCLK by using UCBRSx. I now describe this process for the low-frequency mode.

Example 10.7

Repeat the procedure to find UCBRx and UCBRFx for 9600 baud and $f_{\text{BRCLK}} = 1\,\text{MHz}$ (a decimal megahertz instead of a binary megahertz).

Low-Frequency Mode of the Baud Rate Generator

The low-frequency mode (UCOS16 $=0$) is used when $f_{\text{BRCLK}} < 16\,f_{\text{baud}}$. An extreme example is communication at 9600 baud with BRCLK taken from ACLK and a watch crystal, in which case $f_{\text{BRCLK}} = 32\,\text{KHz}$. In this case $f_{\text{BRCLK}}/f_{\text{baud}} = 3.41$ so 16-times oversampling is clearly out of the question.

Only a single divider and modulator is used in the low-frequency mode, as shown in Figure 10.20(b). It is no longer possible to make the length of each cycle of BITCLK correct. For example, it should be 3.41 cycles of BRCLK in the example. Instead we have to be less ambitious and the USCI_A aims for the best average value of the frequency over eight cycles of BITCLK. (There is no point in using 16 cycles because frames are shorter than this.) The divider UCBRx and the modulation index UCBRSx are given by

$$\text{nint}\left(\frac{8 \times f_{\text{BRCLK}}}{f_{\text{baud}}}\right) = (8 - \text{UCBRSx})\text{UCBRx} + \text{UCBRSx}(\text{UCBRx} + 1)$$

$$= 8\,\text{UCBRx} + \text{UCBRSx}. \tag{10.2}$$

This replaces equation (10.1). The left-hand side is 27, which gives UCBRx = 3 and UCBRSx = 3. Thus each set of eight cycles of BITCLK contains 5 with a duration of $3\,T_{\text{BRCLK}}$ and three with a duration of $4\,T_{\text{BRCLK}}$. The cycles with different durations are distributed to minimize the accumulated error in BITCLK. This gives a fairly small error in the average frequency because $8\,f_{\text{BRCLK}}/f_{\text{baud}} = 27.3$, which is reduced by about 1% on rounding to the integer 27. Unfortunately we now have to worry about the variation in length of each cycle of BITCLK and how this affects the sampling of a received signal.

Figure 10.21 shows how USCI_A samples the start bit and the first data bit (high) in the low-frequency mode with $f_{\text{BRCLK}} = 32$ KHz and 9600 baud. The first two periods of BITCLK have durations of $3\,T_{\text{BRCLK}}$ and $4\,T_{\text{BRCLK}}$ so the first is too short and the second is too long, although the errors have largely canceled after 2 bits. The lengths of the high and low parts of each cycle of BITCLK are also multiples of T_{BRCLK} so the first cycle of BITCLK is divided into a low part of T_{BRCLK} and a high part of $2\,T_{\text{BRCLK}}$. This is important because the central sample is taken on the middle edge of BITCLK, which is $\frac{1}{2}T_{\text{BRCLK}}$ too early if T_{BITCLK} is an odd multiple of T_{BRCLK}. The other two samples are $\frac{1}{2}T_{\text{BRCLK}}$ on either side.

The synchronization error also becomes more significant when $f_{\text{BRCLK}}/f_{\text{baud}}$ is small because the uncertainty in the starting time, $\pm\frac{1}{2}T_{\text{BRCLK}}$, is a larger fraction of T_{baud}. Figure 10.21 shows the two extreme cases as input 1 and input 2. The synchronization

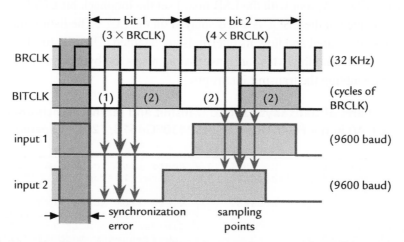

Figure 10.21: Clocks and reception of data in USCI_A using low-frequency mode with BRCLK at 32 KHz and data at 9600 baud. The two inputs show the possible synchronization error.

error interacts with the errors in the sample timing so that the sampling points are not always near the middle of the received bits. The maximum errors are tabulated in the user's guides. For the example, the time at which the central sample is taken may be wrong by -44% or $+21\%$ in units of T_{baud}. The ideal time is 50% (halfway) through each bit so the samples are taken in the worst cases at 6% or 71%. The figure of 6% is worrying because it is so close to the start of the bit but I deliberately chose an extreme (although common) example. You can see the deviations in Figure 10.21. The errors in transmission are smaller because there is no synchronization error; they arise only from the variation in T_{BITCLK}.

10.13.2   Operation of the USCI_A

Once it is configured, the USCI_A handles asynchronous communication in a very similar way to SPI. A character for transmission is written to UCA0TXBUF after checking that the flag UCA0TXIFG is raised. The data then are sent automatically. Similarly, the flag UCA0RXIFG is raised when a new character has been received and is available in UCA0RXBUF. There is usually no way of telling the remote transmitter to pause so it is important to read the buffer promptly before it is overwritten. The overrun flag UCOE is raised if this happens. These operations are almost always triggered by interrupts, as in Listing 10.5 for the SPI.

I provide yet another thermometer as an example in Listing 10.13. This uses the ADC12 in the FG4618 to read the temperature using code from Listing 9.9. The result is transmitted in 2 bytes using USCI_A, sent with the LSB first. I set the loopback bit UCLISTEN so that they are also received in the USCI_A and displayed on the LCD. The listing shows how the USCI_A0 is configured and how its interrupts are handled. The ISR for the ADC12 loads the temperature into the bytes for transmission, resets the byte counter and starts transmission by enabling the transmit interrupts.

Listing 10.13: Parts of `usciasync1.c` to transmit and receive asynchronous data using the USCI_A0 in the FG4618 on an MSP430FG4618/F2013 Experimenter's Board.

```
    P2SEL = BIT4 | BIT5;                // Route pins to USCI_A for comms
// Defaults: no parity, lsb first, 8-bit data, 1 stop bit, UART, async
    UCA0CTL0 = 0;
    UCA0CTL1 = UCSSEL_1 | UCSWRST;      // Clock from ACLK; hold in reset
    UCA0BR1 = 0;                        // Upper byte of divider word
    UCA0BR0 = 3;                        // Divider from ACLK to BITCLK
    UCA0MCTL = UCBRS_3;                 // Low-frequency mode, modul'n = 3
    UCA0STAT = UCLISTEN;                // Loopback mode
    UCA0CTL1 &= ~UCSWRST;               // Release from reset
    IE2_bit.UCA0RXIE = 1;              // Enable interrupts on RX all time
    for (;;) {                          // Transmissions triggered by BT
```

```
          __low_power_mode_3();
     }
}
//-----------------------------------------------------------------
// ISR for USCI_A,B0 TX: load next byte for TX, halt interrupts at end
//-----------------------------------------------------------------
#pragma vector = USCIAB0TX_VECTOR
__interrupt void USCIAB0TX_ISR (void)    // Acknowledge by write to TXBUF
{
     UCA0TXBUF = TXdata[TXcount++];  // Send next byte, update counter
     if (TXcount >= N_MESSAGE) {     // Sent complete message?
         IE2_bit.UCA0TXIE = 0;       // Disable further interrupts
     }
}
//-----------------------------------------------------------------
// ISR for USCI_A,B0 RX: store data, process when all bytes received
//-----------------------------------------------------------------
#pragma vector = USCIAB0RX_VECTOR
__interrupt void USCIAB0RX_ISR (void)    // Acknowledge by read of RXBUF
{
     int16_t Temperature;               // Value received over SPI

     RXdata[RXcount++] = UCA0RXBUF;  // Store recd data, update counter
     if (RXcount >= N_MESSAGE) {     // Received complete message?
         Temperature = (RXdata[1] << 8) | RXdata[0]; // Assemble word
         DisplayDeciCels (Temperature);  // Display temperature to 0.1oC
         RXcount = 0;                // Reset byte counter
         LED = 0;                    // Show end of activity
     }
}
//-----------------------------------------------------------------
// Interrupt service routine for basic timer, requested every 1s
// Enable ADC12 reference, allow time to stabilize using watchdog timer
//-----------------------------------------------------------------
#pragma vector = BASICTIMER_VECTOR
__interrupt void BASICTIMER_ISR (void)   // Acknowledged automatically
{
     LED = 1;                           // Show start of activity
     ADC12CTL0_bit.REFON = 1;           // Turn ADC12 reference on
// Start watchdog as interval timer, 512 cycles of ACLK, clear
     WDTCTL = WDTPW | WDTTMSEL | WDTCNTCL | WDTSSEL | WDTIS1;
}
//-----------------------------------------------------------------
// Wait for reference voltage to stabilize
//-----------------------------------------------------------------
// ISR for watchdog interval timer: start ADC12 conversion
//-----------------------------------------------------------------
#pragma vector = WDT_VECTOR
__interrupt void WDT_ISR (void)          // Acknowledged automatically
{
     WDTCTL = WDTPW | WDTHOLD;          // Stop watchdog timer
     ADC12CTL0_bit.ADC12ON = 1;         // Turn ADC12 on
     ADC12CTL0 |= (ENC|ADC12SC);        // Enable, start single conversion
}
//-----------------------------------------------------------------
// Wait for ADC12 to complete conversion
```

```
//----------------------------------------------------------------
// Interrupt service routine for ADC12, called when conversion complete
// Calculate temperature in tenths of celsius
//----------------------------------------------------------------
#pragma vector = ADC12_VECTOR
__interrupt void ADC12_ISR (void)      // Flag NOT cleared automatically
{
    uint32_t temperature;              // Converted value of temperature

    ADC12CTL0_bit.ENC = 0;             // Disable ADC12 before turning off
    ADC12CTL0_bit.ADC12ON = 0;         // Turn ADC12 off
    ADC12CTL0_bit.REFON = 0;           // Turn ADC12 reference off
    if (ADC12IV == 0x0006) {           // No other value should happen
        temperature = ADC12MEM0;       // Raw converted value
        temperature *= 4200;           // Scaling factor (Vref/temp coeff)
        temperature += 2048;           // Allow for rounding in division
        temperature /= 4096;           // Divide by range of ADC10
        temperature -= 2780;           // Subtract offset to give celsius
        TXdata[0] = (uint8_t) temperature;          // Extract LSB
        TXdata[1] = (uint8_t) (temperature >> 8);   //    and MSB
        TXcount = 0;                   // Initialize byte counter
        IE2_bit.UCA0TXIE = 1;          // Enable interrupts to start TX
    }
}
```

Very little work is needed to configure the USCI_A because most of the defaults are appropriate. The exception is the baud rate, for which I use the values calculated in the previous section. It would probably be better to call the function to display the temperature from the main function so that it could be interrupted by further communications but I left this in the ISR for reception to keep the program simple.

Figure 10.22 shows traces of the output using Listing 10.13. These are taken from the pin of the FG4618, not from the RS-232 driver, and are therefore at the usual logic voltages. The top trace shows data from the thermometer. The more significant byte is 0 for the temperature in my room (this is Scotland) so there are no features between the falling edge at the beginning of the start bit and the rising edge of the stop bit. There is no easy way of telling where frames begin and end in this format. The lower signal looks like a clock, which of course it is not—I edited the program to send bytes of 0x55 instead of the real data. The variation in width of the bits due to modulation is evident. Despite this, the FG4618 is able to receive the data without problems.

A more searching test is to transmit the data to another system. I connected the MSP430FG4618/F2013 Experimenter's Board to an elderly computer with a serial port, configured it for 9600 baud and 8-N-1 format, and it works. The fluctuations in the widths of the bits do not disrupt the transmission. This connection worked much better if the data are sent in ASCII format, which I leave as an exercise.

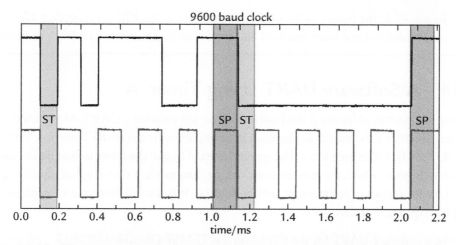

Figure 10.22: Two traces, each showing 2 bytes transmitted by USCI_A using low-frequency mode with BRCLK at 32 KHz and data at 9600 baud. The upper trace shows data from the thermometer while the bytes are both 0x55 in the lower trace to show every bit. I highlight the start and stop bits and the ticks along the top axis show the ideal width of bits at 9600 baud.

Example 10.8

Extend Listing 10.13 to transmit the data in ASCII format instead of binary. Here are a few suggestions:

- Use `UintToBCD()` to transform the computed value of temperature from binary to BCD. You may need to check that the value is within range and I suggest that you assume a positive temperature.

- Convert each BCD digit to ASCII and store the result in `TXdata[]`. This is simple: Just add (arithmetically) the digit to 0.

- It would be better to send the most significant digit first, which is the reverse of the order in Listing 10.13.

- Send a further 2 bytes with the end-of-line characters `\r\n` if your computer uses Windows (just `\n` for unix). This means that you send six characters in all (or five). Remember to enlarge the arrays.

- The ISR for transmission needs no changes at all.

- Reassemble the binary value of the temperature in the ISR for reception after all characters have been received and send it to `DisplayDeciCels()` as before.

10.14   A Software UART Using Timer_A

You have to resort to software if the hardware does not provide a UART. Many code examples illustrate this and the application note *Implementing a UART Function with Timer_A3* (slaa078 illustrate this). The general idea is to use the timer to handle the input and output directly, which ensures accurate timing, and process the sampled input or set up the next bit for output in an interrupt service routine. It is a good demonstration of the principles explained in Chapter 8.

I describe a software UART for the F2013 on the TI MSP430FG4618/F2013 Experimenter's Board because a pair of wires allows it to communicate with the FG4618 and avoids any need for an RS-232 interface. My approach is different from the code examples but closer to the library described in the application note *Using the Timer_A UART Library* (slaa307). The idea is to provide a "virtual peripheral" that works like the USCI_A as closely as possible. All the low-level communication is handled by Timer_A but this should be hidden from the user, who sees an interface with the usual registers, status flags, and bits to enable interrupts. It is a full-duplex UART and therefore needs two channels of Timer_A, one for reception and the other for transmission. (Many of the code examples are half duplex so that they need only one channel.) I use a speed of 9600 baud and the 8-N-1 format for each frame.

Once it has been configured, communications with USCI_A are mainly a matter of writing data to UCA0TXBUF when its flag is raised, usually in response to an interrupt, and similarly reading data from UCA0RXBUF. Unfortunately we cannot imitate these with ordinary registers. For example, writing to UCA0TXBUF does not just store the value but also clears the flag and starts the transmitter if necessary. We must use a function instead. This has the further advantage that the buffer can be hidden from the user by putting the functions and variables for the UART in a separate source file.

The baud rate must be accurate, as was seen in the preceding sections. The F20xx has calibrated values for f_{DCO}, which are specified within a few percent and should be good enough. The VLO is too slow and inaccurate. Older devices lack calibrated frequencies and need a different approach. One possibility is to use a high-frequency crystal to provide MCLK. Alternatively, a software FLL can be used to calibrate the frequency of the DCO against a crystal or possibly the 50- or 60-Hz mains (see the section *Measurement of*

Frequency: Comparison of SMCLK and ACLK on page 312). Slow communication could be performed with a watch crystal and a baud clock modulated in software. The idea was described in the section "Generation of a Precise Frequency" on page 326 and acts like the hardware modulator in the USCI_A.

We also have to find some spare interrupts that can be taken over for the UART. The port interrupts are the obvious choice and I use those for P2.6 and P2.7 on the F2013. It would be more convenient if we could use the flags for the missing pins P2.0–P2.5 but they are not implemented in the F20xx according to the data sheet. (This approach works for the unbonded pins P2.6 and P2.7 in the F1121A.) You must ensure that there is no activity on the pins that could trigger interrupts as well.

10.14.1 UART Library

With these choices, here is the header file `uartlib.h` for the library. There is an interrupt flag and enable bit for transmission and reception, the pair of functions to write and read data, and a function to initialize the UART. The argument to `WriteTXBUF()` is declared `const` to show that the function does not modify it.

```
#define RXIFG    P2IFG_bit.P2IFG_6    // Flag for receiver interrupts
#define TXIFG    P2IFG_bit.P2IFG_7    // Flag for transmit interrupts
#define RXIE     P2IE_bit.P2IE_6      // Enable receiver interrupts
#define TXIE     P2IE_bit.P2IE_7      // Enable transmit interrupts
void UARTSetUp (void);               // Function to initialize UART
void WriteTXBUF (const uint8_t TXData); // Write byte for UART to send
uint8_t ReadRXBUF (void);            // Read received byte from UART
```

The code for the library is shown in Listing 10.14. It includes the three functions declared in the header file and the interrupt service routines for Timer_A, which perform the communication. The external variables `RXBUF` and `TXBUF` are the buffers between the functions called by the user and the ISRs. They provide double buffering like that in USCI_A.

Listing 10.14: Library `uartlib1.c` for a software UART in the F2013 on the TI MSP430FG4618/F2013 Experimenter's Board.

```
// uartlib1.c - software UART using Timer_A2, running from SMCLK at 1MHz
// J H Davies, 2008-02-02; IAR Kickstart version 4.09A
#include <io430x20x3.h>              // Header file for this device
#include <stdint.h>                  // Integers of defined sizes
#include "uartlib.h"                 // Library for software UART
#define LED      P1OUT_bit.P1OUT_0   // Output pin for LED (active high)

uint8_t RXBUF;                       // Buffer for received byte
uint8_t TXBUF;                       // Buffer for byte to transmit
```

```
//------------------------------------------------------------------
// Set up UART: Timer_A2 continuous mode; assume 9600 baud, SMCLK = 1MHz
// TA0 for TX, init to high, flag set to request interrupt when enabled
// TA1 for RX, capture falling edge of start bit, interrupt enabled
//------------------------------------------------------------------
#define BITTIME  ((1000000 + 4800)/9600) // Cycles of TACLK per bit, 1MHz
#define HALFTIME   (BITTIME/2)            // Cycles for half a bit

void UARTSetUp (void)
{
    TACTL = TASSEL_2 | MC_2;            // TACLK = SMCLK = 1MHz
    TACCTL0 = OUT | OUTMOD_0 | CCIFG;
    TACCTL1 = CM_2 | CCIS_1 | SCS | CAP | OUTMOD_0 | CCIE;
    RXIFG = 0;                          // Buffer empty, no byte received
    TXIFG = 1;                          // Buffer empty, ready to accept TX
}
//------------------------------------------------------------------
// Write byte to TXBUF; clears flag, starts transmitter if necessary
//------------------------------------------------------------------
void WriteTXBUF (const uint8_t TXData)
{
    TXBUF = TXData;                     // Transfer data to buffer
    TXIFG = 0;                          // Clear interrupt flag
    TACCTL0_bit.CCIE = 1;              // Enable interrupts if inactive,
}                                      //   no effect if already running
//------------------------------------------------------------------
// Read byte from RXBUF; clears flag
//------------------------------------------------------------------
uint8_t ReadRXBUF (void)
{
    RXIFG = 0;                          // Clear interrupt flag
    return RXBUF;                       // Read data from buffer
}
//------------------------------------------------------------------
// ISR for Timer_A2, channel 0: transmission
//------------------------------------------------------------------
#pragma vector = TIMERA0_VECTOR
__interrupt void TIMERA0_ISR (void) // Acknowledged automatically
{
    static uint8_t TXBitCount = 0;     // Count bits transmitted
    static uint16_t TXShiftReg;        // Shift register for transmission

    LED = 1;                           // Show start of activity
    if (TXBitCount == 0) {             // Ready to start new byte
        if (TXIFG == 0) {              // Byte available to transmit
            TXShiftReg = TXBUF;        // Load data from buffer
            TXShiftReg |= BIT8;        // Add stop bit after msb
            TXIFG = 1;                 // Show that buffer is available
            TXBitCount = 9;            // Number of bits including stop bit
            TACCR0 = TAR + BITTIME;    // Delay until start of start bit
            TACCTL0 = OUTMOD_5 | CCIE; // Clr output on compare for ST
        } else {                       // No data available, shut down
            TACCTL0 = OUT | OUTMOD_0 | CCIFG;  // Idle: output high,
        }  //    disable interrupt, set flag for int'pt when reenabled
    } else {                           // Send next bit
        if (TXShiftReg & BIT0) {       // Send 1 next
```

```
            TACCTL0 = OUTMOD_1 | CCIE;   // Set output on compare
        } else {                         // Send 0 next
            TACCTL0 = OUTMOD_5 | CCIE;   // Clear output on compare
        }
        TACCR0 += BITTIME;               // Delay until next bit
        TXShiftReg >>= 1;                // Shift right, remove lsb
        --TXBitCount;                    // Update bit counter
    }
    LED = 0;                             // Show end of activity
}
//-------------------------------------------------------------------
// ISR for Timer_A2, channel 1 and timer core (inactive): reception
//-------------------------------------------------------------------
#pragma vector = TIMERA1_VECTOR
__interrupt void TIMERA1_ISR (void) // NOT acknowledged automatically
{
    static uint8_t RXBitCount = 0;   // Count bits received
    static uint8_t RXShiftReg;       // Shift register for reception

    LED = 1;                             // Show start of activity
    TACCTL1_bit.CCIFG = 0;               // Acknowledge interrupt, clear flag
    if (TACCTL1_bit.CAP) {               // Capture mode, start bit detected
        TACCTL1_bit.CAP = 0;             // Switch to sampling (compare) mode
        TACCR1 += HALFTIME;              // Wait for half a bit time
        RXBitCount = 9;                  // Bits to receive including ST, SP
    } else {
        switch (RXBitCount) {
        case 9:                          // Start bit
            if (TACCTL1_bit.SCCI) { // Error: start bit should be low
                TACCTL1_bit.CAP = 1;     // Abandon reception
            } else {                     // Correct: proceed to receive data
                TACCR1 += BITTIME;       // Wait for complete bit time
                --RXBitCount;            // Update bit counter
            }
            break;
        case 0:                          // Stop bit
            if (TACCTL1_bit.SCCI) { // Correct: stop bit should be high
                RXBUF = RXShiftReg; // Store data into buffer
                RXIFG = 1;               // Raise flag to show new data
            }                            //   (discard byte if stop bit low)
            TACCTL1_bit.CAP = 1;         // Return to capture mode
            break;
        default:                         // Data bit (lsb first), cases 1-8
            RXShiftReg >>= 1;            // Shift data, clear msb for new bit
            if (TACCTL1_bit.SCCI) {
                RXShiftReg |= BIT7; // Set received bit if high
            }
            TACCR1 += BITTIME;           // Wait for complete bit time
            --RXBitCount;                // Update bit counter
            break;
        }
    }
    LED = 0;                             // Show end of activity
}
```

Set up Timer_A

Timer_A2 runs in a continuous mode, which makes it easy to calculate the delays between bits. I assume that SMCLK runs at 1 MHz; the constants must be changed if this is not true. The calculation of BITTIME for a baud rate of 9600 looks horrible but is done by the compiler, not the MSP430.

I would have preferred to use channel 0 for reception because of its higher priority but that is not possible. The most convenient pins on the F2013 on the MSP430FG4618/F2013 Experimenter's Board are P1.4–P1.7, which are connected to Header 1 (Figure 10.6). The functions available on these pins impose the choice of P1.5 for TA0 out and P1.6 for CCI1B in. I therefore set up channel 0 for output. It is initially in output mode 0, where the output is controlled by the OUT bit. This is set so that the output idles high. Interrupts are not enabled yet because there are no data to transmit. However, I have set the flag CCIFG so that an interrupt is requested as soon as interrupts are enabled. This starts the transmitter.

Channel 1 is set up in the capture mode to detect a falling edge on its input CCI1B. This marks the beginning of a start bit. In this case I enable interrupts so that the UART is ready for data to arrive.

Finally, the interrupt flags are initialized. RXIFG is cleared because there is nothing valid in RXBUF while TXIFG is set to show that data may be written to TXBUF. This mimics the behavior of USCI_A.

Pins P1.5 and P1.6 must be routed to Timer_A by setting their bits in P1SEL. This is done in the main function, which should also enable the interrupts if desired.

Write to the Transmission Buffer

The function WriteTXBUF() copies its argument to the buffer for transmission, TXBUF. There is no check that the buffer is available; a previous value that has not been transmitted is simply overwritten. The function clears TXIFG to show that the buffer is occupied. Finally, it enables interrupts on channel 0 of Timer_A. This causes an interrupt to be requested immediately if the transmitter was idle but has no effect if it is already active.

Read from the Reception Buffer

This is simple: ReadRXBUF() returns the value in the buffer and clears the flag. Again there is no check that the data are valid.

Interrupt Service Routine for Transmission

This carries out the detailed work for transmission. There are local variables for the shift register and bit counter, required for any serial communication. Each bit is placed on the output by a compare event. As a reminder, it works like this:

- The output is set or reset at a compare event, when the value in the counter TAR matches that in TACCR0, so its timing is precise.

- An interrupt is requested at the same time. The time and value for the *next* output are set up in the ISR.

Here the actions taken in the ISR depend on the bit counter, TXBitCount. This is 0 in the ISR that follows the rising edge of a stop bit or if the transmitter has just been started by WriteTXBUF(). In either case, TXIFG is tested to check whether a byte is waiting for transmission. If a byte is available it is copied to the shift register, the ninth bit is set to provide the stop bit, TXIFG is set to show that the buffer is available again and the bit counter is initialized. Channel 0 is set up for the start bit by configuring it to reset the output on compare. The delay is added to the current value in TAR because the transmitter may be starting, in which case the previous value in TACCR0 is useless.

If no byte is waiting for transmission, interrupts are disabled and the channel is returned to its idle state.

A nonzero value of TXBitCount shows that the transmitter is sending data. Channel 0 is configured for a set or reset event according to the next bit to be sent and the duration of the bit is added to TACCR0. Data are sent with lsb first so the shift register is shifted right to bring the next bit into position 0.

Interrupt Service Routine for Reception

This uses two modes of Timer_A to receive each byte:

- It is in the capture mode when idle, waiting for a falling edge to signal a new start bit.

- The "sampling" mode is used to receive each bit. This is my name for the use of the compare mode to read the input and latch the value into SCCI when TAR matches TACCRn. The times are arranged to sample the middle of each bit.

The ISR first checks whether it is in the capture mode. If so, it switches to the compare (sampling) mode, adds a delay of half a bit to TACCR1, and initializes the bit counter

RXBitCount. The action in the sampling mode depends on which bit has just been read. All 10 bits are tested including the start and stop bit:

- The start bit should be low. If not, the reception presumably was triggered by noise and should be abandoned. The channel is therefore returned to capture mode. Otherwise, a delay is added to wait for the first bit of data.

- Bits of data are rotated into the shift register from the msb and a further delay of 1 bit is added.

- Finally, the stop bit is tested. This should be high and the data are discarded if not. If it is correct, the received byte is written to the buffer and the flag is raised to show that new data are available. There is no check for an overrun. Finally the channel returns to the capture mode to await the next frame.

10.14.2   Using the Software UART

I first tested the USCI_A by using its internal loopback mode. Much the same can be done for the software UART by joining pins P1.5 and P1.6 externally. Of course there is no display so you must use the debugger to check that the value is received correctly.

A more appealing test is to receive data from the FG4618 and echo it back. The USCI_A can be routed to pins P4.6 and P4.7, which are not otherwise used, to avoid the RS-232 transceiver. I edited the program in Listing 10.13 slightly to display the temperature after it has been echoed back by the F2013. The initialization must be done carefully or it is possible to put a falling edge onto P4.6, which fools the software UART into receiving a byte and transmitting it back. This is disastrous because it causes the received bytes to be one out of step with the transmitted bytes: The receiver misinterprets them and displays nonsense on the LCD. The problem is that my standard PortsInit() function for the FG4618 drives unused outputs low, which creates a falling edge if there is previously a 1 in the output buffer. I got around this by routing the pins to the USCI_A before initializing the ports.

Figure 10.23 shows oscilloscope traces of the UART receiving a single byte of 0x55 from the FG4618 and echoing it back. It is amusing to compare the irregular widths of the bits produced by the FG4618 using a 32 KHz clock with the uniform bits from the software UART. Reception and transmission are reliable despite the different clocks.

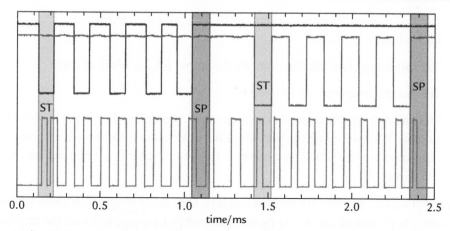

Figure 10.23: Reception followed by echo of a single byte (0x55) by the software UART with data at 9600 baud and MCLK at 1 MHz. The bottom trace shows when the interrupt service routines of the UART are active.

It is important to know how much load the software UART places on the CPU. How much time remains for other functions, for instance? A simple way of testing this is to toggle the voltage on a free pin to show when functions in the UART are active. Ganssle [59] gives further suggestions for this sort of instrumentation. I use the pin that drives the F2013's LED on the MSP430FG4618/F2013 Experimenter's Board and the waveform is shown in the bottom trace of Figure 10.23. The active pulses are very narrow with MCLK running at 8 MHz so I reduced it to 1 MHz for this test. Remember that the interrupt service routines are called just *after* a capture or compare event. The interrupt latency is clear on the very first edge of the input, which begins the start bit. These are the events that call the ISRs:

- The first is caused by the receiver capturing the initial edge of the start bit.

- The next occurs half a bit time later, which should put it in the middle of a bit, and samples the input to check for a valid start bit.

- The following eight interrupts follow samples of the input. They remain safely near the middle of each bit in time, despite the modulation from USCI_A and a possible error in the frequency of the receiver due to calibration of the DCO.

- The final sample is to check the stop bit. Its ISR writes the received byte to the buffer, raises the flag, and returns the receiver to capture mode. This calls a receiver interrupt, which is not shown in the trace. It writes the received data to the transmit buffer, which echoes it. The WriteTXBUF() function enables interrupts on channel 0 of Timer_A, which starts transmission.

- The first transmitter interrupt sets up the falling edge of the start bit after a delay of the bit time.

- The following eight interrupts set up the output for the 8 bits of data.

- The next interrupt sets up the stop bit.

- The final interrupt follows the compare event that drives the output high for the stop bit. It would set up the start bit of the next frame if there were data waiting. Here there is none so the transmitter is stopped by disabling interrupts on channel 0 of Timer_A.

The interrupt service routines for the receiver take noticeably longer than those for the transmitter. They could be shortened by using assembly language to rotate the received bit into the shift register, an operation that is clumsy in C. I originally thought of writing the complete UART in assembly language but it was not clear that much efficiency would be gained elsewhere.

The interrupt service routines use up about 50% of the time during reception so it would not be possible for the device to do much else unless MCLK were faster. The most critical point is at the start of reception, where the first sample must be set up for half a bit time after the capture of the falling edge. There is just enough time in Figure 10.23.

This is not a stringent test of the UART under load because it is either receiving or transmitting at any time—never both. Figure 10.24 shows similar traces for a pair of bytes but with MCLK running at 8 MHz. I should have used a value other than 0x55 because you have to count where one byte ends and the other begins. There are now pairs of interrupts during the overlap between transmission and reception. Interrupts are crowded together after the first byte is received, the second byte begins to arrive, and the first byte is transmitted back. There is also the receiver interrupt, which is not shown. This test required $f_{\mathrm{MCLK}} \geq 4\,\mathrm{MHz}$; it failed if the speed was reduced to $2\,\mathrm{MHz}$.

Example 10.9

What events cause the crowd of five interrupts around 1.1–1.2 ms?

The software UART has plenty of deficiencies. There is no provision for low-power modes, for instance, nor are there flags for overrun and framing errors. These could be provided by adding a status register but this would aggravate one of the UART's major faults, which is the high overhead required to run it. This is the price to pay for general-purpose

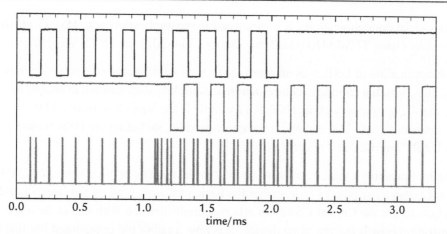

Figure 10.24: Reception and echo of 2 bytes (each 0x55) by the software UART with MCLK at 8 MHz. The bottom trace shows when the interrupt service routines of the UART are active.

software. The code examples are much more compact and efficient because they are specialized. Some more subtle problems come with the current software as well. What would happen if another module requested an interrupt during a call to ReadRXBUF() or WriteTXBUF(), for instance? Hardware cannot be interrupted but software can, with potentially nasty side effects; see the section "The Shared Data Problem" on page 197.

Example 10.10

The F20xx has only Timer_A2 so the software UART uses all of its channels. Would it be simpler to use the Up mode rather than the Continuous mode in this case?

10.15 Other Types of Communication

At present the MSP430 needs assistance from another IC to handle more complicated types of communication. This includes USB, ethernet, and most wireless protocols. The communication IC often uses SPI to exchange data with the MSP430. It would take a great deal of space to explain any of these, and the book is too long already, so I just provide a few pointers to further information.

USB is now the standard interface for most peripherals on personal computers and your JTAG pod probably uses it. Texas Instruments offers the TUSB3410 as a bridge between

USB and serial communications. It is described in the application note *MSP430 USB Connectivity Using TUSB3410* (slaa276). The eZ430–F2013 uses a TUSB3410.

Another application of USB is as an interface to removable storage devices—thumb drives. This is such a common requirement that complete solutions are available to handle the file system as well as USB. An example is the Vinculum from FTDI (www.vinculum.com). You can buy a complete module, including the USB socket, with an SPI or UART interface to a microcontroller.

The Internet presents a major step upward in complexity. The application note *MSP430 Internet Connectivity* (slaa137) describes the basic protocols used on the Internet and shows how to use the Crystal CS8900A ethernet controller in a Web server. Several demonstration boards use the same device. It is now a rather old component but this has the advantage that plenty of software is available. A more modern choice might be the Microchip ENC28J60, which is much smaller and has an SPI interface for the microcontroller, but the manufacturer's software is licensed for use only on its products.

Finally, wireless radio communication is perhaps the most rapidly growing area. A simple application is described in *Implementing a Bidirectional Wireless UART Application with TRF6903 and MSP430* (swra039). Recent protocols such as ZigBee are designed for low power and Texas Instruments acquired Chipcon a few years ago to increase its range of products in this area. An example is described in the application note *IEEE 802.15.4™ and ZigBee™ Hardware Platform Using MSP430F1612* (slaa264). The MSP430FG4618/F2013 Experimenter's Board has a header for wireless evaluation modules and a low-cost tool to demonstrate wireless applications, the eZ430–RF2500, uses a rather simpler protocol called SimpliciTI. See application note *Wireless Sensor Monitor Using the eZ430–RF2500* (slaa378) for an example. The topic of RF applications could fill a whole book and Baugh has written one already [1].

CHAPTER 11

The Future: MSP430X

There are many more peripherals in the MSP430 and applications that it would be fun to explore but a book has to come to an end and we are nearly there. However, I cannot finish without saying a little about the MSP430X.

I was originally attracted to the MSP430 for teaching because of its uniform 16-bit architecture: Both addresses and data are 16 bits wide, which makes it trivial to use any of the general-purpose registers in the CPU for either purpose. Calculations likewise are the same for data and addresses. This is undoubtedly elegant but limits the size of the address space to $2^{16} = 64$K bytes. That is typical of an 8-bit microcontroller and it was inevitable that the demand for evermore functions in embedded applications would necessitate an expansion. The result is the MSP430X, which can address a megabyte of memory. I explain in this chapter how the additional memory is organized and describe some of the changes to the instruction set that were made to accommodate it.

As I mentioned in the section "Where Does the MSP430 Fit?" on page 16, there is no separate family for devices that contain the MSP430X CPU. There isn't even an "X" in the part number to distinguish them! You have to look at the brochure and deduce the presence of the MSP430X from the size of the memory. This seems a curious approach in this publicity-conscious era and was presumably chosen to emphasize the transparent, upward compatibility from the MSP430 to the MSP430X.

11.1 Architecture of the MSP430X

Several major changes are needed to handle the enlarged memory in the MSP430X. I describe the memory and CPU in this section and examine the extended instruction set afterward.

11.1.1   Memory Organization

Memory can be expanded in several ways beyond the number of locations that can be addressed directly. One approach is to divide the memory into *banks*, only one of which is accessible to the CPU at a time. This will be familiar to anybody who has programmed a PIC16 in assembly language, where you need instructions of the form bsf STATUS, RP0 to switch between banks.

Of course the PIC has a Harvard architecture, unlike the MSP430. Perhaps a better comparison would be with another 16-bit, von Neumann microcontroller such as the Freescale HCS12. This can address 64 KB of memory directly, which includes some flash memory that is always available. The rest of the flash memory is divided into *pages* of 16 KB, only one of which is visible at a time. A page register selects the current page. Effectively the page register provides the most significant bits of an overall 24-bit address, while the CPU provides the less significant 16 bits. (This is not like paging in a computer with virtual memory, where data must be copied whenever a new page is required. Changing a page in the HCS12 is like changing a pointer.) This approach has the advantage that no changes are needed to the CPU, which sees only a 64-KB address space. A memory-mapping module prepends the contents of the page register to an address issued by the CPU to form the full address required by the memory. This approach is convenient for the designer of the CPU because no changes are needed to the instruction set for handling longer addresses, but it is less attractive to the programmer or compiler, which must arrange the program and data into pages so that they are available when required.

The MSP430X uses neither of these approaches but retains a simple, linear address space when the memory is expanded from 64 KB to 1 MB. This is much more attractive to a programmer but means that major changes must be made to the CPU so that it can address the enlarged memory and calculate the longer addresses. The description of the MSP430X CPU takes 170 pages in the family user's guides, which gives you an idea of the scale of the redesign. I cannot cover the details but try to give you a feeling for the way in which the changes have been implemented. Most of the particulars will not concern you if you program in C except that a memory model should be specified; see the section "Using C with the MSP430X" on page 613.

Figure 11.1 shows the memory map for the FG4618. This has 116 KB of flash memory and 8 KB of RAM, which makes it one of the largest devices currently available. It is useful to compare this with the memory map for the F20xx, one of the smallest devices, in Figure 2.4. The crucial feature is that the lowest 64 KB of memory, which can be

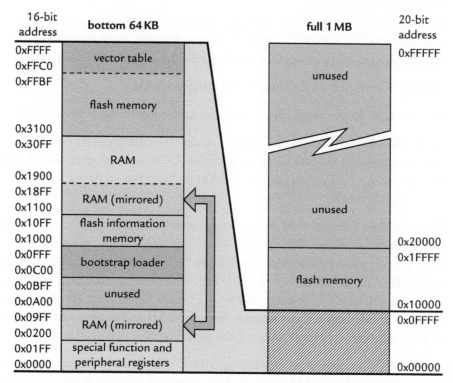

Figure 11.1: Memory map of the FG4618 with 116 KB of flash memory and 8 KB of RAM. The bottom 64 KB, which constitutes the address space of the MSP430, is enlarged on the left with the full 1 MB address space of the MSP430X on the right. The map is not drawn to scale.

accessed using 16-bit addresses, has the same layout in the MSP430 and MSP430X. The extra memory beyond 0xFFFF is all flash (at least for the time being). Inevitably there are a few complications so we look at them in a little more detail, working our way up the addresses:

- The special function and peripheral registers remain at the lowest addresses.

- Normally the RAM lies in the next zone before the bootstrap loader. Unfortunately there is space for only 2.5 KB, which is unlikely to be sufficient for a device that needs an MSP430X. It is inconvenient to have the RAM split into separate regions because this makes it difficult to handle large tables or other blocks of data. The 2 KB block of RAM in this zone is therefore not "real" memory but is *mirrored* from the lowest 2 KB of the larger block of RAM at higher addresses. This means that the addresses point to the same physical memory. For example, both of the addresses 0x0200 and 0x1100 refer to the first location in the RAM. The same feature is found on other devices with large RAM, such as the F1611.

- There is a small unused region after the mirrored RAM.

- The bootstrap loader and flash information memory are in their usual places.

- Next comes the RAM, with 8 KB in this device. The lowest 2 KB is mirrored to lower addresses in the traditional location for RAM, as I described already. There is no difference between the mirrored RAM and the rest—it forms a single, uninterrupted block.

- The main flash memory starts immediately after RAM and continues to 0x1FFFF, into the range that needs 20-bit addressing. Although it is a single block in principle, the table for interrupt and reset vectors lies within it at the same place as in the MSP430, immediately below 0xFFFF. The flash memory is by far the largest block in the lowest 64 KB range, occupying nearly 52 KB of it. There is a further 64 KB of uninterrupted flash memory from 0x10000–0x1FFFF. Therefore 128 KB of the 1 MB memory map is utilized, the higher addresses being unused.

The memory is packed as tightly as possible into the lowest 64 KB, with only one small, unused gap. If the size of the RAM is changed, the boundary with the flash memory moves so that no addresses are wasted.

The layout of the memory is designed so that a program written using only instructions for the MSP430 runs perfectly on a MSP430X, provided of course that all addresses remain below 0x10000. This ensures backward compatibility.

An aspect of the backward compatibility is that the vector table is unchanged. It is located at the same place in the memory map and each vector remains a 16-bit word. This means that all interrupt service routines and the main function (or at least their entry points) must reside in the first 64 KB of memory. This is no problem because they can in turn call functions that lie anywhere, including the extended memory.

11.1.2   Central Processing Unit

The width of the memory address bus (MAB) is increased to 20 bits in the MSP430X to address the extended memory but the memory data bus (MDB) remains 16 bits wide. Most of the registers in the CPU of the MSP430X can be used for either data or addresses and have therefore been enlarged to 20 bits as well. The exception is the status register (SR).

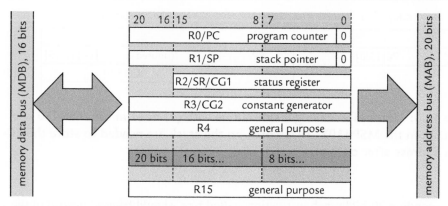

Figure 11.2: Registers in the central processing unit of the MSP430X.

Only 9 of its 16 bits are used in the MSP430 but the unused bits are exploited in the MSP430X, as we shall see.

The registers in the CPU of the MSP430X are shown in Figure 11.2. Both the program counter and stack pointer are used only as addresses and are therefore always treated as 20-bit registers. In contrast, the status register has only 16 bits. The constant generator and general-purpose registers can handle 8, 16, or 20-bit numbers. This raises a problem with the terminology. A "word" usually means 16 bits, which remains applicable for data, but an "address word" now contains 20 bits. I usually refer to the number of bits for clarity.

The general functions of the registers are unchanged from the MSP430. For example, the stack pointer should be initialized to the top of RAM before any functions are called. However, some details of their usage change to accommodate 20-bit values. For example, the current value in the program counter must be stacked when a subroutine is called, as we saw in the section "Automatic Control: Use of Subroutines" on page 99. This requires a single 16-bit word in the MSP430 following the `call` instruction, as shown in Figure 11.3(a). The 20-bit PC in the MSP430X requires *two* words of the stack to hold it, although 12 bits of one 16-bit word are wasted, as shown in Figure 11.3(b). A different instruction, `calla`, must therefore be used unless the whole program fits in the lowest 64 KB. (In this case you could probably save money with a smaller device.) There is a corresponding return instruction, `reta`, that retrieves a 20-bit address from the stack rather than 16 bits. I mentioned this in the section "Conversion from Binary to BCD in Assembly Language" on page 272. Be sure to pair `call` with `ret` and `calla` with `reta` or you will wreck the stack.

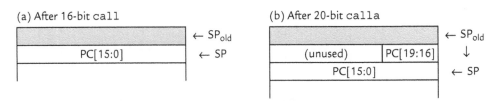

Figure 11.3: Stack after calling a subroutine with (a) `call` in the MSP430 and (b) `calla` in the MSP430X. Two words on the stack are needed to store the 20-bit return address after `calla`.

A corresponding 20-bit branch instruction, called `bra`, should be used instead of the 16-bit `br`. Many processors use the mnemonic `bra` for their standard branch so I wish that they were called `bra` and `braa` on the MSP430(X).

Figure 11.3(b) shows that two 16-bit words are needed to store a 20-bit address on the stack and the same is true for main memory: Address words require 4 bytes of storage, 12 bits of which are unused. This wastes memory and reduces the effective speed of the processor because two cycles are needed to fetch a complete 20-bit address from memory. In most cases these addresses are parts of instructions and the extended instruction set is designed to reduce these overheads.

Interrupts are handled in a slightly different way from subroutines. Both the status register and program counter are stacked in this case, as shown in Figure 6.5 for the MSP430. The upper 7 bits of the status register are not used and the MSP430X therefore takes over the space for the top 4 of them to hold the extra 4 bits of the program counter. Figure 11.4 shows this. It saves both memory and time, which is particularly important for interrupts. The `reti` instruction has been updated to match. An interrupt service routine must be

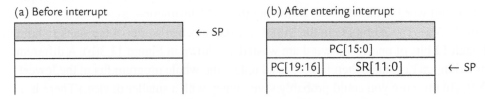

Figure 11.4: Stack before and after entering an interrupt service routine in the MSP430X. The most significant 4 bits of the return address (from the PC) are stored instead of the top 4 bits of the status register (SR), which contain no meaningful data.

situated in the lowest 64 KB of the address space because vectors hold only 16-bit addresses but the full 20-bit address is stacked to ensure that execution can resume at any address after the interrupt.

11.2 Instruction Set of the MSP430X

The instruction set of the MSP430X is extended beyond that of the MSP430 to handle the larger address space. The new instructions are clearly needed in two situations:

- The operands are 20-bit data, which usually arises when computing addresses.

- 20-bit constants are needed to compute the addresses of the operands.

In fact the MSP430X has three sets of instructions although the third, intended for common operations on addresses, contains only a few. For example, there are now three instructions for addition:

add.w, add.b: The original MSP430 instructions for words and bytes, which continue to work in most cases throughout the 20-bit address space.

addx.a, addx.w, addx.b: The extended MSP430X instructions for address words, words, and bytes, which allow any form of 20-bit address to be used.

adda: A more compact and therefore faster version of addx.a but with greatly restricted addressing modes.

The original MSP430 instructions remain useful because they are *not* restricted to 16-bit addresses. Let us look at this in more detail.

11.2.1 Addressing Modes

Here is a quick reminder, condensed from the section "Addressing Modes" on page 125:

Register mode: The operand is in one of the CPU's registers so there are no problems with the address. The most significant bits of the destination are cleared if the operation produces 8-bit or 16-bit data.

Indirect register mode: One of the CPU's registers contains the address of the operand, which is always treated as a 20-bit value in the MSP430X.

Indirect autoincrement register mode: Again a 20-bit address is taken from one of the CPU's registers. The address is automatically incremented by 1 for a byte,

2 for a word, or 4 for an address word. There is one special case: immediate data is implemented as autoincrement addressing on the program counter.

Indexed mode: The address is given by adding the contents of a register to a constant, which follows the instruction in the program memory. The constant is a 16-bit value in the MSP430 but may have 20 bits in the MSP430X, which requires a distinct instruction. There are three special cases of indexed mode:

- The Symbolic mode is indexed on the program counter (PC-relative).

- The Absolute mode is indexed on the status register, which behaves as though it contains 0.

- The SP-relative mode is indexed on the stack pointer and is not considered a separate mode by TI.

This summary shows that the only addressing modes that may need modification for 20-bit addresses are the indexed modes. We next look at how these operate with the original MSP430 instructions before examining the extended instructions.

11.2.2 Original MSP430 Instructions with Indexed Addresses

An MSP430 instruction in the MSP430X forms an indexed address from the sum of a 20-bit register and a 16-bit constant. This operation is handled differently, depending on the value in the register:

- If the value in the register is 0xFFFF or below, so that it lies in the 16-bit addressing range, the sum always wraps around to produce a 16-bit result, exactly as in the MSP430. This ensures backward compatibility for the lowest 64 KB of memory.

- If the register points to an address above the lowest 64 KB, the 16-bit constant is treated as a signed integer and added normally to the 20-bit value in the register. Thus the final address can lie roughly in the range ±32 KB from the value in the register, anywhere in the 20-bit address space.

These two cases are illustrated in Figure 11.5. Look next at what this implies for the various indexed addressing modes:

SP-relative mode: The stack is in RAM, which all lies in the lowest 64 KB, so this works exactly as in the MSP430.

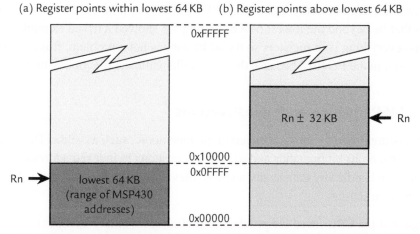

(a) Register points within lowest 64 KB   (b) Register points above lowest 64 KB

Figure 11.5: Ranges accessible by indexed addressing for an MSP430 instruction using a 16-bit constant in the MSP430X, where the register Rn points to an address (a) within the lowest 64 KB and (b) above the lowest 64 KB.

Absolute mode: The effective content of the register (SR) is 0 so this allows access to the lowest 64 KB only. This sounds restrictive but it includes most of the locations for which absolute addressing is usually used: the special function and peripheral registers and static and global variables in RAM. It can also be used for calling functions within the lowest 64 KB.

Symbolic mode (PC-relative): This mode seemed fairly pointless in the MSP430 because absolute addressing could almost always be used instead, but it becomes much more useful in the MSP430X. This is because it allows access to addresses within ±32 KB of the program counter, even when PC points above the lowest 64 KB. It can therefore be used for constants associated with a function, provided that they are stored in memory close to that function. Similarly, it can be used to call another function within the ±32 KB limit.

General indexed mode: This mode is typically used to access tables, in which case the 16-bit constant is the address of the first element of the array and the register contains the index of the element. This mode is therefore restricted to the lowest 64 KB in most cases.

Thus it is possible to write programs for the MSP430X that use its full memory space while employing mainly MSP430 instructions, despite their limited addressing capabilities. Any data should be kept in the lowest 64 KB unless they can be accessed using a symbolic mode. However, it is necessary to use the MSP430X instructions `calla` and `reta` to call functions that lie beyond the lowest 64 KB. Listing 7.8 showed a trivial example of this approach—everything is in registers so the addressing is not a problem. It is essential to end with `reta` in the FG4618 because the C compiler calls the function with `calla`.

11.2.3   MSP430X Extended Instructions

These are distinguished by an **x** appended to the mnemonic, such as `addx`. They can process 8-, 16-, or 20-bit operands and use 20-bit constants in indexed addresses. In other words, they can handle any data at any address. Inevitably this incurs a cost in the size and speed of the instructions.

The designer of the CPU faced the problem of finding extra binary opcodes to accommodate the new instructions. There are some unused values in the MSP430 but nowhere near enough to double the size of the instruction set. The only solution was a longer instruction to accommodate the extra opcodes. The binary form of MSP430X instructions is therefore distinguished from those of the MSP430 by including an *extension word* in front of the usual instruction word, thus doubling its length. The format of the extension word varies but this is its content for a Format I instruction (two operands) where both the source and destination are indexed:

- The most significant 5 bits are 00011 to identify the extension word.

- Four bits are used for the most significant bits [19:16] of the source. These may be part of immediate data or an index (including an absolute address).

- Four bits are used for the most significant bits [19:16] of the destination, which must be an index.

- The A/L bit specifies the length of the operand in conjunction with the B/W bit of the main instruction.

- The two remaining bits are unused (and reserved).

The extension word therefore contains the extra bits needed for the longer addresses as well as the distinguishing opcode, which saves storage and time for fetching the addresses.

Other information is included in the extension word where both operands are in registers or for Format II instructions (single operand). The user's guide says that the four least significant bits "set the repetition count" but this intriguing statement is not explained further.

How much of a penalty is incurred for using 20-bit data and addresses? The longest and slowest types of instruction have two indexed operands. An example is addx.a S(Rs),D(Rd), where S and D are 20-bit constants. This requires 10 cycles of MCLK for the following operations:

1. Fetch the extension word, which includes the most significant bits [19:16] of S and D.

2. Fetch the instruction.

3. Fetch the low 16 bits of S, which are appended to bits [19:16] from the extension word to form the complete 20-bit base for the indexed address.

4. Fetch the first word of source operand, S(Rs).

5. Fetch the second word of 20-bit source operand.

6. Fetch the low 16 bits of D.

7. Fetch the first word of destination operand, D(Rd).

8. Fetch the second word of 20-bit destination operand.

9. Write the first word of result to destination operand, D(Rd).

10. Write the second word of 20-bit result.

In comparison, the MSP430 needs only six cycles for add.w S(Rs),D(Rd) with 16-bit quantities, as analyzed in Figure 5.5. It would take only one cycle instead of two to read and write the operands if they were 16-bit words rather than 20-bit address words, so addx.w S(Rs),D(Rd) requires only seven cycles. This is a smaller penalty than might have been expected. You might have worried that it would take two cycles to fetch each of the 20-bit bases for the indexed addressing, S and D. In fact only one is needed because the most significant 4 bits are included in the extension word. Thus the 20-bit addresses are almost free—the major cost is loading and storing 20-bit operands.

11.2.4 MSP430X Address Instructions

Although the extra cost of extended instructions is smaller than might have been expected, it is not zero. The designers of the MSP430X CPU therefore took advantage of some unused opcodes to add a limited set of address instructions, which cover the most common operations used to manipulate addresses. These operate only on 20-bit address words and have limited addressing modes. The restrictions allow all the information required, including the four most significant bits of constants, to be embedded in the instruction itself so that the extension word is not needed. For example, the adda instruction for addition of 20-bit operands has only two addressing modes:

```
adda    Rsrc,Rdst    ; add Rsrc to Rdst, both operands in registers
adda    #imm20,Rdst  ; add 20-bit immediate value imm20 to Rdst
```

There is no need for a suffix on adda because only 20-bit operands are permitted. The address instructions cmpa and suba are similar but mova is used so often that it has been given more addressing modes. There are also emulated instructions based in the usual way on these four.

I already mentioned calla, reta (emulated) and bra. These are a bit different because calla and bra can use the full range of addressing modes. Again the most significant 4 bits of 20-bit constants are embedded in the instruction for economy. These instructions should be used in the MSP430X in place of the corresponding MSP430 instructions.

You might now be thinking: How do I choose the correct instruction—add, addx, or adda? It is even more perplexing for a move. The rules are fairly straightforward but this is the sort of decision best done automatically by the compiler. In some ways the range of general and restricted instructions seems like a return to an old-fashioned complex instruction set, but this is the price that has to be paid for extending the linear address space without an excessive cost in speed or extra code.

11.2.5   New Instructions in the MSP430X

There are also some completely new instructions in the MSP430X, mostly to speed up repeated operations:

Stack operations: pushm and popm transfer multiple words or address words between the CPU's registers and the stack. A single instruction can save or restore the complete set of registers, which makes it more efficient to call functions. This saves time and space compared with individual instructions but it still takes a cycle to transfer each word or two cycles for each address word. Thus it requires 26 cycles to transfer the full 20-bit values of all 12 general-purpose registers. This may increase the interrupt latency seriously.

Logical shift: I mentioned in the section "Shift and Rotate Instructions" on page 137 that the MSP430 does not provide a logical shift right, where a 0 is inserted into the msb (a logical shift left is the same as an arithmetic shift left because both insert a 0 into the lsb). It has been added to the MSP430X with the name *rotate right unsigned* and the mnemonic `rrux`. The name is misleading because it is a shift rather than a rotation but is consistent with the arithmetic shifts.

Multiple shifts: The shifts and rotations in the MSP430 move the operand only 1 bit to the left or right. Four multiple shifts through one to four places have been added to the MSP430X. The mnemonics are `rrcm`, `rrum`, `rram`, and `rlam`. There is no `rlcm`.

11.2.6   Using C with the MSP430X

You might hope to avoid all the complications of the extended instruction set if you program in C. This is almost true but not quite. To see where problems arise, suppose that you define a pointer `cPtr` with a statement such as `char *cPtr`. How much storage should be reserved for `cPtr`? There is no question in the MSP430 because all addresses contain 16 bits so a word of memory is enough. Of course this no longer holds for the MSP430X with its 20-bit addresses. A 16-bit word is sufficient if `cPtr` points to a variable because all the RAM lies in the lowest 64 KB but `cPtr` might instead point to a table of constant data, which could be located anywhere in the full 1-MB range. Two words should be allocated for the 20-bit address in this case. What should the compiler do?

The usual approach is to specify a *memory model*, which provides overall guidance, and to use key words or intrinsic functions to override the default where necessary. Details vary between development environments and will inevitably change as the software evolves. EW430 currently offers a choice of three *data models*:

Small: All data must reside in the lowest 64 KB, so that only 16-bit pointers are needed. This is the default model.

Medium: The default is that data resides in the lowest 64 KB but this can be overridden by using the key word `__data20` for large data structures that require the extended memory.

Large: The default is to use 20-bit addresses, which you can override with `__data16`.

Extra functions are declared in `intrinsics.h`, such as `unsigned char __data20_read_char (unsigned long address)`, which can be used to read or write data from locations that need 20-bit addresses. A problem with these functions is that they translate into multiple instructions of assembly language and problems could arise if this sequence is interrupted. You are therefore advised to disable interrupts when calling these functions or to use 20-bit pointers instead.

11.3   Where Next?

Suppose that the MSP430 or MSP430X does not provide sufficient power for your application. Where should you look next? The MSP430x5xx family has been advertised with clock frequencies up to 25 MHz, up to 256 KB of memory, and built-in USB, and radio communication. No data sheets are available at the time of writing so we have to wait.

There are two routes to greater processing power within TI's current portfolio of microcontrollers and related products: The TMS470 provides a general-purpose, 32-bit processor, while the C2000 is a more specialized device with a digital signal processor.

11.3.1   TMS470 and the ARM7TDMI

The TMS470 family is based on the 32-bit ARM7TDMI processor core, which I mentioned several times. The company that develops and sells the ARM range (www.arm.com) makes no silicon itself. Instead it licenses its designs of processors to other companies which embed them in their own products. Most of these products are for a particular purpose, such as mobile (cell) phones, but there is also a selection of general-purpose processors based on the ARM7TDMI and other cores. This is where the TMS470 fits in.

The ARM7 is a particular generation of the ARM architecture, which is described in the excellent book by Furber [36]. The extra letters show that the ARM7TDMI includes the Thumb instruction set, which I mention shortly, a debug module, an enhanced hardware multiplier, and in-circuit emulation. Apart from the debugging module it is only a CPU and needs substantial support from other modules to turn it into a complete microcontroller. To give just one example, the ARM7 core has only two interrupts, normal and fast. A vectored interrupt controller must therefore be added to provide the support expected for numerous, distinct interrupts in a microcontroller.

Some features of the ARM7 are familiar from the MSP430. Both have a von Neumann architecture (see the section "Harvard and von Neumann Architectures" on page 13) with 16 registers in the CPU and a small instruction set. Many of the instructions are similar, not just the obvious ones. For example, the MSP430X can save multiple registers with a single instruction like the ARM (although the ARM can save any selection of registers, not just a contiguous block).

However, major differences become apparent on closer inspection. The ARM7 can address 2^{32} bytes or 4 GB of memory, although only a tiny fraction of this is used in practice. The CPU has a load–store architecture, which means that arithmetic and logical instructions work only on the registers in the CPU, as explained in the section "Is the MSP430 a RISC—and Should It Be?" on page 154. This is appropriate for applications that require more processing of the data than control of peripherals. Separate instructions are needed to load the operands and store the results. This separation, combined with the luxury of a 32-bit instruction word, means that the instructions are far more capable than those in the MSP430. For example, the ARM can perform operations with three operands such as a = b + c, while the MSP430 is restricted to a += b. Moreover, one of the operands can be shifted in the ARM so that a = b + (c << d) can be computed in a single clock cycle. A clever feature is that any instruction can be executed conditionally on the state of the flags in the status register. This applies only to jumps in the MSP430. Conditional execution avoids the need for jumps over an instruction or two, which are a common and annoying feature of most assembly languages.

The ARM7 gains significantly higher performance by having a three-stage pipeline. This means that three actions take place simultaneously during a typical clock cycle:

- An instruction is fetched from memory.

- The previous instruction is decoded, which prepares the ALU to perform the operations requested.

- The instruction before that is executed by the ALU and the result written to one of the registers.

The MSP430 handles only one instruction at a time, which means that time must be allowed for the signals to pass through all the logic in the CPU. Splitting the process into three stages means that each can be performed more quickly and the clock frequency can be raised. The pipeline enables the ARM7 to execute one instruction per clock cycle provided that all results remain within the CPU. An extra cycle is needed if a result must be written back to the main memory. Thus the number of clock cycles is ultimately determined by access to memory, just like the MSP430.

The TMS470 does not just have a more powerful CPU than the MSP430: The same is true of its peripherals. For example, the high end timer (HET) has 32 channels, its own memory, processor, and programming language. All TMS470s include hardware for the controller area network (CAN), which is widely used in automobiles and for industrial control. It is considerably more complicated than the protocols handled by the MSP430.

The current range of TMS470 includes devices with 64 KB–1 MB of flash memory and 8–84 KB of RAM. Packages have 80–144 pins. (Other companies make much smaller microcontrollers based on the ARM7TDMI. For example, the LPC2101 from NXP contains only 8 KB of flash memory and 2 KB of RAM, which would be modest for a MSP430.)

Although all ARMs have a 32-bit CPU, their 32-bit instructions consume a lot of memory. The ARM7TDMI therefore offers an alternative, 16-bit *Thumb* instruction set, which was selected to include the most common operations. Thumb instructions typically require only about two thirds of the program memory with some penalty in performance, although this depends on details of the memory (Thumb instructions can instead be faster if the data bus for program memory is only 16 bits wide). The two sets may be mixed at will.

The ARM7TDMI is now a relatively old design and the latest offering from ARM for smaller embedded applications is the Cortex-M3 processor. This includes vectored interrupts and sleep modes and generally looks more like a conventional microcontroller than the ARM7TDMI. The CPU of the Cortex-M3 has a Harvard architecture and uses only 16-bit instructions, with an enhanced Thumb-2 instruction set. Products based on the Cortex-M3 may well be in competition with the MSP430X.

11.3.2   Digital Signal Controller: The C2000

Digital signal controllers offer an alternative approach to higher performance in applications that require extensive computation. Essentially they combine a microcontroller with a digital signal processor. Some MSP430s include a hardware multiplier, which can be combined with direct memory access to act as a basic digital signal processor, although I do not describe this. The TMS320C2000 family provides much higher performance by combining a pipelined processor with separate buses for instructions and data. These features are typical of a dedicated digital signal processor but the C2000 also includes the familiar features of microcontrollers, such as hardware for pulse-width modulation and the handling of interrupts.

Motor control is a common application for digital signal controllers. Simple types, such as brushed DC motors, are fairly straightforward to control with PWM and the MSP430 has sufficient power. Many modern motors are brushless, however, and need much more sophisticated electronics to drive and control them. They have several sets of coils, each of which must be energized with a current that varies with time so that the overall effect is a magnetic field that rotates inside the motor at the desired rate. The magnitudes of the currents are given by formulas that contain trigonometric functions and which are therefore time-consuming to calculate.

Other applications might require extensive processing of inputs. In the past this would have been done with active, analog filters but the modern approach is to sample the signal with an ADC and process it digitally. Here is an example. Modern automobile gasoline engines are controlled by embedded electronic engine management systems of steadily increasing complexity (this is a major application of the TMS470). One function of these systems is to choose the time at which the fuel–air mixture in each cylinder is ignited by the spark plug. Generally speaking, better performance is obtained with earlier ignition. If the spark is too early, however, it causes a sort of small explosion within the cylinder called *knocking*. This damages the engine if it is not corrected and the engine management system therefore needs a knock sensor. The problem is to discriminate between knocking and all the other noise and vibration produced by the engine. One characteristic of knocking is that it excites resonant oscillations within the cylinder at a few kilohertz, which can be detected by filtering the noise from the engine. This is a potential application for the C2000.

11.4 Conclusion

The title of this section is a misnomer because a book like this cannot reach a conclusion. I have not covered all the features of the current range of MSP430s but hope that I provided you with a solid foundation on which you can build. Once you have learned the operation of the basic peripherals and become familiar with the style of the data sheets and user's guides, you should be well equipped to explore new functions. The code examples are helpful too. There has been no space to cover the serious issues of designing software and developing it in a professional manner, so please consult "Further Reading."

For those of you who are new to microcontrollers, I hope you agree that they are versatile, straightforward to use, and provide a remarkable range of capabilities at low cost. Readers who were already familiar with other microcontrollers will probably have migrated to the MSP430 because of its low-power characteristics. These should ensure a successful future for the family in a world where the impact of global warming is leading to continually greater curbs on power dissipation.

Happy programming!

Kickstarting the MSP430

In this Appendix I explain how to use the IAR Embedded Workbench Kickstart for MSP430, or EW430 for short. This is a free integrated development environment (IDE) but is limited to 4 KB of C code. (That is more than enough for the examples in this book.) EW430 is produced by IAR Systems (www.iar.com), which also sells unrestricted versions, but Kickstart is distributed and supported by TI. IAR Embedded Workbench is available for a wide range of processors and I find it to be straightforward to use, predictable, and reasonably reliable if not the most powerful IDE that I encountered. It seems to get better with each release, which is less common than it ought to be. The only problem that I experienced at all frequently is that the emulator loses contact with the device when it enters a low-power mode and it is not hard to see why this might create difficulties.

TI itself developed another IDE, Code Composer Essentials (MSP-CCE430). A free, limited version of this can be downloaded from TI's Web site. The environment is based on the Eclipse open-source platform, with a distinctive look and feel. Some aspects, such as the directives in assembly language, are entirely different from EW430. Unfortunately I lack the space to describe CCE here as well.

A.1 Introduction to EW430

Plenty of documents support EW430. These include a user guide and reference guides for the assembler and the C/C++ compiler. The user guide includes a tutorial but it is of a general nature and not specific to the MSP430. Further information is in the *MSP-FET430 Flash Emulation Tool (FET) User's Guide* (slau138), which is updated for each release of EW430. The "Frequently Asked Questions" are particularly valuable and will solve many of your problems. The first step is of course to install the software from a CD or after you have downloaded it. I have never had any problems with this.

Figure A.1 shows a screenshot of EW430 in use for editing a program. I took this and other images from a screen with a width of only 800 pixels so that the text is legible when reduced to fit on the page. You almost certainly have a larger screen and are able to fit more into the windows. There is the usual menu bar along the top and a toolbar underneath. I describe how to perform actions using the menus and leave you to explore shortcuts. The main part of the screen is divided into three windows.

Workspace window: Has individual tabs for each project in the current workspace with an Overview in the leftmost tab. Make a project active by selecting its tab.

A *project* shows the relations between all the files needed to produce an application. This obviously includes your source code, C or assembly language (or both). The

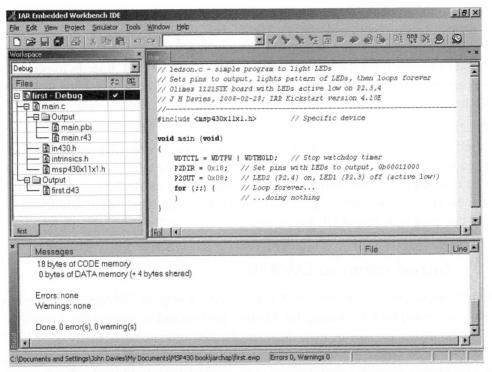

Figure A.1: EW430 while editing a C program. It shows the file `main.c` in the project `first`.

project also shows the header (.h) files that you include and files that provide other information needed to make the application, such as linker scripts (.xcl). Other files are added to the project after the IDE has compiled and linked the code. I show all of these in Figure A.1 but most of the tree is collapsed by default.

There are two columns after the name of each file. The first shows that options have been set for this file. Normally there is a checkmark only against the project itself but sometimes a file needs its own, special settings. The third column contains a red blob if errors were detected when the file was last compiled or if the file has been modified and not yet recompiled.

The layout of files in the workspace window shows their logical hierarchy within the project and bears little relationship to the directory structure on the hard disk.

Each project can have several configurations, shown in the drop-down list at the top of the Workspace window. There are two by default, Debug and Release. Typically the Release configuration omits debugging information and is more highly optimized. I use only the Debug configuration but this would not be a good idea for a version to be released commercially.

A *workspace* contains a set of related projects. For large-scale development the projects might be slightly different versions of the software, which share many files. I simply used a workspace for the programs in each chapter of this book. Figure A.12, shown later, shows a workspace window that contains several projects with the Overview tab on the left, which shows a list of all the projects in the workspace.

Editor window: Where you type and edit your source code, as you might expect. Several files can be edited or viewed by using tabs or multiple windows. It is convenient to see both the source code and header files, for instance. There are the usual features for editing code including syntax coloring and so on, which can be adjusted with the menu item Tools > Options.

Messages window: Shows text output from the tools, such as error messages from the compiler.

A.2   Developing a Project in C

As an example, I work through the simple example in Listing 4.2 to light the LEDs in a constant pattern. Kickstart will inevitably have been updated by the time that you follow these instructions but I hope that the main points remain valid.

1. Before you run Kickstart, create a new directory (folder) on your hard disk to contain the files for the project.

2. Start Kickstart. To begin from a fresh state, select Window > Close All Editor Tabs to empty the editor window and File > Close Workspace to create a new, empty workspace.

3. Create a new project with Project > Create New Project.... You will see the dialog box in Figure A.2. Under Project templates, expand the tree for C and select main. There are no templates for specific devices. The Tool chain at the top should show MSP430 but you should not have to select this unless you have versions of IAR Embedded Workbench for other processors installed. Click OK.

4. You will see the usual Save As dialog box. Navigate to your new directory, give the project a name such as first, and click OK. Your screen will be laid out as in Figure A.1 but the contents of the windows will differ in detail.

5. Configure the project by selecting the name of the project in the Workspace window and choose Project > Options… from the menu. You will be presented with a complicated dialog box, which has a Category list on the left and a set of tabs for each category. We can ignore most of them but some are essential:

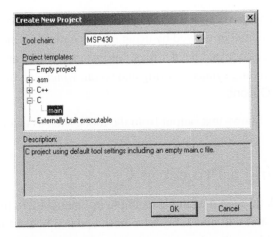

Figure A.2: Create New Project **dialog box for a project using C.**

- Select the category General Options and Target tab if not already shown. Choose your device by clicking on the button and working down the hierarchical pop-up menu as shown in Figure A.3.

- Select the Debugger category and activate the Setup tab if necessary. Choose Simulator from the Driver drop-down list if this is not already shown. See Figure A.4.

- This completes the minimal configuration of the project so click OK. There are plenty of further choices if you wish to explore. I suggest some of these later.

6. Delete the existing text from the editor window for `main.c` and type (or copy and paste) your program instead.

7. Keep the window for `main.c` active and select Project > Compile from the menu. Before EW430 starts the compiler, it presents you with a dialog box for Save

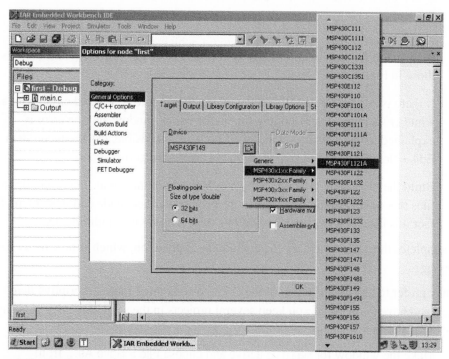

Figure A.3: Project options dialog box, showing the tab and pop-up menu for selecting the target device.

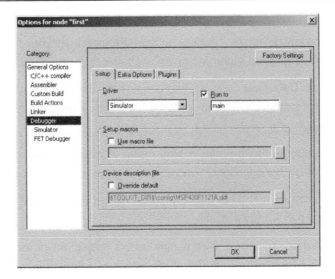

Figure A.4: Project options dialog box, showing the tab for the debugger.

Workspace As. I called it `kickstart`. If all goes well, you will see messages like those in Figure A.1.

8. If there is an error, you will see something like Figure A.5. A red circle with a white cross marks the offending line in the editor window and the error is explained in the messages window. If both are not visible, use the menu item Edit > Next Error/Tag to step through the errors and correct them.

9. Build the project by selecting Project > Make from the menu. This compiles all files that have been modified since the project was last made and links them. In fact it is usually a better idea to do this than compile files individually.

This completes the steps necessary to generate the application, which is now ready for debugging.

You may prefer to change the name of the C file to something more descriptive than `main.c` and have to do this if you keep several projects in the same directory. It is slightly clumsy and takes a few steps:

*   Make the window for `main.c` active and choose File > Save As... from the menu.

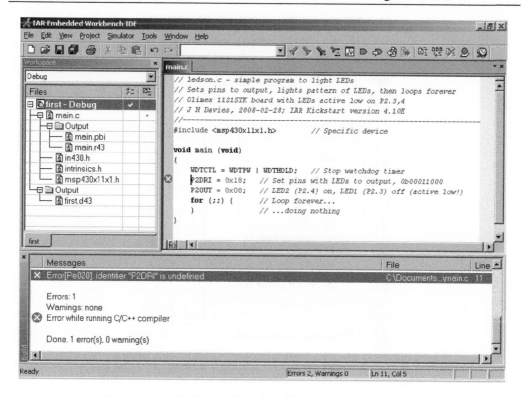

Figure A.5: An error reported when `main.c` is compiled.

- Give the file a more illuminating name, such as `ledson.c`, including the extension `.c`, and save it.

- This changes the name on the tab in the editor window but *the project still contains main.c*, which must also be changed.

- Select the name of the project in the Workspace window, right-click to bring up the contextual menu and move down to the Add item. This expands into a list of files to add, one of which should be Add "ledson.c" as in Figure A.6. Select this.

- Select main.c in the Workspace window and choose Project > Remove from the main menu. You will be asked to confirm that you want to remove the item from the project: Click Yes. The file is not destroyed and you can add it back later if you wish.

- Select Project > Make to rebuild the project with the new file.

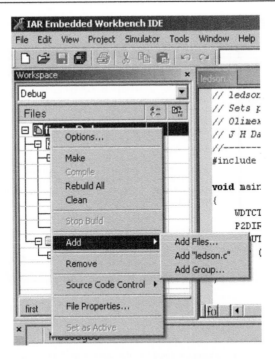

Figure A.6: Contextual menu for adding `ledson.c` to the project.

Congratulations! The debugger is next. The IAR C-SPY Debugger can run in two modes, selected with the Debugger category of the project options:

- The **simulator** is a program that runs entirely on the PC and models the behavior of certain aspects of the MSP430. No hardware is needed (no MSP430) and the simulator gives complete insight into those features that it models. However, it does not include the peripherals and is unsuitable for investigating the interaction between the MSP430 and its external signals.

- The **emulator** looks almost identical on the PC but the MSP430 runs its program itself. The full resources of the microcontroller are available and it can react to all aspects of its working environment. The debugger can stop the CPU and read back the state of any register to the PC so that you can examine any aspect of the program. Unfortunately debugging has some side effects and can be tricky when some peripherals, such as timers, continue to run while the CPU is stopped.

We start with the simulator and explore the emulator later.

A.3   Debugging with the Simulator

The IAR C-SPY Simulator models the CPU but excludes most peripherals. On the other hand, it can simulate interrupts from all sources, which is often enough. Of course `ledson.c` is so simple that we need only the CPU and the input/output ports.

Start the simulator from the menu with Project > Debug. Your project will be rebuilt if it is not already up to date and you will see a screen slightly simpler than Figure A.7. Overall, EW430 looks much the same when debugging as it does for editing the program: It is not like some older debuggers that run as completely separate applications. There is an extra Simulator menu but the most obvious new feature is a Disassembly window, which shows the assembly code that corresponds to each line of C code. Most of this vanishes off the right-hand edge of the window—a large screen is a tremendous boon for debugging. There is also a new toolbar for the debugger.

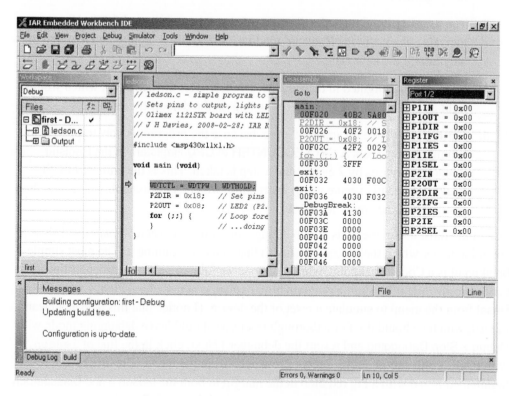

Figure A.7: Simulation of `ledson.c`.

The debugger offers many ways to look inside the processor. The most useful for this simple program is the Register window, which is brought up by choosing View > Register from the menu. The window has a drop-down list at the top to select various sets of registers inside the particular device. By default it shows the registers inside the CPU but I chose Port 1/2 in Figure A.7. This shows that all registers contain 0x00 initially, which is slightly misleading. Some of the registers are initialized this way, such as the direction registers P1DIR and P2DIR. Others, however, are not initialized and may contain anything. These include P1OUT and P2OUT.

This program is so short that it is trivial to step through it line by line. Use Debug > Step Into for this. The green highlight with an arrow in the margin shows the *next* line to be executed. It moves down after each step in both the ledson.c and Disassembly windows:

1. Nothing obvious happens after writing to WDTCTL, which stops the watchdog. You can change the Register window to Watchdog Timer if you want to check that this statement works.

2. The next line configures the direction of port 2 by writing to P2DIR. You can see the value change in the Register window and P2DIR is highlighted in red to draw attention to the new value.

3. P2OUT is highlighted in the Register window after the next step, which would light an LED if the program were running on the demonstration board.

4. Further steps have no apparent effect because the program is now trapped in an infinite, empty loop. A CYCLECOUNTER in CPU Registers in the Register window increases by 2 every time a step is taken. This is the number of CPU cycles needed to execute a `jmp` instruction, which is the assembly language that results from the empty loop `for (;;) {}`.

One of the most useful methods of debugging a program is to set a *breakpoint*. This is a condition under which the processor halts so that you can examine it. The most simple breakpoint is a Code Breakpoint, which causes execution to halt at a particular line. It is hardly worthwhile in so short a program but I explain it nonetheless. First, choose Debug > Reset from the menu to simulate a reset of the device. (I notice that this does not clear P2DIR, which it should do. For a thorough reset you should leave debugging mode with Debug > Stop Debugging and restart the debugger.) Next, click in the line of `ledson.c` with P2OUT and choose Edit > Toggle Breakpoint from the menu. The line becomes highlighted in red with a big red blob in the margin in both the ledson.c and Disassembly windows. Now run the program without stepping with Debug > Go. The simulator runs

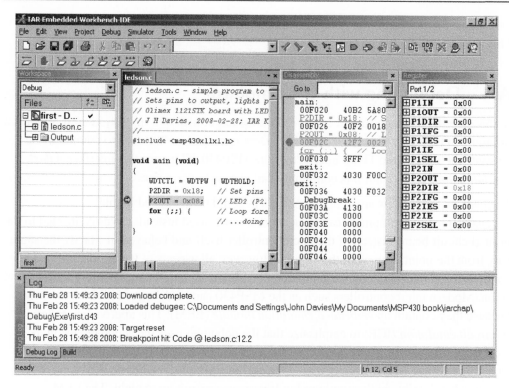

Figure A.8: Simulation of `ledson.c` with a breakpoint.

(for all of two lines) until it hits the breakpoint and stops before executing the highlighted line. This is shown in Figure A.8.

An invisible breakpoint is used whenever you start debugging a C program. This is because the compiler generates code to initialize the processor before it is ready to execute your program. For example, static and global variables must be initialized and the stack must be set up. The debugger therefore sets a breakpoint at the start of your `main()` function so that it stops there after it has performed the initialization. This is the significance of the Run to main item in the dialog box for the Debugger options in Figure A.4.

Another useful trick if you want to examine a particular part of a program is to click in the first line of interest and choose Debug > Run to Cursor. This again sets a breakpoint, runs to the breakpoint and removes it.

Select Debug > Stop Debugging when you have finished. Simulation has the great advantages that you get a perfect view into the device and that no hardware is needed. Unfortunately it is slow, does not include most peripherals, and cannot be used to investigate the interaction between the microcontroller and the surrounding components. This requires emulation instead.

A.4 Debugging with the Emulator

In this case the program runs at full speed on the target MSP430 in its system. It requires a program on the PC, called the C-SPY FET Debugger, and an interface to communicate between the PC and the JTAG interface of the MSP430. TI names this interface a *flash emulation tool* or FET but it is commonly called a *pod* or *wiggler*.

A decade or so ago, few small microcontrollers could be debugged in this way. An expensive piece of equipment called an *emulator* was used instead. It was connected to the printed circuit board instead of the microcontroller itself and behaved in exactly the same way from the point of view of the system. However, all aspects of the emulator could be controlled and interrogated by a computer—breakpoints could be set, memory viewed, and so on. Modern microcontrollers such as the MSP430 can be debugged in the same way because they contain an *embedded emulation module* (EEM). This approach is called *in-circuit emulation* (ICE) to emphasize that the debugger uses the actual device in its working environment: There is no need for a separate emulator in most applications. The capabilities of the EEM vary between devices. For example, smaller devices can set only two breakpoints in the emulator but larger ones permit up to eight. The EEM communicates with the outside world through JTAG, described in the section "Access to the Microcontroller for Programming and Debugging" on page 57.

There are two steps in using the debugger as an emulator: reconfiguring EW430 and connecting the FET or similar interface. First, the software:

1. Open the project options, go to the Debugger category (Figure A.4) and change the Driver to FET Debugger.

2. Select the FET Debugger category and choose the Setup tab, shown in Figure A.9. Several items here may need to be checked:

   - The Connection means the interface between your computer and the JTAG connection. I show this for a TI USB interface. The box with Automatic is to select the virtual COM port used by the FET and should not need to be changed (other USB interfaces may not need this option). The list below it is needed only if your interface connects to a parallel port.

- The Debug protocol applies only to devices that offer a choice of the full JTAG or Spy-Bi-Wire interfaces, such as the F20xx. It is therefore dimmed in Figure A.9. If `ledson.c` were adapted for something like an eZ430–F2013 tool, the dialog box would appear as in Figure A.10. The eZ430–F2013 behaves as a TI USB interface but uses Spy-Bi-Wire. This is the default for a F20xx but can be overridden if you wish to use the four-wire interface instead.

  Click **OK** when you have finished.

Connecting the hardware is easy: Plug in the cables. The only difficulty may be that some demonstration boards use a bare header for the 14-pin JTAG connection, which means that it is possible to connect the plug the wrong way around. The red side of the ribbon cable between the interface and the JTAG plug should be next to pin 1 of the header, which should be marked in some way. Do not forget to apply power to the target board as well, unless it takes it from the JTAG cable.

Back in EW430, select Project > Debug from the menu to start the debugger. If this is the first time that you used the JTAG interface or if you have just updated EW430, you may

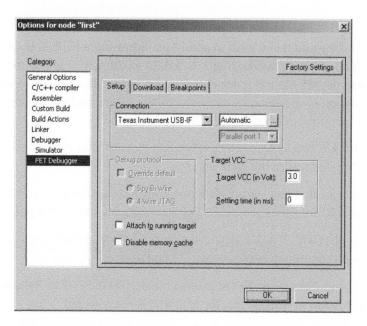

Figure A.9: Configuration of FET debugger for four-wire JTAG and a Texas Instruments USB interface.

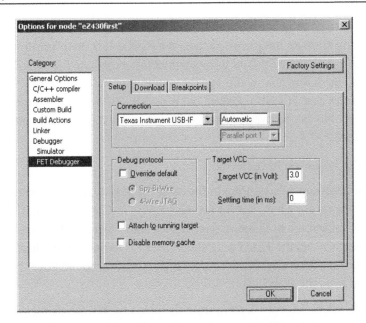

Figure A.10: Configuration of FET debugger for Spy-Bi-Wire and an eZ430–F2013 tool.

see a message asking you if you wish to update the firmware of FET. You should probably agree to this. If all is well, a few messages will flash on the screen as your program is downloaded into the MSP430. If it does not, you are in trouble. (I always dread such problems when running classes with microcontrollers.) Here are few suggestions if something goes wrong:

- A TI FET has two LEDs to show when it is powered and communicating, which might help to localize the problem.

- If you get an alert that starts Fatal error: Failed to initialize you probably forgot to plug the FET into your computer. The alert suggests some other possibilities.

- Another alert with The debugging session could not be started may mean that there is no connection between the FET and device.

- An alert with Fatal error: Failed to write memory at 0xFFFE or Fatal error: External voltage too low may mean that the MSP430 is not powered.

- Check that all connections are secure and correctly oriented, that there is a microcontroller in the target board (has it come loose in a socket, for instance, or been installed the wrong way around?), and that power is applied (use a voltmeter to check—batteries might have run down).

- Consult the "Frequently Asked Questions" in the *MSP-FET430 Flash Emulation Tool (FET) User's Guide* (slau138).

- If your target board is powered by the JTAG interface rather than its own power supply, check the signal connections carefully. The FET uses one wire to sense the voltage of the target board and a separate one to deliver power if desired. There are diagrams in the *FET User's Guide*.

If all these fail, shut down your computer, take a walk to calm down, and check everything again when you return. After that I can only resort to the usual feeble suggestions: try a new MSP430 in case it has been damaged (always a good excuse when students are involved, if often unfair), reinstall the software, and so on.

Assuming that everything works—and it almost always does, despite the formidable list of possible problems—you will see a screen like Figure A.11. It looks almost the same as that for the simulator but the Simulator menu has changed to Emulator. There is also a further toolbar, which I moved next to that from the simulator. You can step through the program, set breakpoints, and so on in exactly the same way. The difference is that the program is now executed on the MSP430 itself. This means that the LED(s) may light up as soon as their pins are configured as outputs by writing to P2DIR. They should show the desired pattern after the write to P2OUT.

Emulation may seem slower than simulation because the contents of all the registers displayed on the screen must be read out through the JTAG interface whenever the CPU stops. You can use the interface in the opposite direction too. Try changing the value of P2OUT in the debugger and you should see the LEDs change in response. The current value of P2OUT is visible in the Register window and you can also check it by making the ledson.c window active and hovering the mouse over P2OUT. For tips on more advanced debugging see the application note *Advanced Debugging Using the Enhanced Emulation Module (EEM)* (slaa263).

A.5 Developing a Project in Assembly Language

You create a project that uses only assembly language in almost exactly the same way. Expand the asm tree in the Create New Project dialog box (Figure A.2) and select asm. A project is created with the file asm.s43 in assembly language. This is set up for

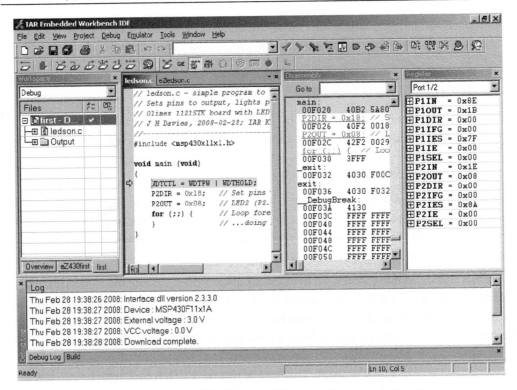

Figure A.11: Debugging `ledson.c` with the emulator.

relocatable assembly, which I do not explain until after the first example in *absolute* assembly. I therefore suggest that you remove this file from the project and replace it with `ledsasm.s43` from Listing 4.3. Set the project options in the same way as for C, which should require only selecting the appropriate target device. Build the project in the same way as well (the menu item is still called Compile even for the assembler).

Simulate the program as before. You may get a Stack Warning error, which can be ignored because this trivial program does not use the stack. Open a Register window, select Port 1/2, and step through it exactly as for the program in C. Change the project options to use the FET Debugger instead and you will find that it works in the same way as well.

The debugger shows a Disassembly window, which might seem rather pointless when the source is in assembly language but in fact has some instructive features. For example, the label `Reset:` appears above the address 0xF000, which is where we directed the assembler to store the program at the start of the flash memory. You can see this explicitly by choosing View > Memory, which opens another window to show the contents of memory. Select FLASH from the drop-down Memory list and 2x Units from the unnamed drop-down list of options. This shows the contents of the memory in 16-bit words, starting from the lowest addresses in flash memory. (Instructions and addresses are stored in 16-bit words, hence this choice.) You will see the binary machine code for the few instructions in the program. All the unused entries memories contain 0xFFFF because this is the erased state of flash memory—not 0.

Now scroll down to the highest addresses in memory. The very last word, at 0xFFFE and 0xFFFF, holds the reset vector—the address at which execution should start. You should see 0xF000 stored here as in Figure A.12. You could try increasing the address in the ORG 0xF000 directive a little and rebuilding the project. This causes the program to be stored at higher addresses and changes the reset vector to match.

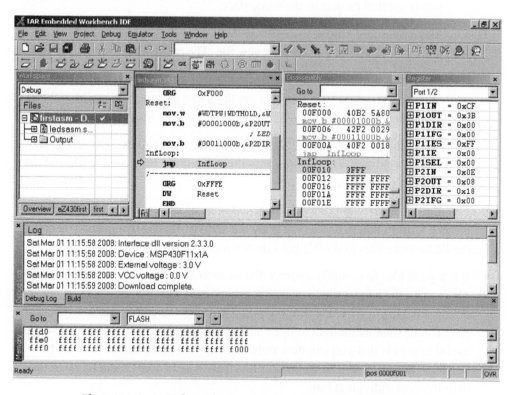

Figure A.12: Debugging `ledsasm.s43` with the emulator.

A.6 Tips for Using EW430

Here are some tips for getting the best out of the current version of EW430.

A.6.1 General

- There is no stationery or "wizard" for constructing projects for a particular target. A new project is almost blank and needs the device defined through the Project Options.... Here is a workaround to avoid some of the tedium.

  — Start with an existing project in the workspace that has the same configuration as the project you wish to create. Duplicate (copy and paste) this project file (extension .ewp) using Windows Explorer and change the name of the new project to something appropriate.

  — Return to EW430 and choose Project > Add Existing Project.... Add the newly created project to the workspace.

  — Make the source file active. This is the file for the project that was copied so use File > Save As... to create a new file. The editing pane shows the new file.

  — Select the name of the project in the Workspace window, right-click, and choose Add > Add "newfile.c", where the name in quotation marks is the newly saved file. It is added to the project.

  — Select the name of the old file in the Workspace window, right-click, and choose Remove. This deletes only the connection between the file and the project, not the file itself.

  You now have a new project that inherited most of its configuration from the old project. Unfortunately it does not copy the settings for the debugger. You could duplicate more files to do this but I do not find it worth the trouble.

- It is a good idea to compile a source file as soon as it contains a "legal" program. This causes the header file to appear in project tree, after which it can be opened by double-clicking or by using the contextual menu in the editor window. There is also a contextual menu item to open a header file in the editor window. The most useful benefit is that you can now right-click on a symbol and choose Go to definition of ... to find it in the header file. This is *very* convenient to get at the definitions of individual bits.

- Beware that the Project > Options... command applies to the file currently selected in the project, not necessarily to the project itself. It is the same as the Options... contextual menu item when you right-click on the name of a file in the Workspace window.

- The editor options are stored only once for each user, not for each file or project. This makes it tedious to use different settings for assembly and C files.

- When you save a new source file, no extension is appended automatically in most cases: You must add `.c` or `.s43` explicitly. This is true even for the Save As... command, which does not copy the extension from the original file.

- To indent a block of code—move it one tab stop to the right—select the text and hit the tab key. There seems to be no way of doing this from the menus. Shift-tab moves the text one tab stop to the left. A more comprehensive menu command, Edit > Auto Indent, lays out the code according to the language, C or assembly.

- Bookmarks are useful to locate points in source files if you have to jump around the file frequently.

- The compiler displays a red circle with a white cross next to lines that produce errors. It is tempting to double-click on these to find the problem but that sets a breakpoint instead. Select Edit > Next Error/Tag instead or hit F4. The warnings are in the Build window but have usually vanished out of sight.

- There is no command to create a new workspace. Just close the existing one, which automatically provides a new, empty workspace. It cannot be saved until a project has been created in it. You will be prompted to do this when you build the project.

- The header (include) files are not protected in any way. It is the original file that you see when you open one within EW430, not a local copy for your project. This means that any accidental edits wrecks your only version of the header file. A simple way of doing this is to drag a word from the header file into your source code. (I find this a natural thing to do, but then I am a Macintosh user by choice.) Unfortunately this deletes the word from the header file. I suggest that you protect the files to avoid such accidents. Navigate to the header files, which are in `430/inc/` within the EW430 directory. Select them all, show their Properties and make them Read-only.

A.6.2   C Language

- There is no notation for binary numbers, such as the common (but nonstandard) format 0b01011010. Use the definitions in the header file for individual bits (BIT0 and so on) or convert the value to hexadecimal.

- It is necessary to launch the debugger to see the disassembly output—the assembly language that corresponds to the C program. This is a limitation of the Kickstart edition, which does not write an assembly file from the C compiler.

A.6.3   Assembly Language

The IAR development environment is primarily intended for the C language and has some minor limitations for assembly language, particularly for absolute assembly:

- With assembly language you are very likely to get a complaint that The stack plug-in failed to set a breakpoint on "main". The stack window will not be able to display stack contents. The problem is that the stack plug-in is enabled automatically by default when the debugger executes the startup code and is ready to begin the user's main() function. This is appropriate for C but the first programs in assembly language do not have a main label. This warning really is no problem and you can safely ignore it. Alternatively you can avoid it by using main instead of Reset as the reset vector. I do not like doing this because it is misleading—main should be after the startup code. Perhaps the best approach is to add a label for main after your basic initialization, which I do in later programs. Alternatively, get rid of the warning by choosing Tools > Options… from the menu, select Stack, and uncheck Stack pointer(s) not valid until program reaches:

- The simulator may complain (rightly) if you do not initialize the stack pointer. You can safely ignore the messages provided that you call no subroutines and should not use interrupts; otherwise, take heed and carry out the initialization.

- The editor highlights assembly *directives* only if they are in uppercase; however it highlights both upper- and lowercase *instructions*. The assembler obeys them all regardless of case. This really is no problem because it is consistent with TI's coding guidelines.

- The header files use a concise but strange notation sfrb and sfrw for defining special function registers.

- The memory map of the device is not defined in the header file; it is in the linker script, which has the extension `.xcl`. This is not useful for absolute assembly, which is not linked. Thus the addresses needed for absolute assembly must be added by hand. Of course a better solution is to use relocatable assembly instead.

- Do not mix C-style directives with their near equivalents in assembly language, such as `#define` and `EQU`. They are processed in slightly different ways, which can lead to strange errors. I suggest that you use the C-style `#include`, which is much clearer than the alternative in assembly language, but stick to the directives for assembly language after that.

- You cannot use the `io430` header files with assembly language, despite what it says in them. This is a minor flaw in the documentation and the files may be updated for assembly language in the future.

A.6.4   Debugger

- It is mildly tedious to switch between the stand-alone simulator and in-circuit emulator, which must be done through the Project Options.... I have not found a shortcut.

- The debugger has two extra toolbars, which appear below the main toolbar. This wastes a lot of the screen so I suggest that you move them to the end of the main toolbar.

- If you get Stack Warning: The stack plug-in failed to set a breakpoint on "main"..., you may have forgotten to select the correct device in the Project Options. This warning is very likely with assembly language, where `main` might not be used—see the preceding section.

- You can set breakpoints in either the editor or the debugger. The editor will let you set as many breakpoints as you want but the emulator can support only a fixed number—two in the smaller devices—so you will be warned when you start the debugger.

- The Breakpoints window, opened by View > Breakpoints, shows only the breakpoints that have been set by the user, not the "system breakpoints" that are set by the debugger. Choose the menu item Emulator > Breakpoint usage… to see *all* the breakpoints in use.

- A strategy for making more breakpoints available is to choose Project Options…> Debugger > Plugins and disable them all. Many are not available in the Kickstart edition but the stack plug-in may set a breakpoint to start monitoring the stack after it has been initialized.

- Another breakpoint is set in case your program exits from `main()`. It can be removed if you are certain that this will never happen. Open the project options, select the category Linker, choose the Output tab, and deselect With runtime control modules.

- A further breakpoint is used temporarily in C to stop execution at the beginning of your `main()` function, after the startup code. This is no problem if you set a breakpoint after the startup code has run but the debugger will want it back when you reset the device.

- The Project > Make & Restart Debugger command is convenient if you want to change the program while debugging. It leaves the debugger, compiles and links the program, and relaunches the debugger.

A.7   Tips for Specific Development Kits

Most of the details in this chapter as well as those in Chapter 4 have been for the Olimex MSP430-1121STK starter kit. Here are some tips for using a few other boards. The more sophisticated ones have lots of jumpers to enable particular features. These are sets of pins, called *headers*, sticking up out of the board, 0.1″ apart, which can be bridged with a small link—a plastic block that contains a wire to connect the two pins. They are usually designated J1, J2, and so on. It is infuriatingly easy to lose the links and there never seem to be any lying around when you need one, despite the billions that must have been manufactured.

Most jumpers have only two pins, in which case you either place the link over them or not, like a switch that is either closed or open (SPST, which stands for single pole, single throw). If the jumper is to be left open, place the link on one of the pins so that it hangs over the side or you will lose it. A few jumpers have three pins, in which case the jumper is used to connect the middle pin to either one of the two end pins like a changeover switch (SPDT, single pole, double throw). Often there are so many jumpers that they are combined into a single block and you may need to puzzle out the numbering if the jumpers are not clearly labeled on the board. Pin 1 should be identified with a 1 or sometimes another mark; if not, look on the bottom of the board, where it usually has a square solder pad rather than a round one. Orient the top of the board so that pin 1 is at the top left. Then pin 2 is on its right, pin 3 is below pin 1 with pin 4 on its right, and so on. I assume that each jumper has only two pins, as in Figure 10.6.

These jumpers must be configured correctly or the MSP430 will not even be powered. This can be the most difficult step when getting started with a new board.

A.7.1 Texas Instruments eZ430–F2013

This has one LED only, connected active high to P1.0. Here is the basic C program to light it:

```
#include <msp430x20x3.h>          // Specific device

void main (void)
{
    WDTCTL = WDTPW | WDTHOLD;     // Stop watchdog timer
    P1DIR = 0x01;    // Set pin with LEDs to output
    P1OUT = 0x01;    // LED on (active high)
    for (;;) {       // Loop forever...
    }                // ...doing nothing
}
```

The JTAG interface inside the eZ430–F2013 looks like a TI FET to the debugger but check that Spy-Bi-Wire is selected, as in Figure A.10. No crystal is installed in the eZ430–F2013, which means that there is no ACLK by default. You can instead take ACLK from the VLO: See the section "Internal Low-Power, Low-Frequency Oscillator, VLO" on page 167. There are no jumpers to worry about and power comes from the USB connection.

A.7.2 Texas Instruments MSP430FG4618/F2013 Experimenter's Board

This provides two MSP430s, a FG4618 and a F2013. They can be used independently or linked to explore communication. Each has its own JTAG header. Some of the LEDS can be disconnected with jumpers and several more control the power supply. All are described in the *User's Guide* (slau213). PWR1 and PWR2 apply power to the FG4618 and F2013 individually. Power for each of them can be taken from either batteries or the JTAG interface, selected with the pair of VCC jumpers. Finally, BATT is used to connect the battery. I suggest that you ensure that jumpers are placed on PWR1 and PWR2, put the VCC jumpers on the left-hand pair of pins labeled LCL, which means the battery, and use the BATT jumper to turn the whole board on and off.

The F2013 has a single LED, labeled LED3, on P1.0 as in the eZ430–F2013 so the same program and debugger settings can be used. If it does not light, check that its jumper JP2 is installed.

The FG4618 has three LEDs. LED4 is connected to P5.1 through jumper JP3, while LED1 (green) and LED2 (yellow) are permanently connected to P2.2 and P2.1. It is simplest to adapt `ledson.c` to use the latter two LEDs. This device uses the full four-wire JTAG.

```
#include <msp430xG46x.h>        // Specific device

void main (void)
{
    WDTCTL = WDTPW | WDTHOLD;    // Stop watchdog timer
    P2DIR = 0x06;    // Set pins with LEDs to output, 0b00000110
    P2OUT = 0x02;    // LED2 (P2.1) on, LED1 (P2.2) off (active high)
    for (;;) {       // Loop forever...
    }                // ...doing nothing
}
```

A.7.3 SoftBaugh ES169 Evaluation System

SoftBaugh makes evaluation boards for many variants of the MSP430 and this is a typical example. The boards have many jumpers, particularly for the sources of power. They are thoroughly described in the user's guide, which you should study carefully. For example, to power the ES169 from a AAA battery, place the following jumpers:

- J20 pins 3 and 4, labeled 3VC, which connect the output of a charge pump to V_{CC}.

- J17, labeled iMSP, which is provided for measuring the current drawn by the MSP430.

- J18, labeled iAAA, which connects the AAA cell to the input of the charge pump.

The board has four LEDs, connected active low, but only D1 is directly wired to the MSP430 at P1.0. The others are brought out to headers and the idea is to use prototyping wires to link them to pins of your choice. This also applies to the two push buttons and reflects the philosophy of making the boards as versatile as possible. Here, for completeness, is the C program but it barely differs from the other examples:

```c
#include <msp430x16x.h>            // Specific device

void main (void)
{
    WDTCTL = WDTPW | WDTHOLD;    // Stop watchdog timer
    P1DIR = 0x01;    // Set pin with LED to output
    P1OUT = 0x00;    // D1 (P1.0) on (active low!)
    for (;;) {       // Loop forever...
    }                // ...doing nothing
}
```

Further Reading

Books and Articles

This includes a wide range of books and material from the Internet that may be useful for embedded systems and microcontrollers. Many are on the interface-to-analog systems because this tends to cause the most trouble. I added a few personal comments. Locations on the Internet are subject to the usual warning about their volatility but a search engine should be able to locate the material unless it has vanished completely. I do not repeat the application notes, which are mentioned throughout the text.

The MSP430

[1] Baugh, Tom. *MSP430 RF Applications with the MRF1611CC1100: and API Reference.* Suwanee, GA: SoftBaugh Inc. 2007. (ISBN 0975475908)

[2] Bierl, Lutz. *Das große MSP430 Praxisbuch.* Poing, Germany: Franzis, 2004. (ISBN 377234299X)

[3] Luecke, Gerald. *Analog and Digital Circuits for Electronic Control System Applications: Using the TI MSP430 Microcontroller.* Boston: Newnes, 2004. (ISBN 0750678100)

[4] Nagy, Chris. *Embedded Systems Design Using the TI MSP430 Series.* Boston: Newnes, 2003. (ISBN 075067623X)

[5] Pereira, Fábio. *Microcontroladores MSP430—Teoria e Prática.* Editora Érica, 2005. (ISBN 8536500670)

[6] Sturm, Matthias. *Mikrocontrollertechnik: Am Beispiel der MSP430-Familie.* Leipzig, Germany: Fachbuchverlag Leipzig, im Carl Hanser Verlag, 2006. (ISBN 3446218009)

[7] There is an interesting description of the CPU and instruction set in the manual for the GCC toolchain for MSP430 at mspgcc.sourceforge.net.

[8] Wikipedia has a good introduction to the MSP430 with further interesting links. (Of course www.wikipedia.org is an excellent place to start learning about anything.)

Embedded Systems

[9] Berger, Arnold S. *Embedded Systems Design: An Introduction to Processes, Tools and Techniques.* San Francisco: CMP Books, 2001. (ISBN 1578200733)

[10] Catsoulis, John. *Designing Embedded Hardware.* Sebastopol, CA: O'Reilly, 2005. (ISBN 0596007558)

[11] Curtis, Keith E. *Embedded Multitasking: With Small Microcontrollers.* Boston: Newnes, 2006. (ISBN 0750679182)

[12] Edwards, Lewin A. R. W. *Embedded System Design on a Shoestring.* Boston: Newnes, 2003. (ISBN 0750676094)

[13] Edwards, Lewin A. R. W. *So You Wanna Be an Embedded Engineer: The Guide to Embedded Engineering, From Consultancy to the Corporate Ladder.* Boston: Newnes, 2006. (ISBN 0750679530)

[14] Ganssle, Jack. *The Art of Programming Embedded Systems.* Boston: Academic, 1991. (ISBN 0122748808)

[15] Ganssle, Jack. *The Art of Designing Embedded Systems.* Boston: Newnes, 1999. (ISBN 0750698691)

[16] Ganssle, Jack. *The Firmware Handbook.* Boston: Newnes, 2004. (ISBN 075067606X)

[17] Ganssle, Jack. *A Guide to Debouncing.* 2004. Available at www.ganssle.com/ debouncing.pdf.

[18] Ganssle, Jack (editor). *Embedded Hardware: Know It All.* Boston: Newnes, 2007. (ISBN 0750685849)

[19] Ganssle, Jack (editor). *Embedded Systems: World Class Designs.* Boston: Newnes, 2008. (ISBN 0750686259)

[20] Kreiman, Michael. "State Machine Shortcuts." *Embedded System Design*, June 23, 2003. www.embedded.com/showArticle.jhtml?articleID = 10700829.

[21] Labrosse, Jean J. *Embedded Systems Building Blocks: Complete and Ready-To-Use Modules in C,* 2nd edition. San Francisco: CMP Books, 1999. (ISBN 0879306041)

[22] Labrosse, Jean J. *MicroC/OS II: The Real Time Kernel.* San Francisco: CMP Books, 2002. (ISBN 1578201039)

[23] Labrosse, Jean J (editor). *Embedded Software: Know It All.* Boston: Newnes, 2007. (ISBN 0750685832)

[24] Mazzocca, Elio. "Contact-Debouncing Algorithm Emulates Schmitt Trigger." *Electronic Design News*, July 7, 2005. www.edn.com/ednmag/article/CA621638.html.

[25] Simon, David E. *An Embedded Software Primer.* Boston: Addison Wesley, 1999. (ISBN 020161569X)

[26] Wilmshurst, Tim. *An Introduction to the Design of Small-Scale Embedded Systems.* New York: Palgrave Macrnillan, 2001. (ISBN 0333929942)

[27] Wolf, Wayne. *Computers as Components: Principles of Embedded Computing System Design.* San Francisco: Morgan Kaufmann, 2005. (ISBN 0123694590)

Programming

[28] Barr, Michael. *Programming Embedded Systems in C and C++.* Sebastopol, CA: O'Reilly, 1999. (ISBN 1565923545)

[29] Hills, Chris. *Embedded C: Traps and Pitfalls*, 3rd edition. Phaedrus Systems, 2005. Available at quest.phaedsys.org.

[30] Kernighan, Brian W., and Ritchie, Dennis M. *The C Programming Language*, 2nd edition. Englewood Cliffs, NJ: Prentice-Hall, 1988. (ISBN 0131103628)

[31] Koenig, Andrew. *C Traps and Pitfalls.* Boston: Addison Wesley, 1989. (ISBN 0201179288)

[32] Martin, George. "Hello World … Want Cookie." *Circuit Cellar*, no. 198, January 2007, pages 60–64.

[33] Martin, George. "More Hello World." *Circuit Cellar*, no. 200, March 2007, pages 67–71.

[34] Plum, Thomas, and Saks, Dan. *C++ Programming Guidelines*. Plum Hall, 1991. (ISBN 0911537104)

[35] Steele, Guy, McMeniman, Linda, and Harbison, Samuel. *C: A Reference Manual*, 5th edition. Saddle River, NJ: Prentice-Hall, 2002. (ISBN 013122560X)

Computer Architecture

[36] Furber, Steve. *ARM System-on-Chip Architecture*. Boston: Addison Wesley, 2000. (ISBN 0201675196)

[37] Maxfield, Clive, and Brown, Alvin. *The Definitive Guide to How Computers Do Math*. New York: Wiley, 2005. (ISBN 0471732788)

[38] Wakerly, John F. *Digital Design: Principles and Practices*. Saddle River, NJ: Prentice-Hall, 2005. (ISBN 0131863894)

Analog and Mixed-Signal Systems

[39] Baker, Bonnie. *A Baker's Dozen: Real Analog Solutions for Digital Designers*. Boston: Newnes, 2005. (ISBN 0750678194)

This book was written for digital engineers who need to handle the interface to analog electronics. Much of the book is on ADCs and DACs and issues such as noise. The author is now at Texas Instruments, having been formerly at Microchip Technology and Burr–Brown, so she knows what she is writing about. One warning: The author uses *differentiate* where I would write *subtract* (the usage comes from the term *differential amplifier*, which subtracts rather than differentiates).

[40] Baker, Bonnie. *Glossary of Analog-to-Digital Specifications and Performance Characteristics*. Texas Instruments, 2006. Application report sbaa147.

[41] Carter, Bruce. *A Single-Supply Op-Amp Circuit Collection*. Texas Instruments, 2000. Application report sloa058.

[42] Horowitz, P., and Hill, W. *The Art of Electronics*. Cambridge, MA: Cambridge University Press, 1989. (ISBN 0521370957)

No electronic engineer should be without this book, with its lucid coverage of all aspects of electronics. The details are now a little out of date, particularly for digital systems, but the principles are as true as ever. The only problem is that it is *too* easy to read, so that the reader tends to fly through the text too fast to absorb it.

[43] Kester, Walt (editor). *Data Conversion Handbook.* Boston: Newnes, 2004. (ISBN 0750678410)

Nearly 1000 pages on all aspects of data conversion from their history to the design of PCBs. It goes into the theory more deeply than Baker's book but is less coherent because of its edited nature. The authors are from Analog Devices and the examples are taken from their range of products.

[44] Kester, Walt (editor). *Mixed-Signal and DSP Design Techniques.* Boston: Newnes, 2002. (ISBN 0750676116)

[45] Kester, Walt (editor). *Op Amp Applications Handbook.* Boston: Newnes, 2004. (ISBN 0750678445)

[46] Kitchin, Charles. "Avoid Common Problems when Designing Amplifier Circuits." *Analog Dialogue*, volume 41, no. 08, August 2007, pages 1–4. Available on the Web at www.analog.com/library/analogDialogue/archives/41-08/amplifier_circuits.pdf.

An excellent and concise summary of common problems, much of which applies to single-supply circuits. There is an earlier article, "Demystifying Single-Supply Op-Amp Design" in *Electronic Design News*, March 21, 2002, pages 83–90.

[47] Kugelstadt, Thomas. *The Operation of the SAR–ADC Based on Charge Redistribution.* Texas Instruments, 2005, application note slyt176.

[48] Mancini, Ron. *Understanding Basic Analog—Passive Devices.* Texas Instruments, 1999. Application report sloa027.

[49] Mancini, Ron. *Understanding Basic Analog—Active Devices.* Texas Instruments, 2000. Application report sloa026.

[50] Mancini, Ron (editor). *Op Amps for Everyone.* Boston: Newnes, 2003. (ISBN 0750677015)

Despite the different title, this covers much the same material as Baker's book although there is rather more about op-amps. This is a good place to look for information on single-supply op-amps. It suffers a little from having an editor rather than single author but is reasonably coherent. An earlier version can be downloaded from the TI Web site as application note slod006b but it is nearly 500 pages long so the printed version might be cheaper.

[51] Mancini, Ron. "How to Read a Semiconductor Data Sheet." *Electronic Design News*, April 14, 2005, www.edn.com/article/CA514964.html.

[52] Motorola (now Freescale). *M68HC11 Reference Manual.* Motorola, 2002. Available on the Web at www.freescale.com/files/microcontrollers/doc/ref_manual/ M68HC11RM.pdf.

I include this because Section 12 provides the best description of the operation of a successive-approximation ADC that I have come across.

[53] Schreier, R., and Temes, G. C. *Understanding Delta–Sigma Data Converters.* New York: Wiley–IEEE Press, 2004. (ISBN 0471465852)

[54] Smith, Steven W. *Digital Signal Processing: A Practical Guide for Engineers and Scientists.* Boston: Newnes, 2002. (ISBN 075067444X)

This provides a relatively nontechnical introduction to digital signal processing with lots of applications. You might find it interesting for the filters in a sigma–delta converter.

Communications

[55] Axelson, Jan. *Serial Port Complete*, 2nd edition. Lakeview Research, 2007. (ISBN 193144806X)

The author has written several highly regarded books on aspects of communication for embedded systems. Check her Web site at www.lvr.com.

[56] Insam, Edward. *TCP/IP Embedded Internet Applications.* Boston: Newnes, 2003. (ISBN 0750657359)

Miscellaneous

[57] Maxfield, Clive. *Bebop to the Boolean Boogie: An Unconventional Guide to Electronics.* Boston: Newnes, 2003. (ISBN 0750675438). See also www.maxmon. com/booginfo.htm.

This book fully lives up to its subtitle: It is nothing like a conventional textbook (such as mine) and it covers a broad spectrum of electronics. It starts with the relation between analog and digital signals, which is an ideal introduction to embedded systems. The book does not concentrate on microcontrollers but there is a lot of background on primitive and programmable logic devices. The main body of the book concludes with useful material on components and construction. There are some fascinating topics in the appendices followed by a comprehensive glossary of terms commonly used in electronics (and many more that are not). Check the section "How to Become Famous."

[58] Information on crystals and oscillators, Micro Crystal. See www.microcrystal.com.

Newsletters, Magazines, and Journals

Many newsletters, magazines, and journals are devoted to embedded electronic systems. You could probably spend a lifetime reading them. Here are some that I have found illuminating.

Embedded Systems Design

This is a trade journal time aimed at the design of embedded systems and was formerly *Embedded System Programming*. The Web site is www.embedded.com. It is oriented toward software rather than hardware but there have been some excellent articles on the analog interface. These are the columns I found most useful:

- "Embedded Pulse" and "Break Points" by Jack Ganssle.

- "Programming Pointers" by Dan Saks.

Here is a selection of articles I found interesting.

[59] Ganssle, Jack G. "eXtreme Instrumenting." May 29, 2006, www.embedded.com/showArticle.jhtml?articleID=187203692.

[60] Maxfield, Clive. "Humans Thrive on Decimals, but Engineers Love Binary." February 13, 2006, www.embedded.com/showArticle.jhtml?articleID=180200596.

[61] Saks, Dan. A series of "Programming Pointers" dealt with the way in which special function registers and pointers are handled in C. These are the most relevant articles (later ones are directed toward larger processors):

- "Mapping Memory." August 11, 2004, www.embedded.com/showArticle.jhtml?articleID=26807176.

- "Mapping Memory Efficiently." October 20, 2004, www.embedded.com/showArticle.jhtml?articleID=50900224.

- "More Ways to Map Memory." December 14, 2004, www.embedded.com/showArticle.jhtml?articleID=55301821.

- "Sizing and Aligning Device Registers." April 7, 2005, www.embedded. com/showArticle.jhtml?articleID=160502362.

- "Use Volatile Judiciously." September 7, 2005, www.embedded.com/ showArticle.jhtml?articleID=170701302.

- "Place Volatile Accurately." November 16, 2005, www.embedded. com/showArticle.jhtml?articleID=174300478.

- "Volatile as a Promise." January 4, 2006, www.embedded.com/ showArticle.jhtml?articleID=175801310.

Electronic Design News

This is another standard trade journal for the whole range of electronic design. It is on the Web at www.edn.com. These are the columns that I found most useful:

- "Baker's Best" by Bonnie Baker.

- "Design Ideas" contributed by readers, which are often remarkably ingenious.

- "Prying Eyes" describes tear-downs of a wide range of products.

Here is a selection of relevant articles:

[62] Baker, Bonnie. "Delta–Sigma ADCs in a Nutshell." This was published as four articles.

　　1. "Delta–Sigma ADCs in a Nutshell." December 14, 2007, www.edn.com/ article/CA6512148.html.

　　2. "The Modulator." January 24, 2008, www.edn.com/article/CA6518678.html.

　　3. "The Digital/Decimator Filter." February 21, 2008, www.edn.com/ article/CA6531581.html.

　　4. "Noise Versus Data Rate." March 20, 2008, www.edn.com/article/ CA6541374.html.

[63] Maxfield, Clive. This is a characteristically entertaining set of articles on BCD. I did not realize that there was so much to it:

- "Mind-Boggling Math: BCD (Binary Coded Decimal)." September 7, 2005, www.edn.com/article/CA6254628.html.

- "More Mind-Boggling Math: Adding and Subtracting Unsigned BCD." September 21, 2005, www.edn.com/article/CA6258681.html.

- "Working with Signed BCD." October 5, 2005, www.edn.com/article/CA6263546.

[64] Passmore, Loren. "Single Microcontroller Pin Senses Ambient Light, Controls Illumination." *Electronic Design News*, October 26, 2006, www.edn.com/article/CA6382648.html.

A similar program by Andreas Dannenberg (slac136) was used as a demonstration at the "430 day" in 2007. It is available on the Web page for the F2013.

More

[65] *Circuit Cellar* is a monthly magazine devoted to embedded systems. Its subtitle is *The Magazine for Computer Applications* but the computers are almost always embedded rather than on a desktop. It is published in the United States and the postage makes the printed version expensive elsewhere but the online edition at www.circuitcellar.com is affordable (unless you print it all).

[66] A useful Web site for all matters analog is www.analogzone.com, now moved to www.en-genius.net. There is an excellent selection of technical notes, many of which are concerned with data acquisition.

[67] *The Embedded Muse* is a fortnightly newsletter by Jack Ganssle. It contains "hints, ideas and rambles about firmware and hardware in embedded systems." Highly recommended. Subscribe on the Web at www.ganssle.com.

Index

Printed and bound by CPI Group (UK) Ltd, Croydon, CR0 4YY

03/10/2024

01040334-0008